WETTABILITY

SURFACTANT SCIENCE SERIES

1. Nonionic Surfactants, *edited by Martin J. Schick* (see also Volumes 19 and 23)
2. Solvent Properties of Surfactant Solutions, *edited by Kozo Shinoda* (out of print)
3. Surfactant Biodegradation, *by R. D. Swisher* (see Volume 18)
4. Cationic Surfactants, *edited by Eric Jungermann* (see also Volumes 34 and 37)
5. Detergency: Theory and Test Methods (in three parts), *edited by W. G. Cutler and R. C. Davis* (see also Volume 20)
6. Emulsions and Emulsion Technology (in three parts), *edited by Kenneth J. Lissant*
7. Anionic Surfactants (in two parts), *edited by Warner M. Linfield* (out of print)
8. Anionic Surfactants: Chemical Analysis, *edited by John Cross* (out of print)
9. Stabilization of Colloidal Dispersions by Polymer Adsorption, *by Tatsuo Sato and Richard Ruch*
10. Anionic Surfactants: Biochemistry, Toxicology, Dermatology, *edited by Christian Gloxhuber* (see Volume 43)
11. Anionic Surfactants: Physical Chemistry of Surfactant Action, *edited by E. H. Lucassen-Reynders* (out of print)
12. Amphoteric Surfactants, *edited by B. R. Bluestein and Clifford L. Hilton* (out of print)
13. Demulsification: Industrial Applications, *by Kenneth J. Lissant*
14. Surfactants in Textile Processing, *by Arved Datyner*
15. Electrical Phenomena at Interfaces: Fundamentals, Measurements, and Applications, *edited by Ayao Kitahara and Akira Watanabe*
16. Surfactants in Cosmetics, *edited by Martin M. Rieger*
17. Interfacial Phenomena: Equilibrium and Dynamic Effects, *by Clarence A. Miller and P. Neogi*
18. Surfactant Biodegradation, Second Edition, Revised and Expanded, *by R. D. Swisher*

19. Nonionic Surfactants: Chemical Analysis, *edited by John Cross*
20. Detergency: Theory and Technology, *edited by W. Gale Cutler and Erik Kissa*
21. Interfacial Phenomena in Apolar Media, *edited by Hans-Friedrich Eicke and Geoffrey D. Parfitt*
22. Surfactant Solutions: New Methods of Investigation, *edited by Raoul Zana*
23. Nonionic Surfactants: Physical Chemistry, *edited by Martin J. Schick*
24. Microemulsion Systems, *edited by Henri L. Rosano and Marc Clausse*
25. Biosurfactants and Biotechnology, *edited by Naim Kosaric, W. L. Cairns, and Neil C. C. Gray*
26. Surfactants in Emerging Technologies, *edited by Milton J. Rosen*
27. Reagents in Mineral Technology, *edited by P. Somasundaran and Brij M. Moudgil*
28. Surfactants in Chemical/Process Engineering, *edited by Darsh T. Wasan, Martin E. Ginn, and Dinesh O. Shah*
29. Thin Liquid Films, *edited by I. B. Ivanov*
30. Microemulsions and Related Systems: Formulation, Solvency, and Physical Properties, *edited by Maurice Bourrel and Robert S. Schecter*
31. Crystallization and Polymorphism of Fats and Fatty Acids, *edited by Nissim Garti and Kiyotaka Sato*
32. Interfacial Phenomena in Coal Technology, *edited by Gregory D. Botsaris and Yuli M. Glazman*
33. Surfactant-Based Separation Processes, *edited by John F. Scamehorn and Jeffrey H. Harwell*
34. Cationic Surfactants: Organic Chemistry, *edited by James M. Richmond*
35. Alkylene Oxides and Their Polymers, *by F. E. Bailey, Jr., and Joseph V. Koleske*
36. Interfacial Phenomena in Petroleum Recovery, *edited by Norman R. Morrow*
37. Cationic Surfactants: Physical Chemistry, *edited by Donn N. Rubingh and Paul M. Holland*
38. Kinetics and Catalysis in Microheterogeneous Systems, *edited by M. Grätzel and K. Kalyanasundaram*
39. Interfacial Phenomena in Biological Systems, *edited by Max Bender*
40. Analysis of Surfactants, *by Thomas M. Schmitt*
41. Light Scattering by Liquid Surfaces and Complementary Techniques, *edited by Dominique Langevin*
42. Polymeric Surfactants, *by Irja Piirma*
43. Anionic Surfactants: Biochemistry, Toxicology, Dermatology. Second Edition, Revised and Expanded, *edited by Christian Gloxhuber and Klaus Künstler*
44. Organized Solutions: Surfactants in Science and Technology, *edited by Stig E. Friberg and Björn Lindman*
45. Defoaming: Theory and Industrial Applications, *edited by P. R. Garrett*

46. Mixed Surfactant Systems, *edited by Keizo Ogino and Masahiko Abe*
47. Coagulation and Flocculation: Theory and Applications, *edited by Bohuslav Dobiáš*
48. Biosurfactants: Production · Properties · Applications, *edited by Naim Kosaric*
49. Wettability, *edited by John C. Berg*

ADDITIONAL VOLUMES IN PREPARATION

Fluorinated Surfactants, *Erik Kissa*

WETTABILITY

edited by
John C. Berg
University of Washington
Seattle, Washington

CRC is an imprint of the Taylor & Francis Group,
an informa business

Library of Congress Cataloging-in-Publication Data

Wettability / edited by John C. Berg.
p. cm. -- (Surfactant science series ; v.49)
Includes bibliographical references and index.
ISBN 0-8247-9046-4 (alk. paper)
1. Wettability. I. Berg, John C. II. Series
QD506.W42 1993
541.3'3--dc20 93-12059
CIP

The publisher offers discounts on this book when ordered in bulk quantities. For more information, write to Special Sales/Professional Marketing at the address below.

This book is printed on acid-free paper.

MARCEL DEKKER, INC.
270 Madison Avenue, New York, New York 10016

Current printing (last digit):
10 9 8 7 6

PRINTED IN THE UNITED STATES OF AMERICA

Preface

Wettability refers to the response evinced when a liquid is brought into contact with a solid surface initially in contact with a gas or another liquid. Many possibilities exist. The liquid may move out over the solid, displacing the original fluid, and finally coming to a halt when the angle between the liquid–fluid and solid–liquid interfaces reaches a certain value, a *contact angle*. On the other hand, the liquid may spread without limit, displacing the original fluid from the entire solid surface area available, a situation evidently corresponding to a contact angle of 0°. The importance of wettability in coating operations, adhesion, detergency, lubrication, and other operations in which liquids are applied directly to solid surfaces is self-evident. It also controls the spontaneous imbibition of liquids into porous media, the separability of particulate solids by flotation, the efficacy of many supported catalysts, and other phenomena of widespread importance.

Although the significance of wettability has never been questioned, its quantifiability in terms of measured contact angles has a long and tortured history. Thomas Young, as long ago as 1805 [1], was able to relate the contact angle to the surface energies of the interfaces meeting at the three-phase junction, and J. Willard Gibbs [2] in 1878 expressed the relationship in mathematical form, thus conferring upon the contact angle the status of a thermodynamic property. It failed to attain commensurate respect, however, through generations of researchers, owing to its notorious irreproducibility and its apparent dependence on many "non-thermodynamic" factors. Its first convincing application as a measure of surface energetics was made by W. A. Zisman and his coworkers [3] in work done during and after the Second World War at the Naval Research Laboratory. Zisman avoided most of the pitfalls of irreproducibility by carefully restricting his systems and procedures. For the most part, he confined his attention to clean, smooth, "low-energy" (thus not easily contaminated) solid surfaces in contact with pure, single-component liquids which were incapable of penetrating, swelling, or chemically reacting in any way with the solid. Contact angles were measured and reported for the situation in which the liquid was advanced over the solid at an infinitesimal rate (producing "static advancing" contact angles). Under such conditions, not only were contact angles reproducible, but for a given solid, their cosines produced linear plots against the surface tensions of the liquids used. Such "Zisman plots" permitted linear extrapolation of the surface tension to the condition of $\cos \theta = 1$, yielding the "Zisman critical surface tension," which is still today the most commonly used measure of the "surface energy" of a given solid.

The major contact angle demon, which Zisman so adroitly avoided—i.e., the generally observed hysteresis between advancing and receding angles and its relationship to surface chemical and morphological heterogeneity—was faced squarely in the landmark article of Johnson and Dettre appearing in 1969 [4]. In my view, it was this article more than any other that salvaged the reputation of the contact angle. It opened with these confident words: "The contact angle is a common and useful measure of wettability. It gives information about surface energetics, surface roughness and surface heterogeneity. It is a sensitive measure of surface contamination, and it controls many technical processes. . . ." It is thus fitting that the present volume opens with a chapter by Johnson and Dettre. In it the authors focus primarily on low-energy surfaces, of the type studied by Zisman, in contact with gases and with other liquids. They incorporate recent developments in the understanding of dispersion forces and reexamine Zisman plots as well as the concepts of Girifalco and Good [5], Fowkes [6], and Neumann [7] in this light. Also, currently used experimental techniques for measurement of contact angle are evaluated, and particular attention is given to the effects of "surface solubility."

The most important developments concerning wettability over the past fifteen years may be divided into three broad categories: (1) increased knowledge of the intermolecular interactions, both within liquids and across the liquid–solid and liquid–liquid interfaces, responsible for wettability, (2) increased knowledge of the *dynamics* of wetting, and (3) increased recognition of the importance of wettability phenomena in processes and products for which such recognition was previously absent or minimal. All three aspects of these recent advances are treated in this book. Chapter 2, by Berg, deals with the current view that intermolecular forces within condensed phases and across condensed-phase interfaces may generally be divided into Lifshitz–van der Waals forces and forces due to acid–base interactions. Current methods for expressing the acid–base contributions to wettability are reviewed and assessed, together with the wide variety of experimental techniques (including many nonwettability measurements) that are used for determining the acid–base effects. Chapter 3, by Bose, deals with wettability effects that are peculiarly related to the fact that the wetting (and/or displaced) fluid may be multicomponent. Both the thermodynamics and the kinetics of multicomponent wetting are considered. Vogler, in Chapter 4, deals with wetting phenomena in systems of biological interest and applies the fundamentals of current wetting theory to these types of systems. The particular problems in characterizing the biomaterial surfaces and in relating the practitioner's concepts of "biocompatibility" to the fundamentals of wetting are considered.

The dynamic aspects of wetting have been treated in recent years from both molecular and macroscopic hydrodynamic points of view. Blake, in Chapter 5, develops and applies the molecular approach and shows that the formats it suggests are useful in describing and correlating the kinetics of interline movement. Kistler, in Chapter 6, considers wetting dynamics primarily from a macroscopic point of view and reviews, assesses, and extends the numerous hydrodynamic models that have been put forth for describing such phenomena. He observes, however, that "all these scenarios for different local wetting regimes suggest that the concept of a moving wetting line is merely a macroscopic substitute for more intricate submicroscopic physics." Hydrodynamic phenomena not associated with events in the immediate micro-vicinity of the moving interline may also play a role in determining apparent dynamic contact angles, and these too are considered. Both Blake and Kistler explicitly consider regimes of very rapid interline movement, relevant to high-speed coating operations.

Chapter 7, by Ko, considers the now-recognized but inadequately reported importance of wetting phenomena in determining the effectiveness of many supported catalysts. Although such catalysts are generally multiphase solids under ambient conditions, their ultimate morphology and surface characteristics are governed by the mobility of the materials during typical thermal treatments. Finally, Chapter 8, by Zelinski, Denesuk, Cronin, and Uhlmann, deals with the importance of wetting phenomena in the preparation of ceramic and ceramic composite systems. The implications are that such phenomena are important in the preparation and ultimate properties of all composite systems because of the key role played by adhesion between the phases.

No finite accounting of the many aspects of wettability phenomena can claim to be completely comprehensive, and the claim is not made for this contribution. The book, however, does provide significant examples of recent developments and new understanding in the field and should be useful to teachers and researchers as well to those whose primary interest is in practical applications.

John C. Berg

REFERENCES

1. T. Young, *Phil. Trans. 95*:65 and 82 (1805).
2. *The Collected Works of J. Willard Gibbs*, Vol. 1, Thermodynamics, Yale University Press, New Haven, 1928.
3. Advances in Chemistry Series 43 (F. M. Fowkes, ed.), American Chemical Society, Washington DC, 1964.
4. R. E. Johnson, Jr. and R. H. Dettre, in *Surface and Colloid Science*, Vol. 2 (E. Matijevic, ed.), Wiley-Interscience, New York, 1969.
5. L. A. Girifalco and R. J. Good, *J. Phys. Chem. 61*:904 (1957).
6. F. M. Fowkes, *ACS Adv. in Chem. Series 43*:99 (1964).
7. A. W. Neumann, *Adv. Colloid Interface Sci. 4*:1 (1974).

Contents

Contributors

John C. Berg Department of Chemical Engineering, University of Washington, Seattle, Washington

Terence D. Blake Research Division, Kodak Limited, Harrow, England

Arijit Bose Department of Chemical Engineering, University of Rhode Island, Kingston, Rhode Island

John Paul Cronin* Department of Materials Science and Engineering, University of Arizona, Tucson, Arizona

Matthew Denesuk Department of Materials Science and Engineering, University of Arizona, Tucson, Arizona

Robert H. Dettre Du Pont Chemicals, E. I. du Pont de Nemours & Co., Inc., Wilmington, Delaware

Rulon E. Johnson, Jr.† Central Research Department, E. I. du Pont de Nemours & Co., Inc., Wilmington, Delaware

---

Current affiliations:
* Advanced Technology Center, Donnelly Corporation, Tucson, Arizona
† Retired

Stephan F. Kistler Magnetic Media Technology Center, 3M Company, St. Paul, Minnesota

Edmond I. Ko Department of Chemical Engineering, Carnegie Mellon University, Pittsburgh, Pennsylvania

Donald R. Uhlmann Department of Materials Science and Engineering, University of Arizona, Tucson, Arizona

Erwin A. Vogler Polymer Chemistry Department, Becton Dickinson Research Center, Research Triangle Park, North Carolina

Brian J. J. Zelinski Department of Materials Science and Engineering, University of Arizona, Tucson, Arizona

WETTABILITY

1

Wetting of Low-Energy Surfaces

RULON E. JOHNSON, JR.* Central Research Department, E. I. du Pont de Nemours & Co., Inc., Wilmington, Delaware

ROBERT H. DETTRE Du Pont Chemicals, E. I. du Pont de Nemours & Co., Inc., Wilmington, Delaware

* Retired.

I. INTRODUCTION

Wetting involves the interaction of a liquid with a solid. It can be the spreading of a liquid over a surface, the penetration of a liquid into a porous medium, or the displacement of one liquid by another. It can help to characterize surfaces and to determine solid/liquid interactions.

Wettability is most often described by a sessile or resting drop. A schematic diagram is shown in Fig. 1. The contact angle (θ) is a measure of wettability. A low contact angle means high wettability and a high contact angle means poor wettability. Zero contact angles are possible but they are always less than 180°. (The highest commonly observed angle, mercury on glass, has been reported to be as high as 148° [1].) Systems having more than one stable contact angle are said to show contact-angle hysteresis.

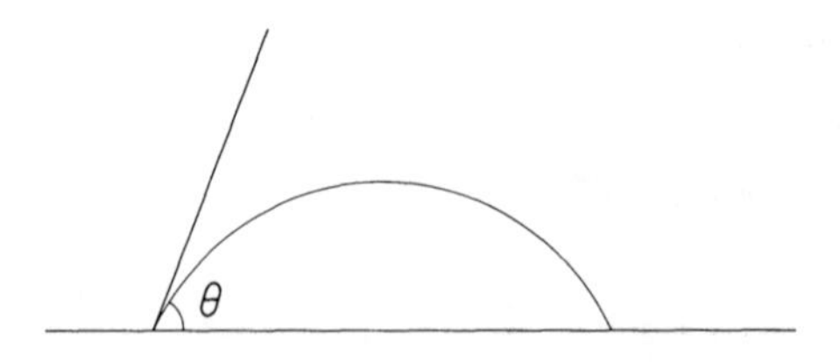

FIG. 1 Sessile drop on an ideal solid surface.

This chapter is an extension of a chapter on wettability written earlier [2]. Our focus is the thermodynamics of wetting, the wetting of low-energy surfaces, and interfacial wettability. This is not a comprehensive review of the literature. We have selected areas of interest where advances have been made, where misunderstanding or contention exists, or areas that have been neglected. It presents our approach to wettability.

Low-energy surfaces interact with liquids primarily through dispersion (van der Waals) forces. Hamaker–Lifshitz [3,4] theory provides a precise understanding of dispersion forces. Two important papers, one by Hough and White [5] and the other by Israelachvili [6], have applied Lifshitz theory to wetting. Their work allows us to evaluate the concepts and theories of Zisman, Girifalco and Good, Fowkes, and others. It will be found that Hamaker–Lifshitz theory requires modification of the concepts of Zisman concerning the effects of molecular packing, surface structure, and the chain length of monolayers on wetting. On the other hand, current Hamaker-Lifshitz theory ignores the possibility of the penetration of molecules of wetting liquids into the surface of a substrate. This "surface solubility" needs to be considered explicitly.

Zisman plots ($\cos\theta$ vs. the surface tension of the wetting liquid) have long been used to characterize the wettability of low-energy surfaces. These plots still give the best empirical fit of experimental data. The intercept of these curves with the $\cos\theta = 1$ axis is known as the critical surface tension of that surface. It is frequently used to describe the wettability of a surface. The slopes of Zisman plots, on the other hand, have been virtually ignored. These slopes contain much valuable information, particularly about surface solubility.

The most controversial addition to the literature of wetting has been the surface equation of state of Neumann [7]. Some effort will be spent in discussing problems associated with this theory.

Our thermodynamic analysis will be more commentary than derivation. Many of the problems we have noticed in the literature are caused by applying equilibrium equations to nonequilibrium systems. Other problems occur when the contact angle approaches zero. Some confusion still exists between surface stress and the free energy of forming a surface. These will each be discussed in turn.

We will discuss experimental measurements primarily as they relate to analysis of the data. Emphasis will be on systems showing hysteresis where poor technique makes analysis difficult. Our discussion of hysteresis will focus almost entirely on hysteresis caused by heterogeneity. Our understanding of hysteresis caused by roughness is not substantially different from that presented in Ref. 2.

We will finish with discussions of the wettability of liquid surfaces and interfacial wettability. The wettability of liquid surfaces provides the most direct test of theories of wetting and interactions between phases. This is because all surface tensions, interfacial tensions, and spreading pressures can be measured independently. Historically, the calculation of liquid/liquid interfacial tensions from theory has provided the best guidance for the development of wetting theory. Unfortunately, Li et al. [33e] have suggested that liquid/liquid/vapor systems do not follow the same laws followed by solid/liquid/vapor systems. We address this issue and provide some examples where liquid/liquid/vapor analysis gives insight into processes occurring in solid/liquid/vapor systems.

Interfacial wettability (solid/liquid/liquid systems) is important in such processes as detergency, secondary oil recovery, and many biological processes. The Bartell–Osterhof equation relating interfacial contact angles to angles measured in air is our starting point. We include the analysis of Melrose who expanded the Bartell–Osterhof equation to include spreading pressures at both the solid/air and solid/liquid interfaces. A new way to handle the complexities of ever-present hysteresis is given. Experimental examples are presented to illustrate the concepts offered. Unless otherwise specified, all data were obtained at 25°C.

II. COMMENTARY ON THE THERMODYNAMICS OF WETTING

The key to understanding wettability is the recognition that it is determined by the balance between adhesive forces between the liquid and solid and cohesive forces in the liquid. Adhesive forces cause a liquid drop to spread. Cohesive forces cause the drop to ball up. The contact angle is determined by competition between these two forces.

We will use the following notation. The subscripts s, v, o, and l refer to solid, vapor, vacuum, and liquid, respectively. An interface or a surface will be represented by a slash (/). Thus, $\gamma_{s/l}$ is the interfacial tension at the solid/liquid interface. A double slash (//) indicates the interface is in equilibrium with the component following the slashes. An asterisk (*) denotes systems not in equilibrium.

A. Ideal Wetting Model

Nearly all theories of wetting use the same ideal model surface. This surface has zero hysteresis. It is smooth, homogeneous, and nondeformable. When a drop of liquid is placed on such a surface, it assumes a characteristic angle as shown in Fig. 1. If the drop is disturbed, it will return to its original shape. The contact angle is independent of gravity [8].

B. Solid Surface and Interfacial Tensions

1. Solid/Vacuum Surface Tension

$\gamma_{s/o}$, commonly called the surface tension, is the reversible work of forming a unit area of surface of the solid in a vacuum. There is usually little difference between a solid surface tension in a vacuum and one in air. $\gamma_{s/o}$ is defined by Eq. (1) [9]

$$\gamma_{s/o} = \left(\frac{\partial G_{s/o}}{\partial \Omega_{s/o}}\right)_{T,P,N_1,\ldots,N_j} \tag{1}$$

$G_{s/o}$ is the Gibbs excess free energy of the solid/vacuum interface, T the absolute temperature, $\Omega_{s/o}$ the solid surface area, P the pressure, and N_j the number of surface excess moles of the jth component. The subscripts T, P, and $N_1, \ldots, N_j$ in Eq. (1) indicate that the derivative is taken at constant T and P and number of surface excess components.

The work of forming a surface should not be confused with surface stress (work of stretching). (See Appendix I for Gibbs' description of the difference.) The surface stress and surface tension are the same for pure liquids but not solids. Unfortunately, the term "surface tension" has usually been applied to the $\gamma_{s/o}$. We will continue with this notation but emphasize that $\gamma_{s/o}$ cannot be interpreted as a stress in the solid surface.

As solid surface tensions are difficult to measure, it is fortunate that a knowledge of solid surface tension is not required to understand wettability. Conversely, wettability measurements cannot be used to determine solid surface tensions.

(a) Solid/Vapor Surface Tension. $\gamma_{s/v}$, the solid/vapor surface tension, is the reversible work associated with forming a unit of surface when the surface is in equilibrium with the saturated vapor of a liquid. It is defined by Eq. (2). $\pi_{s/v}$ is the spreading pressure of the vapor on the solid. It can be calculated from the adsorption isotherm of the vapor of the liquid on the

$$\gamma_{s/v} = \gamma_{s/o} - \pi_{s/v} = \left(\frac{\partial G_{s/o}}{\partial \Omega_{s/o}}\right)_{T,P,N_1,\ldots,N_j} - \pi_{s/v} \tag{2}$$

solid by the Gibbs adsorption equation, Eq. (3)

$$\pi_{s/v} = RT \int_0^{P_0} \frac{\Gamma_{s/v}}{P}\, dP \tag{3}$$

$\Gamma_{s/v}$ is the Gibbs surface excess of the vapor on the solid, P the pressure, P_0 the vapor pressure of the liquid, and R the gas constant.

2. Solid/Liquid Interfacial Tension

$\gamma_{s/l}$, the liquid/solid interfacial tension, is the reversible work of forming a unit of solid/liquid interface. It is defined by Eq. (4)

$$\gamma_{s/l} = \left(\frac{\partial G_{s/l}}{\partial \Omega_{s/l}}\right)_{T,P,N_1,\ldots,N_j} \tag{4}$$

$G_{s/l}$ is the Gibbs excess free energy of the solid/liquid interface and $\Omega_{s/l}$ the solid/liquid area. As with $\gamma_{s/o}$, $\gamma_{s/l}$ needs to be distinguished from the stress in the interface. See Appendix I.

C. Young and Modified Young Equations

Equation (5) was first stated in words by Young [10]. It was advanced on a rigorous and useful thermodynamic basis by Gibbs [11]

$$\cos\theta = \frac{\gamma_{s/v} - \gamma_{s/l}}{\gamma_{l/v}} \tag{5}$$

θ is the contact angle defined in Fig. 1. Gibbs derived this equation by minimizing the free energy of the system in a virtual displacement of all the surfaces and interfaces [8]. Gibbs emphasized the difficulty of measuring $\gamma_{s/v}$ and $\gamma_{s/l}$ and pointed out that they do not have to be known independently. Only the free energy of interaction between the liquid and solid is needed. Gibbs emphasized this idea by deriving the Gibbs-modified Young's equation, Eq. (6). He considered this equation to be both simpler and more general than Young's equation

$$\cos\theta = \frac{\pi_{s/l} - \pi_{s/v}}{\gamma_{l/v}} \tag{6}$$

where

$$\pi_{s/l} = \gamma_{s/o} - \gamma_{s/l} \tag{7}$$

and

$$\pi_{s/v} = \gamma_{s/o} - \gamma_{s/v} \tag{8}$$

The Gibbs derivation is given in Appendix II.

$\pi_{s/l}$ measures the tendency of the liquid to spread. It is the negative of the quantity Gibbs calls the *superficial tension of the fluid in contact with the solid*. It has the units of force per unit length. $\pi_{s/v}$, defined by Eqs. (3) and (8), opposes the spreading of the liquid. $\gamma_{l/v}$ measures the tendency

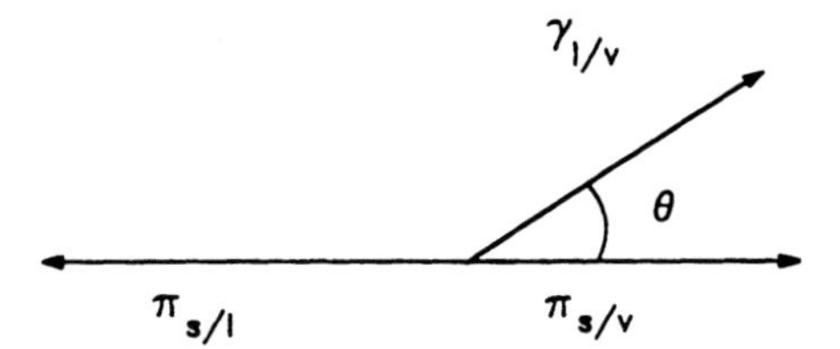

FIG. 2 Vector notation representing forces on the liquid at the three phase boundary.

of the drop to contract under surface tension forces. The interactions of these surface pressures and the liquid surface tension can be represented by vectors. This leads to the familiar diagram shown in Fig. 2. The vectors in Fig. 2 are often shown as surface or interfacial tensions. This is permissible as long as it is recognized that the vectors do not represent stresses in the solid surface, but forces on the fluid at the three-phase boundary.

For low-energy surfaces, $\pi_{s/v}$ is zero or nearly zero when θ is greater than zero. This is observed experimentally. It is usually explained by assuming that a liquid will not adsorb on a solid having a lower surface tension than the liquid. In this case, Eq. (6) simplifies to Eq. (9). Equation (9) clearly shows the competition between adhesive forces $\pi_{s/l}$ and cohesive forces $\gamma_{l/v}$

$$\cos\theta = \frac{\pi_{s/l}}{\gamma_{l/v}} \tag{9}$$

It is sometimes said that the only requirement for good wetting is that the solid surface tension is larger than the liquid surface tension. From Young's equation, it is clear that this is a necessary but not sufficient condition.

D. Works of Adhesion and Cohesion

1. Equilibrium Work of Adhesion

The equilibrium work of adhesion W_a is the reversible work required to separate a unit area of liquid from a solid. It is defined by Eq. (10)

$$W_a = \gamma_{l/v}(1 + \cos\theta) \tag{10}$$

Equation (10) is derived by noting that when a unit area of liquid is separated from a solid, a unit of solid/vapor surface area and a unit of liquid/vapor surface are created, and a unit of solid/liquid interface dis-

TABLE 1 Comparison of Spreading Pressures (20°C) and Works of Adhesion for Several Solid Surfaces

Solid	Liquid	W_a ($mJ\ m^{-2}$)	$\pi_{s/v}$ ($mJ\ m^{-2}$)
Graphite	n-Heptane	40.8	55.2
	Benzene	57.8	76.2
Anatase	n-Heptane	40.8	58.2
	Benzene	57.8	84.2
	n-Propanol	47.6	114.4
	Water	145.6	300.4
Silica	n-Heptane	40.8	59.2
	Benzene	57.8	80.2
	Acetone	65.4	90.6
	Propanol	47.6	134.4
	Water	145.6	316.4
SnO_2	n-Heptane	40.8	57.2
	Benzene	57.8	76.2
	n-Propanol	47.6	104.4
	Water	145.6	292.4

appears. The separation procedure leaves an adsorbed vapor layer on the solid surface that is in equilibrium with the liquid. The free energy change associated with this pressure is $\gamma_{s/v} + \gamma_{l/v} - \gamma_{s/l}$, which combined with Young's equation yields Eq. (10).

2. Nonequilibrium Work of Adhesion

The nonequilibrium work of adhesion is defined by Eq. (11). It is deduced by the same conceptual process used in deriving Eq. (10), except that no adsorbed layer exists after separation of the phases. When $\theta = 0$, spreading pressures are often larger than W_a, the equilibrium work of adhesion. Some examples of $\pi_{s/v}$ and W_a, calculated from data given by Fowkes [12], are shown in Table 1

$$W_a^* = W_a + \pi_{s/v} = \gamma_{l/v}(1 + \cos\theta) + \pi_{s/v} \tag{11}$$

The nonequilibrium work of adhesion is important because this quantity links thermodynamics to molecular theory. In both the theories of Hamaker and Lifshitz (which are reviewed in Sec. IV.B), the free energy of interaction, per unit area, required to bring parallel plates of solid and liquid to a separation D is given by Eq. (12)

$$E = -\frac{1}{12\pi}\frac{A_{s/l}}{D^2} \tag{12}$$

The Hamaker constant at the solid/liquid interface is defined in Sec. IV. If the plates are brought to a separation D_0, the equilibrium separation, then E becomes the negative of W_a^*.

3. Work of Cohesion

The work of cohesion W_c is a good measure of molecular interactions for symmetrical molecules. It is defined by Eq. (13)

$$W_c = 2\gamma_{l/v} \tag{13}$$

Equation (13) is deduced by separating a column of liquid into two parts, creating two liquid/vapor surfaces. The free energy of this process is $2\gamma_{l/v}$. Orientation of molecules may occur during this process. We do not define a nonequilibrium work of cohesion because surface orientation is usually very rapid. This may not be true for highly viscous liquids such as molten glasses or polymers.

Equation (14) is another useful equation illustrating the competition between adhesive forces W_a and cohesive forces W_c

$$\cos\theta = \frac{2W_a}{W_c} - 1 \tag{14}$$

E. Adhesion Tension

$\tau_{s/l}^*$, the nonequilibrium adhesion tension, measures the force per unit length on the liquid at the three-phase boundary. It can be positive or negative

$$\tau_{s/l}^* = \gamma_{s/o} - \gamma_{s/l} = \gamma_{l/v}\cos\theta + \pi_{s/v} = \pi_{s/l} \tag{15}$$

The spreading of the liquid is opposed by adsorption at the solid/vapor interface. The net force on the liquid at the three-phase boundary is the equilibrium adhesion tension $\tau_{s/l}$. It is defined by Eq. (16)

$$\tau_{s/l} = \gamma_{s/v} - \gamma_{s/l} = \pi_{s/l} - \pi_{s/v} = \gamma_{l/v}\cos\theta \tag{16}$$

$\tau_{s/l}$ is equal to $\gamma_{l/v}$ for all surfaces having zero contact angles. This is the reason materials of construction have no effect on the capillary rise method for measuring surface tension, as long as the contact angle is zero.

F. Spreading Coefficients

1. Equilibrium Spreading Coefficient

The equilibrium spreading coefficient W_s is defined by Eq. (17)

$$W_s = \gamma_{s/v} - \gamma_{l/v} - \gamma_{s/l} = \gamma_{l/v}(\cos\theta - 1) \tag{17}$$

It can also be defined by Eq. (18)

$$W_s = \pi_{s/l} - \pi_{s/v} - \gamma_{l/v} \tag{18}$$

The equilibrium spreading coefficient can only be negative or zero. Some texts suggest the possibility of a positive equilibrium spreading coefficient. This is caused by its confusion with the nonequilibrium coefficient.

2. Nonequilibrium Spreading Coefficient

The nonequilibrium spreading coefficient W_s^* is defined by Eq. (19)

$$W_s^* = \gamma_{s/o} - \gamma_{l/v} - \gamma_{s/l} = \gamma_{l/v}(\cos\theta - 1) + \pi_{s/v} = \pi_{s/l} - \gamma_{l/v} \qquad (19)$$

When a liquid drop initially contacts a solid surface, W_s^* can be positive. Harkins [13] called the nonequilibrium spreading coefficient the initial spreading coefficient. He called the equilibrium coefficient the final spreading coefficient. As adsorption occurs at the solid/vapor interface, W_s^* decreases until it becomes zero or negative at equilibrium. Systems with an initial positive W_s^* and a negative W_s are termed autophobic. They usually are liquids composed of unsymmetrical molecules that readily adsorb onto the solid. A typical example is molten perfluorodecanoic acid on glass. A drop of this material will first spread and then in time retract to produce a drop with a finite contact angle.

G. When the Contact Angle Approaches Zero

When $\theta = 0$, Young's equation reduces to

$$\gamma_{l/v} = \gamma_{s/v} - \gamma_{s/l} \qquad (20)$$

and the Gibbs-modified Young's equation becomes

$$\gamma_{l/v} = \pi_{s/l} - \pi_{s/v} \qquad (21)$$

Equation (20) is known as Antonoff's [14] rule or equation.

Some have objected to Eqs. (20) and (21) on the basis that Young's equation is not applicable when θ is zero. In fact, Young's equation is valid when θ is zero. Such objections can usually be resolved by distinguishing between equilibrium and nonequilibrium adhesion tensions. The book by Landau and Lifshitz [15] provides an excellent discussion on the physics of this system.

A more serious problem arises when Antonoff's equation is applied to systems having a finite contact angle. Since the criterion for the validity of Antonoff's rule is a zero contact angle, applying it where the contact angle is not zero will lead to absurdities.

A useful picture of wetting with zero contact angle was given many years ago by Harkins [13]. Equation (3) shows that $\pi_{s/v}$ is determined by integrating $\Gamma_{s/v}/P$ vs. P from P to P_0. A criterion for a finite contact angle is that $\Gamma_{s/v}$ be finite at $P = P_0$. The criterion for a zero contact angle is that $\Gamma_{s/v}$ approach the $P = P_0$ axis asymptotically. As P approaches P_0, the adsorbed layer becomes so thick that the outer surface of the adsorbed film has the properties of bulk water. That is, the outermost surface is not affected by the substrate. Harkins called such a film *duplex*. The work of forming a unit area of the entire surface $\gamma_{s/v}$ of a duplex film is the work of forming the liquid surface $\gamma_{l/v}$, plus the work of forming the inner interface $\gamma_{s/l}$

$$\gamma_{s/v} = \gamma_{l/v} + \gamma_{s/l} \tag{22}$$

This is Antonoff's rule.

The above argument is not a rigorous derivation, but a useful description. In summary, Antonoff's rule is correct only when $\theta = 0$. When $\theta = 0$, $\pi_{s/v}$ becomes large and cannot be ignored. These relations are most easily understood by examining liquids spreading on liquids. Examples are given in Sec. VI.

III. CONTACT-ANGLE HYSTERESIS AND EXPERIMENTAL PROCEDURES

Contact-angle hysteresis is caused by the existence of many thermodynamic metastable states for systems having three-phase (solid/liquid/vapor) boundaries. A different contact angle is associated with each metastable state. The maximum stable angle is called the advancing (or advanced) angle. The minimum stable angle is referred to as the receding (or receded) angle. Hysteresis is the difference between the two.

A. Experimental

There are two main techniques for measuring the contact angle. These are the sessile drop and the Wilhelmy plate methods. Figure 3 is a schematic of a sessile drop showing an advancing and receding angle.

In the sessile drop method, a drop is placed on a horizontal surface and observed in cross section through a telescope. A goniometer in the eyepiece is used to measure the angle. The angle of vision is just slightly off horizontal so the edge of the drop and its reflected image are both visible. This allows the tangent to be determined precisely at the point of contact between the drop and the surface. Agreement between different individuals measuring the same drop will be about 1–2°. Several measurements are often made on both sides of the drop and the numbers averaged. This compensates for any deviation of the surface from the horizontal.

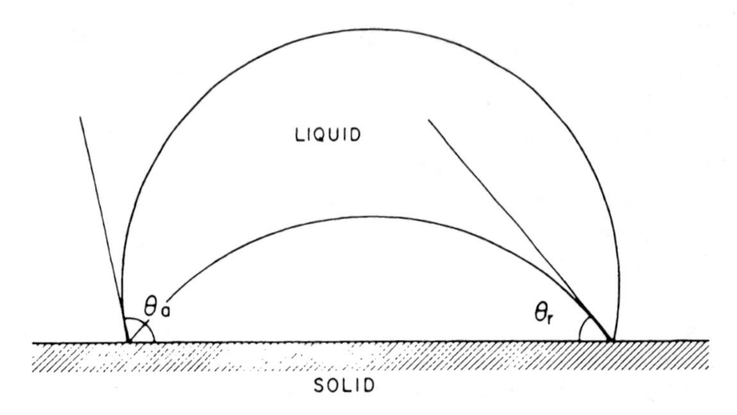

FIG. 3 Schematic of a sessile drop on a surface. Both advancing angles and receding angles are shown.

Major errors do not arise from the measurement itself, but rather not measuring the proper angle. If angles other than the advancing or receding angles are measured, the reproducibility becomes poor and the interpretation of the data becomes difficult.

We have found the following precautions useful:

Advance the angle by adding liquid slowly from a syringe mounted above the drop. Leave the needle in the drop after the addition. Deformation of the drop by contact with the needle will not change the contact angle.
Measure the angle immediately (within 10 sec) after each addition of liquid.
When two consecutive angles are the same, record the angle and repeat the measurement on the opposite side of the drop.
Determine the receding angle by pulling small amounts of liquid back into the syringe.
When two consecutive angles are the same, record the angle and repeat the measurement on the opposite side of the drop.
Do not let the drop become too small. The receding angle is constant only when the size of the drop is much larger than the size of the heterogeneities causing the hysteresis. Typically, a drop should not become smaller than 0.01 ml.

Because the drop is not moving when the measurement is made, these angles are sometimes called advanced or receded angles. We will continue to use the terms advancing and receding, but when applied to sessile

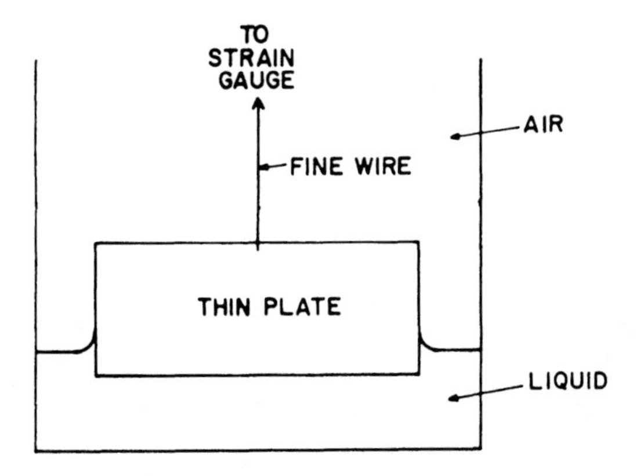

FIG. 4 Schematic drawing of the Wilhelmy plate method for measuring contact angles.

drops, they will refer to the stationary drop after being advanced or receded.

The major advantages of the sessile drop technique are speed and convenience.

The second common method of measuring the contact angle is the Wilhelmy plate technique. This is shown schematically in Fig. 4. A thin plate or cylinder is mounted vertically above the liquid. To begin the measurement, the plate is suspended with the bottom edge nearly touching the surface of the liquid. This is the position of zero force. The plate is lowered (or often the liquid is raised) until it touches the liquid. This is the zero position. The force on the plate is measured as it is cycled slowly down and up. A typical rate of immersion is 2 mm/min. The contact angle is calculated from the equation in Fig. 4 for a plate having a perimeter of 1 cm. The advancing angle is the stable angle as the plate is lowered. The receding angle is the stable angle as the plate is raised. Figure 5 shows typical force curves obtained on a fluoropolymer, a copolymer of tetrafluoroethylene and perfluoropropylene. The top curve illustrates hysteresis loops determined with water and the bottom curve loops determined with hexadecane.

Major advantages of the Wilhelmy plate method are (1) unlike the sessile drop technique, it is independent of the person making the measurement. (2) The conditions of measurement are highly reproducible. (3) There is no problem associated with drop size. (4) The speed of movement

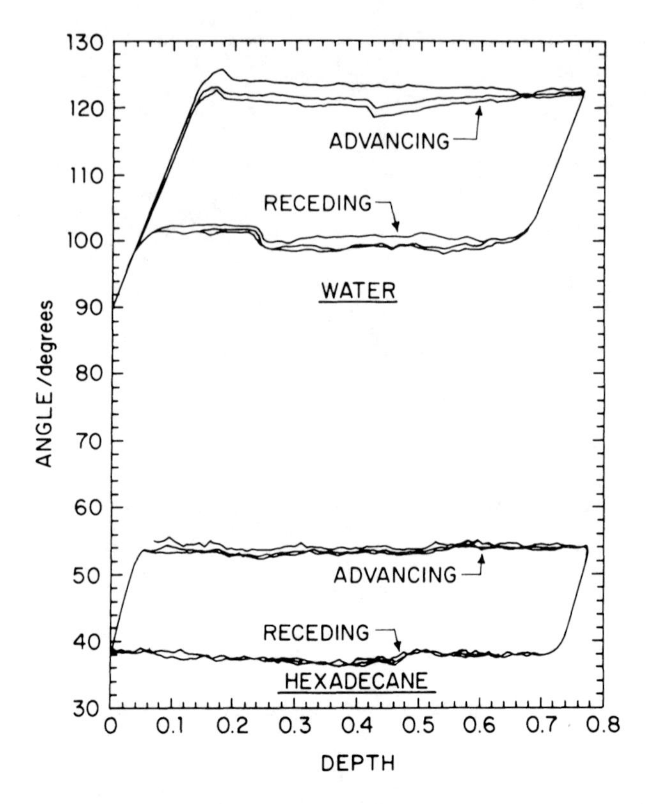

FIG. 5 Wettability of a fluoropolymer by water and hexadecane. Contact angle vs. depth of immersion by the Wilhelmy plate method (three cycles).

of the three-phase boundary is readily controlled. (5) The sensitivity of this technique is very high. Under good conditions changes of as little as 0.5° or less in contact angle can be measured. The major disadvantage is that the wetting surface must be in the form of a sheet, rod, or fiber and all surfaces touching the liquid must have the same wettability. Close agreement can be obtained between the Wilhelmy technique and careful measurements by the sessile drop technique. The plate method is especially useful when kinetic effects (adsorption, desorption, etc.) are important.

B. Hysteresis

Hysteresis is always present. It must be considered when applying theories developed for ideal, hysteresis-free, models to experimental results.

Hysteresis is caused by metastable states at the solid/liquid/vapor interface. Each of these states is characterized by a contact angle. These states can be produced by surface heterogeneity, surface roughness, or surface deformability. Barriers between the metastable states have to be larger than kT. If there is vibration in the system, the energy barriers need to be larger than the energy of the vibration. Surface heterogeneity is the major cause of hysteresis. Small amounts of surface heterogeneity can lead to large hysteresis. Although the heterogeneous regions need to be macroscopic, the regions can be small, perhaps as small as 4 or 5 nm in diameter.

Johnson and Dettre [16] analyzed a model heterogeneous surface made of concentric circular bands. The most important results of their analysis are summarized in Fig. 6. The *X* axis is the fraction of the surface covered with low contact-angle (high-energy) material. The *Y* axis is the calculated contact angle. The upper curve corresponds to the advancing angle. The lower curve corresponds to the receding angle. According to this model, advancing angles are associated with regions of low wettability and the receding angles with regions of high wettability.

Also included in Fig. 6 is a curve calculated for a heterogeneous surface, in which the individual areas of the two surface components are too

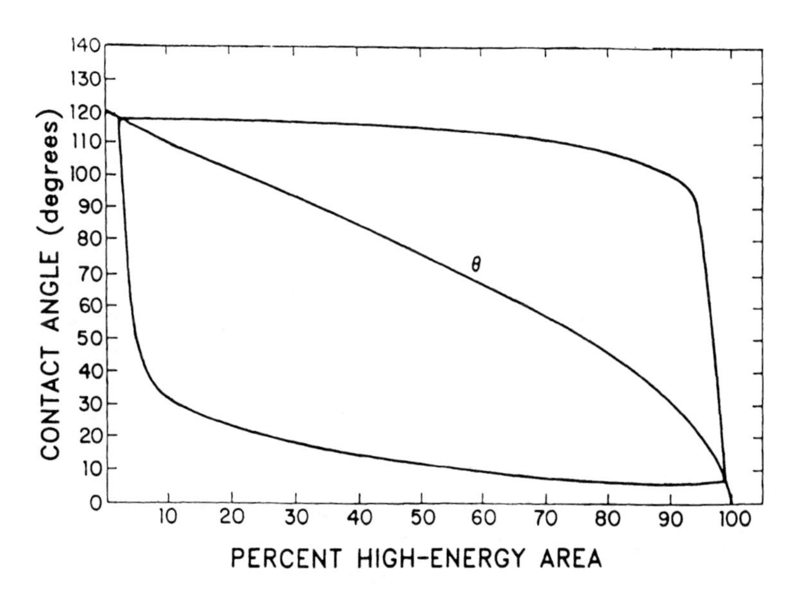

FIG. 6 Contact angles as a function of surface coverage. Concentric circular-band model. (From Ref. 22c by permission.)

small to produce metastable states. This curve, labeled θ, corresponds to a minimum free energy for the system [2]. It is calculated by the Cassie [17] equation, Eq. (23)

$$\cos \theta = f_1 \cos \theta_1 + f_2 \cos \theta_2 \tag{23}$$

θ is the contact angle of a liquid on the heterogeneous surface, θ_1 and θ_2 are the contact angles on each of the two components, and f_1 and f_2 are their respective surface fractions. As the size of the heterogeneous regions become smaller, the outer curves get closer to the Cassie curve.

It has been proposed in a recent paper [18] that the Cassie equation be replaced by Eq. (24) whenever the individual areas of the surface components approach molecular or atomic dimensions

$$(1 + \cos \theta)^2 = f_1(1 + \cos \theta_1)^2 + f_2(1 + \cos \theta_2)^2 \tag{24}$$

In deriving the above equation, the authors have inadvertently made an algebraic error. If Eqs. (6) and (7) in the authors' paper are substituted into Eq. (5) in the same paper, the result is Cassie's equation, Eq. (23), not Eq. (24). Equation (24) cannot be correct physically, because in the limit of heterogeneous regions being atoms (or groups of atoms), it does not reduce to Hamaker's approach of summing the interactions of each group. (See Sec. IV.B.) Cassie's equation does reduce to Hamaker's approach in the limit.

When the model of Fig. 6 applies, theories developed for ideal surfaces can be applied independently to advancing and receding angles. The remainder of this section examines this proposition.

Dettre and Johnson [19] estimated contact angles as a function of surface coverage by making partial coatings of TiO_2 (anatase) on glass. They showed that trimethyloctadecylammonium chloride adsorbed onto the glass, but not the anatase. The surface coverage was estimated from electron micrographs and a reasonable model of deposition. We refer the reader to the original paper for further details. Sessile drop contact angles were measured with water and methylene iodide. Figure 7 shows the curves for water and Fig. 8 the curves for methylene iodide. These data show that advancing angles are nearly independent of coverage by TiO_2 when the coverage is less than about 50%. Receding angles are much more sensitive to coverage. Johnson et al. [20] also estimated hysteresis as a function of surface coverage using radioactive monolayers to monitor the coverage. Radioactive stearic acid and barium stearate monolayers (labeled with carbon 14) were deposited onto glass slides from a Langmuir trough. The monolayers were depleted by immersion in benzene. Contact angles were measured with water and hexadecane by the Wilhelmy plate method. Table 2 shows the changes in contact angle with the number of

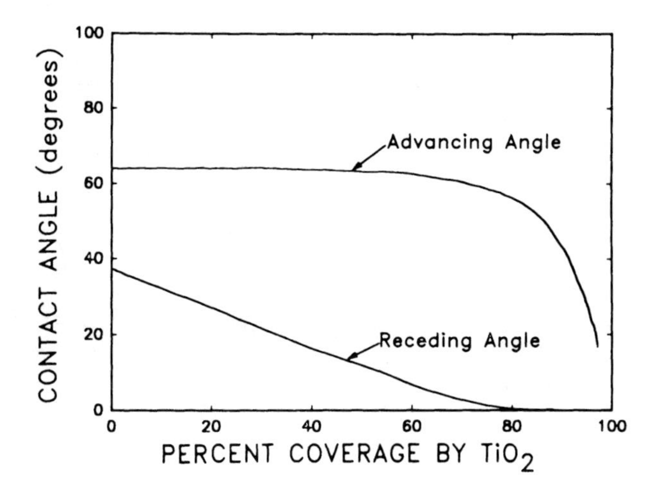

FIG. 7 Water wettability of titania-coated glass after treatment with trimethyloctadecylammonium chloride. (From Ref. 19 by permission.)

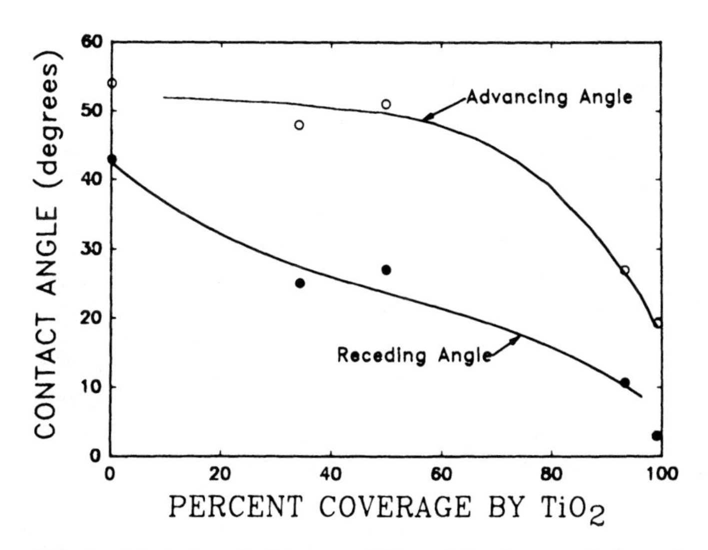

FIG. 8 Methylene iodide wettability of titania-coated glass after treatment with trimethyloctadecylammonium chloride. (From Ref. 19 by permission.)

TABLE 2 Contact Angles of Water and Hexadecane on Complete Monolayers of Stearic Acid (SA) and Barium Stearate (BS)

Cycle number	Monolayer	Water angles (deg)		Hexadecane angles (deg)	
		Adv.	Rec.	Adv.	Rec.
1	SA	107.7	42.2	45.8	43.8
2	SA	97.1	27.8	46.3	44.0
3	SA	90.0	23.4		
1	BS	114.3	28.5	46.3	44.9
2	BS	103.0	12.4	46.9	45.2
3	BS	77.7	12.4		

plate immersion cycles for the monolayers before the depletion measurement. Figures 9 and 10 show plots of contact angle (first immersion cycle) vs. depletion in benzene, as determined from radioactivity measurements.

The data in Table 2 provide a good illustration of the kind of ambiguities that arise with contact-angle measurements. They are usually associated with interactions occurring between the wetting liquid and the solid surface during the measurement. It is apparent that water removed some of the monolayer since the water angle decreased with the number of cycles. Also, the water receding angles were much smaller than expected for a homogeneous Langmuir–Blodgett monolayer.

Although the hexadecane contact angles on the monolayers of Table 2 changed very little and showed low hysteresis, autoradiography measurements indicated that the monolayers were desorbing into the hexadecane during the measurement. Evidently, the hexadecane molecules replaced stearic acid and barium stearate molecules in the monolayer. The hexadecane molecules are much like the molecules of the monolayer in size and shape. If a molecule of hexadecane replaces a molecule from the monolayer, the surface will appear much as it did before the depletion. This mechanism was suggested by Bartell and Ruch [21] to account for some of their wetting observations. When a monolayer is highly depleted, as in the above depletion measurements, adsorption is no longer possible and the contact angles drop rapidly with the additional depletion (see Fig. 10).

An attempt was made in the above work to see if it were possible to determine surface coverage from the contact-angle data of Figs. 9 and 10. Advancing angles alone cannot be used for this purpose. The authors

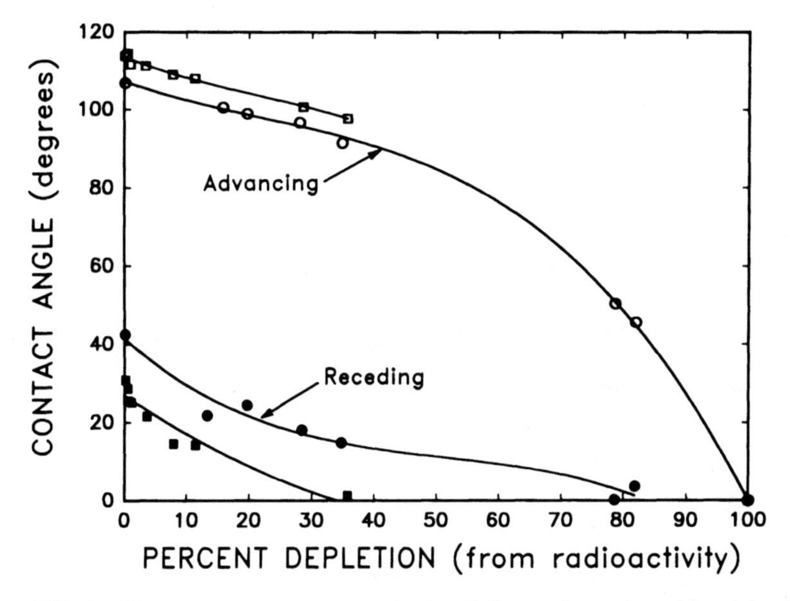

FIG. 9 Water contact angles on depleted films of stearic acid and barium stearate. (From Ref. 20 by permission.)

[20], after labored reasoning that will not be reproduced here, arrived at a relative surface coverage index I defined by Eq. (25)

$$I = 100 \left(\frac{\langle \cos \theta \rangle - \cos \theta_{\min}}{\langle \cos \theta^{\circ} \rangle - \cos \theta_{\min}} \right) \tag{25}$$

where

$$\langle \cos \theta \rangle = (\cos \theta_a + \cos \theta_r)/2$$

$$\langle \cos \theta^{\circ} \rangle = (\cos \theta_a^{\circ} + \cos \theta_r^{\circ})/2$$

θ_a is the experimental receding angle, θ_r the experimental receding angle, $\theta_{\min}$ the angle on the completely clean surface, θ_a° the initial advancing angle, and θ_r° the initial receding angle.

When θ is zero, Eq. (25) reduces to Eq. (26)

$$I = 100 \left(\frac{\langle \cos \theta \rangle - 1}{\langle \cos \theta^{\circ} \rangle - 1} \right) \tag{26}$$

Figure 11 shows the relative coverage index plotted against percent coverage as measured by radioactivity. The index roughly correlates with

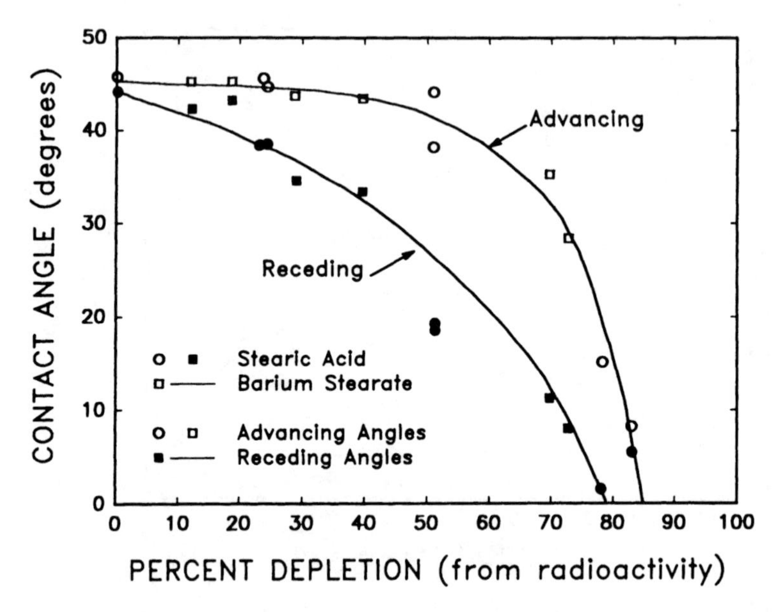

FIG. 10 Hexadecane contact angles on depleted films of stearic acid and barium stearate. (From Ref. 20 by permission.)

the coverage measured by radioactivity even when depletion and/or adsorption were occurring.

Numerous studies of hysteresis have been made [22]. The differences in the studies have existed primarily in the details of the models employed. Both theoretical and experimental models have been used. The conclusions agree qualitatively with our discussion and with each other. It is unlikely that general quantitative models will be developed for two reasons. The first is that the mathematics become very difficult when the geometry becomes more complex than the concentric band model. The second is that to calculate hysteresis quantitatively for a given surface requires more knowledge of the surface than is usually available. The most important result of the studies is that the advancing angle is a good measure of the low-energy portion of a surface.

From the beginning of contact-angle studies, there have been attempts to attribute hysteresis to changes occurring in a surface during the measurement. It was thought that somehow the surface changed as the three-phase boundary advanced and receded over the surface. These ideas have been replaced by the concepts of metastable states and heterogeneity. There have been some suggestions that hysteresis with polymers can be

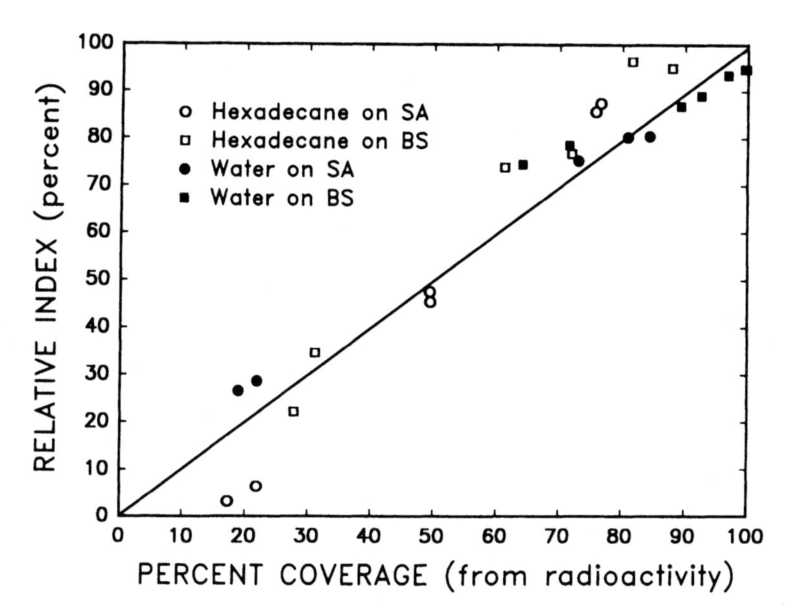

FIG. 11 Surface coverage estimated from wettability compared with coverage determined by radioactivity. Stearic acid and barium stearate on glass. (From Ref. 20 by permission.)

caused by changes in the orientation of a polymer that occur during measurement. Changes of orientation can give the appearance of hysteresis only if the time scale of the changes are the same order as the time scale of movement of the three-phase boundary. If the change of orientation is fast, the system will always be in equilibrium and there no difference will exist between advancing and receding angles. The same will be true if the change of orientation is slow.

The easiest way to check for hysteresis, particularly with the Wilhelmy plate technique, is to do repeat immersion-withdrawal cycles. The repeat cycles will superimpose if hysteresis is present. All the points on both the immersion and withdrawal paths will correspond to stable, reproducible values of the contact angle [23]. Moreover, all points within the hysteresis loop are accessible through control of the position of the plates to give "scanning curves" [23]. Differences in contact angles caused by changes in the surface will be apparent in the curves, which will change with each cycle. A good example of this is the water wettability of the radioactive monolayers discussed above, where contact angles change with each successive cycle (see Table 2.)

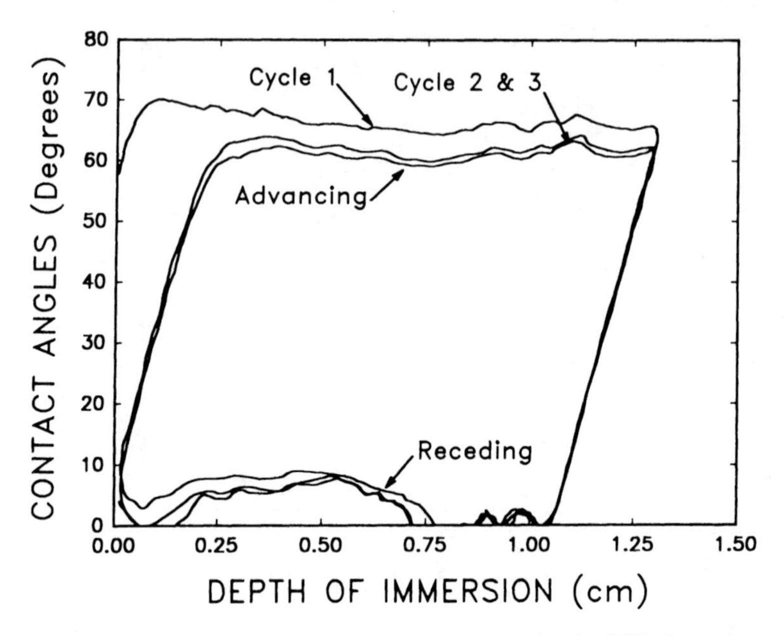

FIG. 12 Hysteresis loops for water on nylon 6 by the Wilhelmy plate method.

If there is still some question whether changes in contact angle are due to hysteresis or changes in the surface caused by interactions with the wetting liquid, one should make measurements at different wetting rates. The Wilhelmy plate method is ideal for doing this because the advancing and receding angles are independent of immersion rate over a wide range [24].

The plate method is particularly useful in studying changes occurring at polymer surfaces. Figure 12 shows a hysteresis curve for nylon in water. There is a slight but significant change in the advancing angle as the film cycles. This is caused by absorption of water into the nylon. Figure 13 illustrates a similar curve for poly(ethylene terephthalate), where there is no change in wettability with an increasing number of cycles.

Most contact-angle data are presented as either the contact angle or the cosine of the contact angle. Either form is useful when geometric considerations such as the penetration of liquids into fabric or porous media are important. When we are interested in gaining information about the nature of the surfaces, however, it is often better to present data as works of adhesion.

Figure 14 is the wettability data of Fig. 5 presented as work of adhesion. Figure 5 gives the impression that the attraction of the water to the surface

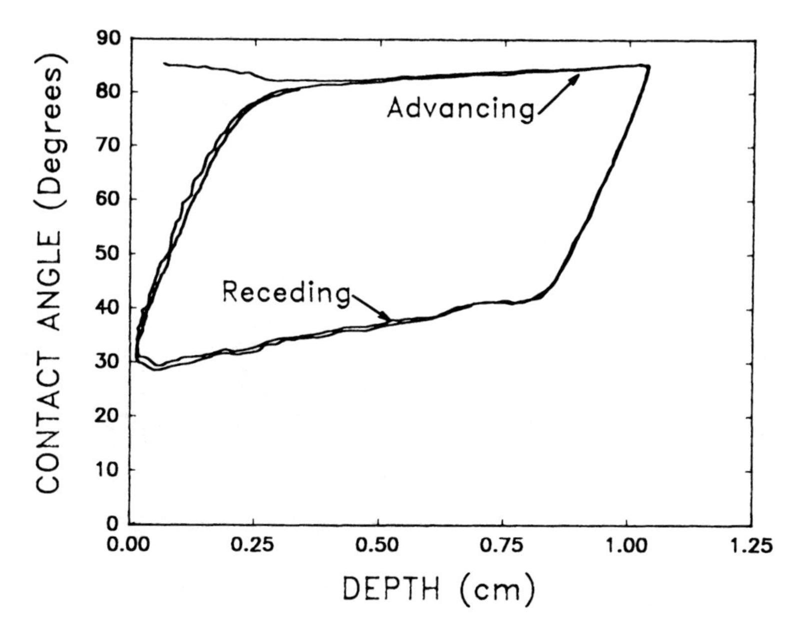

FIG. 13 Hysteresis loops for water on poly(ethylene terephthalate) by the Wilhelmy plate method (two loops).

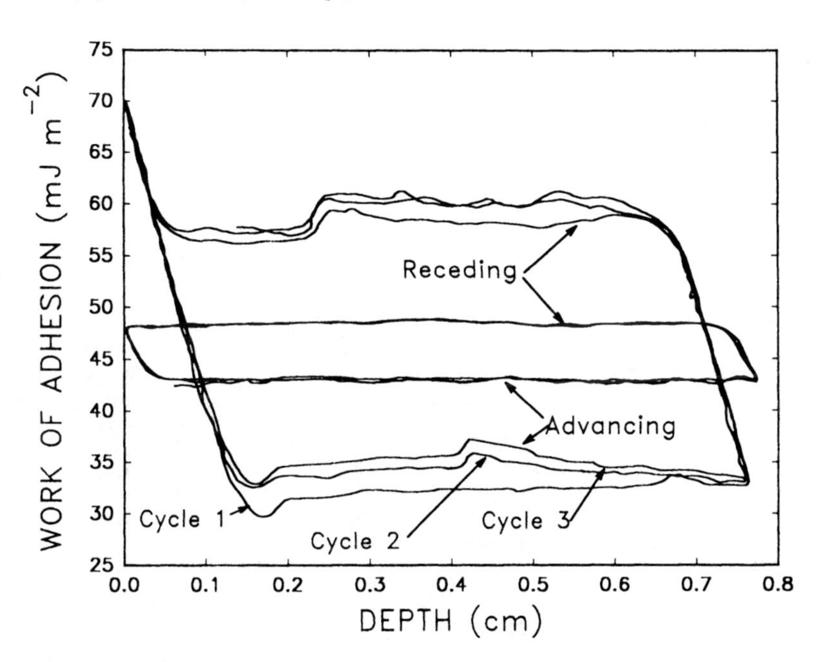

FIG. 14 Wettability presented as work of adhesion (water and hexadecane on a fluoropolymer). Data from Fig. 5.

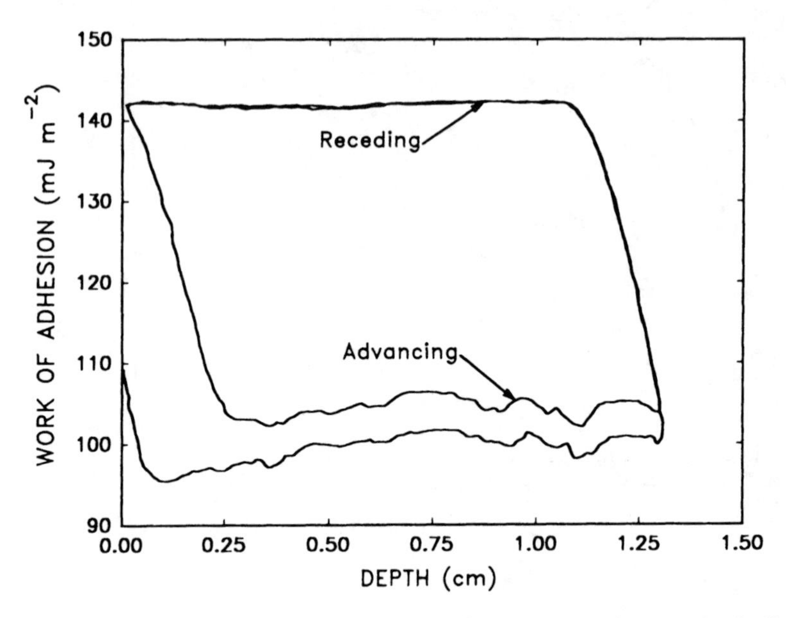

FIG. 15 Hysteresis loops for water on nylon 6 presented as work of adhesion.

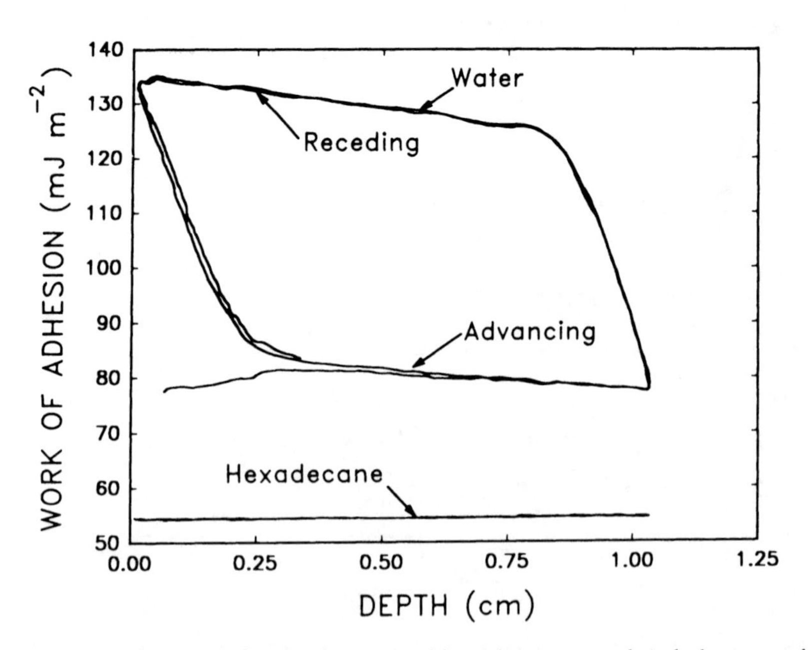

FIG. 16 Hysteresis loops of water and hexadecane on poly(ethylene terephthalate) presented as work of adhesion.

of the fluoropolymer is always less than that of the hexadecane. Figure 14 shows, however, there are sites on the fluoropolymer that interact with the water more strongly than with the hexadecane. This is shown by the large receding work of adhesion. This stronger work of adhesion to water is most probably associated with hydrogen bonding. The hysteresis curve with hexadecane shows that these sites also have a higher dispersion interaction (Hamaker constant) than the predominant low-energy sites.

Figures 15 and 16 illustrate the hysteresis curves of nylon and poly(ethylene terephthalate) in terms of works of adhesion. These data show that the low-energy portion of the nylon (the majority component in the surface) is much more wettable than the low-energy portion of the poly(ethylene terephthalate). On the other hand, the high-energy portions of both polymers have nearly the same wettability.

IV. WETTING OF LOW-ENERGY SURFACES

A. Historical Perspective

Zisman [25] and co-workers provided the impetus for nearly all later wettability studies. Bartell, his co-workers, and a few others had introduced some important concepts, but it was only with Zisman that the work began to flourish. At the time Zisman began his research, the major characteristic of wettability was poor reproducibility. Many papers were being written with titles like "Is the Contact Angle a Thermodynamic Quantity?" Zisman brought reproducibility into the field primarily by working with well-characterized systems and measuring advancing contact angles. A major portion of his work involved fluorinated surfaces. This was advantageous because the contact angles on fluorinated surfaces are easily distinguished from those on hydrocarbon surfaces. Objections related to supposed contamination were almost eliminated.

We can summarize some of the conclusions of Zisman as follows:

When low-energy surfaces are wetted by alkanes, the cosine of the contact angle plotted against the surface tension of the wetting liquid (the Zisman plot) is a straight line. The intercept of this line with the $\cos \theta = 1$ axis is called the critical surface tension or γ_c. Figure 17 shows typical Zisman plots for a series of perfluoroacid monolayers [26].

When contact angles are measured on low-energy surfaces by liquids that can form internal hydrogen bonds or otherwise associate, the Zisman

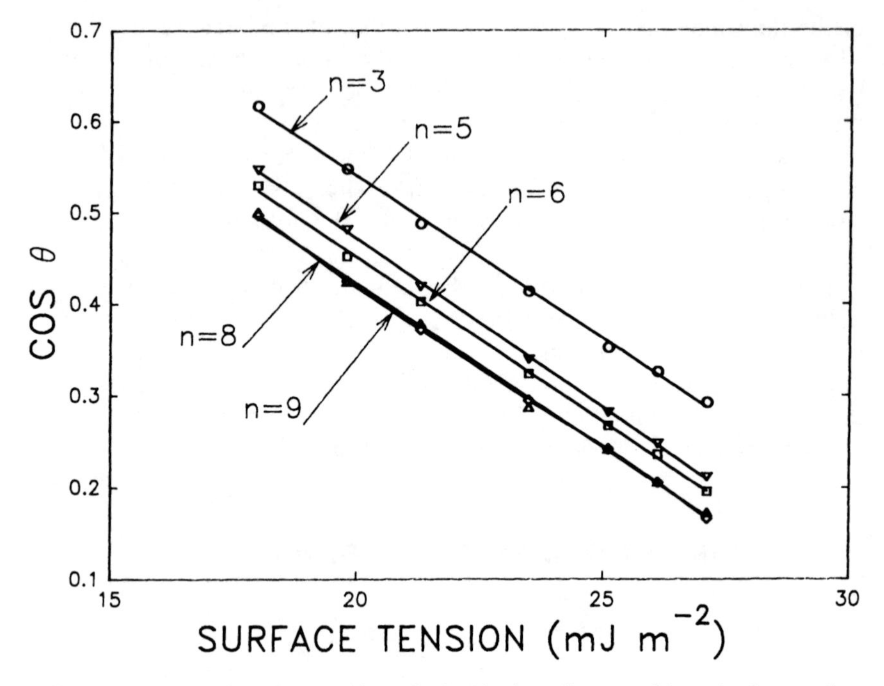

FIG. 17 Zisman plots for a series of adsorbed perfluoro acids. *n* is the number of fluorinated carbons. (From Ref. 26 by permission.)

plot shows curvature and the points no longer fall on a single curve. Zisman included the points in a band.

The outermost chemical groups in the surface determine wettability.

The repellency of fluorocarbon and hydrocarbon groups in the surface decreases in the order $—CF_3 > —CF_2— = —CH_3 > —CH_2—$.

Slight differences between surfaces having similar structures are due to differences in packing. The longer-chain-length acids in Fig. 17 have lower critical surface tensions. Zisman attributed this to closer packing by longer chains.

Branched-chain monolayers have lower contact angles than straight chain because of poorer packing.

Table 3 lists the critical surface tensions of various groups as determined by Zisman.

As soon as sufficient amounts of reliable and reproducible data became available, attempts were made to explain Zisman's results. An intriguing result was that for hydrocarbons on low-energy fluorocarbons, the contact

TABLE 3 Critical Surface Tensions of Selected Groups as Determined by Zisman

Surface constitution	γ_c (mJ m^{-2}, 20°C)
$—CF_3$	6
$—CF_2H$	15
$—CF_2—$	18
$—CH_3$ crystal	22
$—CH_3$ monolayer	24
$—CH_2—$	31
Polymers	
Poly(tetrafluoro ethylene)	18.5
Polyethylene	31
Poly(ethylene terephthalate)	43
Poly(hexamethylene adipamide)	46

Source: Ref. 25 by permission.

angle was a function only of the surface tension of the liquid. This meant that for these liquids and solids, the interfacial tension was a function of only the solid and liquid surface tensions. This became the focus of much of the initial theoretical work. Girifalco and Good [27] wrote seminal papers. They argued that if the surface tensions (surface-free energies) are a result of only dispersion force interactions, then the combining rules for the surface and interfacial tensions should be similar to those for van der Waals interactions between dissimilar molecules in the gas phase. Utilizing the Bertholet [28] relation for interactions of molecules, they developed Eq. (27)

$$\gamma_{s/l} = \gamma_{s/o} + \gamma_{l/v} - 2\sqrt{\gamma_{s/o}\gamma_{l/v}} \tag{27}$$

where $\gamma_{s/l}$ is solid/liquid interfacial tension, $\gamma_{s/o}$ solid surface tension in a vacuum, and $\gamma_{l/v}$ the liquid surface tension. The Bertholet relation states that the dispersion interaction constant between two different gas molecules is the geometric mean of the interaction constants of these molecules. They emphasized that Eq. (27) can be applied only when (1) the surface has the same structure as the bulk. (2) The molecules of both phases are isotropic, that is, the orientation of the molecules at an interface does not affect any interactions. (3) The molecules interact only by van der Waals forces. These conditions are so stringent that no solid at room temperature meets them. Those solids that come closest are linear polymers such as polyethylene and poly(tetrafluoroethylene) (PTFE).

Combining Eq. (27) with Young's equation and setting $\pi_{s/v} = 0$ lead to Eq. (28)

$$\cos\theta = 2\sqrt{\frac{\gamma_{s/o}}{\gamma_{l/v}}} - 1 \tag{28}$$

where θ is the contact angle measured through the liquid.

For a solid surface meeting the criteria of Girifalco and Good, θ is a function only of the liquid surface tension $\gamma_{l/v}$. This was a major step toward understanding Zisman's results. Equation (28) does not predict a linear relationship between θ and $\gamma_{l/v}$, however.

Solving Eq. (28) for $\gamma_{s/o}$ leads to Eq. (29)

$$\gamma_{s/o} = \frac{\gamma_{l/v}(1 + \cos\theta)^2}{4} \tag{29}$$

For hexadecane on PTFE, $\theta = 46°$, $\gamma_{l/v} = 27.2$ mJ m^{-2}, leading to $\gamma_{s/o} = 19.5$ mJ m^{-2}. This is close to the critical surface tension of PTFE. Although there was some speculation that the critical solid surface tension itself was equal to the critical surface tension, Zisman himself never supported the idea.

Girifalco and Good recognized the limitations they had imposed and introduced an adjustable parameter Φ, shown in Eq. (30), to allow for deviations from Eq. (27). Φ differed from 1 when interactions occurred within the liquid or solid phase, but did not interact across the interface. These interactions could be caused by hydrogen bonding, metallic bonds, acid–base interactions, and unsymmetrical molecules

$$\gamma_{s/l} = \gamma_{s/o} + \gamma_{l/v} - 2\Phi\sqrt{\gamma_{s/o}\gamma_{l/v}} \tag{30}$$

A major advance in interpretation was made by Fowkes [29]. He recognized that the critical factors controlling wettability were interactions between the phases across the interfaces. He suggested that these interactions were related to the kinds of forces acting between the molecules of each phase. He assumed these forces to be independent of each other. Thus, if both hydrogen bonding and dispersion forces were acting across the interface, the total work of adhesion would be the dispersion work of adhesion plus the hydrogen bond work of adhesion. Fowkes' focus on interactions rather than surface and interfacial tensions was subtle but important.

With Fowkes' approach, Eq. (28) can be rewritten in terms of interactions rather than solid surface tensions. This interaction is defined by the dispersion surface tension components (γ_s^d and $\gamma_{l/v}^d$) of the total surface tension. For $\gamma_{l/v} = \gamma_{l/v}^d$, Eq. (28) becomes

$$\cos \theta = 2\sqrt{\frac{\gamma_{s/o}^{d}}{\gamma_{l/v}}} - 1 \tag{31}$$

and Eq. (29) becomes

$$\gamma_{s/o}^{d} = \frac{\gamma_{l/v}(1 + \cos \theta)^2}{4} \tag{32}$$

The total solid surface tension is no longer required to determine a contact angle. It should also be apparent that solid surface tensions cannot be determined from contact angles. Fowkes also showed that Hamaker constants can be estimated from contact angles and adsorption and calorimetric measurements.

For systems interacting only by dispersion and hydrogen-bonding forces,

$$W_a = W_a^d + W_a^h \tag{33}$$

where W_a^d is the work of adhesion due to dispersion forces and W_a^h the work of adhesion due to hydrogen-bonding forces.

Owens and Wendt [30], Kaelble [31], and others were soon applying Fowkes' concepts with equations like Eq. (34)

$$W_a = W_a^d + W_a^h + W_a^p \tag{34}$$

where W_a^p is a "polar" work of adhesion.

This approach is useful only as long as the additional terms are recognized as adjustable parameters. A major error that became associated with these equations, however, was the idea that the hydrogen-bonding and polar terms combined in Bertholet-type relations

$$\gamma_{s/l}^{h} = \gamma_{s/o}^{h} + \gamma_{l/v}^{h} - 2\sqrt{\gamma_{s/o}^{h}\gamma_{l/v}^{h}} \tag{35}$$

$$\gamma_{s/l}^{p} = \gamma_{s/o}^{p} + \gamma_{l/v}^{p} - 2\sqrt{\gamma_{s/o}^{p}\gamma_{l/v}^{p}} \tag{36}$$

Another error that was introduced about this time was the idea that the solid surface tension is simply the sum of the interactive terms. That this error is so common provides another good reason for writing the interactions in terms of works of adhesion rather than surface tensions.

Fowkes et al. [32] has since shown that all the interactions across an interface can be reduced to just two types: dispersion and acid–base. (Even hydrogen bonding is an acid–base interaction.) This requires two terms in the work of adhesion equation because for each acid group there must be a corresponding basic group. This leads to Eq. (37)

$$W_a = W_a^d + W_a^{A/B} + W_a^{B/A} \tag{37}$$

where A/B represents an acid–base interaction and B/A a base–acid interaction.

Concurrent with these developments were continuing attempts to calculate solid surface tensions from contact angles. Some attempts were incorrect extensions of the approach of Girifalco and Good. Most were related to attempts to equate Zisman's critical surface tension to the surface tension of the solid. The new approaches differed only in the sophistication used for the extrapolation. These theories all require that the solid/liquid interfacial tension go to zero at some point. These endeavors have culminated in the surface equation of state theory of Neumann et al. [33].

Neumann's surface equation of state theory goes farther than most. It attempts to prove thermodynamically that all liquids having the same surface tension will have the same contact angle on a given solid. This is equivalent to saying that the interfacial tension between any two phases having the same surface tensions will always be the same. This is simply not true.

This theory is readily tested by measuring interfacial tensions in liquid/liquid systems where all surface tensions, interfacial tensions, and spreading pressures can be independently measured. The theory fails to meet this test. (See Table 14 and the discussion in Sec. VI.) Because it fails when tested in liquid/liquid/vapor systems, Li et al. [33e] state that the same thermodynamics do not apply to liquid/liquid/vapor and solid/liquid/vapor systems. Morrison [34] has shown, however, that the liquid/liquid/vapor measurements do indeed provide valid tests of the surface equation of state.

Gibbs has commented on the relation between the surface thermodynamics of solids and liquids. In Ref. 9 (page 314) he introduces the thermodynamics of solid/liquid interfaces and is about to deduce Young's equation. He discusses how to handle stresses in the solid. Then before deriving the basic surface equations, he states, "The notions of the *dividing surface*, and of the *superficial densities* of energy, entropy, and the several components which we have used with respect to surfaces of discontinuity between fluids, will evidently apply without modification to the present case" (pages 219 and 224). He then derives Young's equation exactly as he derived the equilibrium conditions for liquid/liquid systems. The only difference between the two is deformability. Thermodynamics does not support Neumann's surface equation of state.

B. Molecular Models and Dispersion Theory

1. Hamaker Model

Dispersion forces are caused by fluctuations in electron clouds surrounding atoms. An understanding of van der Waals forces only became pos-

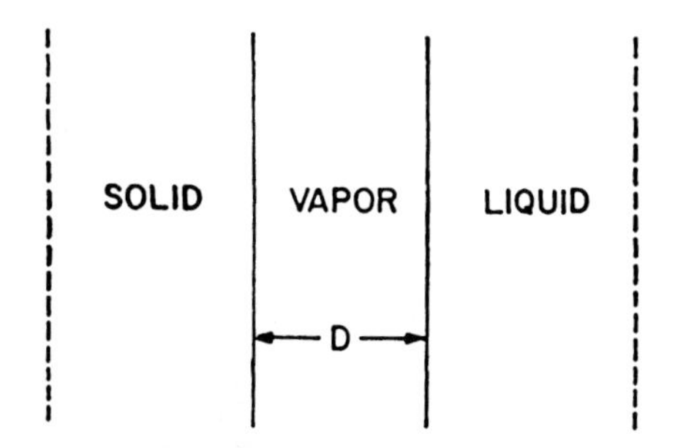

FIG. 18 Hamaker model.

sible with the development of quantum mechanics. They were first shown to account for attractive forces between nonpolar molecules such as nitrogen. Hamaker [3] was the first to show that intermolecular dispersion forces can account for attractive forces between colloids and macroscopic bodies. His model is shown in Fig. 18. The liquid and solid phases have infinitely sharp boundaries separated by a vacuum (or vapor). The energy of interaction between two planes separated by the distance D is given by Eq. (38) of Hamaker

$$E_p = -\frac{1}{12\pi}\left[\frac{A_{s/l}}{D^2}\right] \tag{38}$$

Where $A_{s/l}$ is the solid/liquid Hamaker constant and D the separation distance between the planes at equilibrium. At closest approach, $D = D_0$. This separation is determined by a balance between attractive dispersion forces and repulsive Born forces. Hamaker deduced $A_{s/l}$ by integrating the dispersion interactions between all elements in the two phases as shown in Eq. (39).

$$E_p = -\int_{V_1} dv_1 \int_{V_2} dv_2 \frac{q_1 q_2 \lambda}{r^6} \tag{39}$$

The v's are volume elements of the two phases, the q's are the corresponding densities of atoms per unit volume, λ is the molecular interaction constant, and r is the distance between atoms in the two phases. The integration is over semiinfinite planes. Hamaker actually analyzed interactions of spheres, but his approach is readily applied to planes.

Hamaker constants can be defined for each phase. A joint Hamaker constant can be calculated by the geometric mean rule of Bertholet, Eq. (40)

$$A_{s/l} = \sqrt{A_{s/o}A_{l/v}} \tag{40}$$

Padday and Uffindell [35] calculated $A_{s/l}$ from a molecular theory of Moelwyn–Hughes, and by adjusting D_0, they calculated the surface tensions of alkanes quite precisely. They found that they could fit the data if they calculated interactions between groups rather than between atoms. Their calculated values of $A_{s/l}$ were quite different from those calculated later by the more accurate Lifshitz theory. Even so, their analysis did show that dispersion theory and the Hamaker equation are useful for correlating data and increasing our understanding of interfacial forces.

Fowkes [36] was concerned with Hamaker's use of integration in deriving Eq. (38). He felt that the graininess of matter should be important when the liquid and solid are in contact. Using summation rather than integration, Fowkes found that the form of Eq. (38) stayed the same but K_a in Eq. (41) became 6.1 rather than 12π. Fowkes was quite successful in fitting data using this value and assigning values to D_0 based on geometrical arguments

$$E_p = -\frac{1}{K_a}\left(\frac{A_{s/l}}{D_0^2}\right) \tag{41}$$

Equation (41), where K_a is substituted for 12π, recognizes that $A_{s/l}$, D_0, and K_a have all been used successfully as adjustable parameters. We have examined Eq. (41) using nonlinear regression techniques. The results showed that K_a is hopelessly confounded with D_0. The two cannot be separated with wettability measurements alone. We will use 12π in all our calculations and let D_0 be the adjustable parameter. There is no loss of generality in doing this. Although D_0 is closely related to the separation between planes, it is an adjustable parameter that compensates for any problems that may occur when planes come into molecular contact.

2. Lifshitz Theory

The Lifshitz [4] theory of dispersion forces is one of the more exact theories in surface physics. Its development eliminated many of the problems associated with the Hamaker approach. We will rely heavily on the formulations of Hough and White [5] and Israelachvili [6]. The Lifshitz theory does not present as clear a physical picture of the interface as does that of Hamaker. Accordingly, we will sometimes use the Hamaker model for purposes of discussion, but will use the Lifshitz theory for calculations.

The Lifshitz model is the same as Hamaker's: semiinfinite half-spaces separated by a vacuum. The boundaries between the half-spaces are sharp. The difference between the two theories is in the way the Hamaker constant is determined. The Hamaker method determines the constant by integrating or summing molecular interactions. The Lifshitz method determines the constant from electromagnetic absorption spectra. Equation (38) was also derived by Lifshitz and is the starting point for contact-angle analysis.

A major reason that Lifshitz calculations are so little used today is that the optical measurements must be made on homogeneous, macroscopic objects. They are not directly applicable to monolayers and films that are so important in surface chemistry. Israelachvili [6] helped alleviate this problem by his analysis of surfaces coated with a film.

C. Model of Hough and White

The work of Hough and White [5] plays a major role in our analysis. Their starting point is the fundamental equation of Lifshitz

$$E_{123}(L) = \frac{k'T}{2\pi} \sum_{n=0}^{\infty}{}' \int_0^{\infty} k\, dk \ln(1 - \Delta_{12}\Delta_{32}e^{-2kL}) \tag{42}$$

$$\Delta_{kj} = \frac{\epsilon_k(i\xi_n) - \epsilon_j(i\xi_n)}{\epsilon_k(i\xi_n) + \epsilon_j(i\xi_n)} \tag{43}$$

$$\xi_n = n\left(\frac{4\pi^2 k'T}{h}\right) \tag{44}$$

where h is Planck's constant, k' Boltzmann's constant, and E the energy between plane half-spaces 1 and 3 separated by 2. The prime on the summation indicates that the term $n = 0$ is given half-weight. The prime on Boltzmann's constant is used to distinguish it from the index k employed by Hough and White. $\epsilon(i\xi)$ is closely related to the dielectric response $\epsilon(\omega)$, where ω is the frequency of the electromagnetic radiation.

Hough and White make a change of variable to convert the above equations to a more recognizable form

$$-E_{123} = \frac{A_{123}}{12\pi L^2} \tag{45}$$

where A_{123} is the solid/liquid Hamaker constant defined by Eq. (46) and L the distance between half-spaces 1 and 3

$$A_{123} = -\frac{3k'T}{2} \sum_{n=0}^{\infty}{}' \int_0^{\infty} dx\, x \ln(1 - \Delta_{12}\Delta_{32}e^{-x}) \tag{46}$$

where $x = 2kL$. A_{123} is dependent on the material properties of the system through the function $\epsilon_j(i\xi)$.

We note that the calculation of the Hamaker constant does not depend on any of the approximations or uncertainties discussed earlier. It is as accurate as the optical data allow. The uncertainty lies in determining L in Eq. (45).

We can calculate contact angles from Eq. (45) if we recognize that

$$-E_{123} = W_a^* = \gamma_{l/v}(1 + \cos\theta) + \pi_{s/v} \tag{47}$$

Solving Eq. (47) for cos θ yields Eq. (48)

$$\cos\theta = \frac{A_{s/l} - 12\pi L^2 \pi_{s/v}}{12\pi L^2 \gamma_{l/v}} - 1 \tag{48}$$

where $A_{s/l} = A_{123}$.

For a simple liquid whose molecules do not orient at the surface

$$\gamma_{l/v} = -\frac{1}{2} E_{l/v} = \frac{A_{l/v}}{24\pi L^2} \tag{49}$$

where L is the separation in the liquid.

Substituting $\gamma_{l/v}$ from Eq. (49) into Eq. (42) yields Eq. (50)

$$\cos\theta = \frac{2A_{s/l} - 24\pi L^2 \pi_{s/v}}{A_{l/v}} - 1 \tag{50}$$

Hough and White assume that $\pi_{s/v}$ is zero, which leads to the approximate equation

$$\cos\theta = \frac{2A_{s/l}}{A_{l/v}} - 1 \tag{51}$$

Table 4 shows the Hamaker constants calculated by Hough and White for alkanes and water. These are particularly useful. In all tables, the column headed with "Calc." shows interfacial Hamaker constants calculated by the geometric mean approximation for water and alkanes. These values may be compared with constants calculated from full Lifshitz theory. The differences are a few percent.

Table 5 shows the Hamaker constants calculated for several different materials including poly(tetrafluoroethylene) (PTFE). Table 6 gives the Hamaker constants, surface tensions, densities, and cutoff distances for the alkanes. Table 7 provides theoretical calculations made for the wetting of PTFE by Hough and White using Eq. (51). The experimental results were taken from Fox and Zisman [37].

D. Model of Israelachvili

The equations of Hamaker and Lifshitz apply only to homogeneous solids. Israelachvili [6] extended Lifshitz analysis to systems where the solid is

TABLE 4 Hamaker Constants Calculated by Hough and White (C = Alkane; A = Air; W = Water)

	Hamaker constant ($\times 10^{-20}$ J)					
Alkane	C—A—C	C—W—C	C—W—A	C—A—W	C—A—W calc.	W—C—A
Pentane	3.75	0.336	0.153	3.63	3.72	0.108
Hexane	4.07	0.360	−0.004	3.78	3.88	0.285
Heptane	4.32	0.386	−0.118	3.89	4.00	0.423
Octane	4.50	0.410	−0.200	3.97	4.08	0.527
Nonane	4.66	0.435	−0.275	4.05	4.15	0.624
Decane	4.82	0.462	−0.344	4.11	4.22	0.719
Undecane	4.88	0.471	−0.368	4.14	4.25	0.751
Dodecane	5.04	0.502	−0.436	4.20	4.32	0.848
Tridecane	5.05	0.504	−0.442	4.21	4.32	0.855
Tetradecane	5.10	0.514	−0.464	4.23	4.34	0.886
Pentadecane	5.16	0.526	−0.490	4.25	4.37	0.923
Hexadecane	5.23	0.540	−0.519	4.28	4.40	0.964
Water	3.70					

Source: Ref. 5 by permission.

TABLE 5 Hamaker Constants for Miscellaneous Materials Calculated by Hough and White (M = Material; A = Air; W = Water)

	Hamaker constant ($\times 10^{-20}$ J)				
Material	M—A—M	M—W—M	M—W—A	M—A—W	M—A—W calc.
Fused silica	6.55	0.85	−1.03	4.83	4.92
Calcite	10.10	2.23	−2.26	6.00	6.11
Calcium fluoride	7.20	1.04	−1.23	5.06	5.16
Sapphire	15.60	5.32	−3.78	7.40	7.60
Poly(methyl methacrylate)	7.11	1.05	−1.25	5.03	5.13
Poly(vinyl chloride)	7.78	1.30	−1.50	5.25	5.37
Polystyrene [35]	6.58	0.95	−1.06	4.81	4.93
Polystyrene [32]	6.37	0.91	−0.97	4.72	4.85
Poly(isoprene)	5.99	0.74	−0.84	4.59	4.71
Poly(tetra flouroethylene)	3.80	0.33	0.13	3.67	3.75
"Teflon" FEP	2.75	0.38	0.69	3.12	3.19

Source: Ref. 5 by permission.

TABLE 6 Cutoff Distances ($L_c = D_0$), Hamaker Constants and Surface Tensions, and Densities for Alkanes

Alkane carbon number	A_{LAL} (10^{-20} J)	$\gamma_{l/v}$ (20°C, mJ m^{-2})	ρ (20°C, 10^{-3} kg m^{-3})	$L_c'(D_0)$ (nm)
5	3.75	16.1	0.6262	0.1785
6	4.07	18.4	0.6603	0.1712
7	4.32	20.1	0.6838	0.1685
8	4.50	21.6	0.7025	0.1660
9	4.66	22.9	0.7176	0.1644
10	4.82	23.8	0.7300	0.1637
11	4.88	24.7	0.7402	0.1618
12	5.03	25.4	0.7487	0.1622
13	5.04	26.0	0.7564	0.1604
14	5.05	26.6	0.7628	0.1594
15	5.1	27.1	0.7685	0.1588
16	5.23	27.4	0.7733	0.1507
Water*	3.70	73.0	0.997	0.142

Source: Ref. 5 by permission.

TABLE 7 Theoretical and Experimental Contact Angles on PTFE; Theoretical Predictions Assumed Equal Cutoff Distances (S = PTFE; L = Alkane; A = Air)

	Hamaker constants (10^{20} J)					
Alkane	A_{SAL} Lifshitz	A_{SAL} calc.	A_{LAL} Lifshitz	Cos Θ theory	Θ° theory	Θ° experiment
Pentane	3.75	3.77	3.74	1.006	wets	wets
Hexane	3.91	3.93	4.06	0.924	22	12
Heptane	4.03	4.05	4.31	0.869	29	21
Octane	4.11	4.13	4.49	0.831	34	26
Nonane	4.18	4.21	4.66	0.798	37	32
Decane	4.25	4.28	4.81	0.768	40	35
Undecane	4.28	4.30	4.87	0.758	41	39
Dodecane	4.35	4.37	5.03	0.730	43	42
Tridecane	4.35	4.38	5.04	0.728	43	
Tetradecane	4.38	4.40	5.09	0.719	44	44
Pentadecane	4.40	4.42	5.15	0.709	45	
Hexadecane	4.43	4.45	5.22	0.698	46	46

Source: Ref. 5 by permission.

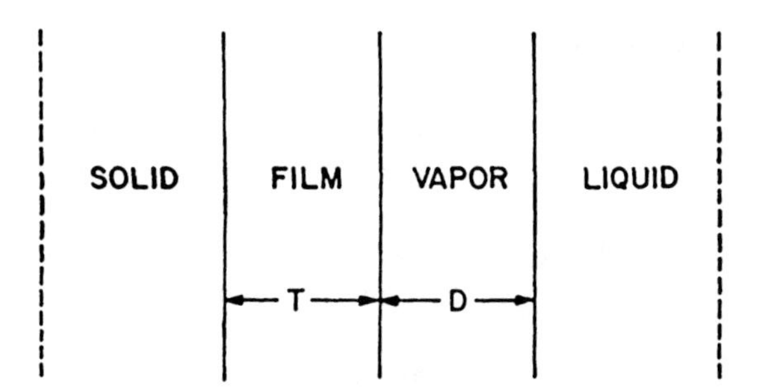

FIG. 19 Israelachvili model.

coated with a film or monolayer. Figure 19 shows his model. T is the film thickness. It is similar to the Hamaker–Lifshitz model, except that a thin film is coating the solid. The boundaries are sharp. Using Lifshitz theory, Israelachvili derived Eq. (52). He also showed that Eq. (52) could be derived from Hamaker theory with an error of less than 20%. D_0 is again an adjustable parameter, closely related to the distance between the plates but containing all the uncertainties associated with the close approach of two materials.

$$W_a = E_p = \frac{1}{K_a}\left[\frac{A_{l/f}}{D_0^2} + \frac{A_{l/s} - A_{l/f}}{(D_0 + D_f)^2}\right] \tag{52}$$

$A_{l/s}$ is the liquid/substrate Hamaker constant and $A_{l/f}$ the liquid/film Hamaker constant. They are often calculated by the geometric mean approximations $A_{l/s} = (A_{l/o}A_{s/o})^{1/2}$ and $A_{l/f} = (A_{l/o}A_{f/o})^{1/2}$. $A_{f/o}$ is the Hamaker constant of the film, $A_{s/o}$ the Hamaker constant of the substrate, and $A_{l/o}$ the Hamaker constant of the liquid. We use ''substrate'' to mean the solid on which the monolayer is adsorbed, plus the inhomogeneous region between the solid and the monolayer.

E. Applications of Dispersion Force Theory

In this section we use the previous theory to investigate the following questions:

What problems are associated with the infinitely sharp boundary of the Hamaker–Lifshitz model?
Why is a Zisman plot linear when theory predicts a curve?
Why does a vapor adsorb on a solid when the contact angle is greater than zero?

How deep is the surface layer? Is it true that only the outermost layer affects wettability?

How does packing of molecular groups in the surface affect wettability? Does increasing the number of —CF_3 groups in a surface really increase repellency?

How does penetration of molecules into the surface (surface solubility or coadsorption) affect wettability? We address this question more fully in Sec. V.

Can we determine the surface tension of a solid surface from contact-angle measurements alone?

1. Zisman Plot

(a) Hough and White Correlation. The work of Hough and White makes a critical analysis of the Zisman plot possible. PTFE and the alkanes are well characterized. Hamaker constants have been calculated by Hough and White from optical measurements by Lifshitz theory.

Figure 20 shows three Zisman plots for alkanes on PTFE. The first plot, labeled "experimental," shows Zisman's experimental data [37]. Statistical analysis detects no sign of curvature. The plot labeled "theoretical" is the plot calculated by Hough and White from optical data. The curvature is apparent.

The theoretical equation of Hough and White ignores the spreading pressures of the alkanes on the PTFE. Graham [38] estimated the spreading pressure of hexane on PTFE due to dispersion forces to be about 1 mJ m^{-2}. The solid diamond in Fig. 20 was calculated from Eq. (6) using this spreading pressure ($\pi_{s/v}$) and the calculated $\pi_{s/l}$ of Hough and White. Since spreading pressures for higher alkanes diminish with increasing chain length, including spreading pressures takes the curvature out of Zisman plots. The modified theoretical plot in Fig. 20 was determined by linear regression using the Hough and White values for decane and higher alkanes for which spreading pressures can be ignored. Spreading pressures also cause Zisman plots for liquids spreading on liquids to be linear. (See Sec. VI.) Figure 20 also illustrates the high sensitivity of contact angles to small changes in the surfaces. Thus, the spreading pressure of 1 mJ m^{-2} resulted in a change in $\cos\theta$ of 0.10 or in θ of 27°.

Including spreading pressures in the analyses increases linearity, but also increases the difference between theory and experiment. The difference between the modified theoretical and experimental values becomes larger as θ approaches zero, thereby causing the slopes of the Zisman plots to be different.

The most probable explanation of these observations is penetration of the molecules into the surface region. This is called surface solubility.

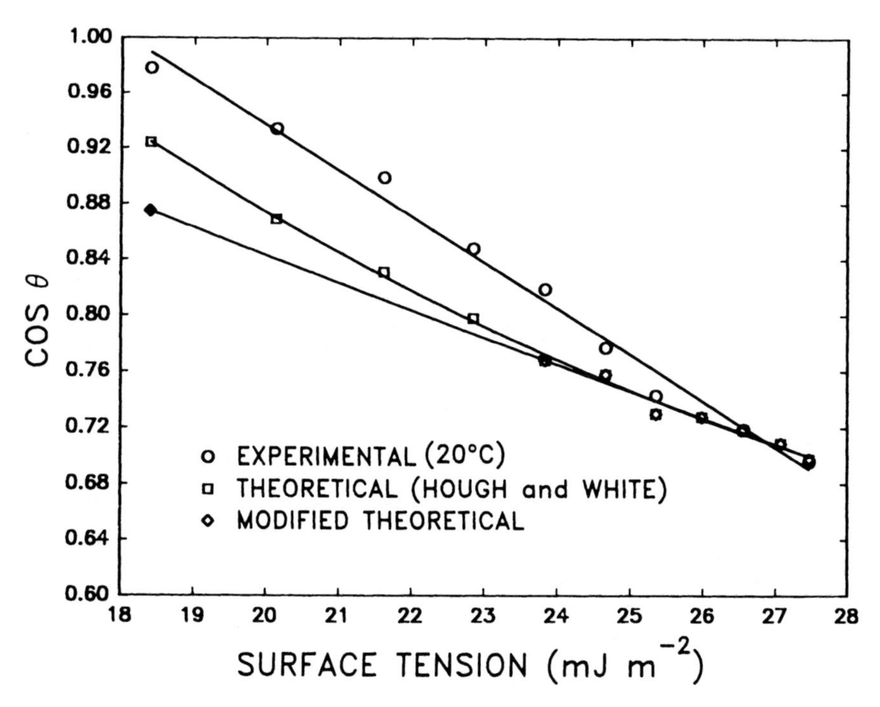

FIG. 20 Zisman plots for alkanes on poly(tetrafluoroethylene) (PTFE) (20°C). (From Ref. 5 by permission.)

The molecules of hexane, for example, can penetrate the surface region, thereby increasing the interaction above that predicted by the Hamaker model. Equation (53) shows an additional term W_a^{SS} added to the work of adhesion given by Eq. (37). W_a^{SS}, the work of the adhesion component associated with surface solubility, is not directly associated with intermolecular forces

$$W_a^d = W_a^{HL} + W_a^{SS} \tag{53}$$

where W_a^d is the conventional work of adhesion due to dispersion forces, W_a^{HL} is the work of adhesion calculated with the Hamaker–Lifshitz model and W_a^{SS} the work of adhesion due to surface solubility. This will be discussed further in Sec. V.

2. Outermost Layer

Zisman emphasized that the outermost molecular layer of a surface (or monolayer) determines the contact angle. This is a useful approximation.

TABLE 8 Wettability of Three Monolayers: Effect of End Groups and Structure Underneath the End Groups

Monolayer	Hexadecane contact angle (deg)	Critical surface tension ($mJ\ m^{-2}$)	Zisman slope ($m^2\ mJ^{-1}$)
$CH_3(CF_2)_{11}COOH$	58.7	15.4	−0.0415
$CF_3(CF_2)_8COOH$	80.6	3.8	−0.0356
$CH_3(CH_2)_{16}COOH$	46.6	21.0	−0.047

It is not true, however, that the underlying structure does not affect wettability. This is illustrated in Table 8.

Table 8 shows data for two monolayers terminated with —CH3 groups and one terminated with $—CF_3$. One $—CH_3$ is anchored onto a fluorocarbon chain, the other onto a hydrocarbon chain. These monolayers are shown schematically in Fig. 21. The critical surface tension is highest for

CF_3 CF_3 CF_3 CF_3 CF_3 CF_3
$(CF_2)_n$ $(CF_2)_n$ $(CF_2)_n$ $(CF_2)_n$ $(CF_2)_n$ $(CF_2)_n$

F F F F F F
$(CF_2)_n$ $(CF_2)_n$ $(CF_2)_n$ $(CF_2)_n$ $(CF_2)_n$ $(CF_2)_n$
$(CH_2)_m$ $(CH_2)_m$ $(CH_2)_m$ $(CH_2)_m$ $(CH_2)_m$ $(CH_2)_m$

CH_3 CH_3 CH_3 CH_3 CH_3 CH_3
$(CF_2)_n$ $(CF_2)_n$ $(CF_2)_n$ $(CF_2)_n$ $(CF_2)_n$ $(CF_2)_n$

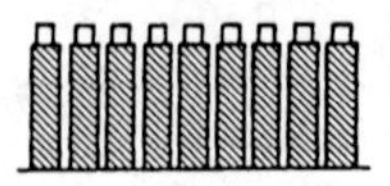

FIG. 21 Schematic diagrams of adsorbed monolayers. The first is a perfluoro monolayer. The second is a segmented monolayer. The third shows a perfluoro chain terminated by a $—CH_3$ group.

the —CH_3 group on the hydrocarbon chain. The underlying structure obviously has an effect. The surface concentration of —CH_3 groups is lower on the film having the higher contact angle. Shafrin and Zisman [39] had earlier suggested that a large dipole between the —CH_2— and —CF_3 groups near the surface of a monolayer of Ω,Ω,Ω trifluoro stearic acid was responsible for significantly higher critical surface tension than that of a monolayer of a fully fluorinated acid such as perfluoro decanoic acid. This is not consistent with Table 8, where the monolayer with the dipole as near the surface as possible has the higher contact angle. How then do we account for these differences?

According to Hamaker theory, the attractive forces in a surface are a product of the attractive force surrounding each molecular group and the density of packing of the groups in the surface. Hoernschemeyer [40] pointed out that the differences in repellency between hydrocarbons and fluorocarbons are not due to differences in the dispersion field surrounding hydrocarbon and fluorocarbons. The field around each —CF_3, —CF_2—, —CH_3, and —CH_2— group is approximately the same. Earlier, Reed [41] had shown that when the molar energy of surface formation is plotted against the molar energy of vaporization, both hydrocarbon and fluorocarbon alkanes fall on the same line. This is consistent with the ideas of Hoernschemeyer. The difference in repellency is due to the smaller concentration of fluorocarbon groups in the surface. It is just not possible to pack as many large fluorocarbon groups as hydrocarbon groups into the same area. This also explains why the surface tensions of fluorocarbon liquids have surface tensions lower than those of analogous hydrocarbons [41].

A vertically oriented hydrocarbon chain in a close packed monolayer occupies about 21 $Å^2$. The area of an analogous fluorocarbon chain is about 29 Å. The attractive force surrounding a hydrocarbon monolayer is therefore about 29/21 times that for a fluorocarbon. This difference can account for nearly all the difference in wettability of hydrocarbons and fluorocarbons. (See Fig. 32 of Ref. 2 for examples.) The difference in wettability of PTFE and polyethylene can also be determined from the ratio of cross sections of —CF_2— and —CH_2— groups [2].

According to Table 8, the wettability of the fluoromonolayer terminated with a —CH_3 group is greater than that of the fluoromonolayer terminated with a —CF_3 group. The surface density of —CF_3 and —CH_3 groups is the same in both. The arguments given above combined with the Hamaker–Lifshitz sharp boundary model would predict about the same wettability. Thus, dispersion theory predicts that a —CH_3 on a fluorinated chain should have a higher contact angle than a —CH_3 on a hydrocarbon chain. That the wettability of a —CH_3 group on a fluorocarbon chain is less than a corresponding —CF_3 is not explained by dispersion theory. This difference will be discussed in Sec. V.

3. Chain Length and Packing

According to dispersion theory, increased packing cannot explain the increasing contact angles seen with monolayers of increasing chain length. Israelachvili analysis shows that dispersion interactions with underlying substrate can quantitatively explain it. Table 9, calculated with Eq. (52), shows how the Hamaker constant of the underlying substrate affects the contact angles of fluorinated monolayers. Column 1 gives the number of fluorinated carbons in the monolayer, column 2 the length of the fluorinated part of the monolayer in angstroms. The next six columns give calculated hexadecane contact angles for monolayers. Each column is headed by the Hamaker constant of the substrate on which the monolayer is adsorbed. The Hamaker constant of the monolayer is assigned the value 1.69×10^{-20} J. In Israelachvili's formulation, the Hamaker constant of the monolayer is the Hamaker constant of a thick layer having the same structure.

Studies of the wettability of perfluoro acids and segmented fluoroacids have been made as a function of the number of fluorinated carbons. Figure 22 is a schematic representation of an adsorbed segmented acid monolayer. The segmented monolayers have the structure $F(CF_2)_N—(CH_2)_{16}—COOH$. Table 10 compares critical surface tensions and Zisman slopes for the adsorbed perfluoro acids and segmented acids.

TABLE 9 Calculated Contact Angles of Hexadecane on a Fluorinated Monolayer (Israelachvili Formulation)

		Substrate Hamaker constants (10^{-20} J)					
N	T (Å)	0.0	1.69	5.22	7.5	10.0	15.0
1	1.25	102.6	82.0	65.8	57.7	49.4	32.2
2	2.50	91.9	82.0	74.4	70.8	67.3	61.3
3	3.75	87.8	82.0	77.6	75.5	73.5	70.1
4	5.00	85.9	82.0	79.1	77.8	76.5	74.3
5	6.75	84.7	82.0	80.0	79.0	78.1	76.6
6	7.50	84.0	82.0	80.5	79.8	79.1	78.0
7	8.75	83.6	82.0	80.8	80.3	79.7	78.9
8	10.0	83.3	82.0	81.0	80.7	80.2	79.6
9	11.25	83.0	82.0	81.3	80.9	80.6	80.0

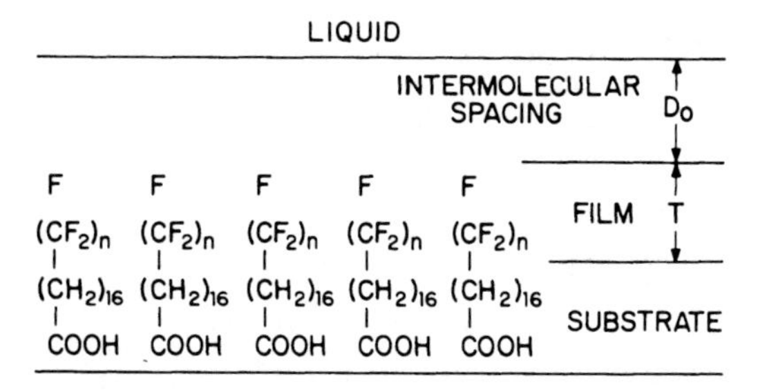

FIG. 22 Schematic representation of a segmented perfluoro acid monolayer.

Table 11 shows hexadecane contact angles on both the perfluoromonolayers and segmented monolayers. Nonlinear regression was used to fit Israelachvili's equation to the data. The standard errors of the fit were 0.5° for the perfluoro acids and 2° for the segmented acids. The substrate Hamaker constants giving the best fits were 10×10^{-20} J for the perfluoro acids and 6×10^{-20} J for the segmented acids. Values of D_0 giving the best fits were 0.16 nm for the perfluoro acids and 0.18 nm for the acids with the segmented chains. The contact angles calculated with these values for the parameters are shown next to the experimental values in Table 11. The change in contact angle with fluorinated chain length is clearly

TABLE 10 Critical Surface Tensions and Slopes of Adsorbed Monolayers of Perfluoro [$F(CF_2)_NCOOH$] and Segmented Perfluoro [$F(CF_2)_N(CH_2)_{16}COOH$] Acids

	Perfluoro acids		Segmented acids	
N	Critical surface tension (mJ m^{-2})	Zisman slope (m^2 mJ^{-1})	Critical surface tension (mJ m^{-2})	Zisman slope (m^2 mJ^{-1})
1				
2				
3	7.2	−0.0357		
4			5.5	−0.0361
5	5.7	−0.0370		
6	4.8	−0.0360	3.8	−0.0368
7				
8	4.0	−0.0360		
9	3.8	−0.0356		

TABLE 11 Contact Angles of Hexadecane on Fluorinated Monolayers

Number of —CF_2— groups	—CF_2— chain length (Å)	Θ Perfluoro acid (deg)		Θ Segmented acid (deg)	
		Calc.	Obs.	Calc.	Obs.
1	1.25			62.4	61.0
2	2.50			64.5	68.2
3	3.75	73.5	73.0	76.7	72.6
4	5.00			78.8	76.7
5	6.75	78.1	77.7	79.7	77.0
6	7.50	79.1	78.7	80.3	79.0
7	8.75				
8	10.00	80.2	80.0		
9	11.25	80.6	80.6		

consistent with the effect of the increasing distance between the substrate and the wetting liquid.

V. SURFACE STRUCTURE

A. Surface Solubility

Zisman did not work within the theoretical framework of Hamaker or Lifshitz. His interpretations were largely intuitive. Penetration of wetting liquids into the surfaces of his monolayers was an important concept. On the other hand, beginning with Girifalco and Good, the concept of penetration of wetting molecules nearly disappeared from theoretical treatments of wetting. Both concepts are needed.

Surface solubility is an analogue of bulk solubility. In this regard, it is well to remember that Gibbs considered an interface to be a region between two phases where properties are continuously changing. He did not consider it to be sharp. Surface excess is the amount of material in the surface region in excess of the amount that would be there if the surfaces were infinitely sharp. When a wetting liquid comes into contact with a solid, there is a normal temperature-driven tendency for the molecules of the wetting liquid to penetrate into the monolayer or polymer. If the wetting molecule has a size and shape compatible with the molecules in the solid surface, solubility will be enhanced. This ability to penetrate the surface also can account for adsorption of the vapors of a liquid whose

Hamaker constant is larger than the solid on which it adsorbs. A phenomenon related to surface solubility is the coadsorption of hexadecane into the radioactive stearic acid molecules described earlier (Fig. 10).

One way of thinking about surface solubility is that the solubility causes an increase in the free energy of interaction. This free energy of surface solubility is then added to the work of adhesion. A second, more mechanistic approach is to note that coadsorption increases the density of surface sites. This effectively increases the Hamaker constant of the surface layer, which in turn increases the wettability. In the Hough and White formulation, where the Hamaker constant of the pure solid is used, the increase in the Hamaker constant appears as a decrease in D_0 (or L_c).

B. Monolayers and Polymers

The effect of surface solubility can be demonstrated by comparing the wettabilities of fluorinated acid monolayers and fluoroalkyl acrylate and methacrylate polymers [26]. The acids in the monolayers have the structure $F(CF_2)_NCOOH$. The polymers have the structure poly[$F(CF_2)_N(CH_2)_2OOCCR{=}CH_2$], where R is either CH_3 (the methacrylate) or H (the acrylate). Figure 23 is a photo of a molecular model of the acrylate with $N = 8$. Figure 24 compares critical surface tensions of the acid monolayers with those of the fluoropolymers as a function of fluorinated chain length. Zisman plots for the monolayers were shown in Fig. 17. It was also shown in Table 11 that the change in hexadecane contact angles with chain length for the acids can be accounted for by the increasing distance between the monolayer and underlying substrate. The changes in critical surface tension for the fluoromethacrylate polymers are too large for this explanation. Moreover, while the slopes of Zisman plots for the acid monolayers remain almost constant at about $-0.036\ m^2\ mJ^{-1}$, the slopes for the fluoropolymers vary with side chain length, changing from $-0.042\ m^2\ mJ^{-1}$ at $N = 4$ to values observed for the acid monolayer for $N > 7$. This is shown in Fig. 25. The higher values for the slopes are a reflection of smaller contact angles exhibited by the lower alkanes. It is proposed that these differences are caused by surface solubility.

If surface solubility is important, we would expect the packing of the side chains of the polymers to affect wettability. X-ray diffraction studies on unsupported films of the fluoropolymers give information on side-chain packing in the polymers. Table 12 shows side-chain areas, calculated from interplanar spacings (by assuming hexagonal packing), and critical surface tensions for these polymers. When the data of Fig. 24 are plotted against the area of the side chain, the points fall on a smooth curve. This is shown

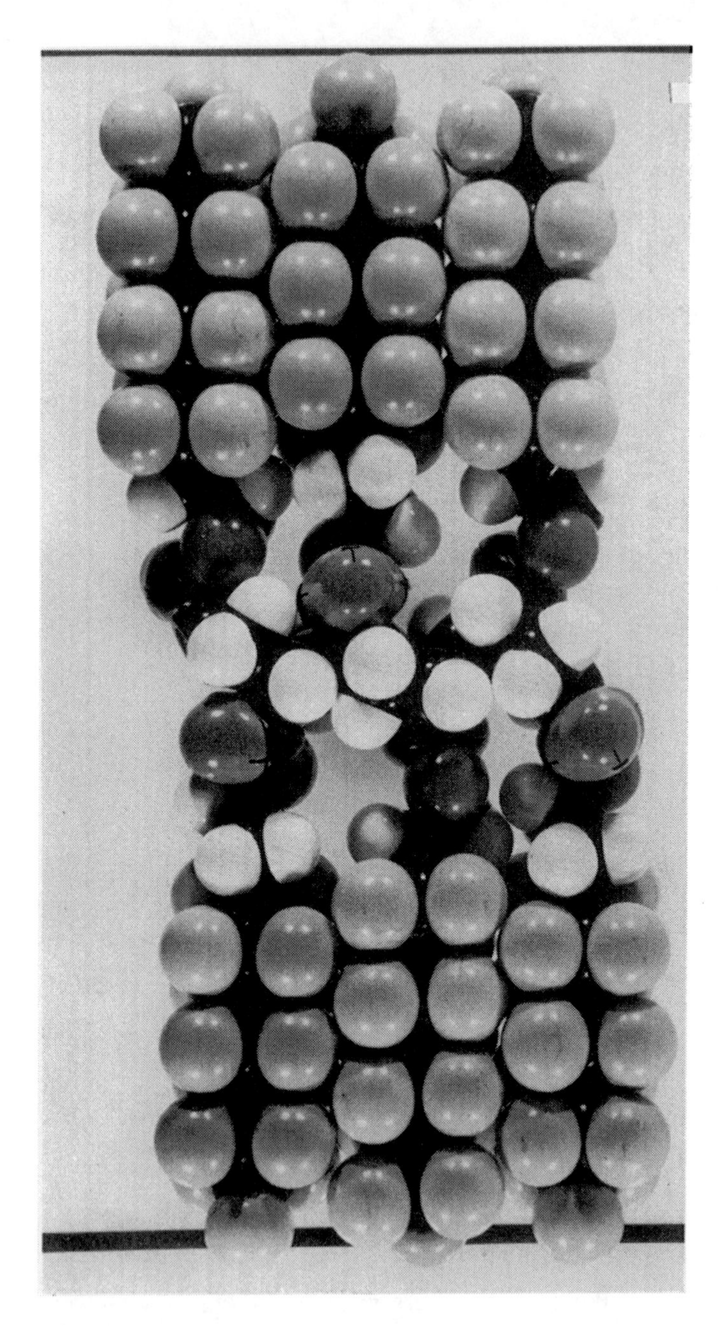

FIG. 23 Photo of a model of a perfluoro acrylate polymer with eight fluorinated atoms in the side chain.

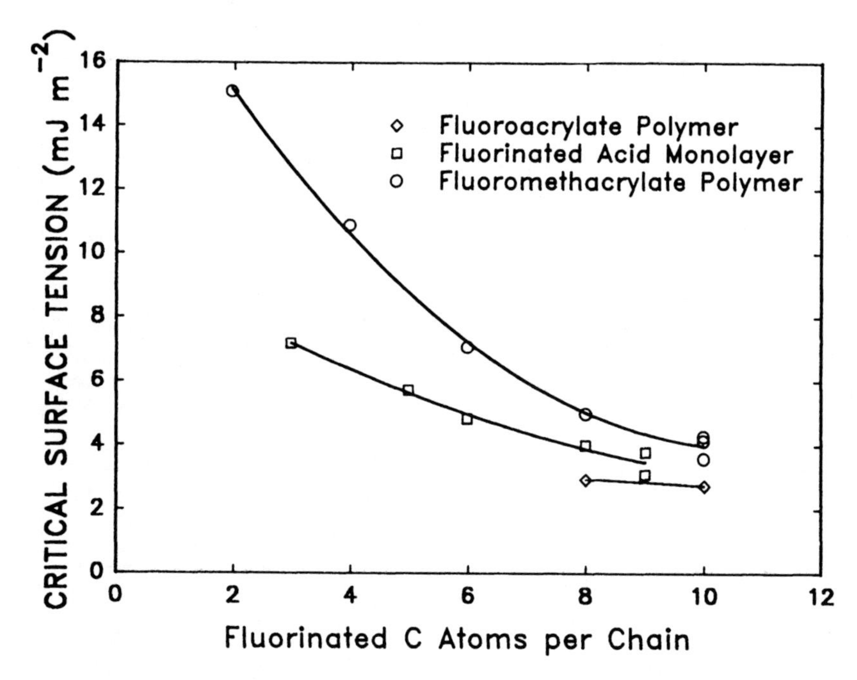

FIG. 24 Critical surface tensions for three systems as a function of fluorinated chain length. (From Ref. 26 by permission.)

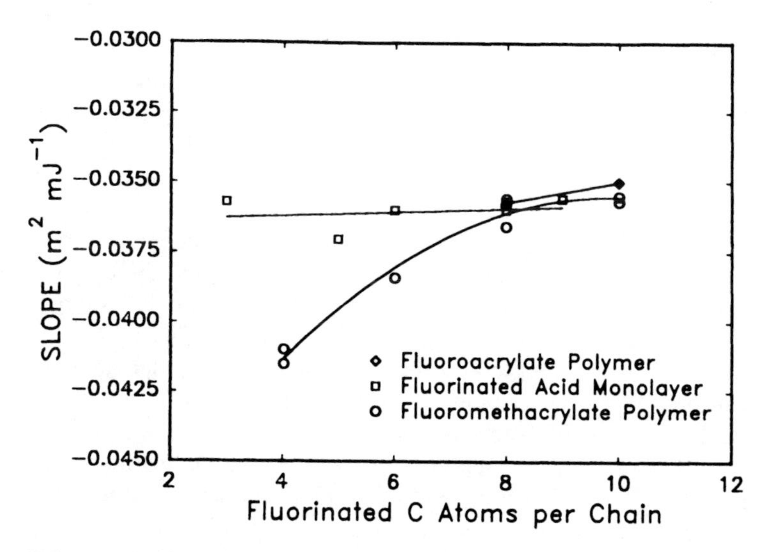

FIG. 25 Slopes of Zisman plots for three systems as a function of fluorinated side-chain length. (From Ref. 26 by permission.)

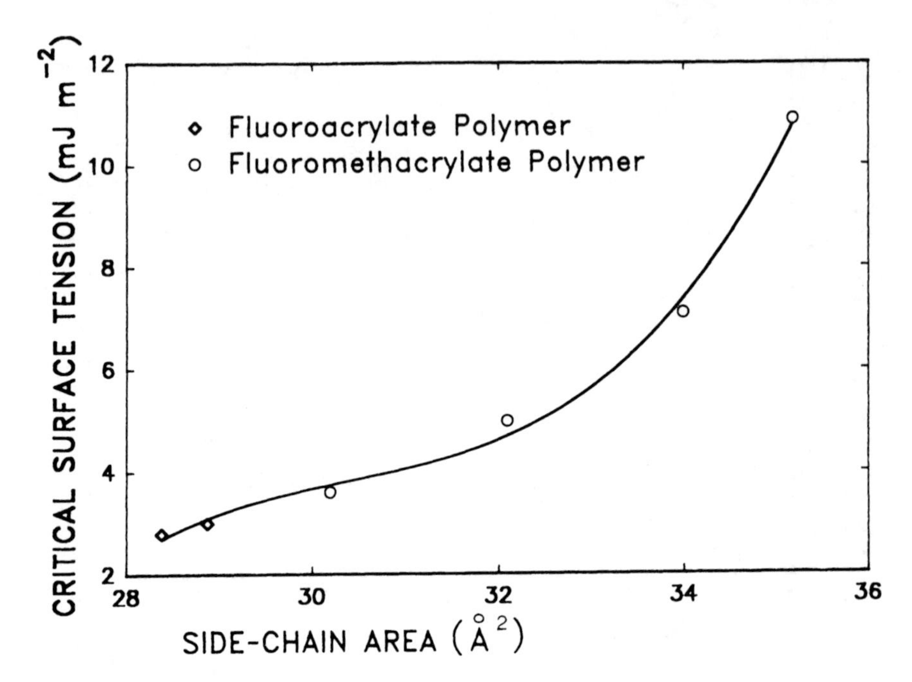

FIG. 26 Critical surface tensions of fluorinated acrylate and methacrylate polymers as a function of fluorinated side-chain area. (From Ref. 26 by permission.)

in Fig. 26. Hamaker–Lifshitz theory predicts that the critical surface tension should increase with increased packing. Table 12 shows the opposite behavior. Surface solubility is a reasonable explanation for the difference. If surface solubility is important, it should show up in the slopes of the Zisman plots.

Slopes of the Zisman plot for each fluorinated polymer are plotted against chain length in Fig. 25. The slopes are plotted against the area of the side chain in Fig. 27. The x-ray diffraction data also show that a large change in the polymer configuration occurs above six fluorinated atoms in the side chain. This is indicated by the appearance of long spacings in the diffraction pattern (not shown here). Similar behavior observed with poly(alkylacrylate) polymers was attributed to a transition from a helical macromolecular configuration to a laminar structure with hexagonal side-chain packing [42]. The increase in the absolute value of the slope with increasing area per side chain is consistent with greater surface solubility for the more loosely packed side chains. The greater slope indicates that the surface solubility is greater for smaller alkanes.

TABLE 12 Areas per Side Chain of Acrylate and Methacrylate Fluropolymers, Poly [$F(CF_2)_n(CH_2)_2OOCCR{=}CH_2$]

n	R	Interplanar spacings (sd = 0.04 Å)	Area per side chain (sd = 0.5 Å^2)	Critical surface tension (adv.) (mJ m^{-2})	Standard error of critical surface tension (mJ m^{-2})
4	CH_3	5.52	35.2	10.9	0.2
6	CH_3	5.43	34.0	7.1	0.3
8	CH_3	5.27	32.1	5.0	0.3
10	CH_3	5.11	30.2	3.6	0.2
8	H	5.00	28.9	3.0	0.4
10	H	4.96	28.4	2.8	0.2

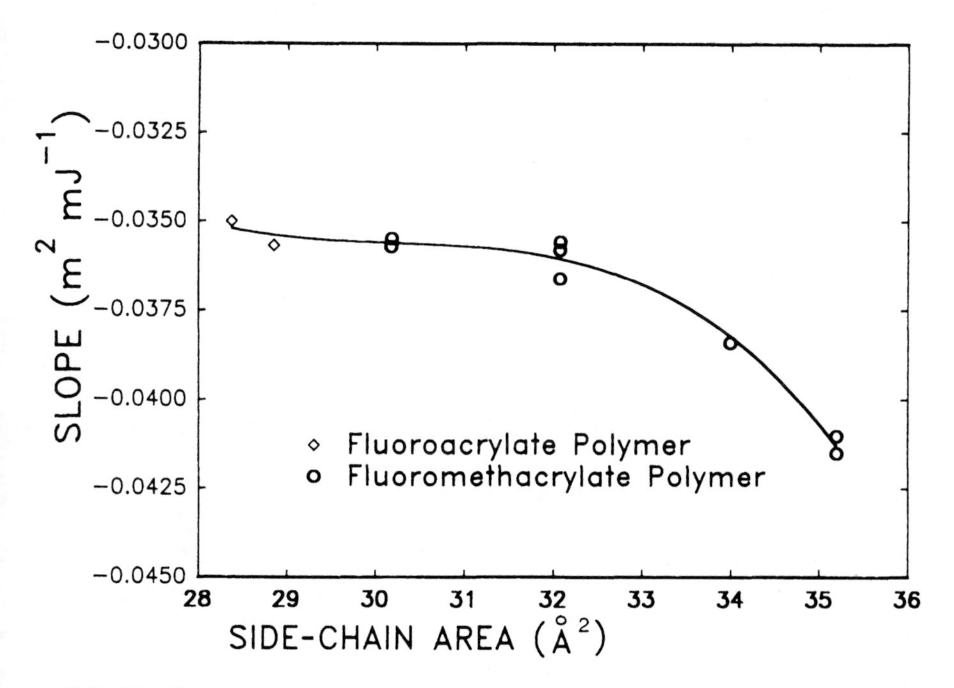

FIG. 27 Slopes of Zisman plots of fluorinated acrylate and methacrylate polymers as a function of fluorinated side-chain area. (From Ref. 26 by permission.)

Another example of the influence of surface solubility on wettability is found with monolayers of perfluoro acids containing terminal branching. Bernett and Zisman [43] measured alkane contact angles on monolayers of acids having the structure $(CF_3)_2CF(CF_2)_NCOOH$. For N in the range 1–11, they found critical surface tensions that are 6–7 mJ m^{-2} higher than monolayers of straight-chain perfluoro acids investigated in earlier work [44]. Our analysis of their data shows that the slopes from the Zisman plots for these monolayers fall in the range of -0.044 to -0.054 m^2 mJ^{-1}, considerably higher than those for the straight-chain acids. (See Table 10.) Bernett and Zisman showed that the smaller contact angles exhibited by the lower alkanes were not due to removal of the monolayers by solvent action.

Surface solubility also can account for the larger critical surface tension and absolute slope for the terminal —CH_3 group on a fluorocarbon chain shown in Table 8. The critical surface tension (15.4 mJ m^{-2}) and slope (-0.0415 m_2 mJ^{-1}) are nearly the same as those for the terminally branched monolayers studied by Bernett and Zisman. The absolute value of the slope is too large to be explained by the Hamaker–Lifshitz model.

There is some correlation between the critical surface tension and the Zisman slope. It is reasonable that as dispersion interactions increase, both wettability and the absolute value of Zisman slope increase. There is not an exact correlation, however, because surface structure influences surface solubility. Figure 28 shows Zisman slopes plotted against critical surface tensions for many polymers and monolayers. Some data come from the literature and some from our own laboratory. It is not our intent to do a complete analysis, but only to show the extent of the scatter.

The solid square is obtained from the theoretical calculations of Hough and White modified to include spreading pressures. The absolute value of every experimental slope was larger than this theoretical value. PTFE is closest in slope to the theoretical. This is reasonable since the crystalline nature of PTFE should show low surface solubility. One liquid, pentafluoropropanol (PPA), is included. It is not far from the regression line. The points having the lowest critical surface tensions are monolayers of perfluoro acids. The points having the highest critical surface tensions are monolayers of stearic acid and stearamine.

If the Hough and White application of Hamaker–Lifshitz theory applied exactly, then all the points should fall on the same curve. The points should also fall on the same curve if the Neumann surface equation of state theory were valid.

C. Summary: Wetting of Low-Energy Surfaces

The most important conclusion to be drawn from the analysis in this section is that more than one number is required to characterize the wett-

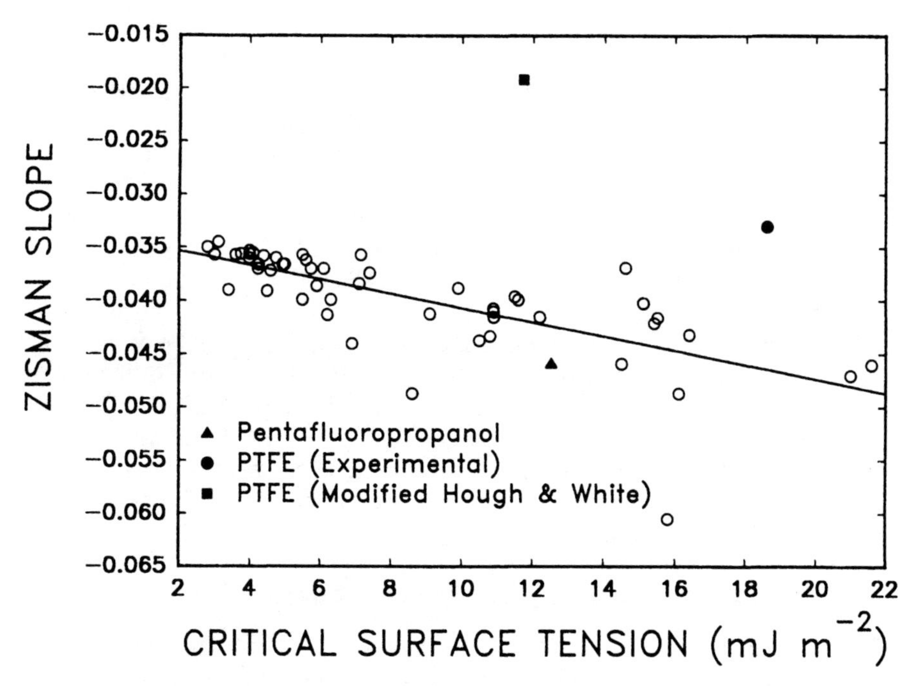

FIG. 28 Zisman slopes vs. critical surface tensions for a wide variety of surfaces. See text for discussion of specific points.

ability of low-energy surfaces. For example, hexadecane contact angles on PTFE, water, and stearic acid are all about the same (45–46°), yet the wetting properties of these substrates are quite different with other liquids.

A characterization adequate for many purposes is to use the critical surface tension and Zisman slope as basic parameters. A more fundamental approach is to use the Hamaker constant of the surface (see Sec. V.E) and free energies of surface solubility with appropriate test liquids. Which test liquids should be used has yet to be determined.

Other conclusions are the following:

The observed decrease in intrinsic repellency of surface groups in the order $—CF_3 > —CF_2— > —CH_3 > —CH_2—$ is due to an increase in the group surface density and not an increase in the force fields around the group.

The repellency of a monolayer increases with increasing chain length of the molecules in the monolayer because the distance between the liquid and substrate increases. This reduces the liquid interaction with the substrate, thereby increasing the contact angle.

The Zisman plot ($\cos\theta$ vs. $\gamma_{l/v}$) is linear because of adsorption of vapor from the liquids onto the solid.

Surface solubility allows adsorption of the vapors even when the Hamaker constant of the liquid is higher than that of the solid. Hamaker–Lifshitz theory does not predict adsorption when the Hamaker constant of the liquid is higher.

Surface solubility also explains why the slopes of the Zisman plots are different from those predicted by Hamaker–Lifshitz theory. A higher negative slope is observed because the smaller alkanes are more soluble than the larger alkanes.

The way in which surface solubility increases wettability can be described in two ways. The first is to say that solution in the surface liberates free energy, which is added onto the work of adhesion. The second is mechanistic. It says that the Hamaker–Lifshitz model is still applicable, but adsorption into the surface region increases the density of groups in the surface, thereby increasing the effective Hamaker constant of the substrate.

D. Hamaker Constants from Contact Angles

The work of Hough and White allows us to estimate Hamaker constants of low-energy surfaces from contact angles of liquids of known Hamaker constant. The basic equation of Hough and White is

$$\cos\theta = \frac{2A_{s/l}}{A_{l/v}}\left(\frac{Lc'}{Lc}\right)^2 - 1 \qquad (54)$$

Lc' and Lc are called "cutoff distances" of the solid and liquid. These have approximately the same meaning as D_0, the distance of closest approach in Eq. (41). If we assume that the Lc's are the same, Eq. (54) simplifies to Eq. (55)

$$\cos\theta = \frac{2A_{s/l}}{A_{l/v}} - 1 \qquad (55)$$

Applying the Bertholet relation (geometric mean approximation)

$$A_{s/l} = \sqrt{A_{s/o}A_{l/v}} \qquad (56)$$

and solving for $A_{s/o}$ yield Eq. (57)

$$A_{s/o} = 0.25A_{l/v}(1 + \cos\theta)^2 \qquad (57)$$

Note the close similarity of Eq. (57) to the original equation of Girifalco and Good [Eq. (29)] and the dispersion energy equation of Fowkes [Eq. (32)].

TABLE 13 Hamaker Constants of PTFE Calculated from Eq. (57) for Three Different Alkanes (units of 10^{-20} J)

Alkane	Hamaker constant (10^{-20} J)
Hexadecane	3.75
Decane	3.97
Heptane	4.02

Table 13 compares the Hamaker constants of PTFE calculated by Eq. (57) for three different alkanes. $A_{l/v}$ was calculated for each alkane by Hough and White from optical data. $A_{s/o}$ for PTFE calculated from optical data is 3.76×10^{-20} J. There is close agreement between the Hamaker constant calculated from the contact angles with hexadecane and the Hamaker constant calculated from optical data. The calculated Hamaker constant for PTFE becomes progressively higher for the lower alkanes. This is consistent with the greater surface solubility of the lower-molecular-weight alkanes. The difference in the Hamaker constant determined with decane and hexadecane is about 6%.

Using hexadecane as a probe liquid rather than a series of alkanes minimizes problems associated with surface solubility. Other liquids, such as methylene iodide, can be used for surfaces where solubility or wetting problems arise. It would be best if the Hamaker constant of the liquid ($A_{l/v}$) were determined from optical data. Otherwise, the Hamaker constant can be estimated from contact angles on PTFE. If it is certain that the liquid surface tension is due only to dispersion forces, then the Hamaker constant can be obtained from Eq. (58) using $A_{s/o} = 3.75 \times 10^{-20}$ J for PTFE

$$A_{l/v} = \frac{A_{s/o}}{0.25(1 + \cos\theta)^2} = \frac{15.0 \times 10^{-20}}{(1 + \cos\theta)^2} \tag{58}$$

Equation (58) is closely related to the equation of Fowkes, Eq. (59)

$$A_{l/v} = 2.8 \times 10^{-21}\gamma^d \tag{59}$$

where γ^d is the dispersion component of the surface tension. Fowkes was handicapped because he had no independent measure of the Hamaker constants of the alkanes, but had to estimate them by estimating D_0 in Eq. (41). The Hamaker constant of hexadecane calculated from Eq. (59) is 7.6×10^{-20} J vs. 5.23×10^{-20} J from optical measurements. This represents close agreement considering the information with which Fowkes had to work.

The Hamaker constant of more interactive liquids (and solids) can also be estimated from contact angles. The error will be larger because D_0 has to be estimated independently. Equation (58) cannot be used for water, for example, because the surface tension of the water is determined by hydrogen bonding, as well as dispersion forces. Equation (61) must be used in such cases.

Equation (60) is derived by equating $-E_p$ to the equilibrium work of adhesion in Eq. (38), setting $D = D_0$ and $A_{s/l} = (A_{s/o}A_{l/v})^{1/2}$

$$\gamma_{l/v}(1 + \cos\theta) = \frac{1}{12\pi}\left(\frac{\sqrt{A_{s/o}A_{l/v}}}{D_0^2}\right) \tag{60}$$

Solving Eq. (60) for $A_{l/v}$ yields Eq. (61)

$$A_{l/v} = \frac{[\gamma_{l/v}(1 + \cos\theta)12\pi D_0^2]^2}{A_{s/o}} \tag{61}$$

Note the fourth-power dependence of $A_{l/v}$ on D_0. If we use the value of $D_0 = 0.15$ nm (the value used for hexadecane), $A_{l/v}$ for water is calculated to be 4.69×10^{-20} J. This is in error by 26%. A value of $D_0 = 0.14$ nm is required to give the correct value.

VI. LIQUIDS SPREADING ON LIQUIDS

Ordinarily, one associates wettability with a solid substrate. However, consideration of the wetting of one liquid by another can be extremely helpful for understanding the thermodynamics and testing the theories of wetting. Liquid/liquid studies have played a dominant role in the development of all theories of wetting. Thermodynamic principles discussed in this chapter can be illustrated with examples from the liquid state. The concepts of this chapter can be applied to liquid/liquid/solid systems if we define θ_E, the equivalent contact angle, by Eq. (62)

$$\cos\theta_E = \frac{\gamma_{L1/v} - \gamma_{L1/L2}}{\gamma_{L2/v}} \tag{62}$$

The subscripts $L1$ and $L2$ refer to the two immiscible fluids. θ_E is called the liquid equivalent contact angle. It cannot be measured directly, but is calculated by Eq. (62) from independent measurements of the surface and interfacial angles. The geometric relationships at the liquid/liquid/vapor three-phase line are defined by the Neumann triangle [45] shown in Fig. 29. The vectors represent the magnitudes of the surface and in-

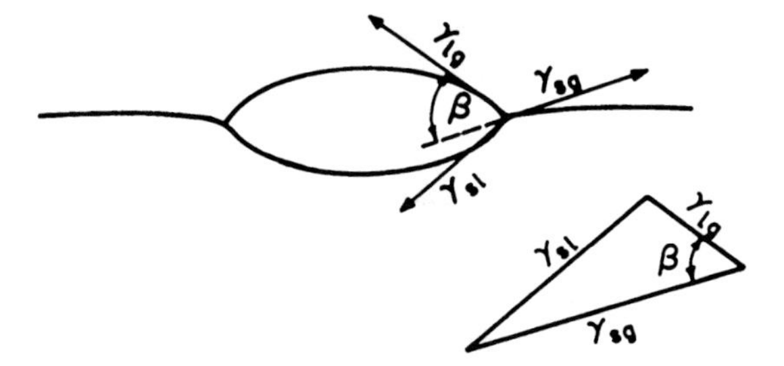

FIG. 29 Neumann triangle for liquid/liquid/vapor systems.

terfacial tensions. The symbol "s" should be interpreted as the other liquid phase. The angle β can be calculated from the triangle, but β is not θ_E.

We [46] evaluated the surface equation of state proposed by Neumann et al. [7] by using liquid equivalent contact angles for alkanes on two liquid substrates, water and pentafluoropropanol. Zisman plots for these two substrates are shown in Fig. 30. These curves will be discussed in more detail later. Table 14 compares measured surface tensions with those

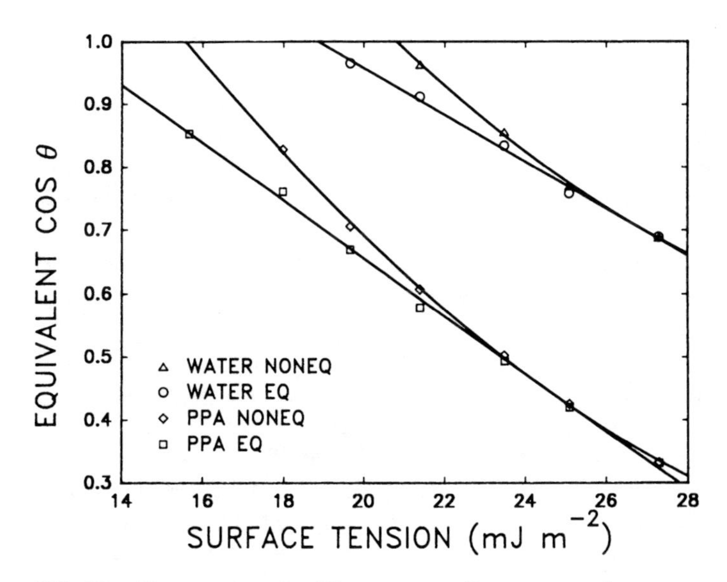

FIG. 30 Zisman plots for alkanes spreading on pentafluoropropanol (PPA) and water.

TABLE 14 Substrate Surface Tensions: Experimental and Calculated by the Method of Neumann

	Surface tensions (mJ m^{-1})		
Liquid	Measured	Calculated	Difference
PPA	17.4	12.8	4.6
Water	72.1	19.7	52.4

Source: Ref. 46 by permission.

calculated by Neumann's equation, Eq. (63). The liquid equivalent contact angles were used for these calculations

$$\cos\theta = \frac{(0.015\gamma_{s/v} - 2.00)(\gamma_{s/v}\gamma_{l/v})^{1/2} + \gamma_{l/v}}{\gamma_{l/v}[0.015(\gamma_{s/v}\gamma_{l/v})^{1/2} - 1]} \tag{63}$$

The differences between the observed and calculated surface tensions are very large, particularly for water. Clearly, Eq. (63) is not valid.

As discussed earlier, Morrison [34] showed that measurements of liquids spreading on liquids prove a good thermodynamic test of Neumann's surface equation of state. The earlier work of Adamson et al. [47] gives strong experimental support for the conclusions of Morrison. He made parallel wetting measurements on water and ice. He found that the contact angle of carbon disulfide on monocrystalline ice at 0°C was 42°. The equivalent contact angle calculated from surface and interfacial tensions for water and carbon disulfide at 20°C was 41°. He concluded, "ice near 0° resembles liquid water." The calculated surface tension of ice using 42° in Eq. (63) is 26 mJ m^{-2}. This is close to the 28 mJ m^{-2} that Adamson found to be the critical surface tension of ice. No other theory of solid surfaces predicts that solid ice could have a surface tension 44 mJ m^{-2} less than liquid water.

Figure 30 presents examples of how spreading pressures cause Zisman curves to become linear. The equilibrium contact angles were calculated using experimentally measured spreading pressures [48]. The curves are similar to the curves shown earlier for PTFE (Fig. 20).

The concept that $\gamma_{l/v} = \gamma_{s/v} - \gamma_{s/l}$ for all systems where $\theta = 0$ is not universally accepted. The data in Figs. 31 and 32, taken from Ref. 48, clearly show the validity of Young's equation when $\theta = 0$ (see Sec. II.G).

These few examples show the power of liquid/liquid studies to illustrate thermodynamic equations and test molecular theories.

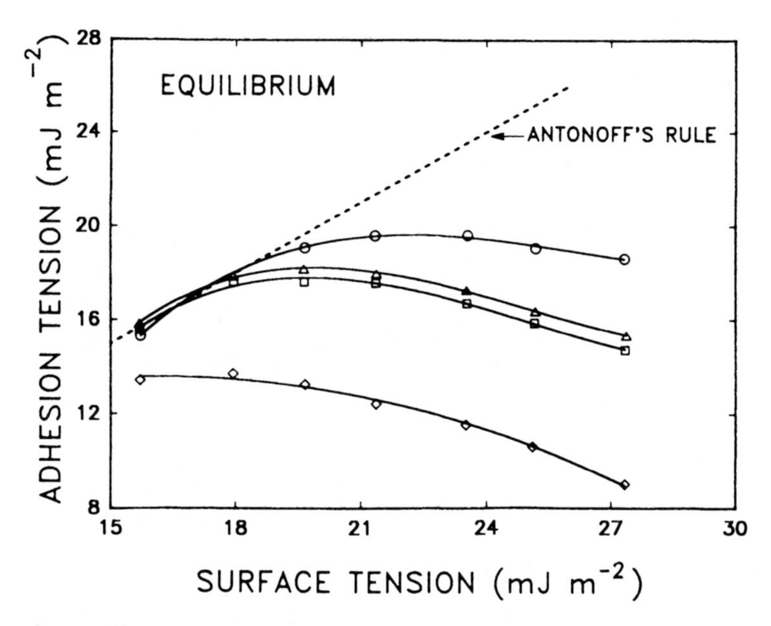

FIG. 31 Equilibrium adhesion tensions of n-alkanes on low-energy surfaces. ○ = water; △ = $H(CF_2)_2CH_2OH$; □ = $H(CF_2)_4CH_2OH$; ◇ = $F(CF_2)_2CH_2OH$. (From Ref. 48 by permission.)

VII. INTERFACIAL WETTABILITY

A. Bartell–Osterhof Equation

Interfacial wettability is the wetting of a solid by one liquid that is surrounded by a second. The starting point for the study of interfacial wettability is usually the Bartell–Osterhof equation [49] [Eq. (68)]. This equation calculates interfacial contact angles from the more readily measurable contact angles at solid/liquid/vapor interfaces. It was derived for ideal surfaces without adsorption.

The notation associated with interfacial wettability becomes cumbersome especially where hysteresis is concerned. We have found it easier to think of water in oil and write the equations accordingly. We use the subscript *w* for the water phase (or liquid 1) and *b* for the oil phase or (liquid 2). We use *b* to prevent confusion with the subscript *o*. θ_L is the contact angle of oil on solid measured through the water phase. The subscript // indicates equilibrium. For example, $\gamma_{s/b//w}$ is the interfacial tension at the solid/oil interface with the interface in equilibrium with bulk water.

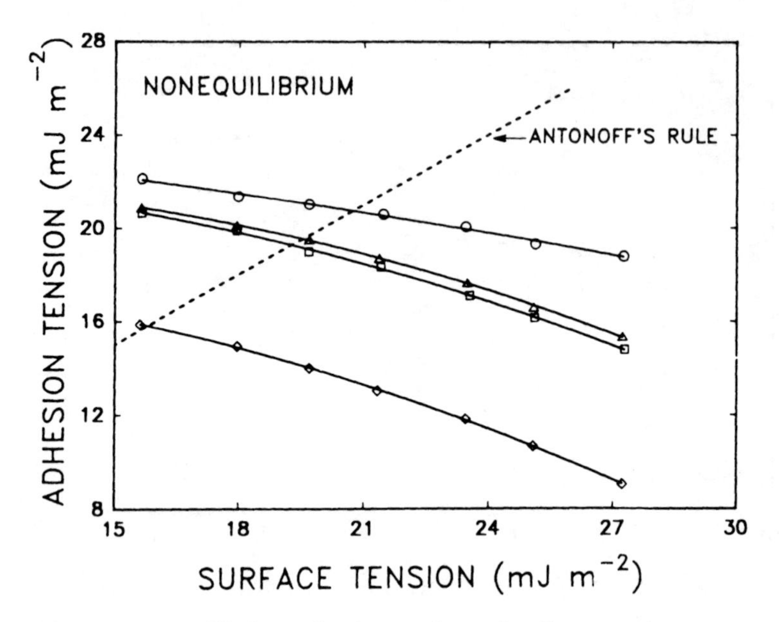

FIG. 32 Nonequilibrium adhesion tensions of n-alkanes on low-energy surfaces. ○ = water. △ = $H(CF_2)CH_2OH$. □ = $H(CF_2)_4CH_2OH$. ◇ = $F(CF_2)_2CH2OH$. (From Ref. 48 by permission.)

The interfacial tension between solid and oil is $\gamma_{s/b}$. It is not in equilibrium with the water. Similarly, $\gamma_{s/w}$ is the interfacial tension between the solid and water not in equilibrium with the oil. By using this notation, Young's equation for a drop of water on a surface with water surrounded by oil is, at equilibrium,

$$\gamma_{w/b} \cos \theta_L = \gamma_{s/b//w} - \gamma_{s/w//b} \tag{64}$$

where the contact angle θ_L is measured through water.

The contact angles in air are measured through the liquid in each case. Equation (65) is Young's equation for the solid/oil/vapor system

$$\gamma_{b/v} \cos \theta_b = \gamma_{s/o} - \gamma_{s/b} \tag{65}$$

Equation (66) is Young's equation for the solid/water/vapor system

$$\gamma_{w/v} \cos \theta_w = \gamma_{s/o} - \gamma_{s/w} \tag{66}$$

The interfacial Young equation, if we do not consider adsorption, is

$$\gamma_{w/b} \cos \theta_L = \gamma_{s/b} - \gamma_{s/w} \tag{67}$$

Substituting Eqs. (65) and (66) into Eq. (67) yields Eq. (68), the Bartell–Osterhof equation

$$\gamma_{w/b} \cos \theta_L = \gamma_{w/v} \cos \theta_w - \gamma_{b/v} \cos \theta_b \tag{68}$$

Melrose [50] extended the derivation to include adsorption. We follow his derivation. The starting point is the complete Young's equation Eq. (64), where all interfaces are in equilibrium. The spreading pressure of water ($\pi_{s/b//w}$) at the solid/oil interface is defined by Eq. (69)

$$\pi_{s/b//w} = \gamma_{s/b} - \gamma_{s/b//w} \tag{69}$$

The spreading pressure of oil at the solid/water interface ($\pi_{s/w//b}$) is defined by Eq. (70)

$$\pi_{s/w//b} = \gamma_{s/w} - \gamma_{s/w//b} \tag{70}$$

Remembering that $\pi_{s/b}$ and $\pi_{s/w}$ are given by Eqs. (71) and (72), respectively, and substituting Eqs. (69–72) into Eq. (68) yield Eq. (73). Substituting Eqs. (71) and (72) into Eq. (64) gives Eq. (73)

$$\pi_{s/b} = \gamma_{s/o} - \gamma_{s/b} \tag{71}$$

$$\pi_{s/w} = \gamma_{s/o} - \gamma_{s/w} \tag{72}$$

$$\gamma_{w/b} \cos \theta_L = \pi_{s/w} - \pi_{s/b} + \pi_{s/w//b} - \pi_{s/b//w} \tag{73}$$

$\pi_{s/w}$ is the spreading pressure of water at the solid/water interface in the absence of oil and $\pi_{s/b}$ the spreading pressure of oil at the solid/oil interface in the absence of water.

Equations (74) and (75) follow from Gibbs modified Young equation [Eq. (6)]

$$\gamma_{w/v} \cos \theta_w = \pi_{s/w} - \pi_{s/v//w} \tag{74}$$

$$\gamma_{b/v} \cos \theta_b = \pi_{s/b} - \pi_{s/v//b} \tag{75}$$

where $\pi_{s/v//w}$ is the spreading pressure at the solid/vapor interface where the vapor is water and $\pi_{s/v//b}$ the spreading pressure at the solid/vapor interface where the vapor is the oil.

Substituting Eqs. (74) and (75) into Eq. (73), we arrive at the Bartell–Osterhof–Melrose equation, Eq. (76)

$$\gamma_{w/b} \cos \theta_L = \gamma_{w/v} \cos \theta_w - \gamma_{b/v} \cos \theta_b + \pi_{s/v//w} - \pi_{s/v//b} + \pi_{s/w//b} - \pi_{s/b//w} \tag{76}$$

$$\Delta\pi = \gamma_{w/b} \cos \theta_L + \gamma_{b/v} \cos \theta_b - \gamma_{w/v} \cos \theta_w = \pi_{s/v//w} - \pi_{s/v//b} + \pi_{s/w//b} - \pi_{s/b//w} \tag{77}$$

These equations are valid for all values of θ including zero.

Of this form of the Bartell–Osterhof–Melrose equation, Eq. (76) is probably the most useful. There is a second form, shown in Appendix III, that may be of interest. The right-hand side of Eq. (77) is called the Δ *film pressure* or $\Delta\pi$. Examples of its use are given in Sec. VII.D.

The equilibrium work of adhesion of water surrounded by oil is given by Eq. (78)

$$Wa_{s/w//b} = \gamma_{w/b}(1 + \cos\theta_L) \tag{78}$$

Equation (78) is derived by noting that when water surrounded by oil is separated from a solid, a unit of solid/oil//water interface and a unit of water/oil interface are created and a unit of solid/water//oil interface is eliminated. In this case, the solid surface is in equilibrium with both water and oil both before and after the separation.

B. Interfacial Contact-Angle Hysteresis

The already complicated set of Bartell–Osterhof equations becomes even more complicated with contact-angle hysteresis. For each set of liquids and solids, it is possible to calculate four Bartell–Osterhof angles. These correspond to the four different combinations of advancing and receding angles shown in Table 15.

Bartell and Zuidema [51] determined the wettability of talc by water immersed in various organic liquids. Their angles measured in air are close to advancing angles, but somewhat less because they removed the pipette from the drops before measurement and allowed the drops to equilibrate. Their interfacial angles were measured by first immersing the talc in water and then advancing a drop of the organic liquid over the surface. Their original data are reproduced in Table 16. The starred points were not calculated originally by Bartell and Zuidema because of the zero contact angles of the organic liquid on the talc. Even when the contact angles are zero, the agreement is good for liquids having surface tensions above 28 mJ m^{-2}.

TABLE 15 Possible Bartell–Osterhoff Calculations Involving Hysteresis

Case	Water in air	Oil in air
1	Adv.	Adv.
2	Adv.	Rec.
3	Rec.	Adv.
4	Rec.	Rec.

TABLE 16 Comparison of Measured Interfacial Contact Angles on Cleaved Talc with Angles Calculated from the Simple Bartell–Osterhoff Equation: Data of Bartell and Zuidema

Liquid	Surface tension ($mJ\ m^{-2}$)	Interfacial tension ($mJ\ m^{-2}$)	Angle in air (deg)	Calculated interfacial angle (deg)	Measured interfacial angle (deg)
Water	72.1		86		
Methylene iodide	50.2	48	53	122	121
α-Bromonaphthalene	44.0	41.6	34	139	141
Bromobenzene	35.9	39.6	12	140	142
Acetylene tetrabromide	49.1	38.3	47	138	139
Chlorobenzene	32.6	37.9	0	137[a]	140
Toluene	28.1	36.1	0	130[a]	129
Ethylene dibromide	38.1	36.0	26	145	145
Benzene	28.2	34.6	0	132[a]	135
Butyl acetate	24.1	13.2	0	180[a]	144
Amyl alcohol	23.4	5.0	0	180[a]	100

[a] These angles were not calculated by Bartell and Zuidema.
Water and oil advancing angles were used to calculate the water receding interfacial angle.
Source: Ref. 51 by permission.

Ray and Bartell [52] studied the wetting properties of some substituted cellulose derivatives. They measured both advancing and receding angles of liquids in air and interfacial angles. They concluded from these measurements that the advancing angle of liquid 1 should be combined with the receding angle of liquid 2 to calculate the advancing angle of liquid 1 immersed in liquid 2. Their rationale was that hysteresis is caused by entrapment of liquid at the interface as a drop moves over the surface. Tables 17 through 19 show three sets of their data. We have used Eq. (68) to calculate the four possible cases shown in Table 15.

Melrose [50] also leaned toward the penetration theory of hysteresis. Considering hysteresis to be caused primarily by heterogeneity gives a different perspective. Figure 6 shows theoretical advancing and receding angles as a function of surface coverage. It should apply as well to solid/liquid/liquid. In air, the advancing angle is characteristic of the low-energy component of the surface and the receding angle is associated with the

TABLE 17 Interfacial Contact Angles Calculated Using the Simple Bartell-Osterhof Equation

Polymer: Ethyl cellulose iodide	Liquids: Water/methylene Measured contact angles (deg)	
	Adv.	Rec.
Water (air)	74.0	44.0
Methylene iodide (air)	59.0	30.0
Interfacial angles (through water)	118	58
Calculated Bartell–Osterhof Angles (deg)		
Case 1 adv./adv. 97		
Case 2 adv./rec. 120		
Case 3 rec./adv. 58		
Case 4 rec./rec. 80		

high-energy fraction. The same is true for interfacial angle measurements. The water advancing angle under oil is associated with the low-energy portion of the surface and the water receding angle is associated with the high-energy portion of the surface. The Bartell–Osterhof equation should apply independently to the high- and low-energy portions of the surface. We would therefore use advancing angles of both oil and water in air to

TABLE 18 Interfacial Contact Angles Calculated Using the Simple Bartell-Osterhof Equation

Polymer: Acetobutyrate	Liquids: Water/bromonaphthalene Measured contact angles (deg)	
	Adv.	Rec.
Water (air)	75.5	23.5
Bromonaphthalene (air)	48.5	11.0
Interfacial angles (through water)	131	86
Calculated Bartell–Osterhof Angles (deg)		
Case 1 adv./adv. 123		
Case 2 adv./rec. 128		
Case 3 rec./adv. 81		
Case 4 rec./rec. 85		

TABLE 19 Interfacial Contact Angles Calculated Using the Simple Bartell-Osterhof Equation

Polymer: Tri-acetate film iodide	Liquids: Water/methylene Measured contact angles (deg)	
	Adv.	Rec.
Water (air)	67.5	52.0
Methylene iodide (air)	47.0	39.5
Interfacial angles (through water)	102	81
Calculated Bartell–Osterhof Angles (deg)		
Case 1 adv./adv. 98		
Case 2 adv./rec. 104		
Case 3 rec./adv. 78		
Case 4 rec./rec. 84		

calculate the interfacial advancing angle (measured through water) and the air receding angles to calculate the water receding angle in oil.

The results shown in Tables 17 through 19 do not clearly show which combination of angles is best. We have found that the combination of angles based on the above heterogeneous surface model (case 1 or case 4) gives reasonable results. These data are presented in Sec. VII.C.

C. Measurement and Data Presentation

The principal methods of measuring interfacial angles are the same as those used in air. These are the sessile drop (or sessile bubble) and the Wilhelmy plate techniques. The major change required for the interfacial sessile drop measurement is that a cell is required to contain the second liquid phase. A complication of the Wilhelmy plate technique is that the wire holding the plate goes through the oil/water interface, and corrections must be made for its buoyancy. This is usually no problem if a fine wire is used.

Figure 33 is a typical example of interfacial contact angles determined by the Wilhelmy plate technique. The film was a sample of commercial poly(ethylene terephthalate). The work of adhesion hysteresis loops for this material by water and hexadecane in air was given earlier in Fig. 16. Figure 34 presents the data of Fig. 33 plotted as works of adhesion. The horizontal straight lines in Fig. 34 were calculated by the simple Bartell–Osterhof equation, Eq. (68). The interfacial advancing works of adhesion were calculated from the average advancing angles in air and the receding

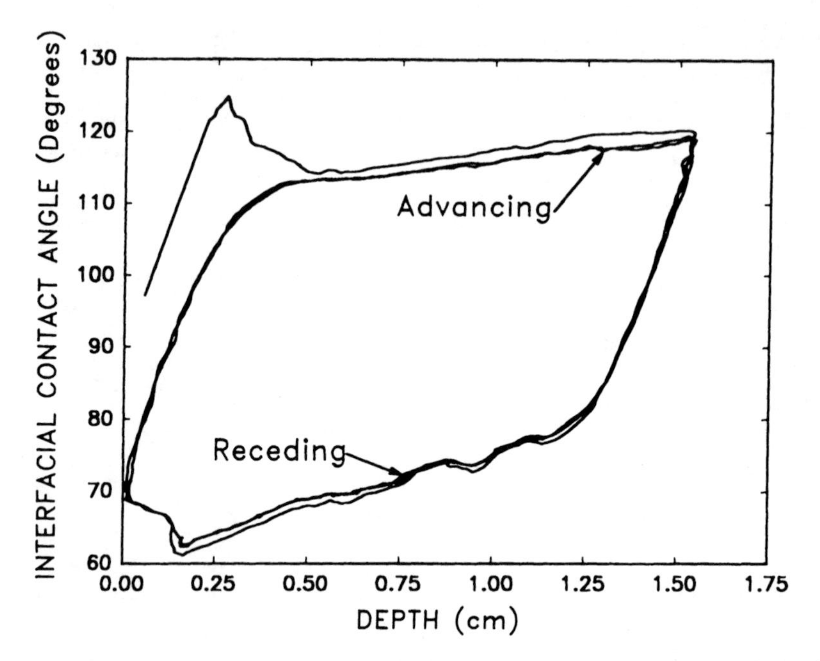

FIG. 33 Interfacial contact angles at the poly(ethyleneterephthalate)/water/hexadecane interface. Measured through the water.

works of adhesion from the average receding angles. The advancing interfacial work of adhesion of water in Fig. 34 is about 31 mJ m^{-2}. It is about 80 mJ m^{-2} in air (Fig. 16).

Table 20 shows calculated and measured interfacial advancing angles and works of adhesion of water on glass, polyethylene terephthalate (PET), and a fluorinated copolymer of perfluoropropylene and tetrafluoroethylene (FEP) in the presence of two oils, hexadecane and decane. Table 21 presents data for receding angles and receding works of adhesion for the same systems. The Bartell–Osterhof equation [Eq. (68)] was used for these calculations. Advancing calculated angles agree quite well for the PET and FEP, but the agreement is poorer for the glass. Table 22 shows Δ spreading pressures ($\Delta\pi$'s) calculated using Eq. (77). These $\Delta\pi$'s are deviations from the Bartell–Osterhof equation. They also represent the differences between the observed and calculated works of adhesion in Tables 20 and 21.

1. Special Cases and Applications

The Bartell–Osterhof–Melrose equation [Eq. (77)] reduces to the Bartell–Osterhof equation [Eq. (68)] when $\Delta\pi$ is zero. Even when it is not zero,

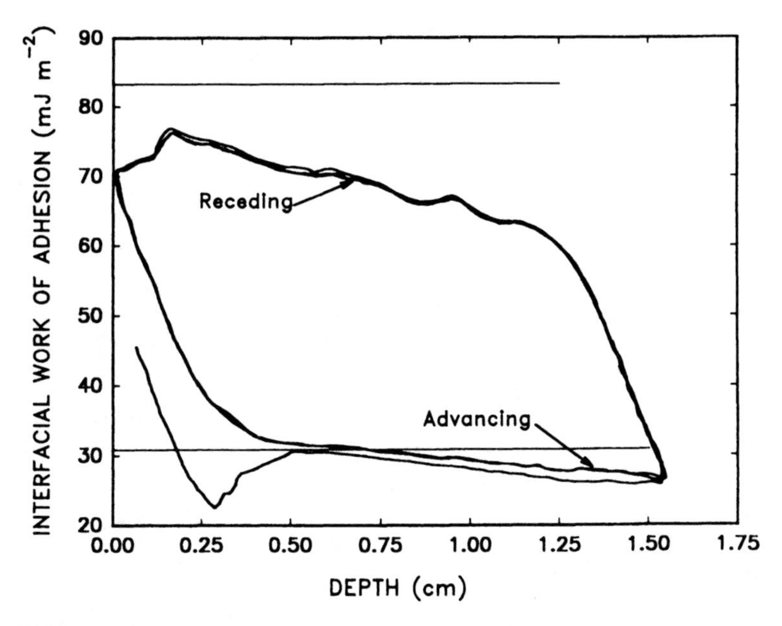

FIG. 34 Interfacial works of adhesion of water at the poly(ethylene terephthalate)/water/hexadecane interface. The angle is measured through the water.

TABLE 20 Bartell-Osterhof Contact Angles at the Water/Hexadecane or Water/Decane Interfaces on Glass, Poly(ethylene terephthalate) (PET), or a Fluorinated Copolymer of Perfluoro Propylene and Tetrafluoroethylene (FEP)

	Hexadecane				Decane			
	Adv. angles (deg)		Adv. works of adhesion (mJ m^{-2})		Adv. angles (deg)		Adv. works of adhesion (mJ m^{-2})	
	Obs.	Calc.	Obs.	Calc.	Obs.	Calc.	Obs.	Calc.
Glass	0	32	105	97	0	32	103	95
PET	117	114	29	31	115	110	30	34
FEP	165	180	2	0	180	180	0	0

TABLE 21 Bartell-Osterhof Contact Angles at the Water/Hexadecane or Water/Decane Interfaces on Glass, Poly(ethylene terephthalate) (PET), or a Fluorinated Copolymer of Perfluoro Propylene and Tetrafluoroethylene (FEP)

	Hexadecane				Decane			
	Rec. angles (deg)		Rec. works of adhesion (mJ m^{-2})		Rec. angles (deg)		Rec. works of adhesion (mJ m^{-2})	
	Obs.	Calc.	Obs.	Calc.	Obs.	Calc.	Obs.	Calc.
Glass	0	30	104	97	0	18	102	100
PET	74	53	66	83	69	47	70	86
FEP	136	125	15	22	149	125	7	22

it is unlikely that all four terms of $\Delta\pi$ will be significantly larger than zero for a given system at the same time. This leads to useful approximations. Melrose has listed the following:

If a vapor-phase contact angle is greater than zero (e.g., $\theta_{s/v//w} > 0$), then $\pi_{s/v//w} = 0$.

If $\theta_L = 0$, then $\pi_{s/w//b} = 0$. For example, a hydrocarbon does not adsorb at the interface between water and a high-energy surface.

It is unlikely that both $\pi_{s/w//b}$ and $\pi_{s/b//w}$ will be significantly different from zero. For example, it is unlikely that water would adsorb at the interface between polyethylene and decane, but decane could adsorb at the water/polyethylene interface. Similarly, it is unlikely that decane would adsorb at the interface between glass and water, but water could adsorb at the interface between decane and glass.

TABLE 22 Δ Spreading Pressures ($\Delta\pi$) of Water in Hexadecane and Decane on Three Different Surfaces: Glass, Polyester (PET), and a Perfluoropolymer (FEP)

	Δ Spreading pressures (mJ m^{-2})			
	Hexadecane		Decane	
	Adv.	Rec.	Adv.	Rec.
Glass	8	7	8	2
PET	−2	−17	−4	−16
FEP	2	−7	0	−15

Melrose applied these approximations to data in the literature for the system PTFE/decane/water. Since the angles in air of both water and decane are finite, he assumed that both $\pi_{s/v//w}$ and $\pi_{s/v//b}$ were zero. Also assuming that water would not adsorb at the PTFE/decane interface, he set $\pi_{s/b//w} = 0$. These approximations reduce Eq. (77) to Eq. (79)

$$\pi_{s/w//b} = -\gamma_{w/v} \cos \theta_{s/w} + \gamma_{b/v} \cos \theta_{s/o} + \gamma_{w/b} \cos \theta_L \tag{79}$$

The spreading pressure of decane at the PTFE/water interface is given by Eq. (79). According to the heterogenous surface model of interfacial wettability, Eq. (79) should apply to the spreading pressure data from advancing angles for PET and PTFE in Table 22. In each case, $\Delta\pi$ is zero within the error of the measurement, which is about 4 mJ m^{-2}. Therefore, there is little tendency for the oils to adsorb under the water.

For the glass data it is probable that both $\pi_{s/v//w}$ and $\pi_{s/b//w}$ are finite and that $\pi_{s/v//b}$ and $\pi_{s/w//b}$ are zero. This leads to Eq. (80)

$$\Delta\pi = \pi_{s/v//w} - \pi_{s/b//w} \tag{80}$$

Applying Eq. (80) to the data in Table 22 gives some useful information. The spreading pressure of water vapor at the glass/vapor interface is about 8 mJ m^{-2} greater than the spreading pressure of water at the glass/hexadecane interface. Similarly, the spreading pressure of water at the glass/vapor interface is about 5 mJ m^{-2} greater than at the glass/decane interface. These values were obtained by averaging advancing and receding $\Delta\pi$'s. By subtracting the two, we find $\pi_{\text{glass/decane//water}} - \pi_{\text{glass/hexadecane//water}} = 3$ mJ m^{-2}. The free energy of adsorption of water is about the same at the two interfaces. If there is any difference, adsorption is stronger at the glass/decane interface.

The spreading pressures are negative for the receding angles on FEP and PET. Since $\pi_{s/v//b}$ is probably zero, this can only occur if $\pi_{s/b//w}$ is larger than $\pi_{s/v//w} + \pi_{s/w//b}$. This suggests that there are hydrophilic sites on the polymers causing adsorption of water at the polymer/oil interface.

There have been few applications of interfacial wettability measurements and fewer still where spreading pressures have been considered. Schultz et al. [53] have attempted to measure acid–base interactions where contact angles in air are zero by measuring under a second liquid where interactions occur only by dispersion forces. They attribute interactions greater than those calculated from the Bartell–Osterhof equation to acid-base interactions. Unfortunately, they did not address the question of hysteresis at all. They also ignored adsorption of the active liquid at the solid/vapor interface. Since the contact angle is zero, the adsorption is finite and probably large.

APPENDIX I. SURFACE TENSION AND SURFACE STRESS

We here give Gibbs' discussion of the difference between *surface stress* and the work *spent in forming a unit of the surface of discontinuity.* Unfortunately, there is a confusion of nomenclature between Gibbs and current usage. What is commonly called a surface tension γ is termed by Gibbs *the work spent in forming a unit of the surface of discontinuity* and given the symbol σ. What Gibbs calls a surface tension is now usually referred to as a surface stress. It is clear that the quantity σ used by Gibbs throughout his papers is the work of formation and not that of stretching the surface. Gibbs states,

> As in the case of two fluid masses, we may regard σ as expressing the work spent in forming a unit of the surface of discontinuity—under certain conditions, which we need not here specify—but it cannot properly be regarded as expressing the tension of the surface. The latter quantity depends upon the work spent in *stretching* the surface, while the quantity σ depends upon the work spent in *forming* the surface. With respect to perfectly fluid masses, these processes are not distinguishable, unless the surface of discontinuity has components which are not found in the contiguous masses, and even in this case (since the surface must be supposed to be formed out of matter supplied at the same potentials which belong to the matter in the surface) the work spent in increasing the surface infinitesimally by stretching is identical with that which must be spent in forming an equal infinitesimal amount of new surface. But when one of the masses is solid, and its states of strain are to be distinguished, there is no such equivalence between the stretching of the surface and the forming of new surface.*

APPENDIX II. GIBBS: ANOTHER METHOD AND NOTATION

The following is a direct quotation from Gibbs [9]. In this section, Gibbs derives our Eq. (6) from general principles. Our comments are enclosed in square brackets.

---

* This will appear more distinctly if we consider a particular case. Let us consider a thin plane sheet of a crystal in a vacuum (which may be regarded as a limiting case of a very attenuated fluid), and let us suppose that the two surfaces of the sheet are alike. By applying the proper forces to the edges of the sheet, we can make all stress vanish in its interior. The *tensions* of the surfaces, thus determined, may evidently have different values in different directions, and are entirely different from the quantity which we denote by σ, which represents the work required to form a unit of the surface by any reversible process, and is not connected with any idea of a direction.

Another method and notation.—We have so far supposed that we have to do with a non-homogeneous film of matter between two homogeneous (or very nearly homogeneous) masses, and that the nature and state of this film are in all respects determined by the nature and state of these masses together with the quantities of the foreign substances which may be present in the film. Problems relating to processes of solidification and dissolution seem hardly capable of a satisfactory solution, except on this supposition, which appears in general allowable with respect to the surfaces produced by these processes. [Gibbs is here talking about the difficulty of determining a solid surface tension. See Appendix I.] But in considering the equilibrium of fluids at the surface of an unchangeable solid, such a limitation is neither necessary nor convenient. The following method of treating the subject will be found more simple and at the same time more general.

Let us suppose the superficial density of energy to be determined by the excess of energy in the vicinity of the surface over that which would belong to the solid, if (with the same temperature and state of strain) it were bounded by a vacuum in place of the fluid, and to the fluid, if it extended with a uniform volume—density of energy just up to the surface of the solid, or, if in any case this does not sufficiently define a surface, to a surface determined in some definite way by the exterior particles of the solid. Let us use the symbol (ϵ_s) to denote the superficial energy thus defined. Let us suppose a superficial density of entropy to be determined in a manner entirely analogous, and be denoted by (η_s). In like manner also for all the components of the fluid, and for all foreign fluid substances which may be present at the surface, let the superficial densities be determined, and denoted by (Γ_2), (Γ_3), etc. *These superficial densities of the fluid components* relate solely to the matter which is fluid or movable. All matter which is immovably attached to the solid mass is to be regarded as a part of the same. Moreover, let ζ be defined by

$$\zeta = t(\eta_s) - \mu_2(\Gamma_2) - \mu_3(\Gamma_3), \text{ etc.} \qquad [(81)]$$

[In Gibbs' notation, η is the entropy, t the temperature, and ζ the same as our $-\pi$. After some discussion he derives the equation]

$$d\zeta = -(\eta_s)\, dt - (\Gamma_2)\, d\mu_2 - (\Gamma_3)\, d\mu_3, \text{ etc.} \qquad [(82)]$$

[After a few more comments he states] The quantity ζ evidently represents the tendency to contraction in that portion of the surface of the fluid which is in contact with the solid. It may be called the *superficial tension of the fluid in contact with the solid.* Its value may be either positive or negative.

It will be observed for the same solid surface and for the same temperature but for different fluids the values of σ [our γ] (in all cases to which the definition of this quantity is applicable) will differ from those of ζ by a constant, viz., the value of σ for the solid surface in a vacuum.

For the condition of equilibrium of two different fluids at a line on the surface of the solid, we may easily obtain

$$\sigma_{AB} \cos \alpha = \zeta_{BS} - \zeta_{AB} \qquad [(83)]$$

[This is the same equation we call the Gibbs-modified Young equation. Gibbs used α for the contact angle. The important part of this analysis is that Gibbs focuses on the energetics of the motion of the fluid phase. The solid is important only as it interacts with the liquid and changes its tendency to spread. Note that Gibbs considers this equation to be even more general than the Young equation. Note also that the contact angle is determined by the *interaction* between the liquid and solid, not by the absolute value of the solid surface tension. Note also that the Gibbs-modified Young equation is derived and is not a starting place of analysis. We highly recommend the interested reader to consult this entire section of Gibbs' book.]

APPENDIX III. ANOTHER FORM OF THE BARTELL–OSTERHOF EQUATION

There is another Bartell–Osterhof–Melrose equation that has the same form as the simple Bartell–Osterhof. This formulation leads to a different set of measurements. In general, these other measurements are not so convenient. We include the equation here for completeness.

Let a drop of water on a surface be in equilibrium with the vapors of both itself w and the second liquid b. Young's equation applied to this situation is

$$\gamma_{w/v//b} \cos \theta_{w//b} = \gamma_{s/v//w//b} - \gamma_{s/w//b} \qquad (84)$$

where the term $\gamma_{s/v//w//b}$ is the surface tension of the solid in equilibrium with the saturated vapors of both the water and oil.

A similar equation applies to a drop of oil on the surface in equilibrium with water

$$\gamma_{b/v//w} \cos \theta_{b//w} = \gamma_{s/v//w//b} - \gamma_{s/b//w} \qquad (85)$$

The interfacial Young equation considering adsorption [see Eq. (67)] is Eq. (86)

$$\gamma_{w/b} \cos \theta_L = \gamma_{s/b//w} - \gamma_{s/w//b} \qquad (86)$$

Substituting Eq. (84) and Eq. (85) into Eq. (86) leads to Eq. (87)

$$\gamma_{w/b} \cos \theta_L = \gamma_{w/v//b} \cos \theta_{w//b} - \gamma_{b/v//w} \cos \theta_{b//w} \qquad (87)$$

This has the same form as the original equation of Bartell and Osterhof [Eq. (68)], but the surface tensions and contact angles are those measured in equilibrium with all the components of the system. For example, if we are concerned with the system water/benzene/polyethylene, $\gamma_{w/v//b}$ is the surface tension of water in equilibrium with the saturated vapor of benzene. This surface tension is nearly 30 mJ m^{-2} lower than that of pure water.

REFERENCES

1. T. F. Young, and W. D. Harkins, *Internat. Critical Tables 4:*434 (1928).
2. R. E. Johnson, Jr., and R. H. Dettre, in *Surface and Colloid Science*, Vol. 2 (E. Matijevic, ed.), Wiley-Interscience, New York, 1969.
3. H. C. Hamaker, *Physica 4:*1058 (1953).
4. E. M. Lifshitz, *Sov. Phys. 2:*73 (1956).
5. D. B. Hough, and L. E. White, *Adv. Coll. Inter. Sci. 14:*1 (1980).
6. J. N. Israelachvili, *J. Chem. Soc., Faraday II 69:*1729 (1973).
7. A. W. Neumann, *Adv. Coll. Inter. Sci. 4:*1 (1974).
8. R. E. Johnson, Jr., *J. Phys. Chem. 63:*1655 (1959).
9. J. Willard Gibbs, *The Collected Works of J. Willard Gibbs, Vol. 1, Thermodynamics*, Yale Univ. Press, New Haven, 1928.
10. T. Young, *Phil. Trans. 95:*65, 82 (1805).
11. J. Willard Gibbs, *op. cit.*, p. 326.
12. F. M. Fowkes, in *Chemistry and Physics of Interfaces* (D. E. Gushee, ed.), ACS, Washington, DC, 1965, p. 1.
13. W. D. Harkins, *The Physical Chemistry of Surface Films*, Reinhold, New York, 1952, Chap. 2.
14. G. Antonoff, (a) *J. Chem. Phys. 5:*364 (1907). (b) *J. Chem. Phys. 5:*372 (1907).
15. L. D. Landau, and E. M. Lifshitz, *Statistical Physics*, Pergamon, London, 1958, pp. 471–473.
16. R. E. Johnson, Jr., and R. H. Dettre, *J. Phys. Chem. 68:*1744 (1964).
17. A. B. D. Cassie, *Discuss. Faraday Soc. 3:*11 (1948).
18. J. N. Israelachvili, and M. L. Gee, *Langmuir 5:*288 (1989).
19. R. H. Dettre, and R. E. Johnson, Jr., *J. Phys. Chem. 69:*1507 (1965).
20. R. E. Johnson, Jr., D. A. Brandreth, and R. H. Dettre, *Contam. Contr. 9:* 17 (1970).
21. L. S. Bartell, and R. J. Ruch, *J. Phys. Chem. 60:*1231 (1956).
22. (a) R. E. Johnson, Jr., and R. H. Dettre, *Adv. Chem. Ser. 43:*112 (1964). (b) R. H. Dettre, and R. E. Johnson, Jr., *Adv. Chem. Ser. 43:*136 (1964). (c) R. E. Johnson, Jr., and R. H. Dettre, *J. Phys. Chem. 68:*1744 (1964). (d) R. H. Dettre, and R. E. Johnson, Jr., *J. Phys. Chem. 69:*1507 (1964). (e) J. D. Eick, R. J. Good, and A. W. Neumann, *J. Coll. Inter. Sci. 53:*235 (1975). (f) A. W. Neumann, and R. J. Good, *J. Coll. Inter. Sci. 38:*341 (1972). (g) C. Huh, and S. G. Mason, *J. Coll. Inter. Sci. 60:*11 (1977). (h) J. F. Oliver,

C. Huh, and S. G. Mason, *Coll. Surf. 1:*79 (1980). (i) J. F. Joanny, and P. G. deGennes, *J. Chem. Phys. 81:*552 (1984). (j) L. W. Schwartz, and S. Garoff, *Langmuir 1:*219 (1985). (k) L. W. Schwartz, and S. Garoff, *J. Coll. Inter. Sci. 106:*422 (1985).
23. D. H. Everett, and W. L. Whitton, *Trans. Faraday Soc. 48:*749 (1952).
24. R. E. Johnson, Jr., R. H. Dettre, and D. A. Brandreth, *J. Coll. Inter. Sci. 62:*205 (1977).
25. W. A. Zisman, in *Advances in Chemistry Series 43* (R. F. Gould, ed.), American Chemical Society, Washington, D.C., 1964, p. 1.
26. R. E. Johnson, and R. H. Dettre, *Polym. Preprint* (ACS Division of Polymer Chemistry) *28(1):*48 (1987).
27. (a) L. A. Girifalco, and R. J. Good, *J. Phys. Chem. 61:*904 (1957). (b) R. J. Good, L. A. Girifalco, and G. Kraus, *J. Phys. Chem. 62:*1418 (1958). (c) R. J. Good, and L. A. Girifalco, *J. Phys. Chem. 64:*561 (1960).
28. D. Bertholet, *Compt. Rend. 126:*1703, 1857 (1898).
29. F. M. Fowkes, in *Wetting, SCI Monograph No. 25*, Soc. Chemical Industry, London, 1967, p. 3.
30. D. K. Owens, and R. C. Wendt, *J. Appl. Polym. Sci. 13:*1741 (1969).
31. D. H. Kaelble, *J. Adhesion, 2:*66 (1970).
32. F. M. Fowkes, F. L. Riddle, W. E. Pastore, and A. A. Weber, *Coll. Surf. 43:*367 (1990).
33. (a) J. K. Spelt, and A. W. Neumann, *J. Coll. Inter. Sci. 122:*294 (1988). (b) J. K. Spelt, and A. W. Neumann, *Langmuir 3:*588 (1987). (c) J. K. Spelt, D. R. Absolom, and A. W. Neumann, *Langmuir 2:*620 (1986). (d) A. W. Neumann, *Adv. Coll. Inter. Sci. 4:*1 (1974). (e) D. Li, E. Moy, and A. W. Neumann, *Langmuir 6:*888 (1990).
34. I. D. Morrison, *Langmuir 7:*1833 (1991).
35. J. F. Padday, and N. D. Uffindell, *J. Phys. Chem. 72:*1407 (1968).
36. F. M. Fowkes, in *Wetting, SCI Monograph No. 25*, Soc. Chemical Industry, London, 1967, pp. 3–30.
37. H. W. Fox, and W. A. Zisman, *J. Coll. Sci. 5:*514 (1950).
38. D. P. Graham, *J. Phys. Chem. 69:*4387 (1965).
39. E. G. Shafrin, and W. A. Zisman, *J. Phys. Chem. 61:*1046 (1957).
40. D. Hoernschemeyer, *J. Phys. Chem. 70:*2628 (1966).
41. T. M. Reed III, in *Fluorine Chemistry* (J. H. Simons, ed.), Academic Press, New York, 1964, p. 189.
42. V. P. Shibayev, B. S. Petrukhin, Yu. A. Zubov, N. A. Plate, and V. A. Kargin, *Vysokomol. Soye. A10:*216 (1968).
43. M. K. Bernett, and W. A. Zisman, *J. Phys. Chem. 71:*2075 (1967).
44. E. F. Hare, E. G. Shafrin, and W. A. Zisman, *J. Phys. Chem. 58:*236 (1954).
45. F. Neumann, *Vorlesungen uber die Theorie der Capillaritat*, B. G. Teubner, Leipzig, 1894.
46. R. E. Johnson, Jr., and R. H. Dettre, *Langmuir 5:*293 (1989).
47. A. W. Adamson, F. P. Shirley, and K. T. Kunichika, *J. Coll. Inter. Sci. 34:* 461 (1970).

48. R. E. Johnson, Jr., and R. H. Dettre, *J. Coll. Inter. Sci. 21:*610 (1966).
49. (a) S. Schultz, K. Tsutsumi, and J.-B. Donnet, *Coll. Symp. Mono. 5:*113 (1928). (b) S. Schultz, K. Tsutsumi, and J.-B. Donnet, *Ind. Eng. Chem. 19:* 1277 (1927).
50. J. C. Melrose, in *Wetting SCI Monograph No. 25*, Soc. Chemical Industry, London, 1967, pp. 123–143.
51. F. E. Bartell, and H. H. Zuidema, *J. Amer. Chem. Soc. 58:*1449 (1936).
52. B. R. Ray, and F. E. Bartell, *J. Phys. Chem. 57:*49 (1953).
53. (a) S. Schultz, K. Tsutsumi, and J.-B. Donnet, *J. Coll. Inter. Sci. 59:*272 (1977). (b) S. Schultz, K. Tsutsumi, and J.-B. Donnet, *J. Coll. Inter. Sci. 59:*277 (1977).

2

Role of Acid–Base Interactions in Wetting and Related Phenomena

JOHN C. BERG Department of Chemical Engineering, University of Washington, Seattle, Washington

I. INTRODUCTION

Wetting refers to the macroscopic manifestations of molecular interaction between liquids and solids in direct contact at the interface between them. Such manifestations include the formation of a contact angle at the solid/liquid/fluid interline, the spreading of a liquid over a solid surface (displacing the fluid initially in contact with that surface), and the wicking of a liquid into (and displacement of a fluid from) a porous solid medium. The interactions of interest in wetting exclude the dissolution or swelling of the solid by the liquid or chemical reactions between the phases that change the system composition. The various wetting phenomena may be characterized without ambiguity in terms of macroscopic thermodynamic properties, which we term the "wetting parameters." We thus begin with a set of definitions of these parameters and examine the ways in which they may be determined by direct measurement. Since wetting phenomena are traceable to molecular interactions, we must consider the origin and nature of these molecular properties and the methods by which the macroscopic wetting parameters may be computed in terms of them. It has long been recognized that molecular interactions in condensed phases, or across interfaces between condensed phases, may be split into those of a physical and those of a chemical nature. Although this view is at least as old as the arguments between Dolezalek and van Laar [1], near the turn of the century, regarding the relative importance of these contributions to solution nonideality, recent years have seen refinements in the prevailing view of each of them. We seek to examine these refinements and their implications for the evaluation of the wetting parameters. The physical molecular interactions are those referred to collectively as the "van der Waals" forces and these are dealt with first. Wetting parameters are evaluated for systems subject only to these types of interactions and compared to direct data obtained for systems of this type. It is the thesis of this chapter that most of the chemical interactions relevant to wetting phenomena, as defined above, are effectively described as acid–base interactions. Acid–base chemistry is briefly reviewed, and the methods currently proposed for describing its influence on the macroscopic wetting parameters is considered in detail. Particular attention is given to methods for measuring the strength of the acid–base interactions and correlating the results for different liquids and solids. Finally, the prospects for *predicting* the effect of acid–base interactions on the macroscopic wetting parameters are examined. We conclude by describing some of the experimental techniques used for the acid–base characterization of solid surfaces.

There is now a very large and growing literature dealing with examples and studies of the practical importance of acid–base interactions, including the enhancement of adhesion through various surface treatments or the use of coupling agents [2–4]; improving the adherence of thin films to substrates [5]; increasing or decreasing thin-film permeability [6]; increasing the toughness of fiber-matrix or particulate filler-matrix composite systems [7–9]; the solvent treatment of wood and paper [10,11]; influencing the adsorption of organic molecules, surfactants, or polymers to various surfaces [12]; or improving the absorbency of porous solids. The literature cited is only a small recent sampling of what is currently available.

The literature on both the fundamentals and the practical aspects of acid–base concepts at solid/liquid interfaces has recently experienced a quantum jump in size. *The Journal of Adhesion Science and Technology* has devoted four issues [Vol. 4, Nos. 4, 5 and 8, and Vol. 5, No. 1] to this topic as a four-part Festschrift in honor of Professor Frederick M. Fowkes on the occasion of his 75th birthday. The entire collection of contributions has been published as a single volume: *Acid–Base Interactions: Relevance to Adhesion Science and Technology* (K. L. Mittal and H. R. Anderson, Jr., Eds.), VSP, P.O. Box 346, 3700 AH Zeist, The Netherlands. Professor Fowkes was a pioneer in the development of this subject and, particularly, its application to practical problems of wetting and adhesion. He was one of the first to understand the pervasive importance of acid–base concepts in describing the interactions between molecules and between macroscopic phases. The present review has drawn heavily on the contents of the Fowkes Festschrift, as well as the large body of published work of Professor Fowkes and his co-workers, and it is affectionately dedicated to his memory.

A. Thermodynamic Wetting Parameters

The quantitative measure of the wetting of a solid by a liquid is often taken as the contact angle θ and defined at a point along the solid/liquid/gas interline as the angle between the normals to the liquid/gas and liquid/solid interfaces at that point. This is often equivalent to the angle observable in profile between the tangents to the intersecting liquid/gas and liquid/solid interfaces, drawn in the liquid. If both fluids in contact with the solid are liquids, the one across which θ is defined must be specified. The language used in the present chapter will hereafter assume that we are dealing with solid/liquid/gas systems, unless otherwise stated. As is well known, measurements of contact angles in real systems reveal significant differences, depending on whether the liquid is advancing or receding across the solid surface. Such hysteresis has generally been traced to solid surface roughness and/or chemical heterogeneity [13]. Advancing contact angles are larger than receding angles and thus correspond to the interline locating itself on the less wettable parts of the surface. Both advancing and receding contact angles are also found to depend on the rate of interline movement, as discussed in detail elsewhere in this monograph. Among the various possibilities, the most generally useful contact angle for describing wetting behavior is the static (zero-velocity) advancing angle, and it is this angle that will be implied when simply referring to "the contact angle." When wetting measurements are made for the purpose of characterizing a solid surface, however, it is advisable to measure and report both the static advancing and receding angles [14]. For our present purposes, however, we shall deal only with a single contact angle and assume that it is characteristic of the thermodynamic state of the system under study. Systems in which the liquid exhibits a contact angle of 0° against the solid are those in which the liquid "wets out" the solid. Spontaneous spreading of the liquid over the solid at a finite velocity may be observed. Partial wetting occurs when the measured contact angle is finite. For the latter case, Young [15] was the first to relate the contact angle to the surface energies of the three interfaces meeting at the interline

$$\cos\theta = \frac{\sigma_{SG} - \sigma_{SL}}{\sigma_{LG}} \tag{1}$$

where σ_{SG}, σ_{SL}, and σ_{LG} are the area derivatives of the surface free energy per unit area of the solid/gas, solid/liquid, and liquid/gas interfaces, respectively, often referred to somewhat imprecisely as the "surface free energies" of these interfaces. In general,

$$\sigma_{ij} = \left(\frac{\partial G}{\partial A_{ij}}\right)_{T,p,\text{equil}} \tag{2}$$

where G is the Gibbs free energy of the system and A_{ij} the area of the interface between phases i and j. The subscript "equil" denotes the assumption of adsorption equilibrium, in the case of multicomponent systems. We shall assume for now that there is no practical difference between the free energies associated with the liquid- or solid-phase surfaces against the gas-phase or their own equilibrium vapors and the energies associated with those surfaces against vacuum. Thus, we shall put $\sigma_{SG} \approx \sigma_S$ and $\sigma_{LG} \approx \sigma_L$. Equation (1), known as Young's equation, may be derived from the free energy minimization of a system consisting of phases S, L, and G meeting at a common interline [16] and confers on the contact angle the status of a thermodynamic property. $\sigma_{LG} \approx \sigma_L$ is equivalent to the surface tension of the liquid, which is readily measurable. It is clear from Eq. (2) that a fluid interface system cannot support a negative interfacial tension because under such conditions, the interfacial area would spontaneously extend itself, leading ultimately to the formation of an emulsion or a solution. The "tensions" of the solid/gas and solid/liquid interfaces are not directly measurable, and in fact the quantity defined in Eq. (2) is not, in general, the total tension in an interface bounded by a solid, even if it could be measured. The total tension of a solid/fluid interface, $\sigma_{ij}^{\text{total}}$, includes a contribution σ^s from the elastic energy stored in the tangential strain field in the solid near to the interface, so that

$$\sigma_{ij}^{\text{total}}(\text{solid/fluid interface}) = \sigma^s + \sigma_{ij} \tag{3}$$

The contribution to G in Eq. (2) of the elastic energy stored in the solid is thus omitted. Only for interfaces between fluids, which by definition are incapable of supporting shear stresses in states of stable equilibrium, can the quantity defined in Eq. (2) be properly referred to as an interfacial "tension." The above distinction between fluid/fluid and solid/fluid interfaces has the practical significance that in the latter case, σ_{ij}-values may be negative. Even though a thermodynamic driving force for spontaneous area extension exists under such conditions, the diffusive processes by which the area extension would be accomplished are sufficiently slow to be neglected.

There are three different expressions for the thermodynamic free energy of interaction between a liquid and a solid, depending on how the contacting event is visualized, which we shall refer to as the "wetting parameters." They are summarized in Fig. 1. Each of the parameters represents the reversible work associated with a *de*-wetting process, as pictured. The first is the "work of adhesion," W_a, defined as the work required to disjoin a unit area of the solid/liquid interface, thereby creating a unit area of liquid/vacuum and solid/vacuum interface. It is given by

$$W_a = \sigma_S + \sigma_L - \sigma_{SL} \tag{4}$$

known as the Dupré equation. W_a is thus equivalent to the negative free energy change associated with the contacting event, or adhesion of the two phases, that is, $W_a = -\Delta G_{\text{adhesion}}$.

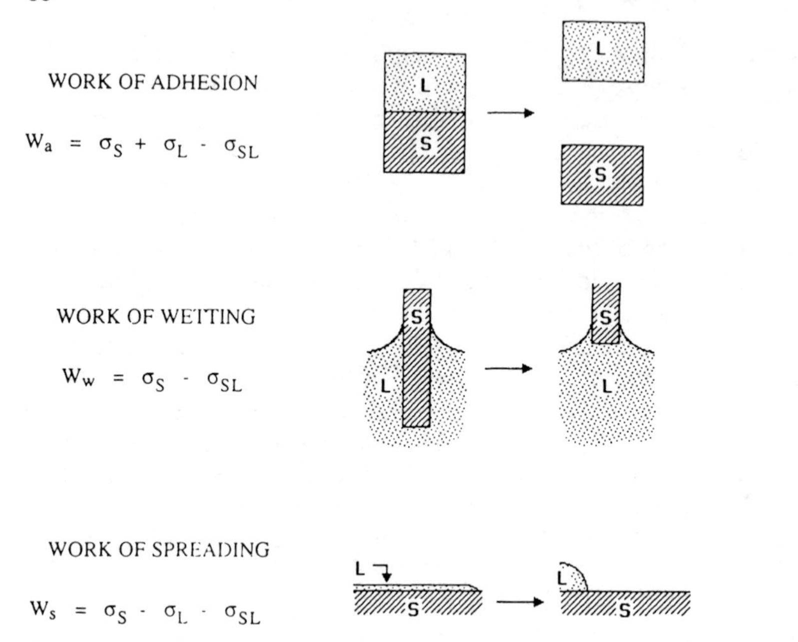

FIG. 1 Thermodynamic expressions for solid/liquid interactions.

The second wetting parameter is termed the "work of wetting," W_w (or more commonly the "wetting tension" or "adhesion tension"), defined as the work expended in eliminating a unit area of the solid/liquid interface while exposing a unit area of the solid/vacuum interface

$$W_w = \sigma_S - \sigma_{SL} \tag{5}$$

The process corresponds to withdrawing a solid material from a pool of liquid. W_w is the negative free energy associated with the process of engulfment of a solid surface by a liquid.

The third wetting parameter is the "work of spreading," W_s (or more commonly the "spreading coefficient"), defined as the work required to expose a unit area of solid/vacuum interface while destroying a corresponding amount of the solid/liquid and liquid/vacuum interfaces

$$W_s = \sigma_S - \sigma_L - \sigma_{SL} \tag{6}$$

W_s thus refers to the retraction of a thin film of liquid from a solid surface. W_s is the negative free energy associated with the process of the spreading of a liquid over a solid surface.

Since W_a, W_w, and W_s are the negative free energies associated with the adhesion, engulfment, and spreading processes, respectively, positive values of them suggest the spontaneity of the corresponding liquid/solid contacting events.

B. Direct Experimental Determination of the Wetting Parameters: Wetting Measurements

Of the quantities defining the wetting parameters, represented on the right-hand sides of Eqs. (4–6), only the liquid surface tension $\sigma_{LG} \approx \sigma_L$ is directly measurable in the laboratory. Substitution of Young's equation into Eqs. (4–6), however, leads to

$$W_a = \sigma_{LG}(1 + \cos\theta) \tag{7}$$

$$W_w = \sigma_{LG}\cos\theta \tag{8}$$

$$W_s = \sigma_{LG}(\cos\theta - 1) \tag{9}$$

which show that all three wetting parameters may be determined in the laboratory by a combination of liquid surface tension and contact-angle measurements. Equation (7) is known as the Young–Dupré equation. The description of these measurement techniques is reviewed elsewhere [17]. We shall refer to these as "wetting measurements." It is clear, however, that wetting measurements may be used to obtain experimental values for the wetting parameters only when the contact angle in finite, that is, when $\theta > 0°$. The above equations suggest that positive adhesion would occur for all $\theta < 180°$, but positive engulfment would occur only when $\theta < 90°$, and evidently both are "optimized" when $\theta = 0°$. Positive spreading is consistent only with $\theta = 0°$.

Considerable care must be exercised in the use of Eqs. (7–9) when the fluid phases are multicomponent. Both the liquid surface tension and contact angles, as used in the equations, refer to conditions of adsorption equilibrium and suggest that the *same* adsorption equilibrium conditions apply during the measurement of the surface tension as during the contact-angle measurement. The particular problems encountered in multicomponent wetting are treated in Chap. 3 of this monograph. There remains a significant complication in the use of Eqs. (7–9), however, even when the fluid phases are nominally pure components, a situation we shall assume to exist, unless otherwise stated, in the present chapter. If the liquid and/or solid phases are volatile, the gas phase at equilibrium will contain their vapors, and these may exhibit finite adsorption at the solid/gas and/or liquid/gas interfaces, respectively [18]. The presence of an adsorbed layer of molecules from the solid phase at the liquid/gas interface would reduce the surface tension of the liquid from its pure-component value by an amount equal to the "equilibrium film pressure" (sometimes called the "equilibrium spreading pressure") of the solid-phase molecules at the liquid/gas interface, $\pi_{e(s/l)}$ [19]. (The term "equilibrium" is used because the adsorbed amount corresponds to a gas-phase partial pressure equal to the equilibrium vapor pressure of the solid.) Correspondingly, the adsorption of vapor molecules of the liquid at the solid/gas interface reduces the solid/gas interfacial free energy by an amount equal to the equilibrium film pressure $\pi_{e(l/s)}$ of the liquid-phase molecules at the solid/gas interface. The adsorption of solid-phase molecules at the liquid/air interface has not generally been considered significant and will hereafter be neglected, but the importance of the equilibrium film pressure of the liquid-phase molecules at the solid/gas interface, hereafter abbreviated as π_e, is still at issue. It has been common to argue that π_e should be negligible for all cases in which the contact angle is finite [20,21], that is, for so-called "low-energy" surfaces. There is increasing evidence, however, that this might not be justifiable [22–24]. The possible significance of the equilibrium spreading pressure at the solid/gas interface is made explicit by rewriting Young's equation in the form

$$\cos \theta = \frac{(\sigma_{SG} - \pi_e) - \sigma_{SL}}{\sigma_{LG}} \tag{1a}$$

Substitution of Eq. (1a) into Eqs. (4–6) for the wetting parameters leads to

$$W_a = \sigma_{LG}(1 + \cos \theta) + \pi_e \tag{7a}$$

$$W_w = \sigma_{LG} \cos \theta + \pi_e \tag{8a}$$

$$W_s = \sigma_{LG}(\cos \theta - 1) + \pi_e \tag{9a}$$

The equilibrium film pressure may be evaluated from the adsorption isotherm for the vapors of the liquid on the solid surface, $\Gamma = \Gamma(p)$, where p is the partial pressure of the vapors of the liquid, using the Gibbs adsorption equation

$$\pi_e = RT \int_0^{p^s} \Gamma \frac{dp}{p} \tag{10}$$

where p^s is the saturation vapor pressure of the liquid. Evaluation of π_e using Eq. (10) is cumbersome and outside the realm of "wetting measurements." A short-cut method, however, using only surface tension and contact-angle measurements has recently been proposed [24].

Whether or not it is necessary to account for the equilibrium film pressure, wetting measurements can be used for direct determination of the wetting parameters only in the event that the measured contact angle is finite. For many practical situations, however, such as adhesion, coating, and wicking, the condition of greatest interest is that of total wetting. The maximization of the various wetting parameters (all within the range of conditions for which $\theta = 0°$) cannot be addressed experimentally with wetting measurements alone involving only the materials of interest. To accomplish the latter, a theoretical framework leading to alternate expressions for the wetting parameters defined in Eqs. (4–6) is required, permitting the interpretation of additional types of measurements. These may be either additional wetting measurements involving probe liquids, or they may be calorimetric, spectroscopic, chromatographic, electrokinetic, or other types of measurements, as detailed below. The interpretive theory in each case must be a model for the molecular interactions within bulk phases and across interfaces between them.

As stated above, it has long been an operational premise that interactions between molecules in bulk phases or across interfaces between them are physical or chemical in nature. Physical interactions are always present, and chemical interactions can arise whenever there is the possibility of donating a proton or an electron pair that is then shared between neighboring molecules or functional groups, leading to complexation, or adduct formation (all describable as acid–base interactions). It is further assumed that when chemical as well as physical interactions are possible, their effect on the macroscopic (configurational) thermodynamic properties of the system is additive. With respect to the wetting parameters, for example, the work of adhesion, this implies that we may write

$$W_a = W_a^{\text{physical}} + W_a^{\text{chemical}} \tag{11}$$

We consider first systems subject only to physical intermolecular forces and the evaluation of the wetting parameters for such systems and next consider those subject to both types of molecular interactions.

II. PHYSICAL MOLECULAR INTERACTIONS

A. Pair Potentials: Lifshitz–van der Waals Theory

Excluding Coulombic interactions (between ions) and metallic bonding, the physical interactions between molecules include the London (dispersion) forces, Keesom (dipole/dipole) forces, and Debye (dipole/induced dipole) forces, known collectively as the van der Waals forces. Forces due to quadrupoles or more complex permanent charge distributions are generally negligible. London forces result from the natural oscillations of the electron clouds of the molecules inducing synchronous oscillations in the neighboring molecules. The resulting temporal dipoles of neighboring molecules have the mutual orientation producing maximum attraction. The London [25] pair potential between molecules i and j is given by (using the form of the equation appropriate to the cgs system of units)

$$\Phi_{ij}^{\text{London}} = -\frac{3}{4}\alpha_i\alpha_j\frac{2I_iI_j}{(I_i + I_j)}\cdot\frac{1}{r_{ij}^6} \tag{12}$$

where α_i and α_j are the molecular polarizabilities and I_i and I_j the first ionization potentials of molecules i and j, respectively, and r_{ij} is the distance between them. The first ionization potentials do not vary substantially from one substance to the next, that is, $I_i \approx I_j = I \approx 10$ eV, so that to good approximation

$$\Phi_{ij}^{\text{London}} \approx -\frac{3}{4}\alpha_i\alpha_jI\cdot\frac{1}{r_{ij}^6} \tag{13}$$

When the molecules possess permanent dipoles, they will exhibit an interaction dependent on the mutual dipole orientation. The Boltzmann-averaged pair interaction potential between molecules i and j having dipole moments of μ_i and μ_j, respectively, has been given by Keesom [26] as

$$\Phi_{ij}^{\text{Keesom}} = -\frac{2}{3}\frac{\mu_i^2\mu_j^2}{kT}\cdot\frac{1}{r_{ij}^6} \tag{14}$$

When both molecules have substantial dipole moments, the Keesom interaction as given by Eq. (14) is comparable in magnitude to the London interaction.

Debye [27] pointed out a third interaction in which a permanent dipole in one molecule induces and interacts with a dipole in the neighboring molecule, and vice versa, to give

$$\Phi_{ij}^{\text{Debye}} = -\frac{(\alpha_i\mu_j^2 + \alpha_j\mu_i^2)}{2}\cdot\frac{1}{r_{ij}^6} \tag{15}$$

The Debye (induced dipole) interaction seldom amounts to more than a few percent of the London interaction, however, and can often be neglected. In any event, since all three of the contributions to the van der Waals interaction vary as the inverse sixth power of the distance of separation, they may be easily summed to give the total interaction*

$$\Phi_{ij}^{\text{total}} = -\left\{\frac{3}{4}\alpha_i\alpha_jI + \frac{2}{3}\frac{\mu_i^2\mu_j^2}{kT} + \frac{(\alpha_i\mu_i^2 + \alpha_j\mu_i^2)}{2}\right\}\frac{1}{r_{ij}^6} \tag{16}$$

* Equations (12–16) are written for interactions between molecules i and j *in vacuo*. For interactions in other media, the right-hand side of each equation should be divided by the square of the static dielectric constant of the medium. For purposes of computing the macroscopic wetting parameters in terms of intermolecular potentials, however, the effect of an intervening medium need not be considered, as shown below.

It is evident that under conditions in which the contribution of permanent dipoles is negligible, the interaction between unlike molecules will be given as the geometric mean of the interactions between like molecules, but not otherwise.

The theory of van der Waals forces represented by Eq. (16) was generalized by McLachlan in 1963 [28,29] to account for the importance of more than a single electron cloud oscillatory frequency of the interacting molecules (more than a single ionization potential). McLachlan's result is

$$\Phi_{ij}^{\text{total}} = -\frac{6\,kT}{r^6}\sum_{0}^{\infty}{}' \alpha_i(i\omega_\nu)\alpha_j(i\omega_\nu) \tag{17}$$

where $\alpha_i(i\omega_\nu)$ and $\alpha_j(i\omega_\nu)$ are the molecular polarizabilities of i and j as functions of the imaginary frequencies $i\omega_\nu$. ν is the quantum number of the relevant oscillation, $\omega_\nu = (4\pi^2kT/h)\nu \approx 2.43 \times 10^{14}\ \nu\ \text{sec}^{-1}$ at room temperature, and h is Planck's constant. The prime on the summation in Eq. (17) indicates that the $\nu = 0$ term is to be multiplied by ½. The molecular polarizability includes an "orientational" or dipole contribution if the molecule has a permanent dipole moment μ. For dilute gases, it is given by [30, p. 76]

$$\alpha(i\omega_\nu) = \frac{\mu^2}{3kT\left(1 + \dfrac{\omega_\nu}{\omega_{\text{rot}}}\right)} + \frac{\alpha_0}{1 + \left(\dfrac{\omega_\nu}{\omega_e}\right)^2} \tag{18}$$

where ω_{rot} is the average rotational relaxation frequency for the molecule and ω_e the classical fundamental frequency of the electron cloud ν_0. It is $h\nu_0$ that London approximated as the first ionization potential I in arriving at Eq. (13). α_0 is the static polarizability. For any $\nu > 1$ and $\omega_\nu \gg \omega_{\text{rot}}$, the first term of Eq. (18) is significant only when $\nu = 0$. It can then be shown that for typical absorption frequencies, substitution of Eq. (18) into Eq. (17) leads to the former result of Eq. (16).

An expression of the type of Eq. (16) may be used to compute the potential energy of a molecule in a macroscopic assembly of other molecules if the composition and structure of the assembly are known (e.g., in terms of the radial distribution function), and one can assume pairwise additivity of the interactions. Hamaker [31] was the first to use such a procedure for computing the "molecular" interaction between macroscopic systems. The results depended on the sizes, shapes, and relative positions of the macroscopic bodies. For example, if the macroscopic systems are semiinfinite "half-spaces" of the same material 1 interacting across a vacuum of constant thickness s_0, one obtains

$$\Phi_{11}^{\text{macro}} = -A_{11}\cdot\frac{1}{12\pi s_0^2} \tag{19}$$

where A_{11} is termed the Hamaker constant for substance 1. It takes the form

$$A_{11} = \pi^2 \rho_1^2 E_{11} \tag{20}$$

where ρ_1 is the molecular density of phase 1, assumed to be uniform, and E_{11} is a constant equal to the quantity in brackets on the right-hand side of Eq. (16), as it would apply to the case of $i = j = 1$.

For the interaction between half-spaces consisting of different substances 1 and 2, one has

$$\Phi_{12}^{\text{macro}} = -A_{12} \cdot \frac{1}{12\pi s_0^2} \tag{21}$$

with the cross Hamaker constant A_{12} given by

$$A_{12} = \pi^2 \rho_1 \rho_2 E_{12} \tag{22}$$

On an ad hoc basis, it was assumed that the geometric mean mixing rule for the cross constant could be applied, that is,

$$A_{12} = \sqrt{A_{11} A_{22}} \tag{23}$$

The structure of E_{12}, from Eq. (16), suggests that only when interactions due to the existence of permanent dipoles are negligible would Eq. (23) be valid. It now appears that in condensed phase media, our principal concern in wetting phenomena, the effect of permanent dipoles, *is* essentially negligible.

It is now recognized [32–34] that the use of expressions of the type of Eqs. (16) and (17), together with the pairwise additivity assumption for the computation of the configurational free energy, or related properties, of macroscopic systems, is valid only when these systems are *dilute gases*. In condensed phase systems, the contribution of dipole interactions cannot be taken as simply the pairwise summation of the Keesom and Debye terms. The essentially vectorial nature of polar interactions, as illustrated schematically in Fig. 2, shows this. The Boltzmann-averaged interaction between *three* dipoles is seen to be far less than the sum of the three Boltzmann-averaged pair interactions. In fact, when multibody permanent dipole interactions are considered, there is a self-cancelling effect that all but removes their influence in determining the thermodynamic properties of condensed phases.

The smallness of the permanent dipole contribution to the relevant thermodynamic properties of condensed phases is arrived at in a more modern (and quantitative) way using the Lifshitz approach [35], in which macroscopic bodies are treated as continua, and the interaction between

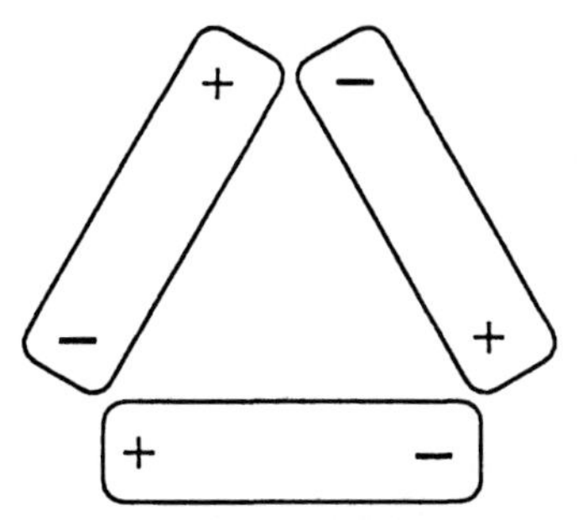

FIG. 2 Schematic of mutual orientation of three permanent dipoles.

them is given in terms of macroscopic properties like dielectric susceptibility and refractive index. The derivation of the Lifshitz approach requires the full quantum field theory, but its implementation may be carried out to good approximation using simplified approaches, as outlined by several authors and summarized by Israelachvili [30]. Without approximation, the Lifshitz approach gives results of the same functional form as the Hamaker approach for the interaction between macroscopic systems, with only the expression for, and in some cases the numerical value of, the Hamaker constant being different. The simplest of the methods for using the Lifshitz method to obtain the Hamaker constant is what Israelachvili terms a "modified additivity approach," in which one begins with the McLachlan expression for the interaction between molecules in medium 1 and those in medium 2 across a vacuum

$$A_{12} = \pi^2 C \rho_1 \rho_2 = 6\pi^2 kT \rho_1 \rho_2 \sum_{0}^{\infty}{}' \alpha_1(i\omega_\nu)\alpha_2(i\omega_\nu) \tag{24}$$

where the "molecular" polarizabilities for 1 and 2 are the excess bulk or volume polarizabilities of these media. These are expressible in terms of the bulk phase dielectric properties, yielding, to good approximation, for materials interacting across a vacuum [34]

$$\begin{aligned} A_{12} &= \frac{3}{2} kT \sum_{0}^{\infty}{}' \left[\frac{\epsilon_1(i\omega_\nu) - 1}{\epsilon_1(i\omega_\nu) + 1}\right] \left[\frac{\epsilon_2(i\omega_\nu) - 1}{\epsilon_2(i\omega_\nu) + 1}\right] \\ &= \frac{3}{2} kT \left\{ \left(\frac{\epsilon_1 - 1}{\epsilon_1 + 1}\right) \left(\frac{\epsilon_2 - 1}{\epsilon_2 + 1}\right) \right. \\ &\quad \left. + \sum_{1}^{\infty}{}' \left[\frac{\epsilon_1(i\omega_\nu) - 1}{\epsilon_1(i\omega_\nu) + 1}\right] \left[\frac{\epsilon_2(i\omega_\nu) - 1}{\epsilon_2(i\omega_\nu) + 1}\right] \right\} \end{aligned} \tag{25}$$

where the $\epsilon(i\omega_\nu)$ are the dielectric susceptibilities of the media as functions of the imaginary frequencies, as specified below Eq. (17), and the ϵ are corresponding static values. The first term on the right-hand side of Eq. (25) (for $\nu = 0$) represents the Keesom and Debye contributions and the remaining terms the London contribution. For many situations, there is a single dominant absorption frequency ω_e so that

$$A_{12} = \frac{3}{2}kT\left\{\left(\frac{\epsilon_1 - 1}{\epsilon_1 + 1}\right)\left(\frac{\epsilon_2 - 1}{\epsilon_2 + 1}\right) + \left[\frac{\epsilon_1(i\omega_e) - 1}{\epsilon_1(i\omega_e) + 1}\right]\left[\frac{\epsilon_2(i\omega_e) - 1}{\epsilon_2(i\omega_e) + 1}\right]\right\} \quad (26)$$

Finally, for the interaction between like materials

$$A_{11} = \frac{3}{2}kT\left\{\frac{1}{2}\left(\frac{\epsilon - 1}{\epsilon + 1}\right)^2 + \left[\frac{\epsilon(i\omega_e) - 1}{\epsilon(i\omega_e) + 1}\right]^2\right\} \quad (27)$$

Evaluation of the Hamaker constant from Eq. (25) requires knowledge of the dielectric susceptibility of the media over the whole frequency range, information that is still not available for many materials, but Eqs. (26) and (27) often give a good approximation. What is important to note is that computations based on these equations (or their more general antecedents) reveal the relative smallness of Keesom and Debye effects in condensed media, and hence the validity of the geometric mean mixing rule for calculating the cross Hamaker constant, that is, Eq. (23).

Now the Hamaker constant can be related directly to the surface tension (or surface free energy) of a substance 1 by considering the work of adhesion between identical phases of 1 across a distance $s_0 = s_m$ corresponding to the effective spacing between molecular planes. In this case, the work of adhesion becomes the work of *co*hesion, that is, $W_{a11} \equiv W_{c1}$ and should be equal to Φ_{11}^{macro} in Eq. (19). Since the interfacial tension between these identical phases is zero, application of Eq. (4) gives

$$W_{c1} = 2\sigma_1 = \frac{A_{11}}{12\pi s_m^2} \quad (28)$$

Israelachvili [36] argues that the distance s_m should be substantially less than the intermolecular center-to-center distance and suggests a "universal" value of approximately 0.165 nm. This gives for liquids whose molecules interact solely through van der Waals forces

$$\sigma_1(\text{mJ/m}^2) \approx 4.76 \times 10^{20} A_{11}\ (\text{J}) \quad (29)$$

The range of applicability of the above approach can be examined by comparing experimental values of surface tension for a variety of liquids with values based on Eq. (29), with the Hamaker constant calculated from Lifshitz theory. Table 1, excerpted from Israelachvili [30, p. 158], shows excellent agreement except for liquids known to self-associate through hydrogen bonding. In those cases, Eq. (29) seriously underestimates the surface tension. The exclusion should extend also to liquids that can self-associate in other ways.

It is important to understand the nature of the limitation to the use of Eq. (29). The Lifshitz approach *is* capable of calculating the appropriate Hamaker constant for a material in terms of the macroscopic dielectric properties of the media involved (if these data are available), even if the molecules of the material exhibit strong self-association (as through hy-

TABLE 1 Comparison of Experimental Surface Energies with Those Calculated on the Basis of Lifshitz Theory

Materials (ϵ)[a] in order of increasing ϵ	Theoretical A (10^{-20} J)	Surface energy σ (m Jm^{-2}) $A/24\pi s_m^2$ (s_m = 0.165 nm)	Experimental[b] (20°C)
Liquid helium (1.057)	0.057	0.28	0.12–0.35 (4–1.6K)
n-Pentane (1.8)	3.75	18.3	16.1
n-Octane (1.9)	4.5	21.9	21.6
Cyclohexane (2.0)	5.2	25.3	25.5
n-Dodecane (2.0)	5.0	24.4	25.4
n-Hexadecane (2.1)	5.2	25.3	27.5
PTFE (2.1)	3.8	18.5	18.3
CCl_4 (2.2)	5.5	26.8	29.7
Benzene (2.3)	5.0	24.4	28.8
Polystyrene (2.6)	6.6	32.1	33
Polyvinyl chloride (3.2)	7.8	38.0	39
Acetone (21)	4.1	20.0	23.7
Ethanol (26)	4.2	20.5	22.8
Methanol (33)	3.6	18	23
Glycol (37)	5.6	28	48
Glycerol (43)	6.7	33	63
Water (80)	3.7	18	73
H_2O_2 (84)	5.4	26	76
Formamide (109)	6.1	30	58

[a] Dielectric constant.
[b] For sources, see Ref. 30.
Source: Ref. 30, p. 158, by permission.

drogen bonding). The computed values of A are validated in other experiments involving the forces between macroscopic bodies at finite distances from one another (i.e., not in direct contact). The failure of Eq. (29) for self-associated materials is in the calculated relationship between the Hamaker constant and the surface tension. It amounts to the failure of the tacit assumption that any special effects of molecular orientation or molecular packing differences near the interface are negligible.

B. Macroscopic Wetting Parameters for Lifshitz–van der Waals Systems

Next it is of interest to consider the computation of the *interfacial* tension (or interfacial free energy) between unlike phases in terms of intermolecular potentials as represented by Hamaker constant. Using the same reasoning as that leading to Eq. (28), together with the geometric mean mixing rule for the cross Hamaker constant, gives for the work of adhesion between phases 1 and 2

$$W_{a12} = \frac{A_{12}}{12\pi s_m^2} = \frac{\sqrt{A_{11}A_{22}}}{12\pi s_m^2} = 2\sqrt{\sigma_1\sigma_2} = \sigma_1 + \sigma_2 - \sigma_{12} \qquad (30)$$

Equation (30) contains the same assumptions as Eq. (29), so that it would not be expected to predict interfacial tensions in terms of Hamaker constants if there is significant self-association in either of the phases. Combining Eq. (30) with Eq. (4) leads to an expression for the interfacial tension (or free energy) in terms of the surface tensions (free energies) of the phases, applicable to van der Waals systems

$$\sigma_{12} = \sigma_1 + \sigma_2 - 2\sqrt{\sigma_1\sigma_2} = (\sqrt{\sigma_1} - \sqrt{\sigma_2})^2 \qquad (31)$$

Equation (31) was found to work well for liquid/liquid systems subject only to London (dispersion) forces, but appeared to fail otherwise, often for systems with large dipole moments. Seeking to develop an equation of the type of Eq. (31) applicable to a wider class of systems, Girifalco and Good [37] and Fowkes [38,39] took two slightly different approaches. Girifalco and Good introduced an "interaction parameter," Φ_{12}, which represented the degree of effectiveness of the geometric mean mixing rule for interactions across the interface

$$\sigma_{12} = \sigma_1 + \sigma_2 - 2\Phi_{12}\sqrt{\sigma_1\sigma_2} \qquad (32)$$

where Φ_{12} was expected to take on values less than unity. In general, measurements of interfacial tension were required to evaluate Φ_{12}, so Eq. (32) was predictive only for so-called "apolar" materials, for which $\Phi_{12} = 1$. Fowkes' reasoning began with expressions for surface tension in terms of the Hamaker constant, Eq. (28). From the form of Eq. (16), Fowkes [38,39] assumed the Hamaker constant could be split into dispersion, dipole, and induced dipole contributions, leading to

$$\sigma = \sigma^d + \sigma^p + \sigma^i \qquad (33)$$

for van der Waals materials, where σ^d, σ^p, and σ^i represent the dispersion (London), polar (Keesom), and induced dipole (Debye) contributions, respectively. If hydrogen bonding, metallic bonding, or other interactions were present, these too could be represented with additional terms in Eq. (33). Fowkes argued that in van der Waals systems, only dispersion forces could effectively operate *across* the interface. Therefore, his result was

$$\sigma_{12} = \sigma_1 + \sigma_2 - 2\sqrt{\sigma_1^d \sigma_2^d} \qquad (34)$$

The dispersion force contributions to the surface tensions of liquids could be obtained using Eq. (34) with interfacial tension measurements against dispersion-force-only probe liquids, such as hydrocarbons, for which $\sigma = \sigma^d$. The value found for water, for example, at 23°C is approximately 21.1 mJ/m^2 [40] (compared with the total surface tension of 72.4 mJ/m^2). The dispersion force contribution to solid surface free energies could be obtained with contact-angle measurements of dispersion-force-only liquids against the solid using Eq. (7)

$$\sigma_S^d = \frac{1}{4}\sigma_L(1 + \cos\theta)^2 \qquad (35)$$

From a practical point of view, this method is limited to low surface energy solids and high surface tension probe liquids in order to obtain finite contact angles. Once a solid was so characterized, contact-angle measurements against it of other liquids could be used to determine the dispersion force contribution to the surface tension of those liquids, etc. A considerable database for the dispersion force components of liquid surface tension and solid surface free energy has been developed [41].

For van der Waals systems, in accord with the Fowkes approach, the work of adhesion should be given by

$$W_a = 2\sqrt{\sigma_S^d \sigma_L^d} \qquad (36)$$

and the interfacial tension by Eq. (34). Nonetheless, there were found to be many systems, particularly those involving water, in which experimental values for the work of adhesion obtained from wetting measurements via Eq. (7) were determined to be considerably greater than predicted by Eq. (36) [44,45]; on the other hand, there were many systems yielding interfacial tensions *less* than predicted by Eq. (34). These observations led to the conclusion that forces other than dispersion forces were effectively operating across the interface (and perhaps within the bulk phases). We know now that these are the forces of acid–base interaction, as suggested by Eq. (11), but it was first speculated by Owens and Wendt [42], Kaelble [43], and others that the discrepancy was due to "polar" interactions. If the "polar" interaction could be computed using the same geometric mean mixing rule* as for the dispersion force interaction, one obtains for the work of adhesion and interfacial tension

$$W_a = 2\sqrt{\sigma_S^d \sigma_L^d} + 2\sqrt{\sigma_S^p \sigma_L^p} \tag{37}$$

and

$$\sigma_{12} = \sigma_1 + \sigma_2 - 2\sqrt{\sigma_1^d \sigma_2^d} - 2\sqrt{\sigma_1^p \sigma_2^p} \tag{38}$$

Using contact-angle measurements and surface tension measurements with pairs (or larger numbers) of probe liquids permitted the determination of the σ^d and σ^p parameters for many liquids and solids [41]. The concept of the "polar fraction" for surface free energy became current, and the principle of "polarity matching," between adhesive and adherend to optimize adhesion, first enunciated by de Bruyne [46], appeared to have been quantified. The practical problem was that different sets of probe liquids led to different polarity splits for a given solid, and the fundamental problem, as pointed out by Fowkes [47], was that computed values of σ^p showed no correlation whatsoever with the dipole moment of the material.

In retrospective view of the relative unimportance of Keesom and Debye, that is, "polar," interactions in determining the thermodynamic properties of van der Waals condensed phases, it can be asserted that the interactions represented by the final terms on the right-hand side of Eqs. (37) and (38) are certainly *not* the result of polarity. The contribution of polarity is not necessarily zero, but is evidently always small. van Oss et al. [48] computed, on the basis of Lifshitz theory, the Keesom–Debye contribution to the surface tension of water at room temperature to be only 1.4 mJ/m^2. They suggest that such polar interactions can be "rolled into" the dispersion force contribution to surface free energy, which can then be renamed the "Lifshitz–van der Waals" contribution, abbreviated LW, that is, $\sigma^d + \sigma^p + \sigma^i = \sigma^{\mathrm{LW}} =$ (for water) $\approx 21.1 + 1.4 = 22.5$ mJ/m^2. Equations (34–36), which remain valid, should be rewritten with d replaced by LW. The remaining contribution to the surface tension of water ($72.4 - 22.5 = 49.9$ mJ/m^2) is due to self-association of the water

* Other mixing rules for the polar interaction have been tried [41], but they have no more general validity than the geometric mean mixing rule.

by hydrogen bonding. In general, contributions to the surface tension, work of adhesion, or other macroscopic wetting parameters in addition to those of Lifshitz–van der Waals (LW) interactions are attributable to chemical interactions, as discussed in the next section.

We may now summarize methods for obtaining the wetting parameters for a system consisting of a solid S and a liquid L subject only to LW interactions, or in other cases, for obtaining the *LW contributions* to those parameters. The LW contributions to the surface tension of the liquid and the surface free energy of the solid may be estimated in terms of the Hamaker constants, computed from Lifshitz theory

$$\sigma_L^{\mathrm{LW}} = \frac{A_{LL}}{24\pi s_m^2}; \qquad \sigma_S^{\mathrm{LW}} = \frac{A_{SS}}{24\pi s_m^2} \tag{39}$$

These quantities can also be obtained in terms of wetting measurements using LW probe liquids. Whether liquid L is a LW liquid or not, the LW component of its surface tension is obtained from the measurement of its interfacial tension against an appropriate LW probe liquid, that is, one with which L is immiscible. The most commonly used probe liquids for this purpose are moderately high-molecular-weight alkanes. Fowkes et al. [40] suggest the use of squalane.

$$\sigma_L^{\mathrm{LW}} = \frac{(\sigma_L + \sigma_{\mathrm{probe}} - \sigma_{\mathrm{probe}/L})^2}{4\sigma_{\mathrm{probe}}} \tag{40}$$

Similarly, a LW probe liquid may be used to evaluate the LW component of the solid surface free energy. The contact angle of the probe liquid against the solid S is measured†

$$\sigma_S^{\mathrm{LW}} = \frac{1}{4}\,\sigma_{\mathrm{probe}}(1 + \cos\theta_{\mathrm{probe}})^2 \tag{41}$$

The requirements for the probe liquid are more stringent in this case. They must be of sufficiently high surface tension to yield a finite contact angle. This means that alkanes, with their low surface tensions, are virtually never suitable, and that for very high surface energy solids, it may not be possible to find a probe liquid at all. Favorite candidates are diiodomethane, with a surface tension of 50.8 mJ/m^2 at 23°C, and α-bromon-

† Equation (41) assumes that the equilibrium film pressure of the probe liquid on the solid π_{ep} is negligible. Otherwise, we must use

$$\sigma_S^{\mathrm{LW}} = \frac{1}{4}\,\sigma_{\mathrm{probe}}(1 + \cos\theta)^2 + \frac{1}{2}\,\pi_{ep}(1 + \cos\theta) - \frac{\pi_{ep}^2}{4\sigma_{\mathrm{probe}}} \tag{41a}$$

aphthalene, with a surface tension of 44.4 mJ/m^2 at the same temperature. We will later show how other types of measurements (particularly inverse gas chromatography) may be used to obtain σ_S^{LW}. The LW component of the interfacial free energy may then computed as

$$\sigma_{SL}^{LW} = (\sqrt{\sigma_S^{LW}} - \sqrt{\sigma_L^{LW}})^2 \tag{42}$$

The LW component of the work of adhesion, wetting, and spreading is then evaluated in accord with

$$W_a^{LW} = 2\sqrt{\sigma_S^{LW}\sigma_L^{LW}} \tag{43}$$

$$W_w^{LW} = 2\sqrt{\sigma_S^{LW}\sigma_L^{LW}} - \sigma_L^{LW} \tag{44}$$

$$W_s^{LW} = 2\sqrt{\sigma_S^{LW}\sigma_L^{LW}} - 2\sigma_L^{LW} \tag{45}$$

We may compute the full values of the above three wetting parameters in cases where there is self-association in either the liquid or solid phase, as long as there is no chemical interaction *across* the interface

$$W_a = \sigma_S + \sigma_L - (\sqrt{\sigma_S^{LW}} - \sqrt{\sigma_L^{LW}})^2 \tag{46}$$

$$W_s = \sigma_S - (\sqrt{\sigma_S^{LW}} - \sqrt{\sigma_L^{LW}})^2 \tag{47}$$

$$W_s = \sigma_S - \sigma_L - (\sqrt{\sigma_S^{LW}} - \sqrt{\sigma_L^{LW}})^2 \tag{48}$$

III. CHEMICAL INTERACTIONS AND THE ACID–BASE CONTRIBUTION TO THE WETTING PARAMETERS

The range of possible chemical interactions that may occur between the molecules of a liquid phase and the surfaces of solids with which they are in contact covers all of reaction chemistry. In view of our definition of wetting phenomena, however, we exclude reactions at the interface that destroy the long-term stability of the interface itself or result ultimately in changing the composition of either of the bulk phases. This leaves for our consideration reactions that are capable of forming complexes between molecules on opposite sides of a solid/liquid interface, as pictured schematically in Fig. 3. Figure 3a shows the situation that exists when molecules in the liquid (above) form 1-1 complexes with a uniform population of a single type of reactive functional group exposed at the solid surface. An example of this might be the interaction occurring between the surface of a solid polyester and chloroform. A hydrogen-bonded complex forms between the hydrogen atoms of the chloroform molecules and the carbonyl oxygens exposed at the surface of the polyester. Another example would be the interaction of liquid pyridine with the solid surface

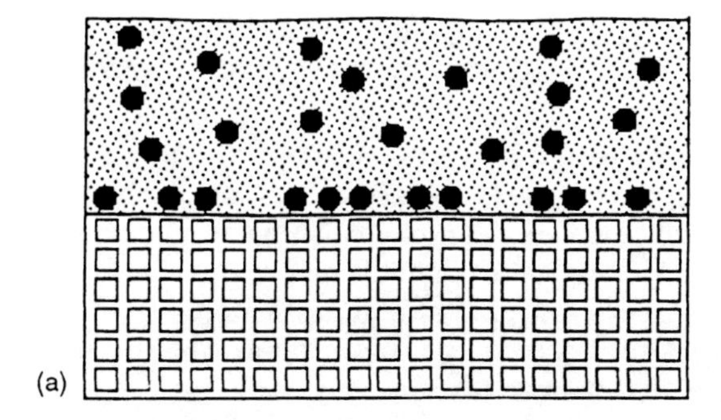

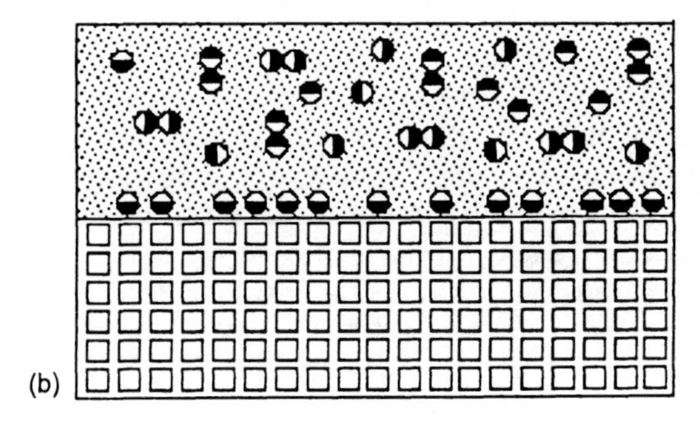

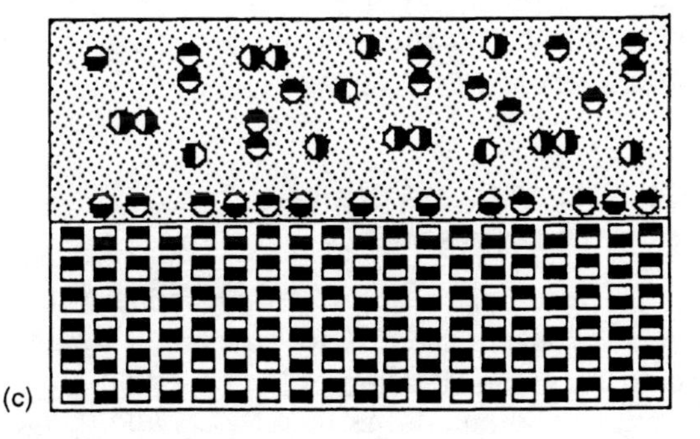

FIG. 3 Schematic of adduct formation at solid/liquid interfaces: (a) monofunctional liquid and monofunctional solid surface of opposite type; (b) bifunctional liquid and monofunctional solid surface; (c) bifunctional liquid and solid surface with two different types of sites.

of polyvinyl chloride. In this case, a covalently bonded complex forms between hydrogen atoms at the surface of the solid and nitrogen atoms of the pyridine molecules. Hydronium ions in liquid water, under conditions of low pH, may interact with hydroxyl (–OH) groups at the surface of a solid material to produce $-OH_2^+ + H_2O$. Most solid surfaces present the possibility for chemical interaction with liquids. Countless other examples could be given. Except for the noble metals, most metallic solids have superficial oxide layers, and after contact with moist air, many or all of the surface oxygens have become hydroxyl groups. The surfaces of ionic crystals expose a regular pattern of their constituent ions. Amorphous minerals present a variety of chemical functionality at their surfaces. Polymeric solids will present at their surfaces the chemistry inherent in their repeat units, as well as possibly more widely spaced functional groups that arise from trace chemicals used in their preparation (e.g., sulfate groups derived from the initiator used for the polymerization) or incorporated as additives. Naturally occurring proteinaceous solids generally contain surface populations of amino and carboxyl groups, and polysaccharides present hydroxyl, carbonyl, and sometimes carboxyl groups. The surface of what is nominally carbon generally presents a variety of oxygen-containing complexes [49]. Even the surfaces of solids like poly (ethylene) or poly (propylene) that are nominally free of any reactive functional groups are often found to contain oxygen or other contaminant atoms. It has been widely contended that nearly all of the chemical reactions between liquids and solid surfaces that are important in wetting phenomena are acid–base reactions [47]. It is therefore reasonable to replace W_a^{chemical} in Eq. (11) with W_a^{AB}. We consider first the direct experimental determination of W_a^{AB} from wetting measurements, followed by a brief review of acid–base chemistry and its relationship to W_a^{AB}. Extensive reviews of acid–base chemistry are given elsewhere [50–55].

A. Determination of the Acid–Base Work of Adhesion from Wetting Measurements

Before attempting to dissect the molecular origins of the chemical contribution to wetting phenomena, it is useful to consider how direct wetting measurements provide a means for the evaluation of the contribution of acid–base interactions *across* the solid/liquid interface, W_a^{AB}, and hence the other wetting parameters. This may be done by combining Eqs. (7), (11), and (43) [56]

$$W_a^{AB} = W_a - W_a^{\text{LW}} = \sigma_L(1 + \cos\theta) - 2\sqrt{\sigma_S^{\text{LW}}\sigma_L^{\text{LW}}} \tag{49}$$

with σ_S^{LW} and σ_L^{LW} evaluated from Eqs. (41) and (40), respectively, using inert probe liquids. Equation (49) is written in the form that assumes the film pressure π_e can be neglected and used only in the situation in which the contact angle formed by the liquid/solid system of interest is finite.

The early published data of the type suggested by Eq. (49) were obtained before the acid–base interpretation was current. It was recognized that there was a significant contribution to the work of adhesion beyond that of dispersion forces, but it was often misinterpreted as a "polar" contribution, or some other contribution of unspecified origin. Thus, Dann [44,45] in 1970 found a significant "nondispersion force" contribution to the work of adhesion between ethanol/water mixtures, mixed glycols, and polyglycols and a mixture of formamide and 2-ethoxyethanol against a variety of solids. Significant nondispersion contributions were found in all cases with poly(styrene), poly(methyl methacrylate), poly(vinyl chloride), poly(ethylene terephthalate), Nylon 11, and Nylon 6.6, but not for poly(ethylene), Teflon, or paraffin, as expected.

Acid–base interactions between so-called "high-energy" surfaces and liquids are generally not accessible through contact-angle measurements in air since for these cases, the contact angle is generally zero. Schultz et al. [57,58] suggested a way of circumventing this problem by performing the wetting measurements not in air but under a second liquid. These angles are generally finite, and the technique is now referred to as the "two-liquid method."

Consider the liquid 1, forming a contact angle θ_{12} against the solid S, under liquid 2. Young's equation for this case gives

$$\sigma_{S1} = \sigma_{12} \cos \theta_{12} + \sigma_{S1} \tag{50}$$

It is unlikely that a film pressure at the $S2$ interface needs to be considered, and it is omitted. Splitting the solid/liquid interfacial energies into Lifshitz–van der Waals contributions and acid–base contributions, represented by I_S, we have

$$\sigma_{S1} = \sigma_S + \sigma_1 - 2\sqrt{\sigma_S^{\mathrm{LW}}\sigma_1^{\mathrm{LW}}} - I_{S1} \tag{51}$$

and

$$\sigma_{S2} = \sigma_S + \sigma_2 - 2\sqrt{\sigma_S^{\mathrm{LW}}\sigma_2^{\mathrm{LW}}} - I_{S2} \tag{52}$$

Substituting Eqs. (51) and (52) into Young's equation and rearranging, we get

$$\sigma_1 - \sigma_2 + \sigma_{12} \cos \theta_{12} = 2\sqrt{\sigma_S^{\mathrm{LW}}}\left(\sqrt{\sigma_1^{\mathrm{LW}}} - \sqrt{\sigma_2^{\mathrm{LW}}}\right) + I_{S1} - I_{S2} \tag{53}$$

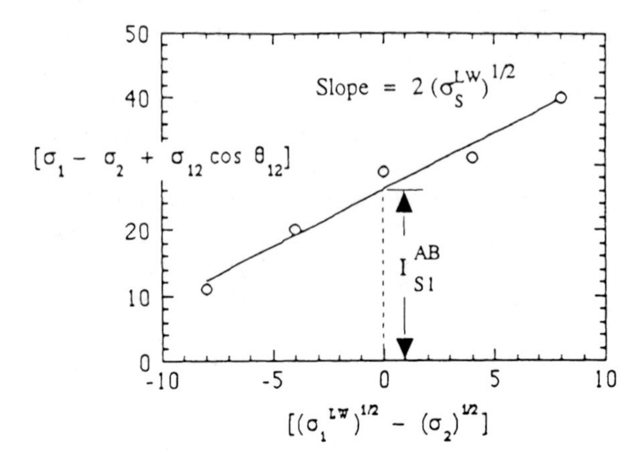

FIG. 4 Schematic of data plot for the two-liquid method.

If we assume that liquid 2 is inert, that is, incapable of entering into acid–base interactions, $I_{S2} = 0$, and $\sigma_2^{LW} = \sigma_2$, so that

$$\sigma_1 - \sigma_2 + \sigma_{12} \cos \theta_{12} = 2\sqrt{\sigma_S^{LW}} \left(\sqrt{\sigma_1^{LW}} - \sqrt{\sigma_2}\right) + I_{S1} \tag{54}$$

For a family of inert liquids L_2, one may plot $(\sigma_1 - \sigma_2 + \sigma_{12} \cos \theta_{12})$ vs.

$$\left(\sqrt{\sigma_1^{LW}} - \sqrt{\sigma_2}\right)$$

to obtain a straight line, as shown schematically in Fig. 4, whose slope is $2\sqrt{\sigma_S^{LW}}$ and intercept is I_{S1}, which is the same as the acid–base contribution to the work of adhesion, W_a^{AB}.

B. Arrhenius and Brønsted–Lowry Theories

The classical view of acids and bases was that of Arrhenius and first presented in 1887 [59]. It dealt exclusively with species and reactions in aqueous media. An acid was identified as any substance that upon reaction with water increases the hydronium ion (H_3O^+) concentration of the solution. A base was defined as a substance whose reaction with water increases the hydroxyl ion (OH^-) concentration of the solution. An acid–base reaction was a neutralization producing a salt and water. The definitions of acids and bases were broadened in 1923 by Brønsted [60] and independently by Lowry [61]. An acid was defined as a substance capable of giving up a proton, and a base was defined as a substance capable of accepting a proton. The central role of the proton was retained from the

Arrhenius definition of an acid, but bases no longer had to be hydroxides, and water was no longer the required solvent. The fundamental acid–base reaction of the Brønsted-Lowry type is written as

$$\underset{\text{acid}}{A} \Leftrightarrow \underset{\text{proton}}{H^+} + \underset{\text{base}}{B} \tag{55}$$

Species A gives up a proton, so that what is left is a base B. Species B is referred to more specifically as the *conjugate base* of the acid A. Conversely, A is the *conjugate acid* of base B. Since it is believed that free protons do not exist in solution, the reaction of Eq. (55) cannot occur in isolation, that is, for the acidity of A to manifest itself, there must be some other substance (a base) present to receive the proton. A more realistic way of representing the Brønsted–Lowry acid–base interaction, termed *protolysis*, is thus that of an acid A_1 of one system reacting with a base B_2 of another system according to

$$\underset{\text{acid}_1}{A_1} + \underset{\text{base}_2}{B_2} \Leftrightarrow \underset{\text{base}_1}{B_1} + \underset{\text{acid}_2}{A_2} \tag{56}$$

where A_1 and B_1 are the conjugate acid and base of one system, and A_2 and B_2 are the conjugate acid and base of the other. The reaction involves the transfer of a proton from A_1 to B_2 in the forward direction, and from A_2 to B_1 in the reverse direction. The solvent in which acid A_1 finds itself may play the role of the base of the other system B_2 if it has sufficient ability to accept a proton, that is, is sufficiently *protophilic*. The more strongly protophilic the solvent, the more Eq. (56) will be driven from left to right. Conversely, in a solvent more prone to furnish a proton, that is, a *protogenic* solvent, Eq. (56) is driven to the left. Solvents, such as alkanes, that neither accept nor furnish protons are termed *aprotic* or inert. Solvents, like water, that both accept and furnish protons are *amphiprotic*. It is clear that an acid will appear stronger in a protophilic or amphiprotic solvent than a protogenic solvent, and conversely for a base, and that neither acids nor bases can manifest their acidity or basicity, respectively, in an aprotic solvent.

The picture underlying Brønsted–Lowry theory is that all substances are imbued with both an acidity *and* a basicity, that is, are to some extent amphoteric. Acids like HCl have a very high acidity and very low basicity, whereas bases like NaOH have a very low acidity and very high basicity. The acid or base strength of a given substance, independent of solvent effects, can be expressed theoretically in terms of the equilibrium constant and its reciprocal, for Eq. (55), termed the "acidity constant" K_A and "basicity constant" K_B.

$$K_A = \frac{C_B a_H}{C_A} \tag{57}$$

$$K_B = \frac{C_A}{C_B a_H} \tag{58}$$

where a_H is the activity of the proton species. K_A expresses the tendency of an acid to give up a proton, and K_B is the tendency of its conjugate base to accept a proton. Thus, the conjugate base of a strong acid is weak, and vice versa. Neither K_A nor K_B can be measured directly since neither acidity nor basicity are manifest in the absence of a proton acceptor or donor, respectively. What *can* be measured is the equilibrium constant for a protolysis reaction of the type of Eq. (56)

$$K = \frac{C_{B1} C_{A2}}{C_{A1} C_{B2}} = K_{A1} K_{B2} \tag{59}$$

The acidity of an acid A_1 can thus be measured only with reference to some base, say, B_2. However, a comparison of K values for a series of acids, all with the same reference base (the solvent), yields a ranking of the acid strengths that should be solvent-independent. The basicities of substances may be similarly ranked with the use of a reference acidic solvent. When attempting to rank a series of acids, it is important that the reference solvent not be too basic. If it is, all acids may be fully dissociated in the protolysis reaction and thereby appear to be equivalently strong. Consider two acids A_1 (with $K_{A1} = 10^5$) and A_2 (with $K_{A2} = 10^{-5}$) in a basic solvent B (with $K_B = 10^5$). The difference in acid strength between A_1 and A_2 is clearly differentiated in this solvent, since the protolysis constant with the first acid is $K_{A1(B)} = 10^5 \times 10^5 \equiv 10^{10}$ (showing complete dissociation) and $K_{A2(B)} = 10^{-5} \times 10^5 \equiv 1$ (showing only partial dissociation). In another more basic solvent B' (with $K_{B'} = 10^{15}$), the protolysis constants are $K_{A1(B')} = 10^{20}$ and $K_{A2(B')} = 10^{10}$, respectively, showing complete dissociation of both acids, which are therefore indistinguishable in terms of strength. It is apparent that highly basic solvents put an upper bound on the acidity of solutions by leveling solute acid strengths to the strength of the conjugate acid of the solvent. A similar statement can be made of the leveling effect of highly acidic solvents on the basicity of the solutions.

The Brønsted–Lowry picture of acids and bases can be conveniently used in the description of the interaction between liquids and solids when the liquid contains molecules that can act as proton donors or acceptors and the solid surface contains functional groups that can act as proton donors or acceptors. This is very often the case and includes all situations

TABLE 2 Examples of Chemical Functional Groups Found on Solid Surfaces

—OH	Hydroxyl
—COOH	Carboxyl
$—NH_2$	Amino
$—NO_2$	Nitro
—CN	Cyano
>O	Ether
>C=O	Carbonyl
$>CONH_2$	Amide

in which the liquid phase is aqueous. Table 2 lists a number of commonly occurring functional groups.

C. Contribution of Brønsted–Lowry Acid–Base Interactions to the Work of Adhesion: Bolger–Michaels Theory

Bolger and Michaels [62] were the first to propose that interactions between liquids and solid oxide surfaces were describable entirely in terms of a combination of dispersion forces and Brønsted–Lowry-type acid–base interactions. They proposed that metal surfaces, as well as silicate and oxide surfaces, could all be described as hydroxylated oxide surfaces for the purpose of describing interfacial forces, and that such surfaces would adsorb and strongly retain several molecular layers of bound water. The hydroxyl groups, —MOH, would take on varying degrees of acidity or basicity depending on the electronegativity of the relevant metal atom M. The same could be said of polymeric solid surfaces with hydroxyl groups, —ROH, in which the influence of the organic radical R on the polarization of the hydroxyl group determines its acid–base character. Their primary focus was the interaction between *aqueous* media and the solid oxide surfaces. Depending on the pH, water could interact with the surface hydroxyl groups in accord with the dissociative reaction

$$—MOH + H_2O \Leftrightarrow —MO^- + H_3O^+ \tag{60}$$

The surface hydroxyl groups need not be ionized in order to render the surface negatively charged. This could also occur through the nondissociative adsorption of hydroxyl ions via hydrogen bonding

$$\text{—MOH} + \text{OH}^- \Leftrightarrow \text{—MOH}\cdots\text{OH}^- \tag{61}$$

At lower pH values, or when the electron density of the surface oxygen atoms is higher, the surface may become positively charged by dissociation

$$\text{—MOH} + \text{H}_2\text{O} \Leftrightarrow \text{—MOH}_2^+ + \text{OH}^- \tag{62}$$

or by hydrogen bonding of hydronium ions

$$\text{—MOH} + \text{H}_3\text{O}^+ \Leftrightarrow \text{—M}\overset{\overset{\text{H}}{|}}{\text{O}}\cdots\text{H}_3\text{O}^+ \tag{63}$$

The pH at which there is an equal surface density of positive and negative charges is the isoelectric point of the surface, IEPS, which can be determined as the pH at which the zeta potential of the solid surface is zero. This is most often determined from measurements of the electrophoretic mobility of the solid in finely divided form or other electrokinetic measurements. IEPS values for a variety of common oxides have been published by Parks [63]. Parks successfully correlated the IEPS with the ratio of the oxidation state (valence) Z of M and its ionic radius r (corrected for crystal field effects), obtaining

$$\text{IEPS} = C_1 - C_2\left(\frac{Z}{r}\right) \tag{64}$$

where C_1 and C_2 are constants characteristic of the electrolyte. With r expressed in angströms, C_1 and C_2 for water were found to be 18.6 and 11.5, respectively, based on IEPS data for 26 oxides.

Bolger and Michaels claim [62]: "Analysis of the factors that influence the IEPS of various oxides in water should . . . serve also as a qualitative guide to the factors that govern the ability of these surfaces to lose, donate, attract, or bind protons in contact with nonaqueous liquids and solids." It should be noted, however, that the surface —OH groups usually have a certain amount of water adsorbed to them, and that the presence of this moisture (since water is both an acid and a base) may significantly reduce their effective acidity or basicity in contact with organic liquids. Thus, the dryness of both the solid surface and the liquid plays an important part in the extent of the net acid–base interaction between the solid and organic liquid.

Bolger and Michaels describe the interaction of the hydroxylated solid surfaces with organic liquids in accord with one or the other of the following reactions:

Type A: Surface interaction with an organic acid

$$—\text{MOH} + \text{HXR} \Leftrightarrow —\overset{\text{H}}{\overset{|}{\text{MO}}}\cdots\text{HXR} \Leftrightarrow —\text{MOH}_2^+\ \bar{\text{X}}\text{R} \tag{65}$$

Type B: Surface interaction with an organic base

$$—\text{MOH} + \text{XR} \Leftrightarrow —\text{MOH}\cdots\text{XR} \Leftrightarrow —\text{M}\bar{\text{O}}\ \text{H}\overset{+}{\text{X}}\text{R} \tag{66}$$

where X may be an oxygen, sulfur, or nitrogen atom, or in Eq. (66), a halogen. Equations (65) and (66) are thus seen to be a generalization of Eqs. (62) and (63). The tendency for either of these reactions to occur can be expressed quantitatively in terms of the size of the relevant equilibrium constant, or its logarithm. For reactions of type A and type B, respectively, we have

$$\kappa_A = \frac{[\text{MOH}_2^+][\bar{\text{X}}\text{R}]}{[\text{MOH}][\text{HXR}]} \quad \text{and} \quad \kappa_B = \frac{[\text{MO}^-][\text{H}\overset{+}{\text{X}}\text{R}]}{[\text{MOH}][\text{XR}]} \tag{67}$$

Bolger and Michaels define $\Delta_A = \log \kappa_A$, and $\Delta_B = \log \kappa_B$. They evaluate these by first writing the expressions for the equilibrium constants for the model surface ionization reactions and considering the positions of these equilibria at the IEPS. These are

$$\text{MO}^- + \text{H}^+ \Leftrightarrow \text{MOH} \quad \text{with } K_1 = \frac{[\text{MOH}]}{[\text{MO}^-][\text{H}^+]} \tag{68}$$

and

$$\text{MOH} + \text{H}^+ \Leftrightarrow \text{MOH}_2^+ \quad \text{with } K_2 = \frac{[\text{MOH}_2^+]}{[\text{MOH}][\text{H}^+]} \tag{69}$$

and combining Eqs. (68) and (69), we get

$$\text{MO}^- + 2\text{H}^+ \Leftrightarrow \text{MOH}_2^+ \quad \text{with } K_1K_2 = \frac{[\text{MOH}_2^+]}{[\text{MO}^-][\text{H}^+]^2} \tag{70}$$

At the IEPS, by definition, $[\text{MOH}_2^+] = [\text{MO}^-]$, so that $K_1K_2 = [\text{H}^+]_{\text{IEPS}}^{-2}$, and Bolger and Michaels assume symmetry with respect to reactions (68) and (69), so that $K_1 = K_2 = 10^{\text{IEPS}}$. Now expressing Δ_A, we obtain

$$\Delta_A = \log \frac{[\text{MOH}_2^+][\overline{\text{X}}\text{R}]}{[\text{MOH}][\text{HXR}]} \equiv \log \frac{[\text{MOH}_2^+]}{[\text{MOH}][\text{H}^+]} + \log \frac{[\text{H}^+][\overline{\text{X}}\text{R}]}{[\text{HXR}]} \qquad (71)$$

The first term on the right-hand side of Eq. (71) is recognized as $\log K_1 = \text{IEPS}$, and the second term is $\log K_A$ (organic acid) $= -pK_{A(A)}$, where $K_{A(A)}$ is the acid dissociation equilibrium constant. Thus, for type A reactions

$$\Delta_A = \text{IEPS} - pK_A \text{ (organic acid)} \qquad (72)$$

and by similar reasoning for type B reactions

$$\Delta_B = pK_A \text{ (organic base)} - \text{IEPS} \qquad (73)$$

where pK_A (organic acid) $\equiv pK_{A(A)}$; pK_A (organic base) $\equiv pK_{A(B)}$. $K_{A(B)}$ is the equilibrium constant for the dissociation of the conjugate base.

Δ_A and Δ_B can be identified with the free energy changes, on a per mole basis, associated with the second of the reactions in Eqs. (65) and (66), respectively

$$\Delta_A = \frac{(-\Delta G)_{\text{type A } rx}}{2.303\ RT} \quad \text{and} \quad \Delta_B = \frac{(-\Delta G)_{\text{type B } rx}}{2.303\ RT} \qquad (74)$$

Thus, if Δ_A or Δ_B is positive, the ionic reactions in Eq. (65) or (66) would be expected to predominate. If they are slightly negative, one would expect the hydrogen bonding shown in the first reactions of Eqs. (65) and (66). On the other hand, if they are large and negative, the ionic reactions would not occur at all, and the hydrogen-bonding reactions would be very weak, so that the only interaction between the liquid and solid surface would be of the Lifshitz–van der Waals type. Tables 3 and 4, taken from Bolger and Michaels [62], show values of Δ_A and Δ_B, respectively, for organic acids and bases interacting with example oxide surfaces. A relatively small number of cases (upper right-hand corner of Table 3 and upper left-hand corner of Table 4) represent predominantly ionic reactions. It is seen that a neutral water interaction is primarily nonionic, although it may act as either a very weak acid or a very weak base with surfaces of high or low IEPS, respectively. This picture changes if the pH is altered by electrolyte addition, shifting the equilibria in Eqs. (62)

TABLE 3 Polar Surface Interactions with an Organic Acid

		$\Delta_A \equiv$ IEPS $- pK_{A(A)}$		
Organic acid	$pK_{A(A)}$	SiO_2 (IEPS = 2)	Al_2O_3 (IEPS = 8)	MgO (IEPS = 12)
Dodecyl sulfonic acid	−1	3	9	13
Trichloroacetic acid	0.7	1.3	7.3	11.3
Chloroacetic acid	2.4	−0.4	5.6	9.6
Phthalic acid	3.0	−1	5.0	9.0
Benzoic acid	4.2	−2.2	3.8	7.8
Adipic acid	4.4	−2.4	3.6	7.6
Acetic acid	4.7	−2.7	3.3	7.3
Hydrogen cyanide	6.7	−4.7	1.3	5.3
Phenol	9.9	−7.9	−1.9	2.1
Ethyl mercaptan	10.6	−8.6	−2.6	1.4
Water	15.7	−13.7	−7.7	−3.7
Ethanol	16	−14	−8	−4
Acetone	20	−18	−12	−8
Ethyl acetate	26	−24	−18	−14
Toluene	37	−35	−29	−25

Source: Ref. 62, by permission.

or (63). The expressions equivalent to Eqs. (72) and (73) for the case of water with varying pH would be

$$\Delta_A = \text{IEPS} - \text{pH} \quad (\text{for pH} < \text{IEPS}) \tag{75}$$

and

$$\Delta_A = \text{pH} - \text{IEPS} \quad (\text{for pH} > \text{IEPS}) \tag{76}$$

Equation (74) allows one to compute the ionic acid–base contribution to the work of adhesion (and other wetting parameters) resulting from the Brønsted–Lowry acid–base interactions described by the Bolger–Michaels theory

$$W_a^{AB} = N(-\Delta G) = \begin{cases} 2.303\ N\ RT\ \Delta_A & (\text{for type A interactions}) \\ 2.303\ N\ RT\ \Delta_B & (\text{for type B interactions}) \end{cases} \tag{77}$$

where N is the number of moles of acid–base bonds that can be formed per unit area of the solid surface. If an organic compound can interact as either an acid or a base, the dominant mode with any oxide will be the one giving the larger value of Δ. The value of N depends on more than

TABLE 4 Polar Surface Interactions with an Organic Base

		$\Delta_B \equiv pK_{A(B)-\text{IEPS}}$		
Organic base	$pK_{A(B)}$	SiO_2 (IEPS = 2)	Al_2O_3 (IEPS = 8)	MgO (IEPS = 12)
Trimethyl dodecyl ammonium hydroxide	12.5	10.5	4.5	0.5
Piperidine	11.2	9.2	3.2	−0.8
Ethylamine	10.6	8.6	2.6	−1.4
Triethylamine	10.6	8.6	2.6	−1.4
Ethylenediamine	10	8	2.0	−2
Ethanolamine	9.5	7.5	1.5	−2.5
Benzylamine	9.4	7.4	1.4	−2.6
Pyridine	5.3	3.3	−2.7	−6.7
Aniline	4.6	2.6	−3.4	−7.4
Urea	1.0	−1.0	−7	−11
Acetamide	−1	−3	−9	−13
Water	−1.7	−3	−9	−13
Tetrahydroduran	−2.2	−4.2	−10.2	−14.2
Dthyl ether	−3.6	−5.6	−11.6	−15.6
t-Butanol	−3.6	−5.6	−11.6	−15.6
n-Butanol	−4.1	−6.1	−12.1	−16.1
Acetic acid	−6.1	−8.1	−14.1	−18.1
Phenol	−6.7	−8.7	−14.7	−18.1
Acetone	−7.2	−9.2	−15.2	−19.2
Benzoic acid	−7.2	−9.2	−15.2	−19.2

Source: Ref. 62, by permission.

just the number of —OH groups per unit area on the solid surface. It depends also on the size and shape of the interacting molecules from the liquid phase. For example, due to steric factors, the number of bonds that might be formed between an acidic or basic polymeric liquid and an oxide surface might be much less than the number that might be formed between water and the same oxide surface. On the other hand, surface roughness may make available more groups than would be assumed on the basis of a flat surface [64]. Using data from Table 3 and assuming one acid–base bond for every 60Å^2 of surface (giving N = 2.77 μmol/m^2), one may estimate, for example, the acid–base contribution to the work of adhesion between acetic acid and a MgO surface at T = 300K. From Table 3, Δ_A = 7.3. Then W_a^{AB} = 116 mJ/m^2. Other values for W_a^{AB} may be calculated using values from Tables 3 and 4, or Eqs. (69) and (70). When the Δ values are negative, the presumption is that W_a^{AB} is zero.

One of the primary objectives of Bolger and Michaels was the determination of criteria under which water (W) would not be able to interpose itself between a solid (S) and some organic (usually polymeric) coating (L), that is, displace L from the solid surface. This will be expected to occur when the ionic attractions at the organic/solid interface are large compared with the ionic attractions at the water/solid interface. This will, in turn, depend on the pH of the water. Resistance to water penetration would be expected only in pH ranges between the relevant pK_A value of L and the IEPS value of S, as follows:

For type A interactions

$$pK_{A(A)} < \text{pH} < \text{IEPS} \tag{78}$$

For type B interactions

$$\text{IEPS} < \text{pH} < pK_{A(B)} \tag{79}$$

As cited by Bolger and Michaels, these equations satisfactorily explain a considerable body of experimental observations.

Certain limitations of the Bolger–Michaels method for characterizing the acidity or basicity of inorganic solid surfaces should be noted. First, it is limited to consideration of surfaces that are populated with hydroxyl groups, and their degree of acidity or basicity is given in terms of the IEPS. The procedure thus cannot account for the existence of other types of chemical functionality or mixed populations of hydroxyl groups of varying acidity or basicity. Iron oxide and similar oxides, for example, are known to possess both acidic and basic sites [65]. Finally, it appears that the procedure may not take fully into account the acid–base contributions due to the nondissociative hydrogen bonding that may occur.

D. Contact-Angle Titrations

Combining relationships of the type of Eqs. (75–77) with the Young–Dupré equation, Eq. (7), suggests that the measurements of the contact angle might be used to follow the state of ionization of functional groups at the solid/aqueous solution interface as a function of pH

$$\begin{aligned} W_a &= W_a^{\text{LW}} + W_a^{AB} \\ &= W_a^{\text{LW}} + 2.303\, N\, RT(\Delta_A \text{ or } \Delta_B) = \sigma_{LG}(1 + \cos\theta) \end{aligned} \tag{80}$$

Since during the variation of the aqueous solution pH, the only quantity expected to change is Δ_A or Δ_B, one might expect to find, for the case of pH < IEPS,

$$\cos \theta = C_1 + C_2(\text{IEPS} - \text{pH}) \tag{81}$$

where C_1 and C_2 are constants. Cos θ would appear in this case (corresponding to the presence of basic hydroxyl groups on the solid surface) to drop linearly with pH until the IEPS is reached and remain at this minimum value as pH is increased further. As the pH is increased from a low value, where there is a preponderance of ionized basic groups, the proportion of ionized basic groups decreases, and the surface becomes less wettable (cos θ decreases, θ increases) until the IEPS is reached, where the extent of ionization is minimized. Such a treatment is an oversimplification because even at the IEPS there may be considerable surface group ionization, and the population of hydroxyl groups may not be of uniform basicity (or acidity), but it suggests the useful idea that the extent of water wettability should depend directly on the degree of ionization of the functional groups at the interface. From a practical point of view, it would be difficult to implement the type of experiment suggested by Eq. (81), because most hydroxylated mineral surfaces will be wet out ($\theta = 0°$) over the whole pH range.

Whitesides and co-workers [66–73] have made extensive studies of protonation and deprotonation reactions at the interface between water and surface-functionalized poly(ethylene) (PE). Although many types of functional groups were examined, most attention was given to carboxyl groups, implanted at high density in the "interphase" region by treating the polyethylene with concentrated chromic acid/sulfuric acid solutions. Among the most fruitful of their experimental techniques for investigating the "proton transfer reactions in the environment of the organic solid-water interphase" was what they termed "contact angle titration." The advancing contact angle against a smooth treated poly(ethylene) sample was measured (goniometrically) as a function of the pH of the water phase, producing results of the type pictured in Fig. 5. At sufficiently low pH values, the carboxyl groups were presumed to be undissociated, and the contact angle was 55°, whereas at sufficiently high pH the groups were dissociated, leading to stronger liquid/solid interactions and a drop in the contact angle to 22°. It is to be noted that even when there was no dissociation, the contact angle was substantially lower than the value associated with unfunctionalized PE (~100°), suggesting the substantial contribution of hydrogen bonding to the undissociated groups. The drop in contact angle produced as the dissociation of the carboxyl groups became more nearly complete reflected the increasing ionic contribution to the

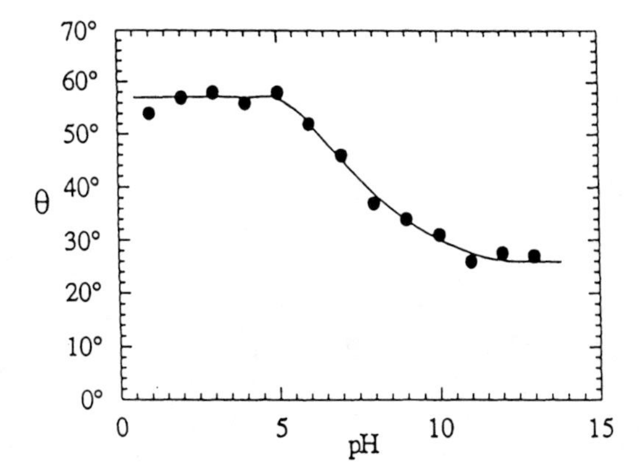

FIG. 5 Contact-angle titration of carboxyl-group-functionalized poly(ethylene). (Redrawn from Ref. 66, by permission.)

work of adhesion. Holmes-Farley et al. [69] related the contact angle directly to the degree of dissociation α, where

$$\alpha = \frac{[CO_2^-]}{[CO_2^-] + [CO_2H]} \tag{82}$$

They assumed that the solid/liquid interfacial free energy σ_{SL} was linearly related to α in accord with

$$\alpha(\text{pH}) = \frac{\sigma_{SL}(\text{pH } 1) - \sigma_{SL}(\text{pH})}{\sigma_{SL}(\text{pH } 1) - \sigma_{SL}(\text{pH } 13)} \tag{83}$$

Substituting for σ_{SL} into Young's equation, Eq. (3), using the fact that the surface tension of the water σ_{LG} was negligibly affected by changing the pH of the water, and assuming that σ_{SG} was similarly unaffected led to

$$\alpha(\text{pH}) = \frac{\cos\theta\,(\text{pH}) - \cos\theta\,(\text{pH } 1)}{\cos\theta\,(\text{pH } 13) - \cos\theta\,(\text{pH } 1)} = 2.83\cos\theta - 1.63 \tag{84}$$

The drop in contact angle was found to occur over a moderately broad range centering on a pH of approximately 10, an interesting result because that value is approximately 4 pH units higher than the value expected for the ionization of carboxyl groups in solution. The midpoint of the pH range over which the contact angle underwent its change was assumed to be the point at which the carboxyl groups at the surface were 50% ionized and was identified as $pK_{1/2}$.

Further contact-angle titrations of untreated PE and PE functionalized with other groups, including —CH_2OH, —$COOCH_3$, —$COOCH_2CH_2Br$, —CH_2OCOCF_3, —NH_2, and others were made by Whitesides and co-workers. The presence of the amino groups led to an increase in contact angle as the pH was increased, whereas all the others showed no change, suggesting that no ionization of the groups occurred. Similar to the results obtained with the carboxyl groups, those obtained with the amino groups suggested that these groups at the surface ionized only at pH values much less than would be required for the ionization in the bulk solution. The increased difficulty of ionization in the interfacial region as compared to the bulk solution was attributed to the decreased effective dielectric constant in that region. Similar results were obtained with more regularly arrayed groups using terminally functionalized self-assembled monolayers on gold [74]. The studies of Whitesides and co-workers have pointed out that the functionalized region between bulk polymer solids and the homogeneous solution must be regarded as an ''interphase region'' rather than a well-defined ''surface'' in the sense of metals and some minerals. They also pointed out the possible importance of lateral interactions between the functional groups in the interphase, both hydrogen-bonding and charge-charge interactions.

Hüttinger et al. [75] have recently reported contact-angle titrations against poly(carbonate) and poly(ether sulfone) polymers, as well as quartz and a variety of carbon fibers. Figure 6 shows results obtained for the polymers, while Fig. 7 illustrates results for three different carbons, expressed as the total work of adhesion, as given by the Young–Dupré equation. The latter shows the presence of several different functional groups on the carbon surface, with different values of $pK_{1/2}$. Contact-angle titration appears to be a powerful technique for the study of wetting phenomena that are influenced by the ionization of functional groups at solid surfaces in contact with aqueous solutions.

Finally, the ionization of surface functional groups leads to electric charge separation at the interface and an electrostatic contribution to the work of adhesion. The magnitude of this contribution may be computed as $W_A^{el} = \psi_0 \sigma$, where ψ_0 is the effective surface potential and σ the surface charge density. If we assume an effective surface potential of 50 mV and a surface charge density corresponding to one electronic charge per 100 Å^2, one obtains $W_A^{el} = 8.0$ mJ/m^2. Thus, the electrostatic contribution to the work of adhesion will in most cases be small, but perhaps not negligible.

E. Lewis Acid–Base Theory

The Brønsted–Lowry theory greatly generalized the older Arrhenius theory of acid-base interactions by not requiring that they occur only in

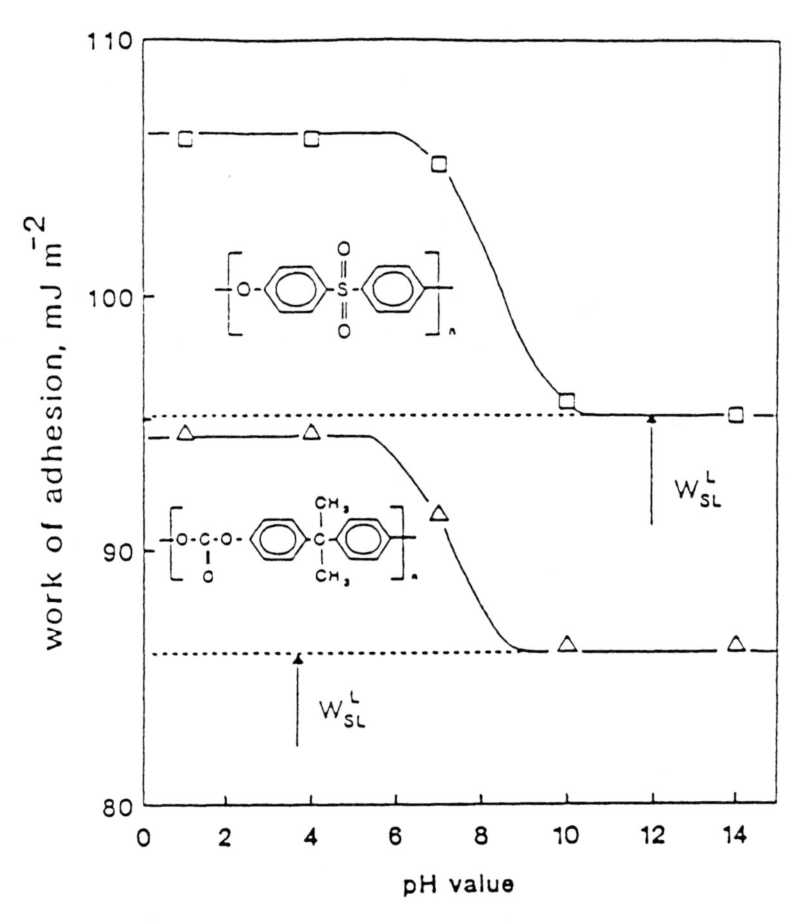

FIG. 6 Work of adhesion as a function of pH for two polar polymers: (–□–) polyethersulfone and (–△–) polycarbonate. (From Ref. 75, by permission.)

aqueous solvents, or in fact in any solvent at all, but its restriction to proton transfer reactions did not allow it to embrace many apparent reactions of the acid–base type that occurred in the absence of protons. A classical example is the neutralization reaction between boron trichloride and triethylamine

$$\underset{\text{acid}}{BCl_3} + \underset{\text{base}}{(C_2H_5)_3N} \Leftrightarrow (C_2H_5)_3N\!:\!BCl_3 \tag{85}$$

This deficiency led Lewis [76] to advance a much broader definition of acids and bases based on his electronic theory of valence. It generalizes

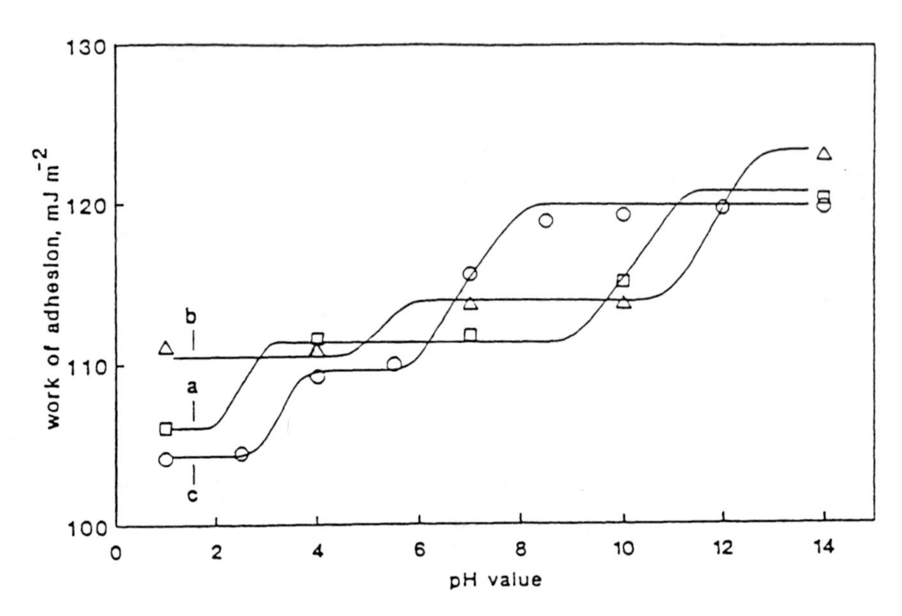

FIG. 7 Work of adhesion as a function of pH for three surface-treated, but unsized HT-type carbon fibers, as received: (a) Grafil XAS; (b) Hercules AS4; (c) Celion G 30-500. (From Ref. 75, by permission.)

the description to include the formation of covalent bonds. Lewis originally defined an acid as any substance that can accept a pair of electrons from a donor substance, whereas a base is any substance capable of furnishing a pair of electrons. The acid–base interaction thus involves the formation of a covalent bond (referred to as a "dative" bond) in which both electrons of the shared pair are provided by the base. In view of molecular orbital theory, the wording of the definitions has changed such that a Lewis base is defined as any substance that is capable of donating electron density to be shared with another substance in a chemical reaction, and conversely, a Lewis acid is any substance capable of accepting electron density from a Lewis base. Reviews of more modern concepts of Lewis acid–base interactions are given elsewhere [77–79]. In any case, the acid–base reaction may be viewed as one in which a simple addition compound (*adduct*) is formed, as exemplified in Eq. (85), or in general

$$A + B \Leftrightarrow A{:}B \tag{86}$$

Consider the types of molecules (and functional groups) that can act as Lewis acids and bases. A Lewis base must thus have one or more pairs of electrons that can furnish electron density to the acid, and for this to be true, the electrons in the base must not all be involved in bonding to other atoms in the molecule, and they must not be so tightly held by the nucleus that there is effectively no tendency for them to be shared. A Lewis acid must be capable of receiving an electron pair from the base. Hydrogen bonding is the most important example of Lewis acid–base interaction, with the hydrogen-containing species serving as the Lewis acid. The definition also includes π bonding, in which the electrons of the π cloud of electrons associated with molecules containing double bonds (such as aromatics) coordinate with an electron-accepting species (acid).

Lewis acid–base adducts themselves can act as acids or bases and undergo further addition, as for example,

$$A{:}B + A' \Leftrightarrow A{:}B{:}A' \tag{87}$$

$$B' + A{:}B \Leftrightarrow B'{:}A{:}B \tag{88}$$

$$A{:}B + A'{:}B' \Leftrightarrow A{:}B{:}A'{:}B' \tag{89}$$

$$A'{:}B' + A{:}B \Leftrightarrow A'{:}B'{:}A{:}B \tag{90}$$

To generalize further, the compounds $A{:}B$, $A'{:}B'$, etc. need not necessarily be the products of Lewis acid–base neutralizations that actually have occurred, but do represent molecules or functional groups that possess both acidic and basic sites. Water is a classical example of such a material. Each water molecule has two acid sites (the hydrogen atoms) and one basic site (the oxygen atom), so that the molecules in liquid water are linked through a three-dimensional network of acid–base hydrogen bonds.

On the other hand, acid–base adducts may undergo displacement reactions in which their constituent acid–base components function as the units of substitution, and one of the original reactant species forming the adduct becomes a leaving group. An acid displacement reaction would be of the type

$$A' + A{:}B \Leftrightarrow A + A'{:}B \tag{91}$$

and a double acid–base displacement reaction would be

$$A{:}B + A'{:}B' \Leftrightarrow A'{:}B + A{:}B' \tag{92}$$

Equation (92) is recognized as the format of the protolytic reaction, Eq. (56), which is the basis of Brønsted–Lowry theory. Brønsted–Lowry acids may thus be Lewis acid–base adducts with the acid component being H^+, and all Brønsted–Lowry bases are also Lewis bases. The Lewis definition of an acid–base reaction thus subsumes the older Brønsted–Lowry definition.

Although extremely broad, Lewis acid–base reactions do not of course cover all chemical reactions. The reaction must involve more than a single electron, and coordination of the acid and base must occur. It thus excludes the simple (single) electron transfer redox reaction, in which the products are ions. Such a reaction may be important in wetting and adhesion first of all because the resulting ionic groups on the solid may interact more strongly with the solvent (usually water) than the original un-ionized groups, as described in the Bolger–Michaels treatment of Brønsted–Lowry interactions between liquids and solids. Second, the resulting ions, on opposite sides of the interface, interact electrostatically. Additional electrostatic effects outside the realm of acid–base interactions may occur due to electron injection from one phase into the other at the interface [80,81]. Covalent reactions, such as free radical combinations, in which both reactants furnish an electron that is shared, are excluded, as well as other reactions involving free radicals. These require initiation through photolysis or other external stimulation, however, and are unlikely to be very important in wetting phenomena.

Nearly all chemical reactions that are important in the wetting of solid surfaces by liquids can be described as Lewis acid–base reactions, as defined above, although when the wetting liquid is an aqueous solution, the Brønsted–Lowry description is often more convenient. When the liquid is a pure component, Lewis acid–base reactions take the form of the adduct-forming reactions of the type of Eqs. (86–90), and they generally involve nonionic species.

By considering the types of acid–base reactions that can occur between pure liquids and solids, that is, Eqs. (86–90), it is clear that we can distinguish between three general types of Lewis acid–base reactants: those compounds that can act exclusively as acids (A, A'), those that can act exclusively as bases (B, B'), and those that can act as either acids *or* bases ($A{:}B$, $A'{:}B$, $A{:}B'$, $A'{:}B'$). van Oss et al. [82] have termed the first two types *monopolar* acids and bases, respectively, and the third category *bipolar* materials. To complete the notation, inert materials, capable of neither acid nor base interactions, are termed *apolar*. This is unfortunate nomenclature because *polarity* has nothing to do with it. We shall therefore use the terminology *monofunctional*, *bifunctional*, and *inert* to distinguish between the above categories of materials. Figure 3b represents the types of interactions that may occur when the liquid phase is bifunctional. The liquid-phase molecules are able to coordinate with either acidic or basic functional groups on the solid surface, and they self-associate in the liquid phase. Figure 3c pictures the case when the liquid is bifunctional and there are two types of functional groups on the solid surface. The picture might be even more complex, since the functional groups on the solid surface may themselves associate or self-associate. It is not always obvious by looking at the structure of a molecule or functional group exactly into which category it falls, and one of the challenges of implementing ideas concerning acid–base interactions at interfaces is to devise experimental tests that will resolve this question. Most substances that are not inert are bifunctional, but bifunctionality is usually a question of degree. Some materials, such as phenol, are strongly acidic, but have a nonnegligible basic functionality as well. Others, such as acetone, are strongly basic, but have nonnegligible acidic functionality. Still others, like water, are equally strongly acidic and basic.

The most important consequence of bifunctionality in a pure liquid is the ability of the molecules to self-associate. These acid–base interactions contribute to the cohesive energy of the liquid, its Hamaker constant, and its surface tension. Fowkes et al. [40] claim that a definitive hallmark of such liquids is their insolubility in inert solvents, such as the higher alkanes, because dissolution requires the dissociation of the acid–base complexes, in contrast to monofunctional acids or bases, which show substantial solubility in these solvents. The criterion must be applied with some caution, however, because the associated complexes themselves may have substantial solubility in inert solvents, particularly those of lower molecular weight. Acetic acid dimers, for example, have substantial solubility in the lower-molecular-weight alkanes. Fowkes et al. [40] have made an extensive study of the miscibility of a variety of liquids with squalane (hexamethyltetracosane), a C_{30}-saturated liquid hydrocarbon (FW 423), with the results shown in Table 5, and have used such information to identify self-association.

Table 5 also shows measured surface tensions and interfacial tensions for all of the liquids against squalane. These values are used to determine the acid–base contribution to the surface tensions of self-associated liquids, making it possible to assess the *degree* of self-association. The surface tension of a self-associated liquid may be expressed as the sum of the Lifshitz–van der Waals and acid–base contributions

$$\sigma = \sigma^{\mathrm{LW}} + \sigma^{AB} \tag{93}$$

and the LW component of it may be evaluated using Eq. (40), with the probe liquid being squalane. The balance of the measured surface tension is thus attributed to acid–base self-association. A measure of the fractional self-association for each liquid in Table 5 is expressed as $(\sigma^{AB}/\sigma)100\%$.

Fowkes [83] proposes a second method for assessing the degree of self-association in liquids by comparing the Lifshitz–van der Waals contribution to the heat of vaporization (computed using Scatchard–Hildebrand

TABLE 5 Solubility and Interfacial Properties of Various Liquids (2) vs. Squalane (1) at 23 ± 0.5°C (surface energies in mJ m^{-2}; dipole moments μ in Debye)

Solvent	σ_2	σ_{12}	σ_2^{LW}	Solubility (%)	Acid–base	% SA
Water	72.4	52.3	21.1	0.01	both	70.9
Formamide	57.3	29.3	28.0	0.01	both	51.1
Methylene iodine	50.8	0	50.8	miscible	neither	0
1,3-Propanediol	45.4	17.3	28.1	0.02	both	38.1
Diethyleneglycol	44.7	12.4	32.3	0.12	both	27.7
Nitrobenzene	43.8	5.7	38.7	0.34	both	11.6
Dimethylsulfoxide	43.5	14.5	29.0	0.01	both	33.3
Salicylaldehyde	42.6	2.2	41.4	2.1	both	2.8
Aniline	42.4	5.6	37.3	0.01	both	12.0
Benzaldehyde	38.3	1.8	37.0	0.32	both	3.4
Pyridine	38.0	0	38.0	miscible	basic	0
Formic acid	37.4	20.7	18.0	0.00	both	51.9
Pyrrole	37.4	4.8	32.6	0.00	both	12.8
Dimethylformamide	36.8	6.5	30.2	0.08	both	17.9
1,4-Dioxane	33.5	0	33.5	miscible	basic	0
Nitromethane	32.5	23.4	19.9	0.03	both	38.9
Dimethyl acetamide	32.4	4.3	28.	0.08	both	13.6
N-Methyl acetamide	32.4	8.1	17.8	0.01	both	45.1
Acetic anhydride	32.4	16.3	17.6	0.00	both	45.7
$POCl_3$	32.2	0	32.2	miscible	acidic	0
cis-Decalin	32.2	0	32.2	miscible	neither	0
1-Nitropropane	30.4	1.7	28.6	0.07	both	5.9
Squalane	29.2	0	29.2	miscible	neither	0
Tetrahydroduran	26.5	0	26.5	miscible	basic	0
Acetic acid	27.6	5.2	22.8	0.01	both	17.4
Chloroform	27.1	0	27.1	miscible	acidic	0
Propionitrile	27.1	4.9	22.6	0.77	both	16.6
CCl_4	27.0	0	27.0	miscible	neither	0
Acetonitrile	27.0	8.6	19.4	0.08	both	28.1
Methylene chloride	26.5	0	26.5	miscible	acidic	0
Cyclohexane	25.5	0	25.5	miscible	neither	0
n-Butyl alcohol	25.4	0	25.4	miscible	weak	0
Ethylacetate	25.2	0	25.2	miscible	basic	0
Methylethyl ketone	24.6	0	24.6	miscible	weak	0
Acetone	23.7	1.4	22.7	0.2	both	4.2
1-Propanol	23.5	0.3	23.3	0.4	both	0.9
Methanol	22.3	6.3	17.4	0.06	both	22.0
Ethanol	22.2	2.7	20.3	0.2	both	8.6
Trifluoroethanol	21.5	10.4	13.5		both (?)	37.2
Triethylamine	20.7	0	20.7	miscible	basic	0
t-Butanol	20.0	0	20.0	miscible	weak	0
Ethyl ether	17.0	0	17.0	miscible	basic	0

% SA = % self-association.
Source: Adapted from Ref. 4, by permission.

TABLE 6 Contributions of Lifshitz–van der Waals and Acid–Base Interactions to Energies of Vaporization of Organic Liquids, in Units of kJ/mol. Listed in Order of Decreasing Self-Association (%SA)

Liquid	ΔU_{vap}	ΔU_{vap}^{LW}	ΔU_{vap}^{AB}	%SA
n-Butanol	51.68	27.7	24.0	46
Acetone	28.72	19.9	8.8	31
Tetrahydrofuran	28.37	20.8	7.6	27
n-Butylamine	22.90	17.8	5.1	22
Ethyl acetate	32.46	26.8	5.7	18
Pyridine	38.85	32.8	6.1	16
Benzene	31.23	28.8	2.4	8
Ethyl ether	24.90	23.3	1.6	6
Chloroform	28.32	27.9	0.46	1.6
Carbon tetrachloride	28.89	29.4	0.46	1.6
Triethylamine	32.35	32.3	0.08	0.1

Source: Ref. 83, by permission.

theory) to the total heat of vaporization. The difference is taken as the contribution due to acid–base self-association, and the ratio of the acid–base contribution to the total value is taken as a measure of the extent of self-association. Fowkes' results for a number of liquids are given in Table 6. The situation with regard to identifying the degree of self-association in liquids is certainly not resolved, as a comparison between results in Tables 5 and 6 indicates. Based on their full miscibility with squalane, pyridine, tetrahydrofuran, and ethyl acetate are identified in Table 5 as monofunctional bases, but based on the acid–base contribution to their heats of vaporization, these liquids are 16, 27, and 18% self-associated, respectively, in Table 6. It is clear that more definitive criteria are required.

Adjacent bifunctional groups on solid surfaces may also exhibit self-association, but another possibility for solid surfaces is that they contain mixed populations of functional groups, some of which may be acidic, some basic, and some bifunctional.

F. Strength of Lewis Acid–Base Interactions

The measure of the strength of Lewis acid–base interactions, or in particular, the quantitative measure of the acidity or basicity of a given material in the Lewis sense, cannot be conveniently expressed in terms of dissociation equilibrium constants as given in Eqs. (57–59) for Brønsted–

Lowry acids and bases. A more appropriate measure of the strength of the acid–base interaction itself is the exothermic molar heat of the adduct-forming reaction, $-\Delta H^{AB}$, a quantity free of the entropy effects that complicate the determination of coordinate covalent bond strength. Various models have been proposed for expressing this enthalpy in terms of the individual contributions of the reactants. These have been reviewed in detail by Jensen [54,77]. The 1960s saw the beginning of what Jensen terms the "quantitative period" for expressing Lewis concepts. In his view, the most important of these were the "hard and soft" acid–base (HSAB) principle of Pearson [55,84], the E & C equation of Drago and Wayland [50,85], the donor and acceptor numbers (*DN* and *AN*) of Gutmann and co-workers [52,86], and the perturbation theory of Hudson and Klopman [87].

Pearson pointed out, following the work of Mulliken [88] and Edwards [89], that there are two contributions to the energy of interaction between a Lewis acid and base: electrostatic ("hard") and covalent ("soft"). Pearson focused on equilibrium constants for reactions of the type of Eq. (86) involving aqueous ions and ligands and proposed using an expression of the form

$$\log K = S_A S_B + \sigma_A \sigma_B \tag{94}$$

where S_A and S_B were "strength factors" for the acid and base, respectively, and σ_A and σ_B the respective "softness factors." H^+ was taken as the arbitrary reference point of zero for the softness parameter. Hardness is associated with low polarizability and high pK_a (for bases), difficulty of oxidation (for bases) or reduction (for acids), small size, and high positive charge density at the acceptor site (for acids) or high negative charge density at the donor site (for bases). Softness is associated with high polarizability and low pK_a (for bases), ease of oxidation (for bases) or reduction (for acids), large size, and low positive charge density at the acceptor site (for acids) or low negative charge density at the donor site (for bases).

1. Drago E & C Equation

Drago and Wayland [85] focused attention on the enthalpy of *AB* complex formation in organic liquids and expressed it in the form

$$-\Delta H^{AB} = E_A E_B + C_A C_B \tag{95}$$

where E_A and E_B represented the "hardness" or electrostatic contribution and C_A and C_B the "softness" or covalent contribution to the exothermic heat of complex formation. It is evident from either the Pearson or Drago formulations that the strongest interactions would be of hard acids with hard bases or soft acids with soft bases. Jensen [77] has pointed out, however, that "in spite of the names assigned to the parameters and their implied basis in bonding theory, there is, to the best of my knowledge, no evidence that they in fact reflect the relative electrostatic and covalent contributions to the bonding in the resulting adducts." This should in no way, however, detract from the operational usefulness of the parameters in predicting the enthalpies of adduct formation.

Drago and Wayland [85] addressed themselves to the assignment of numerical values to the E and C parameters of a variety of materials. Iodine was taken as a reference acid and arbitrarily given values of $E_A = 1$ and $C_A = 1$ (kcal/mol)$^{1/2}$. It was then assumed that for the series of substituted amines NH_3, CH_3NH_2, $(CH_3)_2NH$, and $(CH_3)_3N$, the E_B values should be proportional to the ground-state dipole moment, that is, $E_B = a\mu$, and C_B values should be proportional to the total distortion polarization, that is, $C_B = bR_D$. By using calorimetric measurements of the heat of complexation for all four substituted amines with iodine in an inert solvent (CCl_4), it was found that a single set of constants a and b fit all the data using Eq. (95). The amine constants derived from these iodine data were then used to compute the E and C constants for other acids. The first one tested was phenol. The use of the amine constants with measured enthalpies for the formation of their adducts with phenol reproduced the data with a single set of E and C constants for phenol to within experimental error. In this way, Drago and Wayland produced a table of E and C parameters for 40 different compounds that led to internally consistent values for the calorimetrically determined enthalpies of interaction. Predictions of $-\Delta H^{AB}$ of untried acid–base combinations have generally agreed well with subsequent experiments. The data set has since been expanded and updated [90] to include 42 acids and 55 bases. A representative sample of some of these values is presented in Table 7.

The entries in Table 7 all refer to small molecules, but the presumption is that they may be applied, at least semiquantitatively, to polymeric molecules (in solution) whose repeat units are represented by the appropriate small-molecule analog. Some direct calorimetric measurements of the heat of acid–base interactions in dilute polymeric solution have been reported [83]. The Drago constants may also be applied to functional groups or repeat units at solid polymeric surfaces, but the population of accessible groups at the surface must be determined by independent measurement. The particular problem of the experimental acid–base characterization of solid surfaces is considered further below.

It should be noted that Eq. (95) is limited to adduct formation of a 1:1 stoichiometry. It has also been claimed that "double dative bonding, backbonding, and intermolecular hydrogen bonding effects are recognized

TABLE 7 Drago E and C Parameters for Representative Lewis Acids and Bases, in Units of $kcal/mol^{1/2}$

Acids Compound	E_A	C_A
Iodine [I_2]	1.00	1.00
Iodine chloride [ICl]	5.84	0.54
Phenol [C_6H_5OH]	4.54	0.30
3-Trifluoromethyl phenol [3-$CF_3C_6H_4OH$]	4.80	0.36
tert-Butyl alcohol [$(CH_3)_3COH$]	2.09	0.26
Trimethyl aluminum [$Al(CH_3)_3$]	17.32	0.94
Sulfur dioxide [SO_2]	1.11	0.74
Antimony pentachloride [$SbCl_5$]	7.28	5.09
Chloroform [$HCCl_3$]	2.98	0.07
Isothiocyanic acid [HNCS]	5.70	0.04
Boron trifluoride [BF_3]	12.19	0.81
Water [H_2O]	2.61	0.26
Bases Compound	E_B	C_B
Pyridine [C_5H_5N]	1.30	6.69
Ammonia [NH_3]	1.48	3.32
Methyl amine [CH_3NH_2]	1.50	5.63
Dimethyl amine [$(CH_3)_2NH$]	1.33	8.47
Trimethyl amine [$(CH_3)_3N$]	1.19	11.20
Acetonitrile [CH_3CN]	0.90	1.34
Ethyl acetate [$CH_3C(O)OC_2H_5$]	0.92	1.86
Tetrahydrofuran [$(CH_2)_4O$]	1.06	4.12
Pyridine N-oxide [C_5H_5NO]	1.40	4.40
Benzene [C_6H_6]	0.40	0.85
Dimethyl ether [$(CH_3)_2O$]	1.08	2.92
Piperidine [$C_5H_{11}N$]	1.28	9.00

Source: Excerpted from Ref. 90, by permission.

by experimental ΔH values that do not agree with predictions of the E-C equation'' [51, p. 130]. As Jensen points out [54, pp. 249–251], Eq. (95) is also limited to relatively weak interactions, that is, to $-\Delta H^{AB}$ values of the same order of magnitude as strong intermolecular attractions or hydrogen bonds. Most of the substances for which E and C values are tabulated are discrete neutral molecules, and most of the calculated $-\Delta H^{AB}$ values are less than 200 kJ/mol. Equation (95) fails for many strongly interacting systems, for example, cation/anion and ion/molecule reactions [91]. Kroeger and Drago [92] broadened the applicability of the approach of Eq. (95) to include these types of reactions by introducing a new pair of constants to account for electron transfer effects, yielding

$$-\Delta H^{AB} = e_A e_B + c_A c_B + t_A t_B \tag{96}$$

where the e and c values were recomputed values for the electrostatic and covalent contributions, and the t values were assigned to account for electron transfer effects. Equation (96) was successful at predicting the enthalpies for many reactions, but did so at the expense of two new constants.

Equation (95) also fails to take any explicit account of steric effects, and these might be especially important when dealing with adducts formed at solid surfaces, discussed in more detail below, or those involving functional groups on polymers. It has been suggested that steric effects might also be accommodated by the use of still another pair of parameters, but the value of introducing still more complication is questionable.

Perhaps the most serious present limitation of the Drago E & C formulation is that it treats all compounds as though they were monofunctional acids or bases. In tabulations of the parameters, the same compound never appears in both the acid and base columns. Water, for example, is listed as an acid, but it is known to be equally effective as a Lewis base. What is evidently required for bifunctional compounds or functional groups is a double set of Drago parameters, as might be obtained from calorimetric measurements of interactions with both known monofunctional acids *and* bases.

2. Gutmann Donor and Acceptor Numbers

Gutmann and co-workers [86] introduced a different scale for ranking the strength of Lewis acids and bases. They first addressed the quantification of the Lewis basicity of a variety of solvents by introducing the *donor number*, *DN*. *DN*, for a given basic solvent B, was defined in terms of the molar exothermic heat of its reaction with a reference acid, antimony pentachloride, in a dilute (10^{-3} M) solution in a neutral solvent (1,2-dichloroethane)

$$DN_B = -\Delta H(SbCl_5{:}B) \tag{97}$$

The donor number thus includes in a single parameter both the electrostatic and covalent contributions to the interaction. A representative sample of donor numbers for a variety of bases is presented in Table 8. Credence is lent to the validity of these numbers by noting that they correlate with the enthalpies of interaction in other presumably inert solvents and with other Lewis acids. The database for donor numbers has been sig-

TABLE 8 Donor Numbers (*DN*) for Various Solvents Obtained from Calorimetric Measurements in 10^{-3} *M* Solutions of Dichloroethane with $SbCl_5$ as a Reference Acceptor, in Units of kJ/mol

Solvent	*DN*	Solvent	*DN*
1,2-Dichlorethane (DCE)	—	Ethylene carbonate (EC)	16.4
Benzene	0.1	Phenylphosphonic difluoride	16.4
Thionyl chloride	0.4	Methyl acetate	16.5
Acetyl chloride	0.7	*n*-Butyronitrile	16.6
Tetrachloroethylene carbonate (TCEC)	0.8	Acetone (AC)	17.0
		Ethylacetate	17.1
Benzoyl fluoride (BF)	2.3	Water	18.0
Benzoyl chloride	2.3	Phenylphosphoric dichloride	18.5
Nitromethane (NM)	2.7	Diethyl ether	19.2
Nitrobenzene (NB)	4.4	Tetrahydrofuran (THF)	20.0
Acetic anhydride	10.5	Diphenylphosphoric chloride	22.4
Phosphorus oxychloride	11.7	Trimethyl phosphate (TMP)	23.0
Benzonitrile (BN)	11.9	Tributyl phosphate (TBP)	23.7
Selenium oxychloride	12.2	Dimethyl formamide (DMF)	26.6
Acetonitrile	14.1	*n*-Methyl pyrolidinone (NMP)	27.3
Tetramethylenesulfone (TMS)	14.8	*n*-Dimethyl acetamide (DMA)	27.8
Dioxane	14.8	Dimethyl sulfoxide (DMSO)	29.8
Propandiol-(1,2)-carbonate (PDC)	15.1	*n*-Diethyl formamide (DEF)	30.9
		n-Diethyl acetamide (DEA)	32.2
Benzyl cyanide	15.1	Pyridine (PY)	33.1
Ethylene sulphite (ES)	15.3	Hexamethylphosphoric triamide (HMPA)	38.8
iso-Butyronitrile	15.4		
Propionitrile	16.1		

Source: Ref. 52, p. 20, by permission.

nificantly expanded by inferring the enthalpies of adduct formation from correlations with NMR chemical shifts and infrared spectroscopic measurements [93].

Gutmann and co-workers [94,95] later proposed a function analogous to *DN* for the purpose of ranking the acidity of a solvent, that is, the *acceptor number*, *AN*. It was defined as the relative ^{31}P NMR downfield shift ($\Delta\delta$) induced in triethyl phosphine, $(C_2H_5)_3PO$, when dissolved in the pure candidate acidic solvent. A value of 0 was assigned to the shift produced with the neutral solvent hexane, and a value of 100 to the shift produced by $SbCl_5$. A representative sample of acceptor numbers is

TABLE 9 Lifshitz–van der Waals and Acid–Base Contributions to Acceptor Numbers

Liquid	*AN*	AN^{LW}	$AN\text{-}AN^{LW}$
Dioxane	10.8	10.9	0.0
Hexane	0.0	0.0	0.0
Benzonitrile	15.5	15.3	0.2
Pyridine	14.2	13.7	0.5
Benzene	8.2	7.6	0.6
Tetrahydroduran	8.0	6.1	1.9
Carbon tetrachloride	8.6	6.3	2.3
Diethyl ether	3.9	−1.0	4.9
Ethyl acetate	9.3	4.0	5.3
Methyl acetate	10.7	5.0	5.7
N,N-Dimethylacetamide	13.6	7.9	5.7
1,2-Dichloroethane	16.7	10.3	6.4
N,N-Dimethylformamide	16.0	9.4	6.6
Acetone	12.5	3.8	8.7
Dimethyl sulfoxide	19.3	8.5	10.8
Dichloromethane	20.4	6.9	13.5
Nitromethane	20.5	5.7	14.8
Acetonitrile	18.9	2.6	16.3
Chloroform	25.1	6.4	18.7
tert-Butyl alcohol	27.1	0.6	26.5
2-Propanol	33.6	2.1	31.5
n-Butyl alcohol	36.8	5.1	31.7
Formamide	39.8	7.6	32.2
Ethanol	37.9	2.0	35.9
Methanol	41.5	−0.2	41.7
Acetic acid	52.9	3.6	49.3
Water	54.8	2.4	52.4
Trifluoroethanol	53.3	−2.9	56.2
Hexafluoroo-2-propanol	61.6	−4.7	66.3
Trifluoroacetic acid	105.3	−5.7	111.0

Source: Abstracted from Ref. 96, by permission.

shown in Table 9. Gutmann suggested that the enthalpy of acid–base adduct formation be written as

$$-\Delta H^{AB} = \frac{AN_A DN_B}{100} \tag{98}$$

where the factor of 100 converts the tabulated *AN* to a decimal fraction of the $SbCl_5$ value. Equation (98) can be used in conjunction with calorimetric measurements and a known value of either *AN* or *DN* to obtain the remaining parameter, but has had only limited success in reproducing measured enthalpies of interaction for untried acid–base pairs.

Riddle and Fowkes [96] have recently pointed out that the ^{31}P NMR spectrum is appreciably shifted downward by van der Waals as well as acid–base interactions, and that the former constitute a significant part of many of the tabulated *AN* values. This explains the surprisingly large acceptor numbers for such solvents as pyridine, which are regarded as essentially monofunctional basic. The downfield shift may be written as the sum of acid–base and Lifshitz–van der Waals contribution

$$\Delta\delta = \Delta\delta^{AB} + \Delta\delta^{LW} \tag{99}$$

and it is only the acid–base contribution that should be used in defining an appropriate acceptor number, so that the corrected acceptor number is given by $AN - AN^{LW}$. Riddle and Fowkes correlated the LW contribution to the shift with the LW component of the solvent surface tension σ^{LW} and obtained

$$\Delta\delta^{LW} = \delta_0 + 7.37 - 0.312\sigma^{LW} \tag{100}$$

where δ_0 is ^{31}P NMR peak position for $(C_2H_5)_3PO$ (referenced to diphenylphosphinic chloride, extrapolated to zero concentration, and corrected for volume susceptibility vs. *n*-hexane), as tabulated by Gutmann [52], and σ^{LW} is in units of mJ/m^2. The Riddle–Fowkes corrected values for the Gutmann acceptor numbers are also shown in Table 9. It is noted that 13.7 of the original 14.2 *AN* units for pyridine have been assigned to the LW contribution, so that its true Lewis acidity on this scale is only 0.5, more in line with expectations.

It is to be noted that many of the same compounds appear in both the *DN* and *AN* lists, indicating that the Gutmann formulation, unlike that of Drago, recognizes the bifunctionality of most materials. Its chief weakness is that it is recognizes only single orders of acid and base strength, that is, it does not distinguish between the hard and soft contributions. The agreement found by Gutmann between predictions and measured enthalpies can be traced to the fact that his donor numbers were tabulated almost exclusively for very hard bases. He conceded [97] that the relationship between *DN* values and predicted enthalpies breaks down for soft acids.

G. Experimental Determination of the Acid–Base Properties of Molecules

Except for the Gutmann acceptor number, the acid–base characterizing parameters of either the Drago or Gutmann scales are *defined* in terms of enthalpies of adduct formation. Thus, the direct method for their experimental determination is calorimetry. Unfortunately, the large existing database for the heats of mixing between pure liquid components cannot be used owing to the significant and universally present contribution of van der Waals effects to the measured enthalpy, ΔH. Although measurements in the gas phase would be desirable because they avoid these effects, experimental difficulties and limitations outweigh the advantages [50, Ch. 3]. Many of the adducts of interest do not exist in gaseous form under conditions that can be measured. Instead, measurements are generally obtained with the desired acid and base reactants in dilute solution in an inert solvent, so that the van der Waals contribution remains constant and is subtracted out, and solvation effects (as are present if a non-inert solvent is used) are absent. An important limitation of this method is that some of the desired reactants are not sufficiently soluble in inert solvents. In that case, "polar" solvents must be used, and the appropriate corrections for the solvent effects made. In either case, the standard techniques of batch or flow microcalorimetry can be employed. Alternatively, one can measure the composition changes accompanying the adduct-forming reactions spectrophometrically, and from these the equilibrium constants for the reactions. If done over a range of temperatures, a Gibbs–Helmholtz analysis can be used to extract the desired enthalpy.

The measurement of the enthalpies of adduct formation, either calorimetrically or spectrophotometrically, is a tedious process. Thus, many of the more recently obtained Drago and Gutmann donor constants were obtained not from such data directly, but by making use of the relationship of the desired enthalpy to other more easily determined property changes [98]. Investigators have explored the linear relationship between $-\Delta H^{AB}$ and infrared frequency shifts that accompany acid–base adduct formation [99], Hammet substituent coefficients [99], the Kamlet–Taft β-parameter [100], NMR coupling constants [101], and Doan and Drago [100] have considered the general theoretical validity of such linear free energy correlations. The use of the IR and NMR spectral shifts is discussed briefly below. For the other types of correlations, the reader is referred to the above-cited references.

Consider first the use of the IR spectral shift. Drago and co-workers [101,102] measured and compared the infrared spectra of free phenol *A* and a phenolic adduct *B* in carbon tetrachloride solvent, as shown in Fig. 8 [50, p. 72]. The absorbance at 3609 cm^{-1} in curve *A* is attributed to the stretching vibration of O—H in the free phenol. Curve *B* shows the frequency associated with the same O—H in the acid–base complex, and the shift of 400 cm^{-1} is attributed to the effect of adduct formation. It is this downward frequency shift $\Delta\nu_{OH}$ that correlates with the enthalpy of

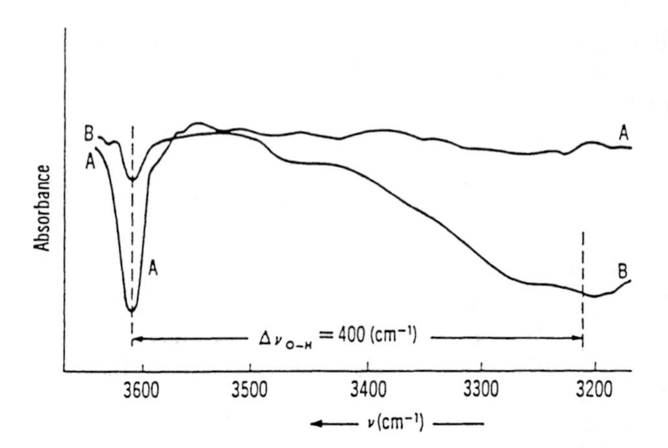

FIG. 8 Infrared spectra of phenol and a hydrogen-bonded phenol: (A) Free phenol in CCl_4 and (B) phenol plus donor in CCl_4. (From Ref. 50, p. 72, by permission.)

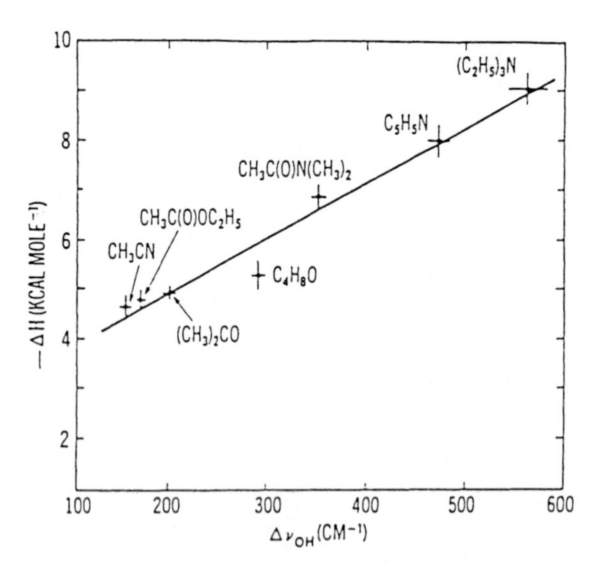

FIG. 9 Correlation of $-\Delta H$ and $\Delta \nu_{OH}$ for phenol adducts. (From Ref. 50, p. 72, by permission.)

adduct formation ($-\Delta H^{AB}$). Figure 9 shows a plot of $\Delta \nu_{OH}$ vs. calorimetrically determined $-\Delta H^{AB}$ values for the formation of seven different phenol adducts, and the linearity of the relationship is evident.

Correlations for the shift in stretching frequency for other bonds undergoing acid–base adduct formation have since been reported. Fowkes et al. [103] in particular have focused on shifts in the carbonyl (C=O) stretching frequency. The carbonyl group is nearly monofunctionally basic and occurs in esters, carbonates, ketones, and amides. The initial studies were carried out with ethyl acetate, and the frequency shift as compared to the vapor state was sought. When ethyl acetate is brought from the vapor state into solution in some solvent, its carbonyl stretching frequency will be lowered, first owing to its van der Waals interaction with the solvent and second owing to any acid–base adduct formation either with the solvent or some acidic species dissolved in the solvent. It was first found that the shift caused by the van der Waals contribution correlated linearly with the surface tension of the solvent (if it was non-self-associating) or more generally to its LW component, σ^{LW}, and the correlating constant was determined from the measured shifts obtained for several different noninteracting (i.e., either inert or basic) solvents. Measurements were then made of the shifts obtained for the C=O stretching frequency in ethyl acetate with acidic solvents or inert solvents containing acidic solutes. These shifts could be corrected by subtracting out the solvent van der Waals contribution, isolating that part of the shift due to the acid–base interaction. Finally, the acid–base shift was correlated with previously published calorimetric values for $-\Delta H^{AB}$ for each case. The present form of the correlation is [83]

$$\nu_{C=O}(\mathrm{cm}^{-1}) = 1764 - 0.714\sigma^{LW}(\mathrm{mJ/m^2}) + 0.99\,\Delta H^{AB}(\mathrm{kJ/mol}) \qquad (101)$$

where the first number on the right-hand side is the frequency in the vapor. It is asserted that parallel equations would hold for other carbonyl-containing compounds, with only the reference (vapor-phase) frequency being different. Thus, Fowkes used measured IR shifts for poly(methyl methacrylate) to compute the $-\Delta H^{AB}$ value for its interaction with chloroform, iodine, antimony pentachloride, and phenol, from which he extracted E_B and C_B values for PMMA, that is, $E_B = 0.45 \pm 0.1$ and $C_B = 1.35 \pm 0.1$ (kcal/mol)$^{1/2}$. Constants were similarly obtained for a variety of polymers and indicated that acidic strength decreases in the order poly(vinyl fluoride) > poly(vinylidine fluoride) > poly(vinyl butyral) > poly(vinyl chloride) [103]. Kwei et al. reported results, similarly obtained, for several additional polymers [104]. All that is needed for the extraction of the E_B and C_B constants for a base is two test acids with significantly different ratios of E_A/C_A, or for the determination of the E_A and C_A constants of an acid, one needs a pair of test bases with different E_B/C_B ratios. Of course, the use of more than two test acids or bases is highly desirable to check the internal consistency of the derived parameters.

The use of measured IR shifts to obtain heats of adduct formation has not only the advantage of convenience over the direct use of calorimetry, but also does not require that all the available reactants be complexed (as is the case for the interpretation of calorimetric measurements). This is especially important when adducts involving polymers are formed or in other situations where steric effects are expected to be significant.

The use of NMR shifts has already been discussed above in connection with the needed correction to the Gutmann acceptor numbers. Riddle and Fowkes [96] have developed a correlation for the ^{31}P-NMR peak position for $(C_2H_5)_3PO$ (relative to the ^{31}P-NMR peak of an external diphenyl phosphinic chloride standard) as follows:

$$\delta(\text{ppm}) = -7.37 + 0.312\sigma^{LW}(\text{mJ/m}^2) - 0.145\,\Delta H^{AB}(\text{kJ/mol}) \qquad (102)$$

The method is generalizable to other compounds, and Fowkes [83] has written that "a series of NMR test probes are under development for determining the hard-soft acid–base (HSAB) properties of solvents, polymers, and inorganic solids. These include a hard and soft base [$(C_2H_5)_3PO$ and $(C_6H_{12})_3PS$], and a hard and a soft acid (triphenyl silanol and diphenyl mercury). With these probes the E and C constants are now being determined for ten polymers."

IV. CORRELATION OF ACID–BASE PROPERTIES OF SOLID SURFACES

Particularly interesting are studies involving the wetting of solids that are presumed to be predominantly acidic or predominantly basic by liquids that are known to be essentially monofunctional acids and bases. Figures 10 and 11 [105] show results of this type, obtained using Eq. (49). Monofunctional acids (bromoform and pyrrole) show virtually no acid–base interaction with the presumably monofunctional acid surface of poly(vinyl chloride) (PVC) (Fig. 10), but significant acid–base interaction with the presumably monofunctional basic surface of poly(methyl methacrylate) (PMMA) (Fig. 11). On the other hand, the predominantly basic liquid dimethyl sulfoxide interacts strongly with the acidic PVC but not the basic PMMA. The bifunctional liquid 2-iodoethanol interacts strongly with both

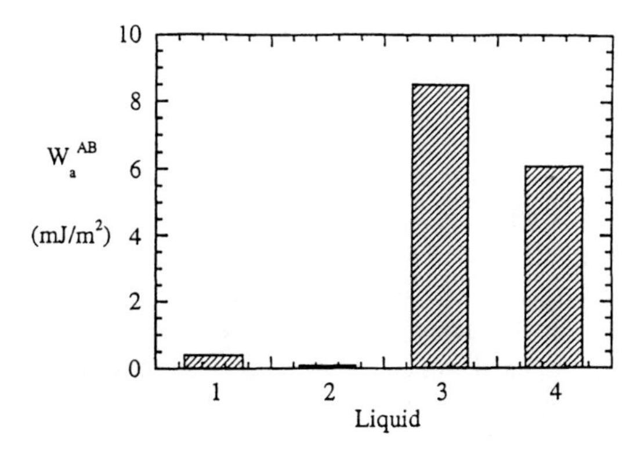

FIG. 10 Acid–base work of adhesion of selected liquids against poly(vinyl chloride): (1) bromoform; (2) pyrrole; (3) dimethyl sulfoxide; (4) 2-iodoethanol. (Redrawn from Ref. 105.)

solid surfaces, as expected. A larger database of this type would be desirable to have, but it will always be limited by the requirement of producing finite contact angles. This usually means that the test liquids used must have a surface tension of at least 30 mJ/m^2. Only a very small number of monofunctional acidic and basic liquids at room temperature meet this requirement. The candidate acids identified so far are bromoform (σ = 40.9 mJ/m^2 at 23°C), antimony pentachloride, $SbCl_5$ (σ = 42.0 mJ/m^2 at

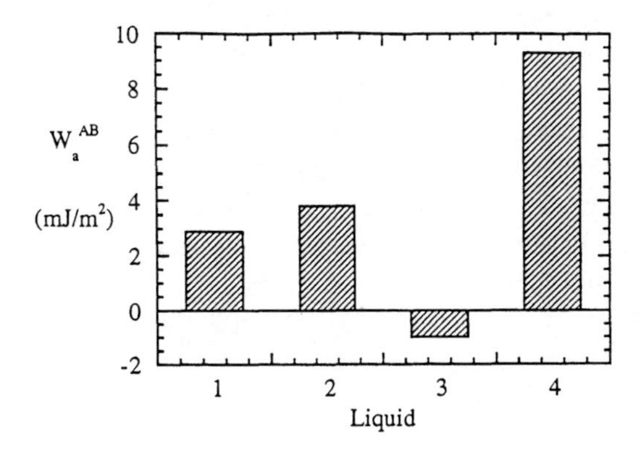

FIG. 11 Acid–base work of adhesion of selected liquids against poly(methyl methacrylate): (1) bromoform; (2) pyrrole; (3) dimethyl sulfoxide; (4) 2-iodoethanol. (Redrawn from Ref. 105.)

23°C),* and phosphorous oxychloride, POCl (σ = 32.2 mJ/m^2 at 23°C). Among these, POCl is awkward to use owing to its toxicity. Pyrrole (σ = 37.4 mJ/m^2 at 23°C) is strongly acidic, but has some measurable basicity. The monofunctional bases appear to include pyridine (σ = 38.0 mJ/m^2 at 23°C) and 1,4-dioxane (σ = 33.3 mJ/m^2 at 23°C). Dimethyl sulfoxide (σ = 43.5 mJ/m^2 at 23°C) is strongly basic, but has a nonnegligible acidic component. Furthermore, the solids that can be tested using wetting measurements must be limited to those of "low surface energy." Otherwise, they are likely to be wet out even by the above test liquids. This excludes from consideration most hard mineral surfaces such as oxides and silicates, as well as most ionic crystalline materials.

The obtainability of quantitative values for the acid–base contribution to the work of adhesion, even if for a rather limited range of systems, suggests that such data be analyzed in terms of the fundamental acid–base properties of the interacting liquids and solid surfaces. The missing component in such an analysis is knowledge of the appropriate properties for the solid surface. Considerable attention has been given to characterizing solid surfaces with respect to their ability to interact with liquids, as manifested through wetting phenomena or adhesion. The foregoing discussion suggests the difficulty of ever being able to make a universal characterization scheme in terms of a small (even manageable) set of parameters. Although inert (i.e., non-acid–base functional) surfaces may be well described in terms of their surface energies, and simple oxide surfaces are sometimes well described in terms their isoelectric points, the general situation involving chemical (acid–base) interactions is much more complex. Nonetheless, two simplified approaches have been strongly suggested in the recent literature. One, propounded by van Oss et al. [106] uses only wetting measurements to evaluate a characteristic acid parameter and base parameter for each solid. The other approach, proposed by Fowkes [107], expresses the acid–base contribution to the work of adhesion in terms of the enthalpy of the interaction between the liquid and solid surface, and expresses this enthalpy in terms of the Drago parameters. In this case, the solid surface (considered either an acid *or* a base in any instance) is characterized in terms of a single set of Drago parameters, together with information concerning the population of accessible functional groups on the solid surface. Both approaches are examined below.

* Measured in the author's laboratory.

A. van Oss–Chaudhury–Good Correlation

Adopting a combining rule for short-range interfacial forces first suggested by Small [108], van Oss et al. [106] expressed the acid–base component of the work of adhesion as

$$W_a^{AB} = 2\sqrt{\sigma_S^{\oplus}\sigma_L^{\ominus}} + 2\sqrt{\sigma_S^{\ominus}\sigma_L^{\oplus}} \tag{103}$$

where $\sigma_S^{\oplus}$ and $\sigma_S^{\ominus}$ are the characteristic acid and base parameters of the solid surface and $\sigma_L^{\oplus}$ and $\sigma_L^{\ominus}$ the characteristic acid and base parameters of the liquid. No distinction between the hard and soft contributions to the acid or base properties of the solid or liquid is made, and no fundamental justification for using a geometric mean type of expression is given (or is needed, if the parameters are simply operational). Equation (103) *defines* the empirical parameters it contains.

The form of Eq. (103) recognizes that both the solid surface and liquid may be acid–base bifunctional. It reduces to the following simpler forms in the event that either or both the solid surface and liquid are monofunctional acids or bases, or are inert. W_a^{AB} is zero if the solid surface is inert ($\sigma_S^{\oplus} = \sigma_S^{\ominus} = 0$) and/or the liquid is inert ($\sigma_L^{\oplus} = \sigma_L^{\ominus} = 0$), or if both the solid surface and liquid are monofunctional acids ($\sigma_S^{\ominus} = \sigma_L^{\ominus} = 0$) or monofunctional bases ($\sigma_S^{\oplus} = \sigma_L^{\oplus} = 0$). If the solid surface is bifunctional or monofunctionally basic, while the liquid is monofunctionally acidic; or if the liquid is bifunctional or monofunctionally acidic, while the solid surface is monofunctionally basic

$$W_a^{AB} = 2\sqrt{\sigma_S^{\ominus}\sigma_L^{\oplus}} \tag{104}$$

With the above roles of acids and bases reversed,

$$W_a^{AB} = 2\sqrt{\sigma_S^{\oplus}\sigma_L^{\ominus}} \tag{105}$$

One may also express the acid–base contributions to the surface tension and interfacial tension in terms of the van Oss–Chaudhury–Good parameters

$$\sigma^{AB} = 2\sqrt{\sigma^{\oplus}\sigma^{\ominus}} \tag{106}$$

and

$$\sigma_{12}^{AB} = 2(\sqrt{\sigma_1^{\oplus}} - \sqrt{\sigma_2^{\oplus}})(\sqrt{\sigma_1^{\ominus}} - \sqrt{\sigma_2^{\ominus}}) \tag{107}$$

van Oss and co-workers were the first to implement a formalism to account for the bifunctionality of liquids and solids in acid–base wetting phenomena, and they also pointed out the important fact that acid–base interactions can, and perhaps often do, lead to negative values of the interfacial free energy σ_{SL} [109].

van Oss and co-workers addressed themselves to the determination of $\sigma_S^{\oplus}$ and $\sigma_S^{\ominus}$ for a solid of interest by making use of measured contact angles against the solid of a pair of liquids, 1 and 2, of known surface tension and known values of $\sigma^{\oplus}$ and $\sigma^{\ominus}$. Using the Young–Dupré equation, in the form of Eq. (49), for each case we get

$$\sigma_1(1 + \cos\theta_{1S}) = 2\sqrt{\sigma_1^{LW}\sigma_S^{LW}} + 2\sqrt{\sigma_1^{\oplus}\sigma_S^{\ominus}} + 2\sqrt{\sigma_1^{\ominus}\sigma_S^{\oplus}} \qquad (108)$$

and

$$\sigma_2(1 + \cos\theta_{2S}) = 2\sqrt{\sigma_2^{LW}\sigma_S^{LW}} + 2\sqrt{\sigma_2^{\oplus}\sigma_S^{\ominus}} + 2\sqrt{\sigma_2^{\ominus}\sigma_S^{\oplus}} \qquad (109)$$

The LW components of the solid surface and both liquids 1 and 2 can be determined using inert probe liquids, as shown in Eqs. (40) and (41). The difficulty in then solving Eqs. (106) and (107) simultaneously for the unknowns $\sigma_S^{\oplus}$ and $\sigma_S^{\ominus}$ is, of course, that the values for $\sigma^{\oplus}$ and $\sigma^{\ominus}$ for the reference liquids 1 and 2 are not known. van Oss et al. chose water as the first reference liquid and glycerol, formamide, and dimethyl sulfoxide as second reference liquids. They obtained the necessary parameters for them in the following way. First, they assumed that for water

$$\sigma_W^{\oplus} = \sigma_W^{\ominus} = \tfrac{1}{2}\sigma_W^{AB} = \tfrac{1}{2}(\sigma_W - \sigma_W^{LW}) = \tfrac{1}{2}(72.8 - 21.8) = 25.5 \text{ mJ/m}^2 \qquad (110)$$

Then they made measurements of the contact angles of the second reference liquid (2) against a variety of *monofunctional solids*, yielding expressions of the type

$$\sigma_2(1 + \cos\theta_{2S}) - 2\sqrt{\sigma_2^{LW}\sigma_S^{LW}} = 2\sqrt{\sigma_2^{\oplus}\sigma_S^{\ominus}}$$

(for S, a monofunctional base) or

$$= 2\sqrt{\sigma_2^{\ominus}\sigma_S^{\oplus}}$$

(for S, a monofunctional acid) (111)

From such measurements, one may obtain the ratio of $\sigma_W^{\oplus}/\sigma_2^{\oplus}$ as follows

$$\frac{\sigma_W^{\oplus}}{\sigma_2^{\oplus}} = \left[\frac{\sigma_W(1 + \cos\theta_{WS}) - 2\sqrt{\sigma_W^{LW}\sigma_S^{LW}}}{\sigma_2(1 + \cos\theta_{2S}) - 2\sqrt{\sigma_2^{LW}\sigma_S^{LW}}}\right]^2 = \beta_2 \qquad (112)$$

By using measurements with 12 different, presumably monofunctional solids, average β values were obtained for glycerol (6.505), formamide (11.2), and DMSO (36.4), leading to the values for $\sigma^{\oplus}$ and $\sigma^{\ominus}$ for these

TABLE 10 Surface Parameters for Some Polymers, in Units of mJ m^{-2}

Solid	σ^{LW}	$\sigma^{\oplus}$	$\sigma^{\ominus}$
Poly(methylmethacrylate), cast film	39–43	(0)	9.5–22.4
Poly(vinylchloride)	43	0.04	3.5
Poly(oxyethylene): PEG 6000	45	(0)	66
Cellulose acetate	35	0.3	22.7
Cellulose nitrate	45	0	16
Agarose	41	0.1	24
Gelatin	38	0	19
Human serum albumin (dry)	41	0.15	18
Poly(styrene)	42	0	1.1
Poly(ethylene) (commercial film)	33	0	0.1

Source: Abstracted from Ref. 109, by permission.

liquids given in Table 10. Then by using water and glycerol as the two reference liquids, the values of $\sigma_S^{\oplus}$ and $\sigma_S^{\ominus}$ for several solids were obtained, as presented in Table 10.

It must be noted that values of $\sigma_S^{\oplus}$ and $\sigma_S^{\ominus}$ for the various solids obtained using other combinations of the reference liquids were not reported. This is a serious omission because such results are needed to check the internal consistency of the method. The number of reference liquids that might be employed for an even greater test of internal consistency is limited by the requirement that they have sufficiently high surface tension as not to wet out the desired solid. The results of Table 10 suggest that most solids appear to be almost totally basic. This is especially surprising for poly(vinyl chloride), which has generally been considered to be monofunctionally acidic.

The most that can be done to check the van Oss et al. results is to examine their internal consistency with wetting measurements, because the underlying equation for expressing the acid–base contribution to the work of adhesion cannot readily be connected to any other types of measurements, such as enthalpy changes. There is no way to meaningfully compare the characteristic parameters $\sigma_S^{\oplus}$ and $\sigma_S^{\ominus}$ with other measures of solid surface acidity or basicity, like those provided by Drago or Gutmann.

B. Fowkes Method

In a series of papers beginning in 1978, Fowkes [107] suggested that the acid–base contribution to the work of adhesion be expressed in the form

$$W_a^{AB} = fN(-\Delta H^{AB}) \tag{113}$$

where ΔH^{AB} is the molar enthalpy of acid–base adduct formation, N the number of moles of interacting functional groups per unit area on the solid surface, and f a conversion factor to correct enthalpy values to free energy values.

Equation (113) establishes the crucial link between the wetting parameters and substantial existing database on the strength of Lewis acid–base interactions and motivates the further extension of that database to include more information on the acid–base properties of solid surfaces. It indicates how wetting measurements may lead to the evaluation of Drago or Gutmann parameters for solid surfaces. In order to implement the linkage between the wetting parameters and database on molar enthalpies of adduct formation, the independent determination of the parameters N and f is required for any specific system. Depending on the particular solid, and if it is available in a form yielding sufficient specific area, N may often be obtained through potentiometric or conductometric titrations. Otherwise, it may be obtained through adsorption measurements as detailed below. With respect to f, Fowkes has insisted, until very recently [83], that it should be essentially unity. Such an assumption denies the probable existence of any entropy changes accompanying the formation of acid–base adducts at the solid/liquid interface. Recent work [105], detailed below, has shown that this assumption cannot be made a priori and that f factors significantly less than unity may be encountered.

Values of the factor f as a function of temperature were determined recently [105] for four different solid/liquid systems using wetting measurements. Careful determinations of surface tension and contact angle were made at closely spaced temperature intervals from 20–60°C, permitting the computation of W_a^{AB} as a function of temperature over that interval using Eq. (49). The enthalpy per unit area was then extracted using a Gibbs–Helmholtz analysis of the temperature dependence of W_a^{AB}

$$N(-\Delta H^{AB}) = T^2\left(\frac{d\,\dfrac{W_a^{AB}}{T}}{dT}\right) = W_a^{AB} - T\frac{dW_a^{AB}}{dT} = \frac{1}{f}\,W_a^{AB} \qquad (114)$$

Since the Gibbs–Helmholtz equation applies to molar rather than per-unit-area quantities, Eq. (114) is written assuming that N is not a function of temperature. It leads directly to an expression for f

$$f = \left(1 - \frac{d\ln W_a^{AB}}{d\ln T}\right)^{-1} \qquad (115)$$

Computed results for the factor f are shown in Fig. 12, and they are seen to deviate significantly from unity.

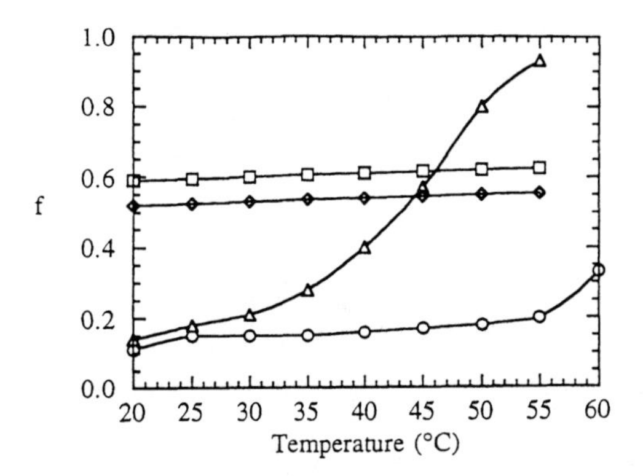

FIG. 12 Enthalpy to free energy correction factor f as a function of temperature for four systems: (–○–) bromoform against poly(methyl methacrylate); (–□–) dimethyl sulfoxide against poly (vinyl chloride); (–△–) dimethyl sulfoxide against copolymer of poly(ethylene)/poly (acrylic acid) (5%); (–◇–) dimethyl sulfoxide against copolymer of poly (ethylene)/poly(acrylic acid) (20%). (Redrawn from Ref. 105.)

The f factors, as well as values of N, are also obtainable from measurements of W_a^{AB} obtained from adsorption isotherms for the desired liquid onto the desired solid surface (either from the gas phase or solution in an inert solvent) as a function of temperature, as detailed below.

An attempt to check Eq. (113) using wetting measurements of W_a^{AB}, together with the appropriate value of f determined as above and N determined by independent titration, has been made for one solid/liquid system, and the results were not conclusive [105]. More studies of this type are needed. Fowkes [83] checked Eq. (113) for the liquid/liquid system of benzene/water, with $-\Delta H^{AB}$ computed from the published Drago parameters for benzene and water. For benzene $E_B = 0.75$ and $C_B = 1.8$ $(\text{kJ/mol})^{1/2}$, and for water $E_A = 5.01$ and $C_A = 0.67$ $(\text{kJ/mol})^{1/2}$, leading to $-\Delta H^{AB} = 5.0$ kJ/mol. If we assume that each benzene molecule occupies 0.50 nm^2, $N = 3.3$ μmol/m^2, and $f \approx 1$, W_a^{AB} is computed at 16.5 mJ/m^2. This is compared with W_a^{AB}, computed as

$$\begin{aligned} W_a^{AB} &= \sigma_b + \sigma_w - \sigma_{b/w} - 2\sqrt{\sigma_b^{LW}\sigma_w^{LW}} \\ &= 28.9 + 72.8 - 35.0 - 2\sqrt{(28.9)(22.0)} = 16.3 \text{ mJ/m}^2 \end{aligned} \tag{116}$$

This represents excellent agreement if benzene has only one bond per water molecule. If benzene bonds with two molecules of water, more likely the case in view of the relative size of the molecules, the implication is that $f \approx 0.5$.

Finally, Eq. (113) must sometimes be rewritten to account for the fact that the solid surface possesses more than one uniform population of functional groups, or groups that have acidic as well as basic sites, that is,

$$W_a^{AB} = \sum_i f_i N_i(-\Delta H^{AB})_i \tag{117}$$

The unambiguous deconvolution of the terms on the right-hand side of Eq. (117) is a challenging task.

V. EXPERIMENTAL TECHNIQUES FOR THE EVALUATION OF ACID–BASE EFFECTS IN SOLID/LIQUID INTERACTIONS

Most of the experimental techniques described above, other than those employing wetting measurements, for the evaluation of enthalpies of adduct formation and the correlating parameters such as those of Drago or Gutmann, pertain to interacting species in the liquid phase. Most of these techniques can be adapted to the study of adduct formation at the solid/liquid interface, and to the determination of Drago or Gutmann parameters for solid surfaces.

A. Calorimetric and Spectroscopic Measurements

The presently available instrumentation for batch calorimetry has sufficient sensitivity to measure the heats of adsorption of acidic or basic molecules to acidic or basic sites on a solid surface, provided the specific area of the solid substrate is moderately high (>20 m^2/g). Fowkes and coworkers [103] have used the Tronac isothermal microcalorimeter to measure the heats of adsorption of several test bases: triethylamine, pyridine, and ethyl acetate, onto acidic silica and titania from solution in dry cyclohexane. The C_B/E_B ratios for the three test bases were 11.14, 5.47, and 1.78, respectively, sufficiently different from one another for an unambiguous determination of the E and C constants for the oxide surfaces. As described by Fowkes [110],

> After thermal equilibration of a well-stirred suspension of 0.3–1 gram of silica (Aerosil 380 or HiSil 233) in 25 ml of cyclohexane, a cyclohexane solution of enough base to neutralize about a third of the surface sites was titrated into the suspension over a fifteen minute period. The heat of adsorption evolved for more than an hour, after which the

> equilibrium concentration of base was determined by UV spectrometry. . . . The resulting heats of adsorption had a standard deviation of about 0.4%, and at the approximately 30% coverage used in these studies we did not find any difference between HiSil 233 and Aerosil 380, even though the HiSil is a precipitated silica and the Aerosil is a pyrolized silica.

The Drago constants for the SiOH surface sites were found to be E_A = 4.39 ± 0.01 and C_A = 1.14 ± 0.01 $(\text{kcal/mol})^{1/2}$. Parallel studies gave for the TiOH sites of rutile: E_A = 5.66 ± 0.05 and C_A = 1.02 ± 0.03 $(\text{kcal/mol})^{1/2}$.

Flow microcalorimetry has sufficient sensitivity to be used for measuring heats of adsorption at solid surfaces of even rather low specific area (>1–4 m^2/g). Using the Microscal flow microcalorimeter, Fowkes et al. [110,111] determined heats of adsorption at different iron oxide surfaces. A flow microcalorimeter is essentially a liquid chromatograph with a sensitive temperature-sensing element in the test chamber, a 0.15-mL adsorption bed of powdered solid, through which the test solution flows at low velocity. Measurement of the flow rate and temperature allows the heat evolution to be computed (as the area under the "adsorption exotherm"), and a downstream concentration detector allows determination of the amount adsorbed. Using the same test bases as in the batch calorimetry described above, Fowkes et al. [111] obtained for α-hematite (an iron oxide) E_A = 3.92 ± 0.08 and C_A = 0.77 ± 0.01 $(\text{kcal/mol})^{1/2}$. The rates and heats of adsorption of polymers onto oxides were also observed using flow microcalorimetry. Specifically, heats of adsorption for poly(vinyl pyridine) onto α-hematite from benzene solution were measured. Time effects for the adsorption were found to be important. It required 26 min in the latter case for the full heat of adsorption to be expended. In all calorimetric techniques for measuring the heat of adsorption, it is important to work at relatively low surface coverages to assure that the van der Waals interactions of the adsorbing species are only with the dispersing solvent and not with each other on the adsorbent surface.

Heats of adsorption of acids or bases onto solid surfaces may also be obtained using the correlations of such heats with IR or other spectral shifts as described earlier. The test solid, for example, silica, may be prepared as a mull in mineral oil. The IR shift is then obtained by measuring the stretching frequency for a given bond, say, the C=O bond in acetone or PMMA predissolved in a small amount in the mineral oil, before and after addition of the silica. Heats of adsorption obtained in this way have shown good agreement with those obtained calorimetrically. More recently, photoacoustic spectroscopy has been employed [110]. Spectral shifts at extended solid surfaces (e.g., cast polymeric films) have also been obtained using attenuated total reflection (ATR) IR [103]. The van der Waals shift must be adjusted by what appears to be an arbitrary constant amount from that found for the solvent. Fowkes claims [83] that the prospect for the use of solid-state NMR spectral shifts looks promising, but results of studies using this technique have not yet been published.

B. Liquid-Phase Adsorption Techniques

Studies that lead to Langmuirian adsorption isotherms for the desired liquid adsorbing onto the desired solid surface from an inert solvent as a function of temperature can be analyzed to obtain the values of W_a^{AB}, N, f, and $-\Delta H^{AB}$. It should be noted that not all systems yield Langmuirian isotherms, as Huang et al. [65] found in their study of the adsorption of several solutes at iron oxide surfaces. Appropriate inert solvents for such studies might be hydrocarbons, carbon tetrachloride, or diiodomethane. If the solid is in the form of a powder with a relatively large specific area ($>20\ m^2/g$), the adsorption isotherm may be obtained stoichiometrically in the usual way by measuring the supernatant concentration of the adsorbate initially and after adsorption equilibrium has been established. If the specific area Σ of the solid powder is determined independently, as by the BET method, the extent of adsorption Γ_2 of solute 2 is determined as

$$\Gamma_2 = \frac{V}{m\Sigma}(C_2^0 - C_2) \tag{118}$$

where V is the volume of the solution, m the mass of the adsorbent powder, C_2^0 the initial concentration of solute in the solution, and C_2 its concentration after adsorption equilibrium has been established. Such equilibration may require several hours or even days. After several such adsorption values are determined, they may be plotted as C_2/Γ_2 vs. C_2 to yield a straight line (if the format of the Langmuir adsorption isotherm is observed) of the form

$$\frac{C_2}{\Gamma_2} = \frac{1}{\Gamma_m K_{eq}} + \frac{1}{\Gamma_m} C_2 \tag{119}$$

where Γ_m is the close-packed monolayer, or saturation, adsorption, and K_{eq} the adsorption equilibrium constant. Γ_m and K_{eq} are evaluated from the slope and intercept of the Langmuir plot. N may be identified as $1/\Gamma_m$.

If the adsorption data were plotted as Γ_2 vs. C_2, the adsorption equilibrium constant K_{eq} would be recognized as the initial, linear slope of the curve. It corresponds to the partition coefficient between the adsorbed layer and bulk solution at low concentration, corresponding to low adsorbent surface coverage where lateral interaction between adsorbed molecules is negligible. Thus, K_{eq} may be identified with the molar free energy change of adsorption and the work of adhesion for the adsorbate liquid on the solid surface

$$-\Delta G_a = \frac{W_a}{N} = \frac{1}{N}\left(W_a^{AB} + 2\sqrt{\sigma_S^{LW}\sigma_2^{LW}}\right) = RT \ln K_{eq} \tag{120}$$

Equation (120) can be solved to yield the acid–base contribution to the work of adhesion between liquid adsorbate 2 and the solid substrate S.

If the adsorption isotherm is obtained at two or more temperatures, Eqs. (114) and (115) may be applied to obtain $-\Delta H^{AB}$ and f.

If the solid surface of interest is not that of a large specific area powder, but rather a flat surface, other methods must be used to obtain the isotherm. One technique that has been employed is that of ellipsometry [112]. This requires that the surface be smooth enough to be sufficiently reflective. Isotherms for pyridine adsorbing onto silica from octane have been reported [113].

An interesting alternative method for determination of the isotherm under these circumstances makes use of contact-angle and surface tension measurements. If the inert solvent used is one of high surface tension, such as diiodomethane, it and its solutions of the adsorbate will produce finite contact angles against the solid surface of interest. If the Wilhelmy technique is used for measuring the contact angles and surface tensions [114], this method may be applied to fine fibers, as well as flat solid specimens. The analysis makes use of the Gibbs adsorption equation [115] for calculating the adsorption from the concentration dependence of the solid/liquid interfacial free energy

$$\Gamma_2 = -\frac{1}{RT}\frac{d\sigma_{SL}}{d \ln C_2} \tag{121}$$

where C_2 is the concentration of the adsorbing solute. Equation (121) is written in the form assuming that the solution of the solute in the inert solvent is ideal dilute, that is, obeys Henry's law. If this is not the case, the concentration must be replaced with the activity. The solid/liquid interfacial free energy σ_{SL} is then expressed in terms of Young's equation

$$\sigma_{SL} = \sigma_S - \sigma_L \cos\theta \tag{122}$$

σ_S is a constant with respect to solute concentration, so the substitution of Eq. (122) into Eq. (121) gives

$$\Gamma_2 = \frac{1}{RT} \frac{d(\sigma_L \cos \theta)}{d \ln C_2} \tag{123}$$

Thus, measurement of the surface tension and contact angle as a function of solute concentration produces the adsorption isotherm. Fowkes [116] reports measurements of this type for phenol and iodine adsorbing onto PMMA surfaces from solution in diiodomethane. Sets of measurements were made at two different temperatures, so using the analysis outlined above, we obtained values of W_a^{AB}, N, f, and $-\Delta H^{AB}$, and the Drago constants for the PMMA solid surface were determined. Incidentally, the value of f obtained was substantially less than unity.

C. Inverse Gas Chromatography

One of the most powerful techniques for experimentally probing the properties of solid interfaces, including their acid–base properties, is inverse gas chromatography (IGC) [117]. It is conventional gas chromatography with respect to the equipment and techniques required but is called "inverse" because it is the solid stationary phase, rather than the gas phase, whose properties are to be investigated. The stationary phase may be in the form of a powder, fiber mass, or thin coating on the wall of the column. Probe gases of known composition and properties are passed by means of an inert carrier gas through the column, and by monitoring the residence time and shape of the elution curve, inferences may be drawn concerning the properties of the solid adsorbent surface. Probe gases that are known to adsorb only through Lifshitz–van der Waals interactions, such as saturated hydrocarbons, may be passed through the column to determine the LW component of the solid surface free energy. Then acidic and basic probe gases may be used to investigate the acid–base properties of the surface.

Inverse gas chromatography may be operated in two different modes: the infinite dilution regime and the finite concentration regime. The infinite dilution regime corresponds effectively to very low surface coverage of the solid by the adsorbate. In this case, the probe gas concentrations are kept very low so that the resulting adsorption is in the linear, or "Henry's law," region. Under these circumstances, results are not affected by any interactions between adsorbate molecules. The fundamental quantity measured is the relative retention volume V_N. It is defined as the amount of carrier gas required to elute the given amount of adsorbate from the column and is measured as

$$V_N = jF_{\text{col}}(t_R - t_{\text{ref}}) \tag{124}$$

where j is a correction factor to account for the difference in pressure between the inlet and outlet to the column [118], t_R is the retention time of the probe gas, and t_{ref} is the retention time for a nonadsorbing probe gas. For nonspecifically adsorbing probe gases, the elution curves are symmetrical, and the retention time is just the time measured to its respective maximum. When specific adsorption occurs, the elution curves usually show some tailing, and the retention time is computed as

$$t_R = \frac{\int_0^\infty t f(t)\, dt}{\int_0^\infty f(t)\, dt} \tag{125}$$

where $f(t)$ represents the elution curve.

The retention volume is directly related to the adsorption equilibrium constant for the partitioning of the adsorbate between the mobile gas and stationary phase K_{eq}

$$V_N = K_{\text{eq}} A \tag{126}$$

where A is the adsorbent surface area in the column. If we assume ideal gas behavior in the column, K_{eq} is given by

$$K_{\text{eq}} = \frac{\Gamma RT}{p} \tag{127}$$

where Γ is the adsorption and p the partial pressure of the adsorbate gas.

The standard molar free energy change upon adsorption of the adsorbate gas is given by

$$\Delta G^0_{\text{ads}} = -RT \ln K_{\text{eq}} + C_1 = RT \ln V_N + C_2 \tag{128}$$

where C_1 and C_2 are constants dependent on the standard states chosen for the gaseous and adsorbed states and the adsorbent surface area. It is not necessary to know their values.

The free energy of adsorption is related to the work of adhesion for the adsorbate on the solid surface [119]

$$W_a = \frac{-\Delta G^0_{\text{ads}}}{a_{\text{mol}}} \tag{129}$$

where a_{mol} is the molar area of the adsorbate on the surface. For nonspecifically adsorbing species, the work of adhesion is given by Eq. (43). Combining Eqs. (43), (128), and (129), we get

$$RT \ln V_N = 2a_{\text{mol}} \sqrt{\sigma_S^{\text{LW}} \sigma_L^{\text{LW}}} + C_2 \tag{130}$$

where subscript L refers to the adsorbate. A plot of $RT \ln V_N$ against $a_{\text{mol}}\sqrt{\sigma_L^{\text{LW}}}$ for a series of nonspecifically adsorbing species, such as a homologous series of n-alkanes, should yield a straight line whose slope is $2\sqrt{\sigma_S^{\text{LW}}}$. The molar area of the adsorbate may be computed from known molecular dimensions and an assumed adsorbed packing configuration. It may sometimes be estimated from the liquid molar volume of the adsorbate v_{mol} by assuming a spherical molecular shape [119]

$$a_{\text{mol}} = 1.33\ N_{Av}^{1/3} v_{\text{mol}}^{2/3} \tag{131}$$

Figure 13, from the author's laboratory, shows a plot (solid line) of the type suggested by Eq. (130) for a series of n-alkanes on a sample of microcrystalline cellulose.

The acid–base contribution to the work of adhesion of an adsorbate capable of acid–base interaction with the solid is obtained by comparing its retention volume with that obtained with an inert probe with the same value of $a_{\text{mol}}\sqrt{\sigma_L^{\text{LW}}}$. Then

$$W_a^{AB} = \frac{RT}{a_{\text{mol}}} \ln \frac{V_N}{V_N^{\text{alkane}}} \tag{132}$$

Values of $RT \ln (V_N/V_N^{\text{alkane}})$ for a number of acid and base probes on a sample of cellulose powder (Whatman CF1) are shown in Fig. 13. The displacement of the acid and base probes above the alkane line indicates that this cellulose has primarily acidic sites on its surface. If IGC runs of the type described are run for at least two different temperatures, Gibbs–Helmholtz analysis can be applied [120–123] in order to extract values for $-\Delta H^{AB}$ and f. These, in turn, may be used to evaluate effective Drago or Gutmann parameters by way of Eqs. (95) [123] or (98) [124]. In evaluating Drago parameters, it is especially useful to use monofunctional acidic or basic probe gases. Many IGC studies are now in the literature, which include applications to carbon fibers, wood pulp fibers, glass fibers, and many synthetic polymers. Much attention has been given to determining the effect of various surface treatments [123–125].

Gas chromatographic determination of the entire adsorption isotherm for the probe gas may also be made [126], and these data (when available for two or more temperatures) may be processed in the same way as isotherms for adsorption from the liquid phase (described earlier) to yield values for N, f, W_a^{AB}, $-\Delta H^{AB}$, and in addition, the equilibrium film pressure π_e. Wesson and Allred [127] have examined the adsorption of various probe gases on graphite fiber surfaces in which the isotherms computed from chromatograms were analyzed with the CAEDMON (computed adsorption energy distribution in the monolayer) algorithm [128] to produce histograms of surface area fraction vs. adsorptive energy. In this partic-

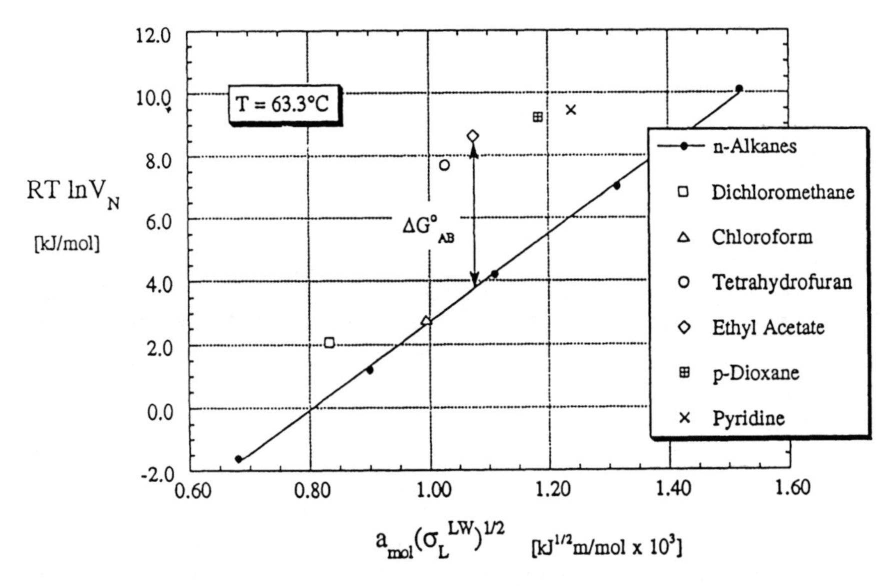

FIG. 13 Inverse gas chromatography data for Whatman CF1 cellulose powder, obtained in author's laboratory.

ular study with Lewis base probe gases, the surface energy distribution was found to be bimodal.

D. Electrokinetic Measurements

The acid–base interaction between liquids and solid surfaces often leads to ionization or charging of the solid surface (as treated in earlier sections), such that it acquires either a positive or negative electrical potential relative to the medium in which it is immersed [129]. The sign of the charged solid surface can be detected by determining the direction of migration of the particles when an electric field is imposed on the system, and the speed of migration per unit electric field strength (electrophoretic mobility) can be related through theory to the zeta potential (an effective measure of the potential near the particle surface relative to the bulk medium). The use of electrokinetic measurements for the determination of the isoelectric point (IEPS) for solid oxide particles was mentioned in the earlier treatment of Bolger–Michaels theory. The IEPS is the pH at which a given solid in contact with water has zero zeta potential and is therefore the point at which the negatively and positively charged sites are equal in number. At pH values below the IEPS of the solid, the solid surface in contact with water is positively charged, whereas above the IEPS it is negatively charged. Acidic surfaces in general in such interactions are proton-donating or electron-accepting, so that they acquire a negative potential when surface charging occurs. Surfaces with basic sites acquire a positive potential. Thus, (acidic) particles of silica in basic solvents such as water (above a pH of 2), ethers, esters, ketones, aromatics, and amines acquire a negative potential and will migrate toward the anode when an electric field is imposed. Conversely, (basic) alumina particles will acquire a positive potential in acidic solvents such as water (below a pH of 9), chlorinated hydrocarbons, phenols, alcohols, and nitro-compounds. Measurement of the zeta potential of solid particles in organic solvents of known acidity and/or basicity can provide important information on the acidity or basicity of the particle surfaces. Labib and Williams [130–132] have shown, for example, that the donicity (donor number) of a solid equals the donicity of that liquid which produces a zero zeta potential. Fowkes et al. [133] have recently characterized the effectiveness of silane coupling agents for modifying the surface acidity or basicity of glass powders by zeta potential measurements in a mixture of methanol and methy isobutyl ketone. The untreated glass had a zeta potential of less than 5 mV. Powders treated with aminosilanes became basic to the extent that the zeta potential became −60 mV, whereas those treated with methacryloxysilane became acidic and registered a zeta potential of +50 mV. An extensive review of the characterization of solid surfaces by electrokinetic measurements has been given recently by Jacobasch [134].

ACKNOWLEDGMENT

This work was supported in part by gifts from the Weyerhaeuser Company and the James River Corporation.

REFERENCES

1. J. J. van Laar, *Z. Phys. Chem. 72*:723 (1910).
2. E. J. Berger, *J. Adhesion Sci. Tech. 4*:373 (1990).
3. M. F. Finlayson, and B. A. Shah, *J. Adhesion Sci. Tech. 4*:431 (1990).
4. H. F. Webster, and J. P. Wightman, *J. Adhesion Sci. Tech. 5*:93 (1991).
5. T. S. Oh, L. P. Buchwalter, and J. Kim, *J. Adhesion Sci. Tech. 4*:303 (1990).
6. A. C. Tiburcio, and J. A. Manson, *J. Adhesion Sci. Tech. 4*:653 (1990).
7. B. J. Briscoe, and D. R. Williams, *J. Adhesion Sci. Tech. 5*:23 (1991).
8. E. Papirer, and H. Balard, *J. Adhesion Sci. Tech. 4*:357 (1990).
9. H. P. Schreiber, and F. St. Germain, *J. Adhesion Sci. Tech. 4*:319 (1990).

10. A. Larsson, and W. E. Johns, *J. Adhesion 25:*121 (1988).
11. D. Ma, W. E. Johns, A. K. Dunker, and A. E. Bayoumi, *J. Adhesion Sci. Tech. 4:*411 (1990).
12. P. M. Costanzo, R. F. Giese, and C. J. van Oss, *J. Adhesion Sci. Tech. 4:*267 (1990).
13. R. E. Johnson, Jr., and R. H. Dettre, in *Surface and Colloid Science*, Vol. 2 (E. Matijevic, ed.), Wiley-Interscience, New York, 1969.
14. R. J. Good, N. R. Srivatsa, I. Islam, H. T. L. Huang, and C. J. van Oss, *J. Adhesion Sci. Tech. 4:*607 (1990).
15. T. Young, *Phil. Trans. 95:*65, 82 (1805).
16. A. Sheludko, *Colloid Chemistry*, Elsevier, Amsterdam, 1966, pp. 90–92.
17. A. W. Neumann, and R. J. Good, in *Surface and Colloid Science*, Vol. 11 (R. J. Good and R. R. Stromberg, eds.), Plenum Press, New York, 1979.
18. R. E. Johnson, Jr., and R. H. Dettre, *J. Coll. Inter. Sci. 21:*610 (1966).
19. R. J. Good, *Nature 212:*276 (1966).
20. F. M. Fowkes, D. C. McCarthy, and M. A. Mostafa, *J. Coll. Inter. Sci. 78:*200 (1980).
21. R. J. Good, *J. Coll. Inter. Sci. 52:*308 (1975).
22. J. Schröder, *Farbe u. Lack. 86:*19 (1980).
23. H. J. Busscher, K. G. A. Kim, G. A. M. van Silfout, and J. Arends, *J. Coll. Inter. Sci. 114:*307 (1987).
24. M.-N. Bellon-Fontaine, and O. Cerf, *J. Adhesion Sci. Tech. 4:*475 (1990).
25. F. London, *Trans. Far. Soc. 33:*8 (1937).
26. W. H. Keesom, *Phys. Z. 22:*126 (1921); *23:*225 (1922).
27. P. Debye, *Phys. Z. 21:*178 (1920); *22:*302 (1921).
28. A. D. McLachlan, *Proc. Roy. Soc. London, Ser. A. 271:*387 (1963); *274:* 80 (1963).
29. A. D. McLachlan, *Mol. Phys. 6:*423 (1963).
30. J. N. Israelachvili, *Intermolecular and Surface Forces*, Academic Press, New York, 1985.
31. H. C. Hamaker, *Physica 4:*1058 (1937).
32. R. J. Good, and M. C. Phillips, in *Proceedings of Conference on Wetting*, Monograph 25, Vol. 44, 1966.
33. M. C. Phillips, R. J. Good, D. A. Cadenhead, and H. F. King, *J. Coll. Inter. Sci. 37:*437 (1971).
34. R. J. Good, and M. K. Chaudhury, in *Fundamentals of Adhesion* (L.-H. Lee, ed.), Plenum Press, New York, 1991, Chap. 3.
35. E. M. Lifshitz, *Sov. Phys. JETP 2:*73 (1956).
36. J. N. Israelachvili, *Q. Rev. Biophys. 6:*341 (1974).
37. L. A. Girifalco, and R. J. Good, *J. Phys. Chem. 61:*904 (1957).
38. F. M. Fowkes, *J. Phys. Chem. 66:*382 (1962).
39. F. M. Fowkes, *Ind. Eng. Chem. 56:*40 (1964).
40. F. M. Fowkes, F. L. Riddle, Jr., W. E. Pastore, and A. A. Weber, *Coll. Surf. 43:*367 (1990).
41. S. Wu, *Polymer Interface and Adhesion*, Marcel Dekker, New York, 1982.

42. D. K. Owens, and R. C. Wendt, *J. Appl. Poly. Sci. 13:*1741 (1969).
43. D. H. Kaelble, *J. Adhesion 2:*66 (1970).
44. J. R. Dann, *J. Coll. Inter. Sci. 32:*302 (1970).
45. J. R. Dann, *J. Coll. Inter. Sci. 32:*321 (1970).
46. N. A. de Bruyne, *The Aircraft Engineer 18:*51 (1939).
47. F. M. Fowkes, *J. Adhesion Sci. Tech. 1:*7 (1987).
48. C. J. van Oss, R. J. Good, and M. K. Chaudhury, *J. Coll. Inter. Sci. 111:* 378 (1986).
49. R. S. Farinato, S. S. Kaminski, and J. L. Courter, *J. Adhesion Sci. Tech. 4:*633 (1990).
50. R. S. Drago, and N. A. Matwiyoff, *Acids and Bases*, D. C. Heath & Co., Lexington, MA, 1968.
51. H. L. Finston, and A. C. Rychtman, *A New View of Current Acid–Base Theories*, Wiley, New York, 1982.
52. V. Gutmann, *The Donor-Acceptor Approach to Molecular Interactions*, Plenum Press, New York, 1978.
53. T.-L. Ho, *Hard and Soft Acids and Bases Principle in Organic Chemistry*, Academic Press, New York, 1977.
54. W. B. Jensen, *The Lewis Acid–Base Concepts: An Overview*, Wiley, New York, 1980.
55. R. G. Pearson, *Hard and Soft Acids and Bases*, Dowden, Hutchinson and Ross. Stroudsburg, PA, 1973.
56. F. M. Fowkes, in *Physicochemical Aspects of Polymer Surfaces* (K. L. Mittal, ed.), Plenum Press, New York, 1983.
57. J. Schultz, K. Tsutsumi, and J.-B. Donnet, *J. Coll. Inter. Sci. 59:*272 (1977).
58. J. Schultz, K. Tsutsumi, and J.-B. Donnet, *J. Coll. Inter. Sci. 59:*277 (1977).
59. S. Arrhenius, *Z. Phys. Chem. 1:*631 (1887).
60. J. N. Brønsted, *Rec. Trav. Chim. Pays-Bas 42:*718 (1923).
61. T. M. Lowry, *Chem. Ind. (London) 42:*43 (1923).
62. J. C. Bolger, and A. S. Michaels, in *Interface Conversion* (Weiss and Cheevers, eds.), Elsevier, New York, 1969.
63. G. A. Parks, *Chem. Rev. 65:*177 (1965).
64. J. F. Douglas, *Macromol. 22:*3707 (1989).
65. Y. C. Huang, F. M. Fowkes, and T. B. Lloyd, *J. Adhesion Sci. Tech. 5:* 39 (1991).
66. G. M. Whitesides, H. A. Biebuyck, J. P. Folkers, and K. L. Prime, *J. Adhesion Sci. Tech. 5:*57 (1991).
67. J. R. Rasmussen, D. E. Bergbreiter, and G. M. Whitesides, *J. Am. Chem. Soc. 99:*4746 (1977).
68. J. R. Rasmussen, E. R. Sterdronsky, and G. M. Whitesides, *J. Am. Chem. Soc. 99:*4736 (1977).
69. S. R. Holmes-Farley, R. H. Reamey, T. J. McCarthy, J. Deutch, and G. M. Whitesides, *Langmuir 1:*725 (1985).
70. S. R. Holmes-Farley, and G. M. Whitesides, *Langmuir 2:*266 (1986).
71. S. R. Holmes-Farley, and G. M. Whitesides, *3:*62 (1987).

72. S. R. Holmes-Farley, C. D. Bain, and G. M. Whitesides, *Langmuir 4:*921 (1988).
73. G. M. Whitesides, and P. E. Labinis, *Langmuir 6:*87 (1990).
74. C. D. Bain, and G. M. Whitesides, *Langmuir 5:*1370 (1989).
75. K. J. Hüttinger, S. Höhmann-Wien, and G. Krekel, *J. Adhesion Sci. Tech. 6:*317 (1992).
76. G. N. Lewis, *Valence and the Structure of Atoms and Molecules*, The Chemical Catalog Co., New York, 1923.
77. W. B. Jensen, *J. Adhesion Sci. Tech. 5:*1 (1991).
78. S. R. Cain, *J. Adhesion Sci. Tech. 4:*333 (1990).
79. L.-H. Lee, *J. Adhesion Sci. Tech. 5:*71 (1991).
80. F. M. Fowkes, *Mater. Res. Symp. 40:*239 (1985).
81. M. E. Labib, *Coll. Surf. 29:*293 (1988).
82. C. J. van Oss, R. J. Good, and M. K. Chaudhury, *Langmuir 4:*884 (1988).
83. F. M. Fowkes, *J. Adhesion Sci. Tech. 4:*669 (1990).
84. R. G. Pearson, *J. Am. Chem. Soc. 85:*3533 (1963).
85. R. S. Drago, and B. B. Wayland, *J. Am. Chem. Soc. 87:*3571 (1965).
86. V. Gutmann, A. Steininger, and E. Wychera, *Monatsh. Chem. 97:*460 (1966).
87. R. F. Hudson, and G. Klopman, *Tetrahedron Lett. 12:*1103 (1967).
88. R. S. Mulliken, *J. Phys. Chem. 56:*801 (1952).
89. J. O. Edwards, *J. Am. Chem. Soc. 76:*1540 (1954).
90. R. S. Drago, N. Wong, C. Bilgrien, and G. C. Vogel, *Inorg. Chem. 26:*9 (1987).
91. R. S. Drago, and D. R. McMillan, *Inorg. Chem. 11:*872 (1972).
92. M. K. Kroeger, and R. S. Drago, *J. Am. Chem. Soc. 103:*3250 (1981).
93. Y. Marcus, *J. Solution Chem. 13:*599 (1984).
94. V. Gutmann, *Chem. Tech. (Leipzig) 7:*255 (1977).
95. U. Mayer, V. Gutmann, and W. Gerger, *Monatsh. Chem. 106:*1235 (1975).
96. F. L. Riddle, Jr., and F. M. Fowkes, *J. Am. Chem. Soc. 112:*3259 (1990).
97. V. Gutmann, *Angew. Chem. Int. Ed. Engl. 9:*843 (1971).
98. R. S. Drago, *Coord. Chem. Rev. 33:*251 (1980).
99. R. S. Drago, N. Wong, and G. C. Vogel, *Inorg. Chem. 26:*9 (1987).
100. P. E. Doan, and R. S. Drago, *J. Am. Chem. Soc. 106:*2772 (1984).
101. T. F. Bolles, and R. S. Drago, *J. Am. Chem. Soc. 88:*5370 (1966).
102. M. D. Joeston, and R. S. Drago, *J. Am. Chem. Soc. 84:*3817 (1962).
103. F. M. Fowkes, D. O. Tischler, J. A. Wolfe, L. A. Lannigan, C. M. Ademu-John, and M. J. Halliwell, *J. Polym. Sci. 22:*547 (1984).
104. T. K. Kwei, E. M. Pearce, F. Ren, and J. P. Chen, *Sci. Poly. Phys. Ed. 24:*1597 (1986).
105. M. D. Vrbanac, and J. C. Berg, *J. Adhesion Sci. Tech. 4:*255 (1990).
106. C. J. van Oss, M. K. Chaudhury, and R. J. Good, *Adv. Coll. Inter. Sci. 28:*35 (1987).
107. F. M. Fowkes, and M. A. Mostafa, *Ind. Eng. Chem. Prod. Res. Dev. 17:* 3 (1978).

108. P. A. Small, *J. Appl. Chem. 3:*71 (1953).
109. R. J. Good, M. K. Chaudhury, and C. J. van Oss, in *Fundamentals of Adhesion* (L.-H. Lee, ed.), Plenum Press, New York, 1991, Chap. 4.
110. F. M. Fowkes, in *Surface and Colloid Science in Computer Technology* (K. L. Mittal, ed.), Plenum Press, New York, 1987.
111. S. T. Joslin, and F. M. Fowkes, *Ind. Eng. Chem. Prod. Res. Dev. 24:*369 (1985).
112. K. Bäckström, B. Lindman, and S. Engström, *Langmuir 4:*872 (1988).
113. L. A. Casper, Ph. D. Thesis, Lehigh University, 1985.
114. J. C. Berg, in *Composite Systems from Natural and Synthetic Polymers* (L. Salmén, A. De Ruvo, J. C. Seferis, and E. Stark, eds.), Elsevier, Amsterdam, 1986.
115. F. M. Fowkes, and W. D. Harkins, *J. Am. Chem. Soc. 62:*3377 (1940).
116. F. M. Fowkes, M. B. Kaczinzinski, and P. M. Kelly, *Langmuir* (1990) (submitted for publication).
117. D. R. Lloyd, T. C. Ward, and H. P. Schreiber, eds., *Inverse Gas Chromatography*, ACS Symposium Series 391, Washington, DC, 1989.
118. J. R. Conder, and C. L. Young, *Physicochemical Measurements by Gas Chromatography*, Wiley, New York, 1979.
119. G. M. Dorris, and D. G. Gray, *J. Coll. Inter. Sci. 71:*93 (1979).
120. G. M. Dorris, and D. K. Gray, *J. Coll. Inter. Sci. 77:*353 (1980).
121. S. Katz, and D. G. Gray, *J. Coll. Inter. Sci. 82:*318 (1981).
122. C. S. Flour, and E. Papirer, *J. Coll. Inter. Sci. 91:*69 (1983).
123. H. L. Lee, and P. Luner, *Nord. Pulp Pap. J. 2:*164 (1989).
124. A. E. Bolvari, and T. C. Ward, in *Inverse Gas Chromatography* (D. R. Lloyd, T. C. Ward, and H. P. Schreiber, eds.), ACS Symposium Series 391, Washington, DC, 1989.
125. L. Lavielle, and J. Schultz, *Langmuir 7:*978 (1991).
126. S. Katz, and D. G. Gray, *J. Coll. Inter. Sci. 82:*326 (1981).
127. S. P. Wesson, and R. E. Allred, *J. Adhesion Sci. Tech. 4:*277 (1990).
128. S. Ross, and I. D. Morrison, *Surface Sci. 52:*103 (1975).
129. F. M. Fowkes, F. W. Anderson, R. J. Moore, H. Jinnai, and M. A. Mostafa, in *Colloids and Surfaces in Reprographic Technology* (M. L. Hair and M. D. Croucher, eds.), ACS Symposium Series 200, Washington, DC, 1982.
130. M. E. Labib, and R. Williams, *J. Coll. Inter. Sci. 97:*356 (1984).
131. M. E. Labib, and R. Williams, *Coll. Polym. Sci. 264:*533 (1986).
132. M. E. Labib, and R. Williams, *J. Coll. Inter. Sci. 115:*330 (1987).
133. F. M. Fowkes, D. W. Dwight, J. A. Manson, T. B. Lloyd, D. O. Tischler, and B. A. Shah, *Mater. Res. Soc. Symp. Proc. 119:*223 (1988).
134. H.-J. Jacobasch, *Prog. Org. Coatings 17:*115 (1989).

3
Wetting by Solutions

ARIJIT BOSE Department of Chemical Engineering, University of Rhode Island, Kingston, Rhode Island

I. INTRODUCTION

The wetting of solid substrates by liquids is a basic element in many natural and commercial processes, and its understanding has been the subject of intensive investigation over the past century. Some common examples include the spreading of liquid droplets on solids such as in the spraying of paint and agricultural chemicals, the penetration of ink in paper, the liquid absorbency or repellency of fabrics, imbibition of fibers in absorbent media, and the displacement of one fluid (oil) by another (water or a complex solution) over a solid in enhanced oil recovery. From a fundamental perspective, a study of wetting behavior is challenging because contact angles are the macroscopically observable consequence of interactions at a molecular level. Although these interactions are generally complex and very system-specific, some basic predictions of static contact angles can be made from a rudimentary knowledge of these intermolecular interactions. For example, using simple arguments, Israelachvili [1] shows that the static contact angle θ formed by material B or C on material A is given by $\cos(\theta) = (B + C - 2A)/(B - C)$, where A, B, and C represent the molecular property of A, B, and C to which the interaction energy between molecules is proportional. Using a Lennard–Jones type interaction potential along with molecular dynamic simulations, Koplik et al. [2] have obtained static contact angles that verify this expression. Conversely, contact angles serve as a very sensitive probe of the "state" of interfaces, since small changes can result in dramatic variations in the local meniscus shapes close to the contact line [3].

Wetting behavior is commonly characterized by the value of the contact angle θ within the liquid, defined as $\theta = \cos^{-1}(\mathbf{n}_l \cdot \mathbf{n}_s)$, where $\mathbf{n}_l$ and $\mathbf{n}_s$ represent the outward pointing unit normals from the liquid and the solid, respectively, at the contact line (Fig. 1). When the speed of the contact line in the direction of its outward normal and along the substrate surface is zero in the frame of reference of the solid, this angle is referred to as a *static* contact angle. For most practical materials, the static contact angle is not unique, but falls within a range defined as contact-angle hysteresis. When the contact line speed is positive (negative), the corresponding angle is called the *dynamic advancing* (*receding*) contact angle. Dynamic contact angles, in general, vary with contact line speed, asymptotically reaching an upper bound representing entrainment [4] and a lower bound representing complete wetting. An understanding of wetting when the contact line is in motion requires delineation of both the physical

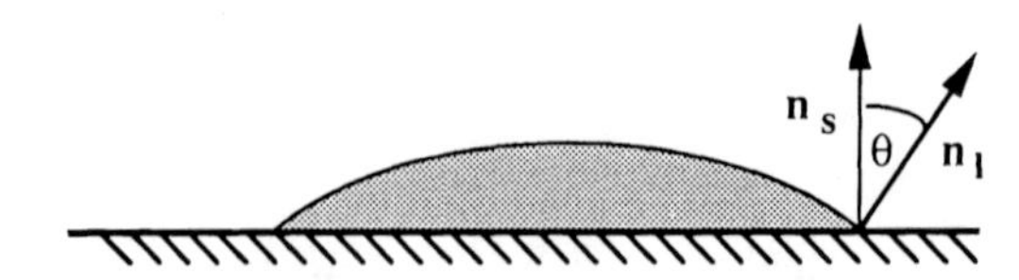

FIG. 1 Contact angle θ in the liquid phase can be defined rigorously as $\cos^{-1}(\mathbf{n}_L \cdot \mathbf{n}_S)$, where $\mathbf{n}_L$ and $\mathbf{n}_S$ represent the unit outward normals from the liquid and solid phases, respectively, at the contact line. Experimentally, this angle is often determined by analyzing the projected image shown in this figure.

chemistry, as well as hydrodynamics in the proximity of the contact line, making it an exciting interdisciplinary challenge. From a transport modeling perspective, contact angles serve as boundary conditions to the equations that govern meniscus shapes, to the well-known Young–Laplace equation for static systems and as an essential condition in fluid flow problems with free surfaces and contact lines. A full understanding of wetting behavior therefore necessarily includes specification of the appropriate contact angles.

The presence of at least one additional component in a liquid phase potentially makes wetting by solutions quite different from that by a pure liquid. For example, static contact angles are related to the free energies of the liquid/vapor, liquid/solid, and solid/vapor interfaces. Additional components in the solution can directly impact these free energies and have concomitant effects on the resulting contact angle. Contact-angle hysteresis is sometimes affected by the presence of solutes within the liquid phase because their adsorption onto the underlying solids changes the nature of the substrate. In wetting with moving contact lines, several dynamic interactions caused specifically by the presence of additional components in the liquid can impact wettability; these include diffusion and convective transport of solute and their rates of adsorption or desorption at the two-phase boundaries.

A. Static Contact Angles

Pioneering work on the understanding of wettability was initiated by Young [5], where the *static* contact angle θ was related to the free energies of the liquid/vapor (γ_{LV}), solid/liquid (γ_{SL}), and solid/vapor (γ_{SV}) interfaces through Young's equation

$$\cos(\theta) = \frac{\gamma_{SV} - \gamma_{SL}}{\gamma_{LV}} \tag{1}$$

This equation can be derived using the principle of energy minimization (see de Gennes [6]), as well as a force balance along the solid surface at the contact line, although the latter approach leaves the forces normal to the solid unbalanced. As pointed out by de Gennes [6], the quantities appearing in Young's equation are all "far-field" free energies, that is, interfacial free energies "sufficiently" far away from a core region near the contact line (this region has molecular dimensions). Therefore, the precise details of the interactions within the core region need not be known.

For both pure and multicomponent liquids, *static* contact-angle data are often interpreted by invoking Young's equation, that is, an effort is made to understand the role played by each of these components, or external influences, on the three interfacial free energies. If those are understood, then the resulting effect on contact angles can be predicted easily. However, several practical difficulties may arise when using this approach. First, although liquid/vapor interfacial free energies are amenable to *direct* measurement, effects on γ_{SL} and γ_{SV} cannot be independently obtained with the same ease (these quantities can possibly be measured using a surface force apparatus (see Israelachvili [1]), or obtained from experimental measurements of contact angles and liquid surface tensions by postulating an additional "equation-of-state" relationship between these energies). Second, for rough or chemically heterogeneous solid surfaces, static contact angles may be not be unique [7]. In multicomponent liquids, differential adsorption onto the substrate can produce chemical heterogeneity, so that Young's equation must be used with care. Third, because this equation is the result of an energy minimization, it is only applicable when systems are at equilibrium; this may be particularly difficult to guarantee for systems that have long relaxation times. When interpreting static contact angles for solutions using Young's equation, these shortcomings must be kept in mind.

B. Contact-Angle Hysteresis

If the underlying solid surface is rough or has impurities, the experimentally observed static contact angle is not unique, but lies within an interval $[\theta_A, \theta_R]$, where θ_A and θ_R are the static advanced and receded contact angles, respectively. Using a series of concentric circular grooves as a model system, Johnson and Dettre [7] show that liquid drops of a given volume can achieve several "metastable" configurations such that the microscopic contact angle is the same, whereas the macroscopically observed contact angles vary. Although there is a global minimum free energy corresponding to a specific macroscopic contact angle, the drop could get trapped in local free energy minima that had different observable contact angles, thus producing hysteresis. de Gennes [6] modeled contact-angle hysteresis by randomly distributing defects (these could be caused either by surface roughness or chemical heterogeneities) on a solid surface and balancing the force exerted by these defects on the contact line with an elastic restoring force proportional to the interfacial tension. For smooth defects, a threshold roughness or chemical heterogeneity has to be crossed before two stable positions of the contact line are possible. This idea is interesting, because it implies that it is not necessary to have a perfectly smooth and homogeneous surface to remove hysteresis. For defects with sharp edges or "mesa" defects, however, *any* heterogeneity or roughness produced more than one stable position for the contact line, implying hysteresis. Knowledge of contact-angle hysteresis is important, not only because it means that even for single-component liquids, a unique static contact angle may be difficult to obtain experimentally, but also because it determines the pressure required to dislodge a drop in a capillary, or the limits to the capillary driving force in wicking, or the maximum substrate tilt angle before a drop slides.

The bounds of contact-angle hysteresis represent the border between static and dynamic states. To ensure the absence of any dynamic artifacts in an experimental determination of the advanced and receded contact angles is a challenging problem. This is especially true in multicomponent liquids, since relaxation times in these materials can often be quite large (of the order of minutes in some systems).

C. Dynamic Contact Angles

In many engineering and other practical situations, the contact line is in motion. The dynamic contact angle then becomes the relevant measure of wettability. It can be obtained experimentally using an indirect technique where a capillary force is measured accurately and the contact angle calculated by knowing the liquid surface tension γ_{LV} and the wetted perimeter, or by direct visualization of the moving meniscus. The latter approach becomes vital if there are uncertainties associated with the value of γ_{LV} at the contact line or the viscous drag on the wetted area of the solid substrate cannot be neglected.

Modeling of fluid flow with dynamic contact lines also poses some challenges. As pointed out by Dussan V. [8], if the liquid is considered Newtonian and the normal continuum "no-slip" boundary condition is applied at the solid surface, the velocity field at the contact line becomes multivalued. A nonintegrable stress is produced at the contact line, leading to an infinite force at that location. This, of course, does not conform to our everyday experience, where, for example, a glass does *not* shatter when being filled with water! This particular difficulty is very serious, since it does not permit the specification of a well-posed boundary-value problem. The aphysicality can be removed and the force made finite by making some ad hoc assumptions: The most common one is to assume that slip occurs in an "inner region" in the immediate vicinity of the contact line. In the outer region, the normal no-slip condition holds. The respective solutions are then matched over the overlap region. One way to do this is to introduce an additional length scale into the problem description, designated as the slip coefficient β, using Navier's condition at the solid surface

$$\mathbf{tn}:\tau = \beta^{-1}\mathbf{t}\cdot(\mathbf{U}_{cl} - \mathbf{U}_s) \tag{2}$$

where τ is the stress tensor, **t** and **n** are the unit tangent and unit normal to the solid surface at the contact line, and $\mathbf{U}_{cl}$ and $\mathbf{U}_s$ represent the liquid velocity at the contact line and the solid velocities, respectively. For $\beta = 0$, no slip holds, whereas $\beta = \infty$ corresponds to perfect slip. Dussan V. [8] has shown that the details of the slip behavior, modeled by varying the rate at which the contact line speed approaches the speed of the solid, has no impact on the flow structure at distances from the contact line that are large compared to the slip length. The only way in which flow in the outer region is affected by the particulars of the flow in the inner region is through the interface shape. Therefore, the status of the slip length as a true material property can only be confirmed by probing meniscus shapes within the inner region. Since accurate free surfaces profiles have to be obtained within 1000 Å of the *actual* contact line, reliable experimental measurements for most currently available geometries in this regime are extremely difficult. An alternate method for removing this singularity is to assume that liquids are shear thinning when the strain exceeds a certain critical value. Although polymeric solutions have been known to display this sort of behavior, this avenue has not been explored analytically.

Under certain circumstances, just the presence of an additional component in the liquid can force the existence of a multivalued velocity field at the contact line; this singularity is not relieved by using Navier's condition. This issue is discussed in Sec. V.

Recently, exciting theoretical work on the simulation of the wetting process using *molecular dynamics* has been initiated [2,9]. The multivalued velocity field at the contact line still persists in their simulation of the displacement of two immiscible fluids in a tube, showing that the singularity was not simply an artifact of making a continuum assumption. Their simulation of a drop spreading on a substrate [9] shows the layered structure observed experimentally by Heslot et al. [10] in their experiments on the spreading of polydimethylsiloxane on silicon wafers. From the perspective of pure liquids, although much still remains uncertain, there appears to be at least a "comfortable" amount of both experiments and analytical work.

D. Experimental Reports of Contact Angles

An important caveat about experimentally reported values of contact angles is illustrated in Fig. 2. If this angle is obtained directly and not from a capillary force measurement, it is usually obtained from a shadow projection of the meniscus using a low-power microscope. Therefore, the number reported depends quite critically on the scale at which the mea-

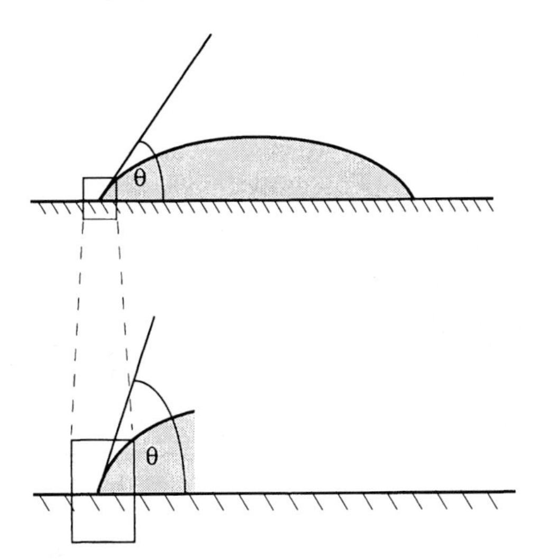

FIG. 2 Experimental value of the contact angle can depend on the image magnification. Large magnification is required if the meniscus is very curved near the contact line.

surement is made or the magnification; obviously, this effect is exaggerated further if the region near the contact line is highly curved. Recent advances in optics, as well as the availability of inexpensive image analysis units, have removed some of the operator bias inherent in direct measurements of contact angles using goniometers. Nevertheless, for most experiments reported in the literature to date, what is being reported is an *apparent* macroscopic angle obtained by extrapolating a meniscus profile to an apparent contact line. Some variations in reported values of contact angles can be attributed simply to different magnifications at which the measurements were made.

E. Contact Angles in Solutions

Several studies have focused on understanding the wetting behavior of *pure* liquids on solid surfaces, and these been summarized in two excellent reviews [6,8]. However, only recently has attention been focused on the problem of how *solutions* wet solids. Because wetting is an *interfacial* phenomenon, attention in this chapter will be focused on how additional components in the solution vary the relevant *interfacial* properties, the most important one, of course, being the free energies of the three interfaces. This consideration drives the discussion naturally toward the effect of *surface-active agents* on wetting, since it is the presence of these constituents that primarily affect surface free energies. These studies are important because natural or commercially used liquids rarely consist of a single component, and if the contaminant is a surfactant, often the largest "effects" may occur at the lowest "impurity" concentrations, since the surface tension of a liquid is most sensitive to concentration in this regime. Some practical examples where specifically understanding the wetting behavior of a *multicomponent* liquid may be important are the recovery of oil trapped in sandstone where surfactant- and/or viscosity-graded aqueous polymer solutions are employed to displace trapped oil ganglia, coating of photosensitive emulsions on polymer substrates for the production of photographic film, deposition of printing inks on paper, and application of thin layers of lubricant on textile fibers. In each of these processes, the liquid is "complex" and its motion must not only be compatible with the conservation of linear momentum, but also must satisfy mass conservation for each of the species. In dynamic systems, different species may be transported at different rates to the three-phase region. Within the simplest context of a *binary* solution containing a solvent A and solute B, the preferential segregation of B at any of the three two-phase interfaces can strongly affect the far-field interfacial free energies with concomitant effects on the contact angle, in most cases producing angles that are considerably different from those of the pure solvent, or with a few exceptions, even another pure liquid with the same surface tension as the solute/solvent system. As we shall show later in this chapter, this is typically the case when the solute is surface-active and has the ability to modify the solid/liquid or solid/vapor interfacial free energy, in addition to the more well-known effect on the liquid surface tension.

If the contact line becomes stationary after being in motion, the contact angle may display transient behavior as the system relaxes to its equilibrium state. The time scales associated with these transients may correspond to diffusion of the solute over characteristic lengths (this length scale is typically associated with the physicochemical properties of the solute and not to the experimental geometry), adsorption or desorption barriers at the various interfaces, or to the rearrangement of adsorbed molecules at the three-phase line. If the contact line is in either spontaneous or forced motion, fluid flow provides an additional mechanism by which concentrations of the solute can become nonuniform. If the solute is surface-active, variations in surface concentration produced by the flow result in surface tension gradients that must be balanced by shear stresses in the underlying fluid. Furthermore, the presence of surfactants can modify the rheology of the interfacial region, producing additional surface forces. These forces can either enhance or retard the motion of the contact line and in some cases render them unsteady. It is apparent from this discussion that a very rich variety of behaviors can then be expected when solutions wet solids.

The purpose of this chapter is to review some of the important concepts involved during the wetting of solids by solutions; attention is primarily given to surfactant solutions because they have the most direct impact on interfacial properties. Solution wetting in the context of *static* contact angles is treated in Sec. II. The effect of having a second component in the solution on contact-angle hysteresis is discussed in Sec. III. Finally, the dynamic wetting of solids by multicomponent solutions is discussed in Sec. IV.

II. STATIC CONTACT ANGLES

Pioneering work on the variation of static contact angles with solute concentration was completed by Zisman and co-workers [11]. Their interpretation was based on the concept of the critical surface tension of a solid γ_c, defined as the surface tension of a liquid that would just wet out the solid. The key issue they dealt with was whether the critical surface tension of a solid was an appropriate indicator of whether a *surfactant* solution would spread spontaneously on the solid. For a variety of aqueous solutions of different concentrations of both anionic and cationic soluble surface-active agents on polyethylene and polytetrafluoroethylene surfaces (both are considered surfaces of low energy), they observed that the critical surface tension for the solid was *nearly* independent of the particular surfactant used, indicating that aqueous surfactant solutions with surface tension below γ_c would spread spontaneously. This conclusion is very important, because it implies that the point of complete wetting of a solid is truly independent of the liquid properties. The observed variation in the critical surface tension was attributed to the difference in the nature and packing of the hydrocarbon groups of the surfactant at the solution/air and solution/solid interfaces. Careful observation of the variation of the contact angle with surface tension for the surfactant solutions reveals only a small spread in regions of incomplete wetting. The implication is that free energy changes at the solid/liquid and solid/vapor interfaces upon exposure to the surfactant solution are negligible, so that for all practical purposes, static contact angles for surfactant solutions on solids can be predicted if *only* the solution surface tension is known. This, as we shall see later, is partly caused by the fact that the surfactants studied were "inert" to the substrate and nonvolatile, so that specific adsorption at those two interfaces was not significant.

The other important observation was the presence of a break in the slope of the curve of cos(θ) vs. solution surface tension at locations corresponding closely to the critical micelle concentration of the surfactant. Such breaks were not observed for surfactants that were incapable of micelle formation. The location of the discontinuity in the slope was independent of the solid, implying that *surfactant* properties were controlling this phenomenon. Since solution surface tension varies only marginally beyond the critical micelle concentration, the rapid increase in the slopes of these plots implies that the presence of micellar aggregates in the solution must modify the energetics associated with adsorption at the solid/liquid interface. At low surfactant concentrations, the bulk solution consists mainly of free monomers that have little specific attraction for the low-energy polyethylene and polytetrafluoroethylene surfaces. At higher concentrations, the bulk solution consists mainly of micellar aggregates, and their interaction with the solid substrate must be substantially different, as illustrated schematically in Fig. 3.

McGuiggan and Pashley [12] measured contact angles of aqueous solutions of dihexadecyl dimethyl ammonium acetate on cleaved mica surfaces in order to eliminate effects of roughness and surface heterogeneities and thus contact-angle hysteresis. The mica surfaces were previously equilibrated with this bitailed surfactant, providing it with a coated surface with the organic tails exposed. At every concentration, three separate measurements are reported: the initial advancing contact angle, obtained immediately after the meniscus had advanced to a fresh section of the surface; the initial receding contact angle; and the equilibrated receding contact angle. The transients (differences between the initial and equilibrated receding angles) were related to the formation of the bilayer at the high concentration of the surfactants. These angles displayed markedly different characteristics as the concentration of the surfactant was varied. The initial advancing contact angle increased then reached a plateau, whereas the initial receding angle went through a maximum, as the concentration of surfactant was increased. These increases are associated with adsorption of the cationic surfactant to the solid substrate. For dilute solutions, the surfactant orientation at the solid/vapor and solid/liquid interfaces is different, but because of the very slow adsorption and reorientation of these surfactants, the contact line can advance without any alteration in the adsorbed state of the molecules. Both the solid/vapor and solid/liquid interfacial free energies are "high," producing large advancing contact angles. By contrast, the receding process exposes high-energy groups (the surfactant head groups) to the air, producing low-equilibrium receding angles that remained invariant with concentration. These surfactants display particularly sluggish desorption characteristics,

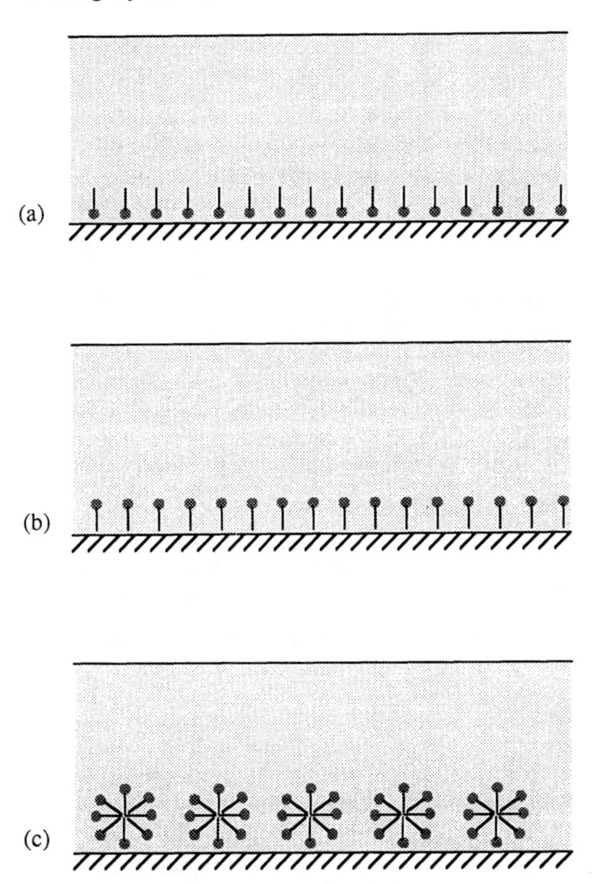

FIG. 3 Schematic representation of differences in adsorption at the solid/liquid interface when the bulk solution contains monomers and micelles: (a) ampiphilic adsorption, (b) ampipathic adsorption, (c) adsorption of micelles or aggregates.

presumably because even at low concentrations they form aggregates in the solution. The dynamics associated with short-chain cationic surfactants such as CTAB are much faster, and such long-time transients are absent in experiments conducted with this surfactant.

Pyter et al. [13] measured contact angles of hydrocarbon (AOT) and fluorocarbon (perfluorooctanoic acid, PFO) surfactant solutions on the low-energy semipolar solid polymethylmethacralate and the nonpolar solid paraffin. For the hydrocarbon surfactants, adsorption at the solid/liquid interface was much more significant for the PMMA than the nonpolar paraffin. On the other hand, the fluorocarbon surfactant solution was a much poorer wetting agent than pure liquids of the same surface tension. These results have been rationalized by invoking differences in the adsorption of surfactants at the solid/liquid interface, producing large differences in the change to the solid/liquid interfacial free energy. This quantity was calculated by combining an equation of state using the harmonic mean equation developed by Wu [14]

$$\gamma_{SL} = \gamma_{LV} + \gamma_{SV} - \frac{4\gamma_{LV}^d \gamma_{SV}^d}{\gamma_{LV}^d + \gamma_{SV}^d} - \frac{4\gamma_{LV}^p \gamma_{SV}^p}{\gamma_{LV}^p + \gamma_{SV}^p} \quad (3)$$

with Young's equation, knowledge of the dispersion and polar contributions to the free energies γ_{ij}^d and γ_{ij}^p, respectively, and the experimentally observed contact angles of surfactant solutions. AOT reduced the solid/liquid interfacial free energy of PMMA only slightly, whereas there was a much greater lowering on paraffin surfaces.

Gau and Zografi [15] have measured advancing contact angles of two nonionic surfactants $C_{12}E_5$ and $C_{10}E_5$ on surfaces of polystyrene and polymethylmethacralate and attempted to relate it to adsorption at the solid/liquid interface. Combining Young's equation and Gibbs absorption isotherm, we get

$$\frac{\partial(\gamma_{LV} \cos \theta)}{\partial \gamma_{LV}} = \frac{\Gamma_{SV} - \Gamma_{SL}}{\Gamma_{LV}} \quad (4)$$

Here Γ_{ij} represents the surface excess concentration of the surfactant at each of the *ij* interfaces. If the surface-active agent does not precoat the solid and it is nonvolatile, $\Gamma_{SV} = 0$, so that a plot of the adhesion tension $\gamma_{LV} \cos \theta$ vs. γ_{LV} should have a slope equal to Γ_{SL}/Γ_{LV}. If such a plot shows a constant slope of -1, $\Gamma_{SL} = \Gamma_{LV}$ for all bulk concentrations, and the surfactant solution will wet just like a pure liquid of the same surface tension as the solution. Deviations from a slope of -1 give clues about specific absorption effects at each of the interfaces. Gau and Zografi [15] obtained Γ_{SL} directly from adsorption experiments done on micron-sized particles that have high surface area, by monitoring the change in solution surface tension on long-term exposure to surfactant solutions. Γ_{LV} was obtained from solution surface tension measurements in combination with the Gibbs adsorption isotherm. If Γ_{SV} is assumed to be zero and the adhesion tension for pure water is known, the adhesion tension for aqueous surfactant solutions can then be *predicted* from the equation

$$\begin{aligned} \gamma_{LV}(c) \cos \theta(c) &= \gamma_{LV}(w) \cos \theta(w) + \gamma_{SL}(w) - \gamma_{SL}(c) \\ &= \gamma_{LV}(w) \cos \theta(w) + \pi_{SL} \end{aligned} \quad (5)$$

where π_{SL} can be obtained from the Gibbs adsorption isotherm

$$\pi_{SL} = RT \int_0^c \Gamma_{SL}\, d\ln(c) \tag{6}$$

By using these ideas, adhesion tensions for aqueous $C_{10}E_5$ and $C_{12}E_5$ solutions to PMMA, as well as $C_{12}E_5$ on polystyrene, have been obtained from adsorption data and shown to compare favorably with experimental values [15].

It is interesting to examine Zisman's [11] data in the context of the analysis proposed above. Figure 4 is a recast of the data from Fig. 15 in Zisman [11] for aqueous sodium dodecylsulfate and cetyltrimethylammonium bromide solutions on polyethylene in the form of the adhesion tension $\gamma_{LV} \cos\theta$ vs. γ_{LV}. The slope is extremely small, indicating negligible adsorption of surfactant at the solid/liquid interface, corresponding to the assumption made in their paper.

An interesting analysis of the contact-angle surface tension plots on low-energy surfaces by Zisman [11] appears in Johnson and Dettre [7]. If we assume no variation in the solid/liquid interfacial free energy with a change in the liquid surface tension, Young's equation is manipulated to give

$$\cos(\theta) = \left(\frac{2\gamma_c - \gamma_{LV}}{\gamma_{LV}}\right) \tag{7}$$

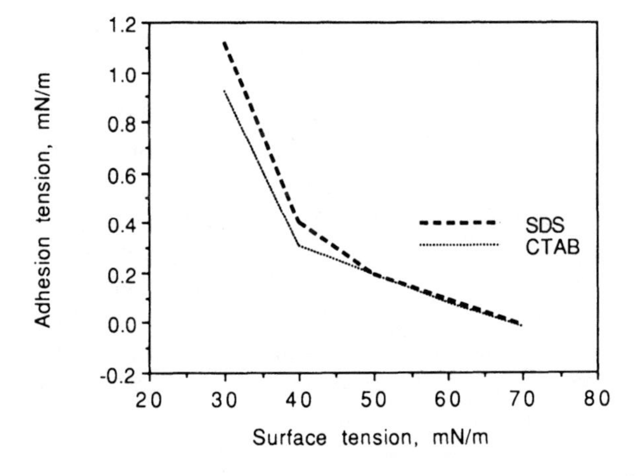

FIG. 4 Adhesion tension $\gamma_{LV} \cos\theta$ plotted vs. the liquid surface tension γ_{LV} for aqueous solutions of sodium laurylsulfate and cetyl trimethylammoniumbromide on polyethylene substrates. (The data are calculated from Fig. 15 in Ref. 11.)

Knowledge of the critical surface tension of the solid γ_c then permits the variation of the contact angle with liquid surface tension to be predicted. This simple prediction closely matched experimental data for the wetting of polyethylene by aqueous solutions of nonfluorinated surfactants. However, data for fluorinated surfactant solutions on polyethylene surfaces were overpredicted by the analysis, indicative of adsorption at the solid/vapor interface.

III. VARIATION IN STATIC CONTACT ANGLES BY SOLID SURFACE MODIFICATION

As shown in the previous section, modifications to the solid surface free energy γ_{SV} can have a fundamental impact on contact angles. This may be achieved in a variety of different ways including grafting of ligands onto surfaces by ion exchange, image adsorption, or chemisorption and physisorption that leave hydrophilic or hydrophobic moieties on the substrates [16]. The presence of even a monolayer of this coated material can drastically alter the wetting characteristics of the underlying solid [11] and this behavior can be usefully exploited. Bain and Whitesides [3] exposed gold substrates to varying ratios of solutions of C_{19} and C_{11} ω-hydroxyalkanethiols. These thiols self-assemble on the solid surface, with the sulfur atoms binding strongly to the gold substrate. The extreme sensitivity of contact angles of water to the "state" of the solid substrate is then exploited to show that both molecules attach to the gold surface with an orientation such that only their hydrophilic hydroxyl groups are exposed. When the gold substrate is exposed to mixtures of the two thiols, they disperse themselves on a molecular scale rather than phase-segregating into separate islands.

One interesting way of modifying a solid surface is to change its electrostatic charge density, by either varying the concentration of the potential determining ion in the surrounding bulk liquid (e.g., the pH in amphoteric solids) or directly controlling it by making it a part of a variable potential capacitance, such as two metallic plates connected through a direct current power supply. Wahal et al. [17,18] used two vertically separated concentric stainless steel disks as parallel plate capacitors and modified the charge density on the substrate by changing the potential difference between the plates. By placing a drop of liquid on the lower plate, the variation of the advancing contact angle with the field strength between the plates is monitored. Their experiments were restricted to field strengths where electrical body forces are negligible. No changes in contact angles are observed for pure water or glycerol drops. However, as shown in Figs. 5 and 6, for anionic (SDS) and cationic (CTAB) sur-

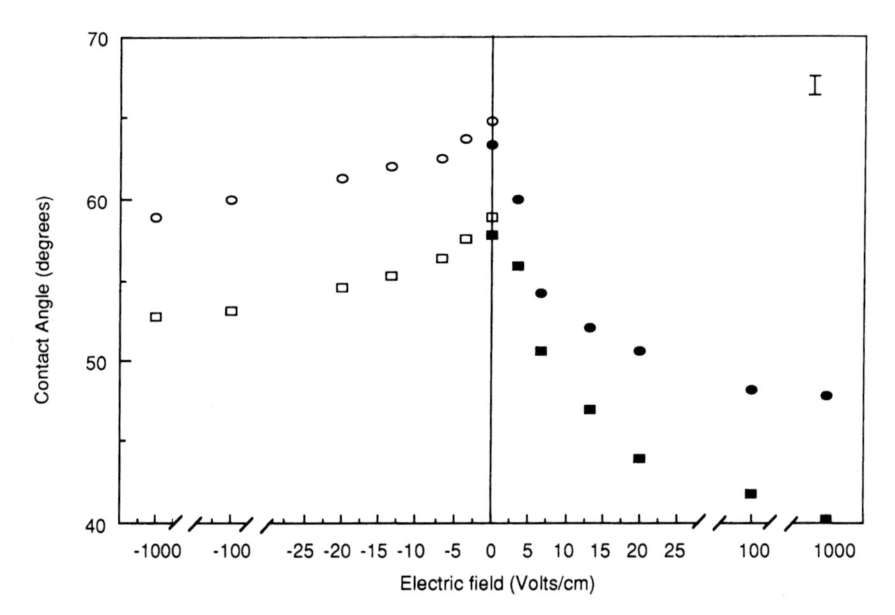

FIG. 5 Static advanced contact angle vs. applied electric field for an aqueous $10^{-3}M$ SDS solution on a stainless steel substrate. The various symbols represent different experimental runs and are indicative of a substantial amount of contact-angle hysteresis. The sample drop is placed on the lower plate. A positive field implies that the upper plate is at a higher potential than the lower plate. The vertical bar represents the maximum error in the contact-angle measurement.

factant solutions, the contact angles *decreased* upon application of the electric field. The magnitude of this decrease depended on not only the nature of the surfactant but also the polarity of the field. The maximum changes in contact angles take place at the lowest field strengths studied. An abrupt change in the field strength induces a response in the solution contact angle that takes approximately 10 min to reach steady state, shown in Fig. 7. This time scale is far higher than what might be expected from diffusion only, indicating that the slow change in contact angle is clearly not diffusion-controlled. By observing changes in sessile drop shapes, variations in liquid/vapor surface tension have been monitored as the magnitude of the electric field is increased. Surface tensions always decreased as the field was increased. However, if Young's equation is assumed to apply, the magnitude of the contact angle change cannot be accounted for just by the change in the solution surface tension; a drop in the solid/liquid interfacial free energy must also be postulated, indi-

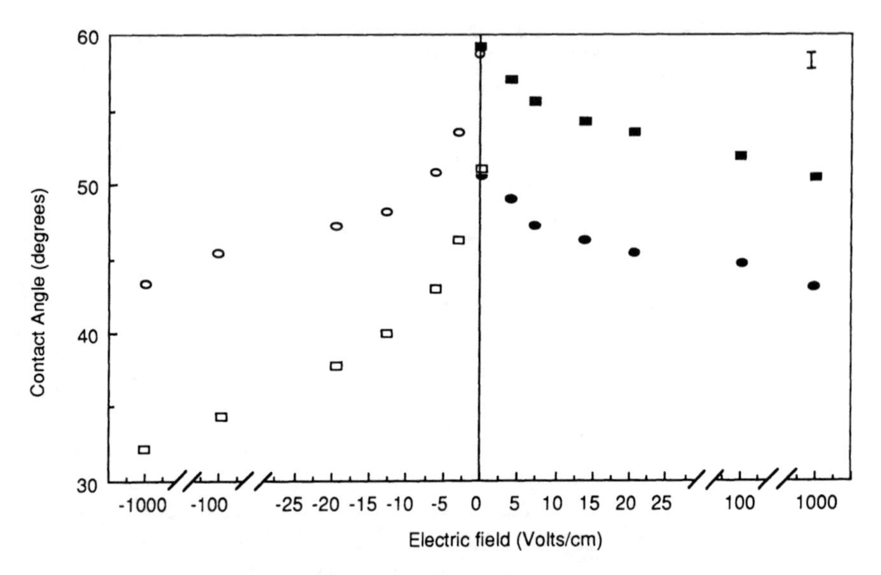

FIG. 6 Static advanced contact angle vs. applied electric field for an aqueous $5.5 \times 10^{-4}M$ CTAB solution on a stainless steel substrate. The various symbols represent different experimental runs and are indicative of a substantial amount of contact-angle hysteresis. The sample drop is placed on the lower plate. A positive field implies that the upper plate is at a higher potential than the lower plate. The vertical bar represents the maximum error in the contact-angle measurement. (Reprinted with permission from Ref. 17.)

cating nontrivial adsorption effects at that interface that are very sensitive to the externally applied electric field [18].

IV. EFFECT OF SURFACTANTS ON CONTACT-ANGLE HYSTERESIS

Contact-angle hysteresis is most easily defined by referring to Fig. 8. The limiting angles produced as the contact line speed approaches zero from the positive (advancing) direction θ_A, and the negative (receding) direction θ_R, are not the same, and the spread is called hysteresis. Almost all known systems display hysteresis, but this effect is enhanced by heterogeneities on the solid surfaces that expose regions of different surface energies to the same liquid. From a practical perspective, contact-angle hysteresis controls the pressure required to dislodge a captive drop in a uniform pore $\Delta P = \gamma_{LV}(\cos \theta_R - \cos \theta_A)/R$, where R is the pore radius, or the

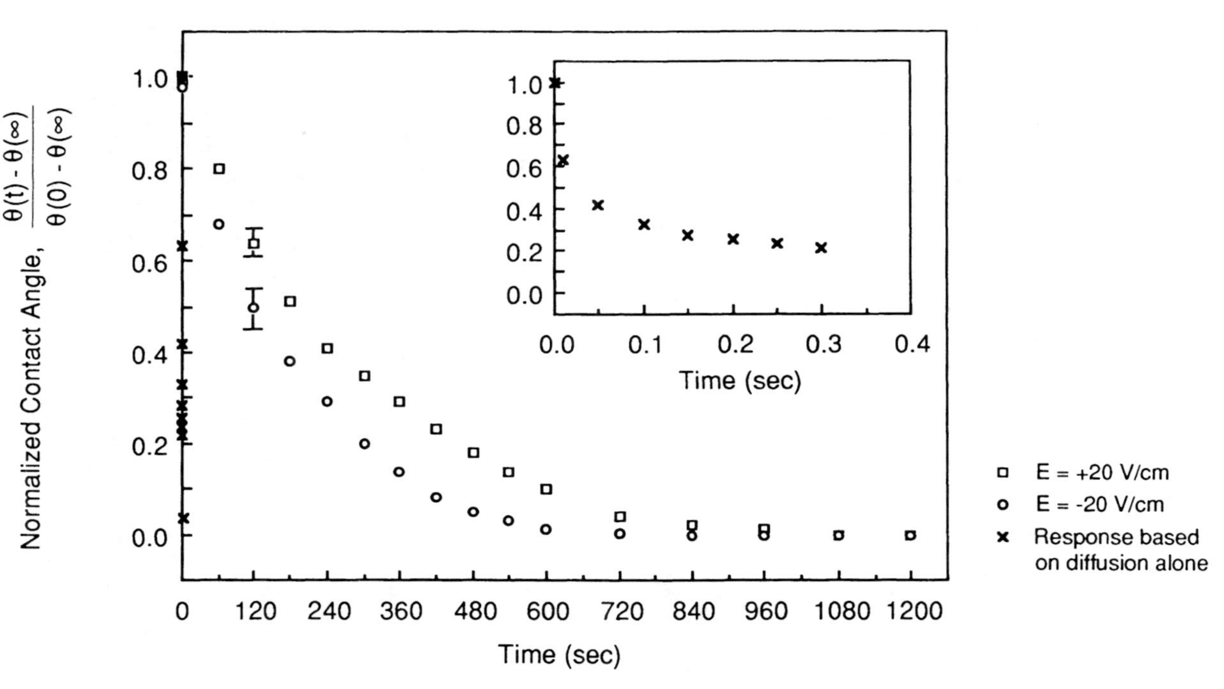

FIG. 7 Variation of the normalized contact angle, $[\theta(t) - \theta(\infty)]/[\theta(0) - \theta(\infty)]$, with time after application of the electric field at time $t = 0$ for an aqueous $10^{-3}M$ SDS solution. The vertical bars represent the error in the contact-angle measurement. The crosses represent the response based on a one-dimensional model if diffusion of the surface active ion is controlling [17,18], clearly indicating that the transient is not diffusion-dominated. The inset shows the diffusion response on an enlarged scale. (Reprinted with permission from Ref. 17.)

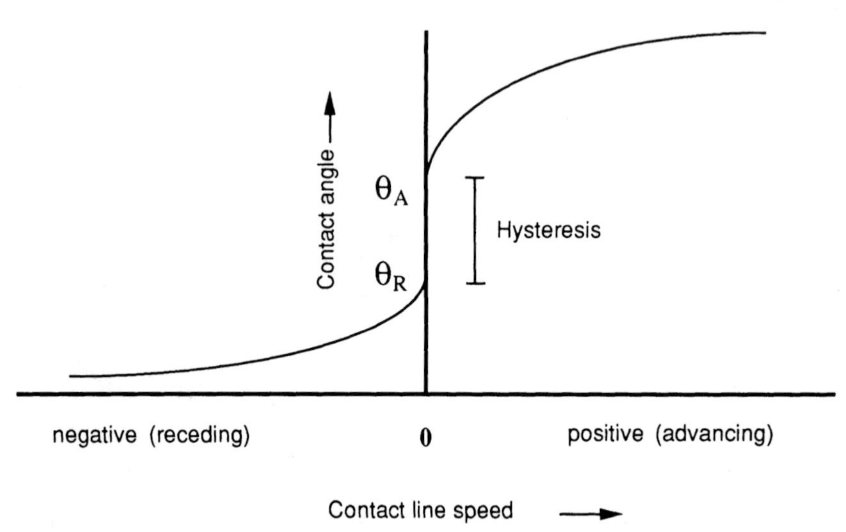

FIG. 8 Illustration of contact-angle hysteresis. The static limit as the advancing (receding) contact line speed approaches 0 is called the static advanced contact angle θ_A (the static receded contact angle θ_R), whereas the difference $\theta_A - \theta_R$ is the contact-angle hysteresis.

gravitational force required to initiate sessile drop motion on a solid substrate.

Measurement of contact-angle hysteresis is usually done by one of two methods. A finite volume of liquid is placed on a substrate, and this volume is differentially incremented until the contact line is observed to advance. The contact angle obtained just before the meniscus moves is called the advancing contact angle. To obtain the receding contact angle, the volume of the drop is differentially reduced until the contact line is again observed to move; the liquid meniscus shape just before the drop moves is used to obtain the receding contact angle. An alternate strategy is to place a sessile drop on a substrate, then tilt the substrate until the drop *just* begins to slide because of an imbalance between the gravitational and capillary forces. The angle subtended at the front of the drop is the advancing contact angle, whereas that at the rear is the receding contact angle. Both these techniques are illustrated in Fig. 9.

Although contact-angle hysteresis is a property that is relevant for static systems, its measurement necessitates locating the border between static and dynamic states. Although this issue may not be very important when measuring hysteresis for pure liquids, it becomes critical when mak-

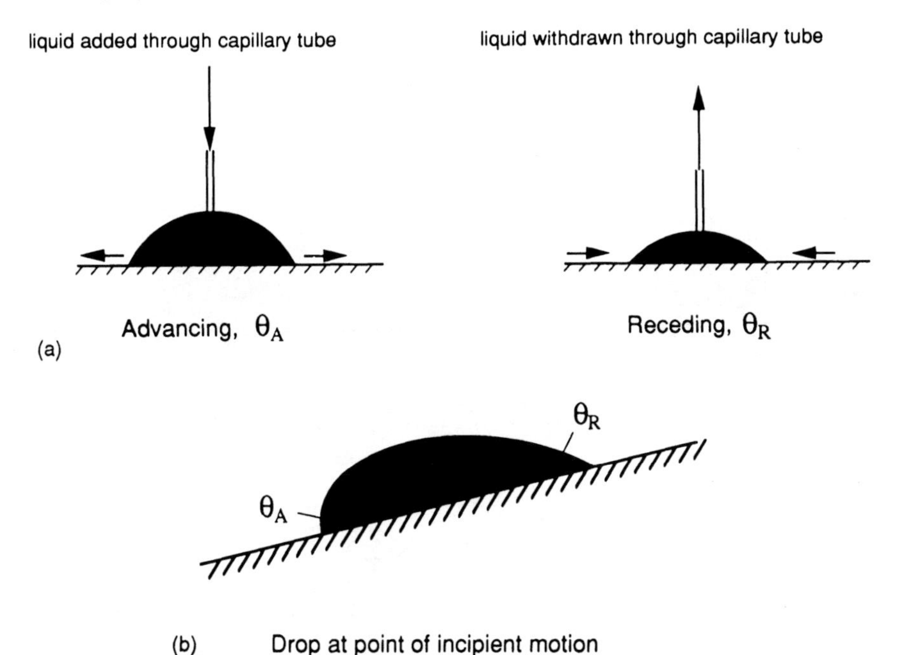

FIG. 9 Techniques for measuring contact-angle hysteresis. (a) Sessile drops volumes are incrementally increased or decreased until the contact line is just set in motion. Images of the drop at these limits are used to obtain θ_A and θ_R. Alternatively, sessile drops are placed on a substrate that is subsequently inclined until the point of incipient drop motion. Drop images at this critical position simultaneously give the advanced and receded contact angles.

ing the same measurements in surfactant solutions. Padmanabhan and Bose [19] used the tilting stage method and showed that the contact-angle hysteresis of the fluorocarbon surfactant solutions FSA™ and FSC™ on teflon substrates was substantially higher than that produced by water, as shown in Fig. 10. They attributed this increased hysteresis to nonspecific adsorption of the surfactant to the solid surface. They also compared experimentally obtained dimensionless drop volumes at the critical configuration to the prediction by Dussan V. [20] and showed that the experimental values were between two and four times those predicted by the analysis. When a "weak" surfactant, such as octanol was used, experimental and analytical predictions matched. These results are intriguing, because the strength of the surfactant $\partial\gamma_{LV}/\partial C$ is a material property that should only be relevant in *dynamic* situations, yet contact-angle hys-

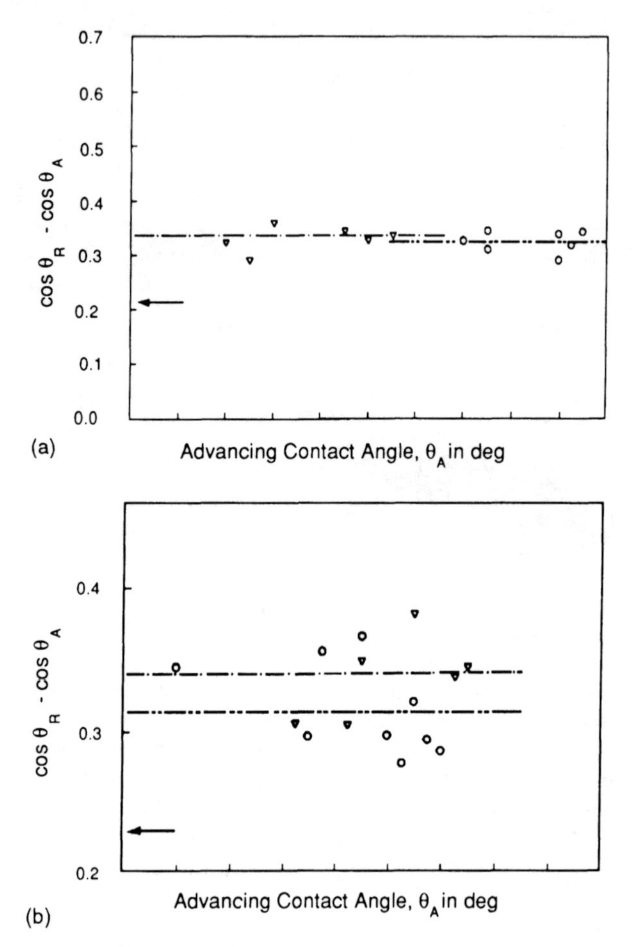

FIG. 10 (a) Hysteresis parameter cos θ_R − cos θ_A vs. the advanced contact angle for 1.32 × $10^{-5}M$(O) and 3.96 × $10^{-5}M$(▽)Zonyl FSA^{TM} aqueous solutions on teflon. The arrow indicates the hysteresis parameter for water. (b) The hysteresis parameter cos θ_R − cos θ_A vs. the advanced contact angle for 1.14 × $10^{-5}M$(O) and 2.85 × $10^{-5}M$(▽)Zonyl FSC^{TM} aqueous solutions on teflon. The arrow indicates the hysteresis parameter for water. (Reprinted with permission from Ref. 19.)

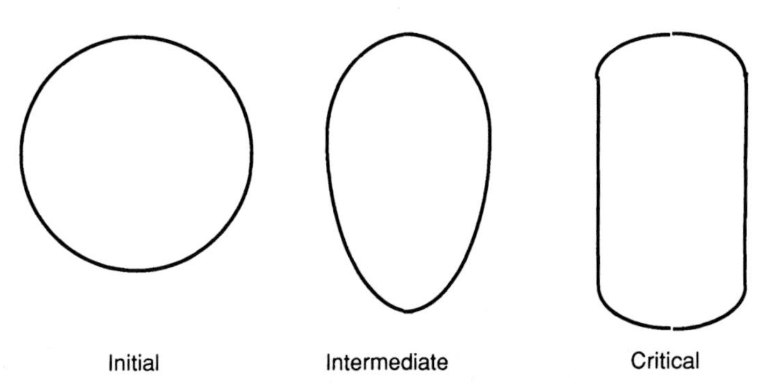

FIG. 11 Sequence of contact line shapes achieved by a sessile drop as a substrate is progressively tilted to the critical drop configuration.

teresis represents a "static" property. In a later paper, Padmanabhan and Bose [21] showed that the source of the large discrepancy was entirely related to the measurement technique. The drops were placed on a substrate that was initially horizontal, so that the contact line was circular. At the critical configuration, the shape of the contact line changes to one with parallel sides (Fig. 11), so that some fluid motion must take place between the starting and final configurations. This motion is always opposed by the hydrodynamic rigidity imposed by the surfactant present at the interface. When the experiments were repeated with the drop placed on substrates that were *initially* tilted, experimental and analytical values converged, as shown in Fig. 12. In order to avoid any dynamic effects from invalidating a nominally static measurement, the recommended method for obtaining hysteresis should clearly be one where the sample is instantaneously at the point of incipient motion. If the incremental drop volume approach is used to measure hysteresis for strong surfactant solutions, the results could easily be dependent on the rate of addition or withdrawal of the liquid.

V. DYNAMIC CONTACT ANGLES

A. Experiments

The most prevalent experimental technique for exploring the wetting behavior of solutions is the use of a dynamic Wilhemy apparatus, where the wetting tension $\gamma_{LV} \cos \theta_{\text{dynamic}}$ is monitored during both immersion and emersion of a solid substrate. If one is interested in the value of the contact angle, some assumption needs to be made about the local surface tension

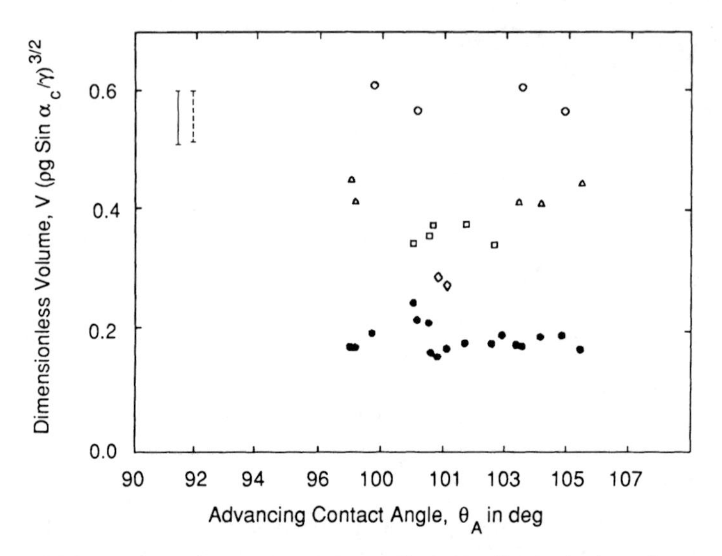

FIG. 12 Experimental and theoretical (●) dimensionless drop volumes vs. the advancing contact angle for varying substrate pretilt angles. V is the liquid volume, ρ the density, g the gravitational acceleration, α_c the critical angle of inclination, and γ the liquid surface tension. Pretilt angles: 0° (O), 2.5° (△), 5.0° (□), and 6.0° (◇). The solid bar represents the maximum error in the analytically predicted critical volume, the dashed bar the maximum experimental error. (Reprinted with permission from Ref. 21.)

at the contact line, and most often, it is assumed equal to the equilibrium value. On the other hand, if there is a direct measurement of the instantaneous dynamic contact angle, the local surface tension at the contact line can be obtained from the capillary force measurement. Padmanabhan and Bose [22] have taken the latter approach. Using a modified Wilhemy apparatus where a camera and light source were attached on the same stage that held the liquid sample and synchronizing the image capture with the force measurement, they were able to obtain direct measurements of the instantaneous contact angles and use the capillary force to obtain the instantaneous surface tension (often referred to as the dynamic surface tension). The capillary force is, of course, equal to the tension on the electrobalance if the viscous drag on the wetted area is negligible and buoyancy effects are properly accounted for. Nondimensionalization of the surfactant flux balance at the liquid surface (if we assume no adsorption/desorption barriers) yields a Peclet number, $-(\partial\gamma_{LV}/\partial C)U/\mathbf{D}RT$, which essentially determines whether the surfactant concentration at the contact line will have a steady value or be time-dependent. Here U is the plate speed, **D** the surfactant diffusion coefficient, and T the temperature. At low values of the Peclet number, diffusional interchange with the bulk solution is fast, and variations in surfactant concentration along the surface cannot be sustained. The surface tension at the contact line will therefore correspond to the equilibrium value. When the Peclet number is high, transients caused by surfactant buildup or depletion at the contact line should be created initially; the surfactant concentration can then be expected to level off to some steady-state value (not necessarily the equilibrium value) after the decay of these transients. These notions were tested by Padmanabhan and Bose [22], by running advancing meniscus experiments over a wide range of Peclet numbers for aqueous solutions of the anionic fluorocarbon surfactant FSA™ on glass microscope slides. The key results are that the contact line appears to be in stick-slip motion during its initial movement when the Peclet number is high ($\sim 10^4$). The dynamic contact angle predicted by assuming equilibrium concentration at the contact line differs significantly from the directly measured value, confirming the departure from equilibrium at the contact line. When the Peclet number is low, the surfactant concentration is equal to the value in equilibrium with the bulk solution and the measured and predicted contact angles converge.

Hodgson and Berg [23] have conducted experiments on the wicking of surfactant solutions into fibrous networks consisting of Whatman No. 40 filter paper and commercial wood pulps and found that in spite of the complex geometry of these materials, they obey the simple Lucas–Washburn rate law, with the penetration distance proportional to the elapsed time $t^{1/2}$. To remove the need to specify any geometrical features of the solid, experiments are conducted using a nonswelling liquid, and the Lucas–Washburn slope of the test liquid is normalized by the slope obtained using the reference liquid. Thus, a wicking equivalent surface tension γ_{LV}^* can be defined as

$$\gamma_{LV}^* = \gamma_{LV\ \mathrm{ref}} \left(\frac{\cos\theta_{\mathrm{ref}}}{\cos\theta}\right)\left(\frac{\mu}{\mu_{\mathrm{ref}}}\right)\left(\frac{k}{k_{\mathrm{ref}}}\right)^2 \tag{8}$$

where $\gamma_{LV\ \mathrm{ref}}$, θ_{ref}, μ_{ref}, and k_{ref} represent the reference liquid surface tension, contact angle, viscosity, and Lucas–Washburn slope, respectively, whereas θ, μ, and k represent the same quantities for the test liquid. All quantities on the right-hand side are measured independently. Any deviation of the wicking equivalent surface tension from the equilibrium value is indicative of a lack of adsorption equilibrium at the contact line. If we use water as the reference liquid, for complete wetting of both the test and reference liquid and for dilute aqueous solutions, Eq. (8) reduces to

$$\gamma_{LV}^{*} = 72 \left(\frac{k}{k_{\text{ref}}}\right)^{2} \tag{9}$$

The results for the aqueous solutions of the three surfactants used 0.005*M* SDS, 0.1*M* CTAB, and 0.01 wt % Triton X-100 illustrate some interesting differences, as shown in Fig. 13. γ_{LV}^{*} is always greater than the equilibrium surface tension, clearly indicating lack of adsorption equilibrium at the penetrating front. This difference is the least for the SDS, intermediate for the CTAB, and largest for the Triton X-100 solutions. Two factors play key roles in determining deviation from equilibrium: the adsorption capability of the surfactant on the solid matrix and its diffusion coefficient. The difference between the results for the SDS and CTAB solutions arises because of differences in adsorption of the anionic and cationic surfactants, whereas the low diffusivity of the larger Triton X-100 is responsible for persistent large deviations from equilibrium for this system. For systems displaying incomplete wetting, wicking equivalent *adhesion tensions* have been similarly defined and show values consistently higher than equilibrium.

Kamath et al. [24] have studied the Marangoni effect during the water wetting of surfactant-coated human hair fibers. They treated human hair with the cationic surfactant Triton X-400 in solutions 20° and 80°C and monitored the dynamic capillary force exerted by the motion of water on these fibers using an electrobalance. A distinct stick-slip behavior was observed as the water first advanced over the treated fiber. For a treatment temperature of 20°C, the coated surfactant dissolved into the liquid at the contact line, so that the subsequent receding wetting force mimicked that of an untreated fiber. For the treatment at 80°C, the cationic surfactant is more strongly adsorbed to the negatively charged fiber. The frequency of the stick-slip phenomenon for the fiber treated at the higher temperature is smaller than the one observed for the 20°C-treated samples. This is related to the difference in the rates of dissolution of surfactant at the contact line: fast for the weakly bound surfactant at 20°C and slow for the tightly held surfactant at 80°C.

B. Analytical and Numerical Solutions

The presence of a layer of surfactant raises additional complications related to the stress singularity at the contact line beyond that present in pure liquids. This may be most easily understood by considering the surfactant to be insoluble and being transported and mapped directly to the

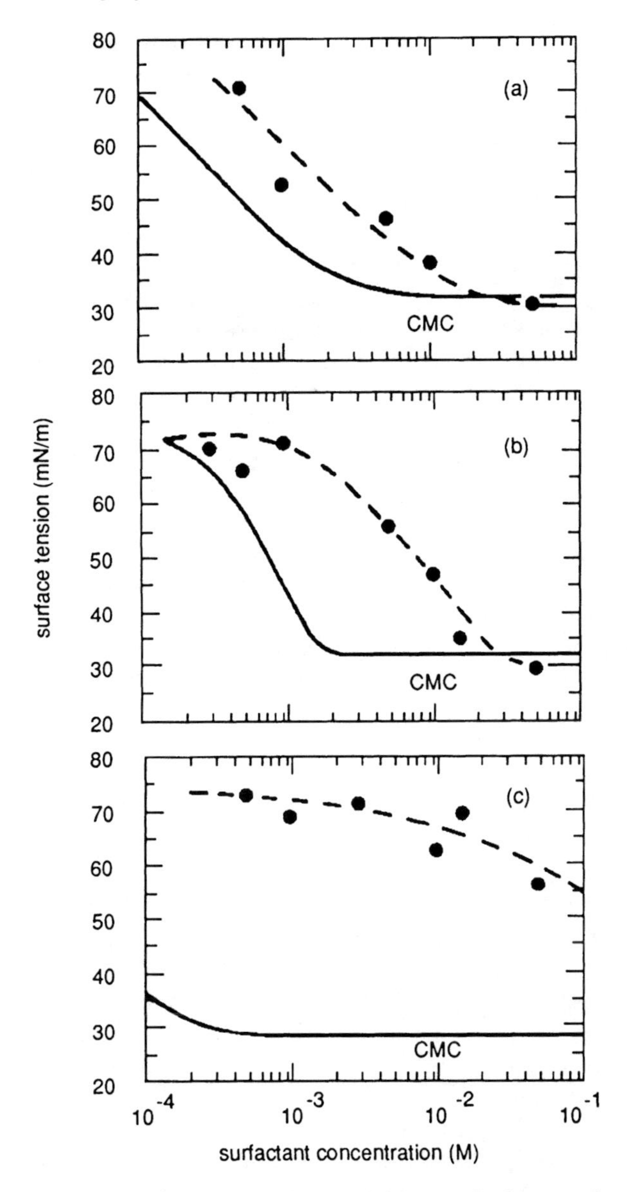

FIG. 13 Comparison of the wicking-equivalent surface tension (---) with the equilibrium surface tension (—) for aqueous solutions of (a) SDS, (b) CTAB, and (c) Triton X-100, penetrating into strips of Whatman No. 40 filter paper strips. (Reprinted with permission from Ref. 23.)

solid surface *only* at the contact line [25]. If surface and bulk diffusion are ignored, the flux of surfactant at the contact line is $\Gamma u|_{cl}$, where $u|_{cl}$ is the mass-averaged velocity of the liquid at the contact line. Because Γ is bounded, $u|_{cl}$ must be nonzero and directed tangent to the liquid/vapor interface. Fluid points along the solid surface must have a velocity tangent to the solid. This implies that the velocity at the contact line must be multivalued, regardless of whether slip occurs in the inner region or not, again leading to the problem of a nonintegrable stress. This singularity can be relieved by posing some additional indirect mechanisms for surfactant transfer, either surface diffusion or imposition of at least a limited amount of solubility, thus providing for dissolution then reabsorption through the bulk phase. In order to make the problem of practical value, Ramé [25] has identified that specification of two *material* parameters are necessary to make the hydrodynamic problem well posed: the dynamic contact angle and the transfer ratio of the insoluble surfactant, both depending on the velocity of the moving contact line, as well as the local concentration of the surfactant in the vicinity of the contact line. These parameters are specified at some (measurable) distance from the contact line, and the fluid mechanical problem solved for *two* cases related to the deposition of Langmuir–Blodgett films: one in which the plate immersion depth is small compared to the container depth and the distance of the float from the contact line (this float moves to maintain a constant surface tension at its location), and the other where the container depth is small compared to the distance of the float from the contact line. From the results of this asymptotic analysis, two experimentally measurable quantities, the force exerted by the moving liquid on the plate as well as the meniscus height, are obtained, which in principle can be used to determine the two *material* properties that crucially affect this process. Cox [26] solved a similar problem, but did not include the transfer of surfactant from the liquid to the solid surface, avoiding the issue of velocity multivaluedness introduced specifically by direct transfer of surfactant.

Joanny [27] studied the spreading of a liquid with an insoluble surfactant onto a solid surface. All cases studied belonged to the category of repulsive solid surfaces, implying a greater affinity of the surfactant for the liquid/vapor rather than solid/liquid interface. This difference gives rise to concentration gradients of surfactant along the liquid/vapor interface as the contact line is approached, causing surface tension gradients that affect spreading. When there is no affinity of the surfactant for the solid, the dynamic contact angle θ_d is simply connected to the dynamic contact angle θ_0 for a pure liquid predicted by the Tanner law by the relationship

$$\theta_d \sim \theta_0 2^{1/3} \tag{10}$$

As the affinity for the solid surface is increased, θ_d decreases and approaches θ_0 when the concentration of surfactant at the solid surface is half that in the liquid surface. Joanny also generalizes the de Gennes rule that relates dissipation in the precursor film to the spreading pressure to a system that contains surfactants; the spreading power is balanced by the viscous force, as well as the Marangoni force in the film.

In order to establish some of the key parameters necessary to understand the wetting of solids by *soluble* surfactant solutions, Padmanabhan [28] has modeled the problem of a plate of liquid entering a bath of soluble surfactant solution at a dynamic contact angle of 90°, shown in Fig. 14. The liquid surface has been assumed flat. Surfactant is transferred from the liquid to the solid surface only at the contact line (the amount is controlled by the transfer ratio T that is assumed independent of immersion speed and local surfactant concentration). Once surfactant is mapped onto the solid, it is not permitted to dissolve and readsorb. Because the surfactant is assumed soluble in the bulk liquid phase, any surfactant that is brought to the contact line and not mapped onto the solid simply dissolves back into the bulk liquid phase. As in pure liquids, slip at the contact line is modeled using Navier's condition that then removes the stress singularity. These results are, of course, dependent on the choice of the slip length, which, as pointed out earlier, is not known. Because the free surface is assumed flat everywhere, the physical domain is known a priori and the normal stress balance is no longer applicable. A two-dimensional Cartesian coordinate system is used with the origin located at point D. Although the problem is inherently unsteady because of the

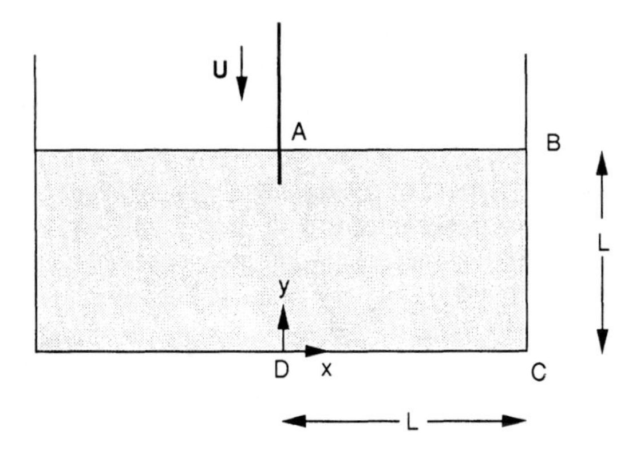

FIG. 14 Domain *ABCD* used for calculation of velocity, pressure, and concentration profiles for a plate entering a bath of a soluble surfactant solution.

changing immersion depth, a quasi-steady-state analysis has been used to examine the roles played by the important parameters that govern the system.

With lengths scaled with the container dimension L, velocities with the plate speed U, surface tension with the far-field equilibrium value $\gamma_{LV\,\text{eq}}$, concentrations with the bulk concentration C_0, and pressure using the inertial scale ρU^2, the governing equations are

$$\nabla \cdot \mathbf{V} = 0 \tag{11}$$

$$\mathbf{V} \cdot \nabla \mathbf{V} = -\nabla P + Re^{-1} \nabla^2 \mathbf{V} \tag{12}$$

$$\mathbf{V} \cdot \nabla C = Pe_b^{-1} \nabla^2 C \tag{13}$$

where Eqs. (11–13) are the continuity, momentum, and diffusion equations, respectively. For the boundary conditions that follow, u and v represent the dimensionless x and y components of the liquid velocity, respectively.

At the contact line, Navier's slip boundary condition reduces to

$$\beta^{-1}(v + 1) = \frac{\partial u}{\partial y} + \frac{\partial v}{\partial y} \tag{14}$$

At all other points on the solid surface,

$$u = 0, \qquad v = -1, \qquad \text{and} \qquad \frac{\partial C}{\partial x} = 0 \tag{15–17}$$

representing the no-slip, kinematic, and no-normal-solute-flux conditions, respectively.

At the plane of symmetry under the plate,

$$u = 0, \qquad \frac{\partial v}{\partial x} = 0, \qquad \text{and} \qquad \frac{\partial C}{\partial x} = 0 \tag{18–20}$$

At the container walls, the kinematic, no-slip, and no-flux conditions give

$$u = 0, \qquad v = 0, \qquad \text{and} \qquad \frac{\partial C}{\partial n} = 0 \tag{21–23}$$

where n is the direction of the unit normal from the respective surfaces.

At the "free" upper surface, the kinematic condition, tangential stress balance (if we neglect surface rheological effects), and solute flux balance give

$$v = 0 \tag{24}$$

$$\frac{\partial u}{\partial y} + \frac{\partial v}{\partial x} = -El \frac{\partial C}{\partial x} \tag{25}$$

$$\frac{\partial C}{\partial x} + C \frac{\partial u}{\partial x} = Pe_s^{-1} \frac{\partial C}{\partial y} \tag{26}$$

At the contact line, the solute flux balance is modified to reflect the "loss" of solute to the solid

$$u \frac{\partial C}{\partial x} + C \frac{\partial u}{\partial x} - Pe_s^{-1} \frac{\partial C}{\partial y} = TC_{\text{contact line}} \tag{27}$$

Far away from the contact line, the concentrations and surface tensions are assumed to be equal to their undisturbed equilibrium values. The equation representing the conservation of surfactant mass within the domain is therefore not applicable.

The concentration and velocity problems are nonlinearly coupled through the shear-stress balance at the liquid/air interface. The important dimensionless groups are the transfer ratio T; the elasticity number $El = (-\partial\gamma_{LV}/\partial C)C_0/\mu U$, representing the ratio of surface tension to viscous forces; the surface Peclet number $Pe_s = Uk/\mathbf{D}$, representing the ratio of

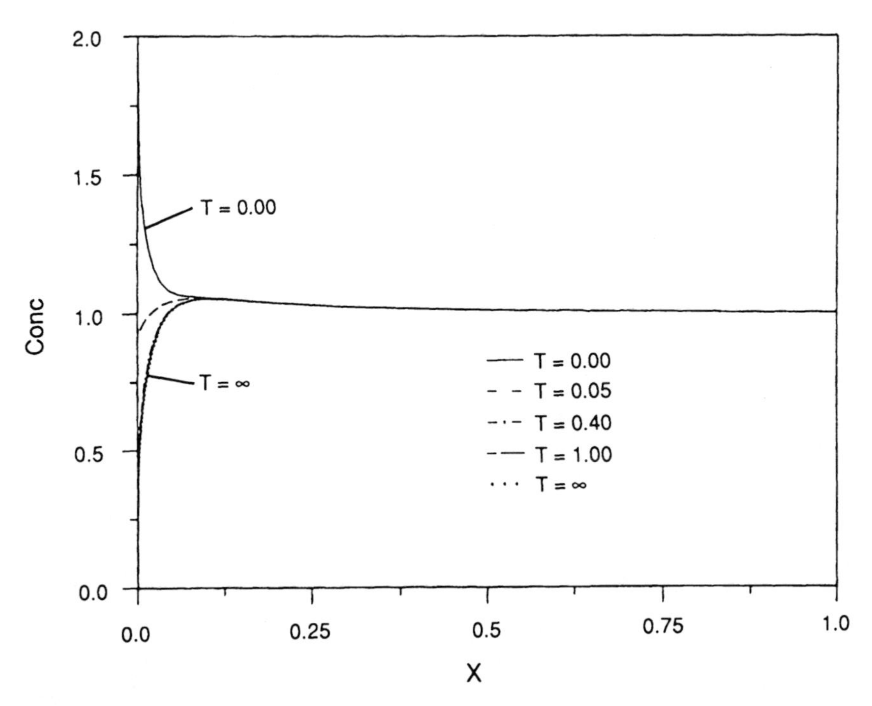

FIG. 15 Surface concentration of surfactant vs. distance from the plate x for different transfer ratios T. $Pe_s = 1.0$, $Pe_B = 100.0$, slip coefficient $\beta = 2.0 \times 10^{-4}$, $El = 1.0$, and $Re = 11.25$. The plate is immersed to $1/10$ of the container depth.

surfactant transported by the flow to that by diffusion, where k is obtained by assuming local equilibrium at the surface so that $\Gamma_s = kC_s$ [by using Gibbs adsorption isotherm, $k = -(1/RT)\partial\gamma_{LV}/\partial C$]; the bulk Peclet number $Pe_b = UL/\mathbf{D}$; and the Reynolds number UL/ν. An additional parameter that must also be specified is the dimensionless slip length β. Some results of the numerical simulation are summarized in Figs. 15 through 17. The most instructive data are the concentration profiles of surfactant along the liquid surface. Values of the dimensionless groups are calculated using physical properties of sodium dodecylsulfate surfactant at a concentration of 0.5 CMC. As shown in Fig. 15, for a transfer ratio of 0 (no surfactant transfer to the solid), the solute concentration builds up monotonically as the contact line is approached. However, for a transfer ratio of 0.05, there is a drop in surfactant concentration as the three-phase line is approached, and all the concentration fields for $T > 0.4$ are nearly identical,

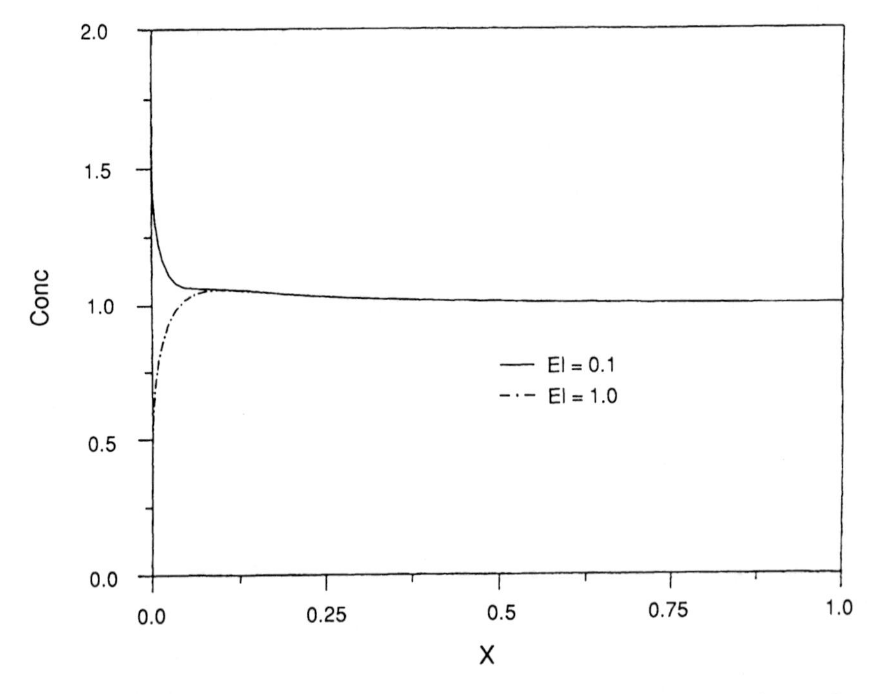

FIG. 16 Surface concentration of surfactant vs. distance from the plate x for different elasticity numbers El. $Pe_s = 1.0$, $Pe_B = 100.0$, slip coefficient $\beta = 2.0 \times 10^{-4}$, $T = 0.4$, and $Re = 11.25$. The plate is immersed to $\frac{1}{10}$ of the container depth.

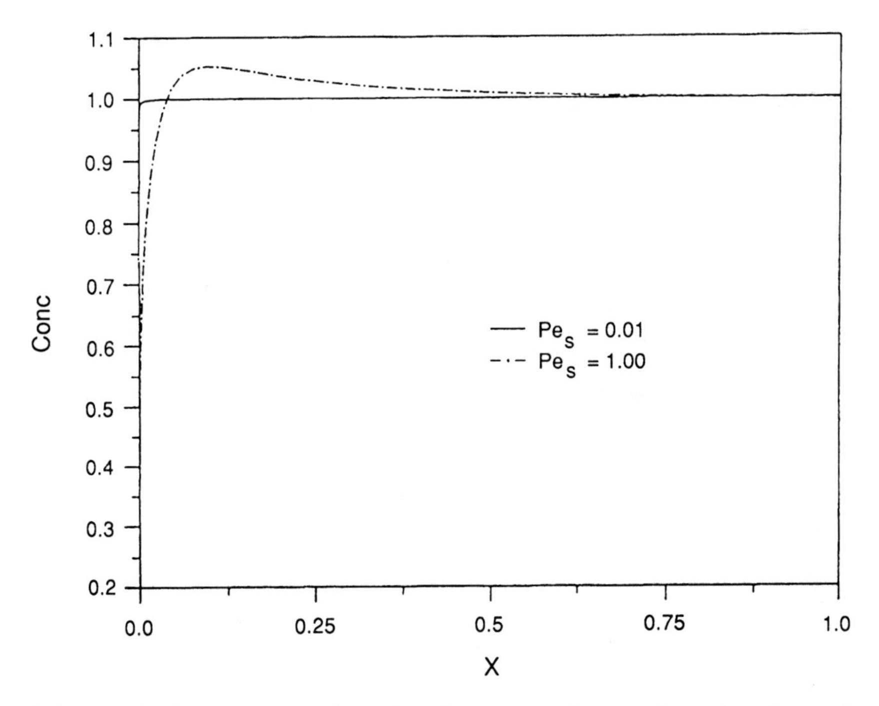

FIG. 17 Surface concentration of surfactant vs. distance from the plate x for different surface Peclet numbers Pe_s. $Pe_B = 100.0$, slip coefficient $\beta = 2.0 \times 10^{-4}$, $El = 1.0$, $Re = 11.25$, and $T = 0.4$. The plate is immersed $1/10$ of the container depth.

showing a steep drop as the contact line is approached. The drop in concentration for a transfer ratio of 0.05 is at first counterintuitive, because the solute must clearly be rejected by the solid at the contact line. This buildup is opposed by both the elasticity of the surfactant, as well as the diffusion of the surfactant into the bulk liquid, resulting in the observed surface concentration variation. The role of surfactant elasticity El is demonstrated in Fig. 16. Clearly, concentration profiles can display drastically different behavior, depending on whether the surfactant is strong or weak. Weaker surfactants generate lower hydrodynamic rigidity at the solution surface to oppose the transport of surfactant by the imposed flow. Surfactant can then build up as the contact line is approached. The role of the surface Peclet number is demonstrated in Fig. 17. Again, small Peclet numbers imply that diffusion into the bulk phase is rapid, so that the solute concentration can be expected to be quite uniform along the liquid surface.

Diffusional interchange with the bulk phase is reduced as the Peclet number goes up, resulting in the excess buildup as the contact line is approached. Of course, because of both the surfactant elasticity as well as the transfer ratio chosen for these runs, the surfactant concentration dips as the three-phase line is approached.

ACKNOWLEDGMENTS

It is a pleasure to acknowledge fruitful collaborations with all our former and current students working on wettability. We have benefited greatly from many discussions with E. B. Dussan V., E. Ramé, and J. C. Berg. The work at URI was financially supported by the donors of the Petroleum Research Fund of the American Chemical Society and the National Science Foundation.

REFERENCES

1. J. N. Israelachvili, *Intermolecular and Surface Forces*, Academic Press, San Diego, CA, 1985.
2. J. Koplik, J. R. Bannavar, and J. F. Williamson, *Phys. Rev. Lett. 60:*1282 (1988).
3. C. D. Bain, and G. M. Whitesides, *Science 240:*62 (1988).
4. T. D. Blake, and K. J. Ruschak, *Nature 282:*489 (1979).
5. T. Young, *Phil. Trans. R. Soc. Lond. 95:*65 (1805).
6. P. G. de Gennes, *Rev. Mod. Phys. 57:*827 (1985).
7. R. E. Johnson, Jr., and R. H. Dettre, in *Surface and Colloid Science*, Vol. 2 (E. Matijevic, ed.), Wiley-Interscience, New York, 1969, p. 85.
8. E. B. Dussan, V., *Ann. Rev. Fluid Mech. 11:*371 (1979).
9. J.-X. Yang, and J. Koplik, Paper O11.5, Materials Research Society Annual Meeting, Boston, MA, 1992.
10. F. Heslot, N. Fraysse, and A. M. Cazabat, *Nature 338:*640 (1989).
11. W. Zisman, in *Contact Angles, Wettability and Adhesion*, Advances in Chemistry Series, Vol. 43 (F. M. Fowkes, ed.), ACS, Washington, DC, 1964, p. 1.
12. P. M. McGuiggan, and R. M. Pashley, *Coll. Surf. 27:*277 (1987).
13. R. A. Pyter, G. Zografi, and P. Mukerjee, *J. Coll. Interf. Sci. 89:*144 (1982).
14. S. Wu, *J. Polym. Sci. 34:*19 (1971).
15. C.-S. Gau, and G. Zografi, *J. Coll. Interf. Sci. 140:*1 (1990).
16. J. C. Berg, in *Absorbency* (P. K. Chatterjee, ed.), Elsevier, Amsterdam, 1985, Chap. V, p. 149.
17. S. Wahal, C. A. Owiti, and A. Bose, *J. Adhesion Sci. Tech.* (1992) (in press).
18. S. Wahal, Ph. D. diss., Univ. Rhode Island, Kingston, 1992.
19. S. Padmanabhan, and A. Bose, *J. Coll. Interf. Sci. 123:*494 (1988).
20. E. B. Dussan V., *J. Fluid Mech. 151:*1 (1985).

21. S. Padmanabhan, and A. Bose, *J. Coll. Interf. Sci. 139:*535 (1990).
22. S. Padmanabhan, and A. Bose, *J. Coll. Interf. Sci. 126:*164 (1990).
23. K. T. Hodgson, and J. C. Berg, *J. Coll. Interf. Sci. 121:*22 (1988).
24. Y. K. Kamath, C. J. Dansizer, and H.-D. Weigmann, *J. Coll. Interf. Sci. 139:*535 (1984).
25. E. Ramé, Ph. D. diss., Univ. Penn., Philadelphia, 1989.
26. R. G. Cox, *J. Fluid Mech. 168:*195 (1986).
27. J. F. Joanny, *J. Coll. Interf. Sci. 128:*407 (1989).
28. S. Padmanabhan, Ph. D. diss., Univ. Rhode Island, Kingston, 1989.

4
Interfacial Chemistry in Biomaterials Science

ERWIN A. VOGLER Polymer Chemistry Department, Becton Dickinson Research Center, Research Triangle Park, North Carolina

I. INTRODUCTION

Biomaterial* scientists have long sought a single, material-related parameter that effectively measures biocompatibility and might serve as a practical design guide. It was quite natural that these investigators looked to theories of surface energetics and wetting for such parameters since surface properties are important determinants of biomaterials function. As examples, Baier pioneered the use of Zisman's critical surface tension as an indicator of blood compatibility [1–4] and bioadhesion [5–8]. Neumann and co-workers employed their "equation-of-state" approach to calculate interfacial tensions from contact-angle measurements [9–12] that, in turn, were used to predict cell adhesion [13–20] and thromboresistance [21,22]. Whereas concepts such as these have served as useful general guidelines or "rules of thumb" for biomaterials design, each has fallen far short of being the desired quantitative predictor of biocompatibility, particularly when applied to proteinaceous environments. Thus, the detailed physicochemical events that link surface chemistry and interfacial properties with the biological response to materials remain obscure.

There has been some progress in the area of polymer surface modification for various target biological responses [23], despite the fact that many of the fundamentals of biomaterials surface science are not yet well understood. The highest-volume commercial example is undoubtedly the treatment of materials used in tissue culture technology. Here, animal cells derived from some source organism are sustained in vitro, and it is found that many types of these cells will proliferate only when attached to a surface (anchorage-dependency). Plasticware, most commonly pol-

---

* The definition of biomaterials subscribed to here was principally enunciated by Baier in 1976 [1] as "any material designed to supplement, store, or otherwise come into intimate contact with living biological cells or biological fluids." It follows that biocompatibility is a measure of the success of the material in the specific biomedical task for which it was designed. See [23] for a recent applications review.

ystyrene, is subjected to some proprietary pretreatment by the manufacturer to render surfaces more wettable and adhesive to cells [24,25]. Although a great deal of work has been aimed at elucidating those surface functional groups responsible for cell adhesion, adhesive mechanisms at the chemical level still remain unclear [24–33]. As a result, the selection of plastics for disposable labware has been based on bulk properties or motivating economic factors such as cost and availability. Development of surface treatments has occurred largely through trial and error.

A more thorough understanding of the interactions and interrelationships between surfaces, interfaces, and the biological environment is required before rational, prospective preparation of biomaterials from the surface chemistry viewpoint can occur. Biomaterials surface science will benefit from a grasp of the fundamentals in the same way that classical organic chemistry has from the various, well-established structure-function relationships that guide synthesis. In the author's view, there are both analytical and conceptual barriers impeding progress toward elucidating these quantitative relationships. Analytical barriers are related to difficulties in characterizing biomaterial surfaces in a manner that provides information relevant to the biological environment encountered in end use. Conceptual barriers are encountered in the attempt to translate the desirable biomedical properties that define the term "biocompatible" into properties of practical use to the materials scientist. These barriers and some avenues around them are the subject of this chapter.

Analytical issues related to surface sensitivity, types of surface information obtained, and relevance of this information to biomedical applications will be discussed. The special roles of interfacial chemistry and wetting phenomena will be emphasized, with the goal of identifying some of the reasons why surface energetics cannot serve as a single material-related parameter that directly measures biocompatibility. Different methods for measuring polymer wetting properties in biological fluids will be compared using a simplified model system to show how time-dependent protein adsorption on solid surfaces can complicate data interpretation. A discussion of conceptual barriers will address the difficulties encountered in factoring complicated biological processes at interfaces into manageable pieces. Utility of surface thermodynamics in the interpretation of wetting data and as a biophysical tool for carrying out the aforementioned biomedical-to-materials property translation step will be emphasized. Finally, prospects for the rational preparation of surfaces for biomedical applications will be discussed. In each of these areas, specific examples drawn from the author's research will be given, along with a brief statement of philosophy.

II. CHARACTERIZATION OF BIOMATERIAL SURFACES

A. Surface Sensitivity

1. Surfaces and Interfaces

The boundary between any two phases is an interface. It is common parlance among surface physical chemists to refer to an interface as a *surface* when one of the phases is a vacuum [34]. The term *interface* is usually applied to the junction between a solid and vapor (*sv*), solid and liquid (*sl*), or two liquids (*ll*), for example. Interface has special meaning when applied in surface thermodynamics (see Sec. III.B.4 for further discussion). On occasion, *interphase* is used in reference to a region separating bulk phases, to stress that a finite volume with varying component densities separates these phases as in, for example, the concentration of protein within an interphase due to adsorption.

The term surface analysis has become strongly associated with one of the many high-vacuum surface spectroscopies, whereas interfacial analysis is typically applied to wetting/contact-measurements or any of the various nonvacuum techniques such as the use of radiotracers, optical and acoustic spectroscopies, microscopies, scattering techniques, etc. There are a number of good reviews that compare different surface spectroscopies and interface analysis techniques (see, e.g., [35,36]). Of particular interest to biomaterial scientists is Ratner's recently edited collection of papers on surface characterization of biomaterials [37] (see also [38–40]). A generality that can be drawn from these literature reviews is that each surface analytical technique has a characteristic *surface sensitivity*. Surface sensitivity effectively measures the number of molecular layers probed by a particular analytical technique. If we think of a book as a metaphor for a surface with pages representing molecular layers that comprise the surface, then each different surface analytical method "reads and reports" on the contents of different portions of the book.

2. Biomaterial Interfaces

Surface sensitivity is of critical importance in biomaterials surface science because only the uppermost layers are in direct physicochemical contact with the biological environment. Consequently, only the upper few molecular layers determine biocompatibility. The highly magnified, cross-sectional view of a polymer surface in contact with a generalized biological fluid diagrammed in Fig. 1 is intended to illustrate this important point and provide a sense of scale for interfacial phenomena. Chemical events such as acid–base reactions, hydrogen bonding, and ion exchange occur over atomic bond-length distances. Longer-range "hydrophobic forces"* can extend up to about 10 nm and are responsible for nonspecific adsorption, adhesion [41–47], and surface-induced water structure within this zone [48–51]. Surface-adsorbed proteins can have exposed regions that are specific for receptors on biological cells and can greatly influence cell attachment, proliferation, and the response of tissue to biomaterials [52–56]. Thus, the interaction of a material with the biological environment occurs at or through the narrow region termed the interface. Deeper molecular layers are effectively shielded from intimate physicochemical interaction with the liquid phase. Since these molecules are not in direct chemical contact with the external world, no primary influence on biomaterial surface properties is anticipated.

Thus, the interfacial chemistry of concern to biomaterial scientists is determined by material composition within the upper nanometer or so. Interpreted in terms of the surface-book metaphor mentioned earlier, any surface analytical technique that reads more than the first few pages effectively dilutes interfacial information. This dilution can be quite misleading unless one can be certain that subsequent pages have the same chemical information as the first few.

Generally speaking, polymer surfaces are not homogeneous and are more accurately viewed as having chemically distinct strata that blend into bulk composition with depth. The first molecular layers can be quite different from subadjacent layers due to surface oxidation, contamination, or migration of incompatible components such as oils, processing aids, or low-molecular-weight oligomers. These strata are not necessarily fixed, again particularly for polymer surfaces, and molecular reconfiguration can lead to time- or environmentally-induced exchange between the interface and subadjacent zones. Migration of polar head groups to solid/liquid interfaces [57–66] is a particularly pertinent example because biomaterial interfaces are always aqueous.

B. Kinds of Surface Analytical Information

Another generality that can be drawn from a survey of the different surface characterization techniques is that each gives a different kind of information. Broadly speaking, spectroscopies provide some form of compositional information, microscopies yield morphology, and results of ten-

---

* Hydrophobic force is a rather ill-defined term frequently found in the literature collectively referring to dispersion forces related to the interaction of momentary dipoles about molecules comprising matter. These dipoles, in turn, are due to rapid fluctuations in the electron density within molecular orbitals (see [41] and [128]).

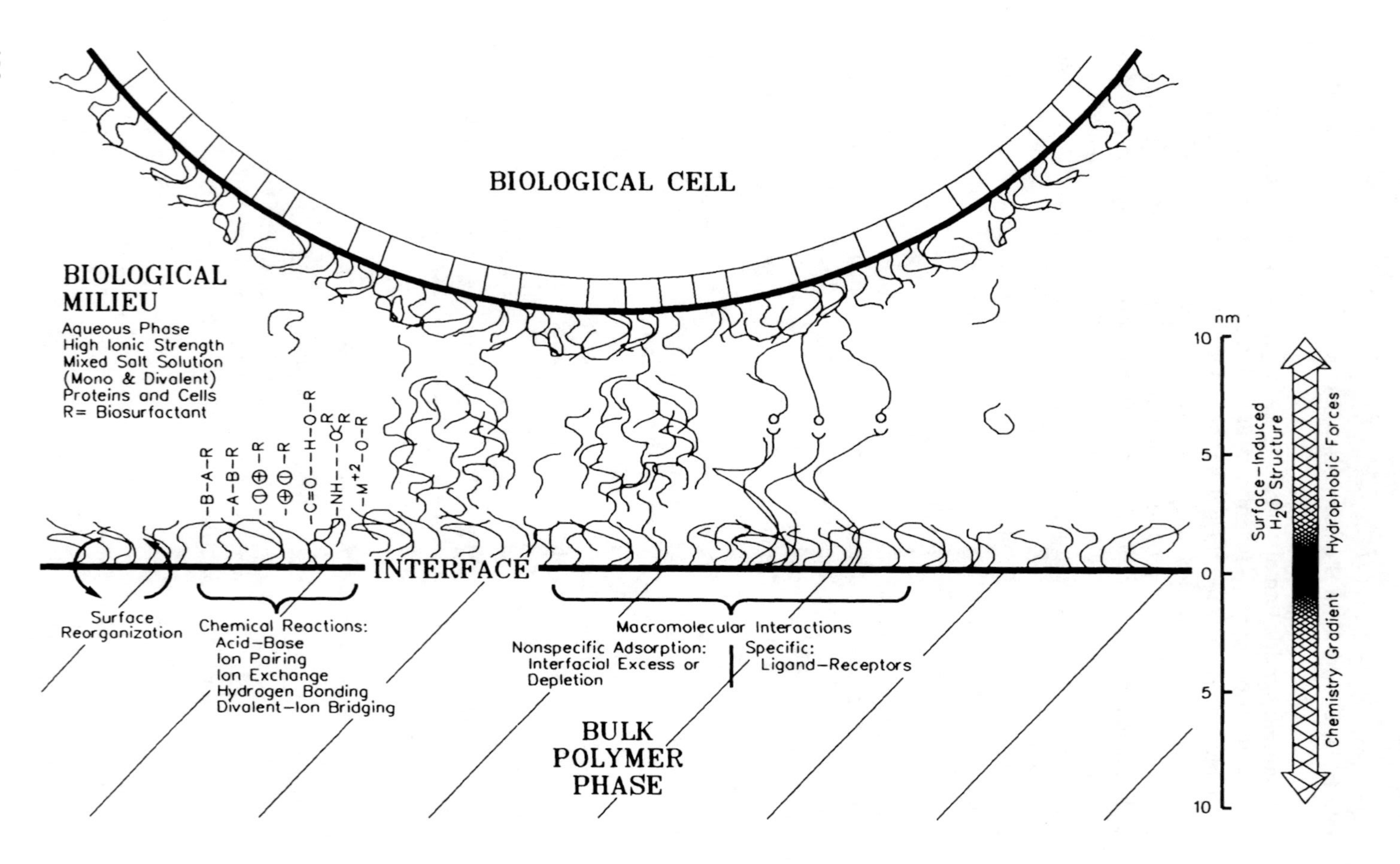

BIOLOGICAL CELL
BIOLOGICAL MILIEU
Aqueous Phase
High Ionic Strength
Mixed Salt Solution
(Mono & Divalent)
Proteins and Cells
R= Biosurfactant
-B-A-R
-A-B-R
-⊖⊕-R
-⊕⊖-R
-C=O---H-O-R
-NH---O-R
-M+2-O-R
INTERFACE
Surface Reorganization
Chemical Reactions:
Acid-Base
Ion Pairing
Ion Exchange
Hydrogen Bonding
Divalent-Ion Bridging
Macromolecular Interactions
Nonspecific Adsorption:
Interfacial Excess or Depletion
Specific:
Ligand-Receptors
BULK POLYMER PHASE
nm
10
5
0
5
10
Surface-Induced H_2O Structure
Hydrophobic Forces
Chemistry Gradient

siometric techniques (wetting and contact angles, see the following sections) are in the form of surface energetics. Within each of these broad categories, there are typically differences between specific techniques. For example, the surface spectroscopy ESCA (electron spectroscopy for chemical analysis) essentially counts different atoms within the upper 10 nm or so of a polymer and results are reported as a kind of surface stoichiometry in atom percent (excluding hydrogen, which is not detected). SSIMS (static secondary ion mass spectroscopy) is another kind of surface spectroscopy that is gaining considerable popularity among biomaterial scientists; it generates mass spectra from molecules sputtered from the upper 1 nm of a surface by energetic ion beams. Thus, each analytical technique provides a different view of a surface with a characteristic surface sensitivity. Not only are different numbers of pages of the surface book read, but also the contents are reported in different "languages."

This richness of analytical detail is actually a kind of problem. In many ways, the "art" in modern biomaterials surface science is a sensible combination of these different views into a coherent and meaningful interpretation of interfacial chemistry.

C. Relevance to the Biomedical Environment

The third surface characterization issue that biomaterial scientists encounter is relevance of the analytical information to biomedical applications. The important question here asks whether the surface that the microscope, spectrometer, or tensiometer "sees" is the same surface that the biological environment confronts. All biomaterial interfaces are aqueous in nature. However, many surface analysis techniques require dry-state, if not high-vacuum, sample preparation.* This introduces the relevance question since, as mentioned previously, it is known that sur-

* Dry-state or high-vacuum (solid/air or solid/vacuum) interfaces are "hydrophobic," whereas aqueous interfaces are "hydrophilic."

FIG. 1 High-magnification, diagrammatic view of a polymer interface (10^7 × magnification) in contact with a generalized biological milieu, providing a sense of scale for interfacial phenomena at biomaterial surfaces. So-called "hydrophobic forces" emanating from the upper 1 nm of a surface extend into solution about 10 nm. These forces are responsible for adsorption and unfolding of proteins at surfaces that can reveal domains that are ligands or receptors for other biological molecules. Chemical reactions such as ion exchange and hydrogen bonding occur over single-atom distances. A chemistry gradient usually exists within the bulk polymer phase and polymeric surfaces can reorganize under hydrating conditions.

faces can reorganize in aqueous environments. More important, perhaps, is the effect of protein adsorption. Proteins adsorb nearly instantaneously and tenaciously to most polymers, obscuring the original surface chemistry from important secondary processes such as cell/tissue adhesion that determine the chronic behavior of biomaterials in end use [1–8,67].

That every tool has a limited range of application is particularly true in the surface characterization of polymeric biomaterials because of the confounding complexities of surface analysis and biology at interfaces. There are both practical and fundamental surface science problems in biomaterials and no single surface analytical tool suffices in every application. An old adage asserts that if the only available tool is a hammer, every problem becomes a nail. Lest biomaterial investigators fall victim to this sometimes not so obvious pitfall, surface and interface analytical tools must be carefully applied, keeping in mind which layer or layers of the material surface are being analyzed and how those layers interact with the biological environment.

D. Tensiometry

Tensiometry encompasses a broad range of related "wetting" techniques that measure surface energetics. These include the observation of contact angles, which is perhaps the most familiar and widely applied method. Tensiometric methods have singular potential in biomaterials surface science based on the criteria of surface sensitivity, kind of analytical information obtained, and relevance of that information to biomedical problems. First, with respect to surface sensitivity, wetting measurements are sensitive only to the upper 0.5 nm or so of a surface [68,69] and are, therefore, among the most surface-sensitive techniques available. Second, tensiometry directly measures the fundamental energetics at an interface that drive important processes such as adsorption and adhesion. This kind of information must be particularly pertinent to biomaterial problems because of the overwhelming importance of protein adsorption and cell/tissue adhesion. Third, wetting measurements can be made using proteinaceous saline solutions that are particularly relevant to biomedical applications. Special high-vacuum preparation techniques that might introduce experimental artifacts are not required.

Many modifications and extensions of the basic tensiometric methods developed in the previous century have been made [70–80]. Underlying principles remain unchanged, however, and results of tensiometric measurements are best interpreted within the framework of thermodynamics, the language of energetics. This introduces certain conceptual problems in application to biomaterial problems because complex biological process must be broken into understandable, yet realistic, pieces and interpreted in physicochemical terms that reveal the role of interfacial energetics.

E. Example

A simple study comparing ESCA and cell adhesion serves to illustrate the point that biological events such as cell attachment are sensitive to interfacial phenomena that certain surface analytical techniques may not directly detect.

Table 1 summarizes semiquantitative surface composition of four different ethylene-based polymer films obtained by ESCA over the binding energy region 1100-0 ev (see [81] for details; PE = polyethylene, E/MA = ethylene methacrylic acid copolymer, E/MA/NA, Zn = E/MA neutralized with sodium or zinc). Figure 2 shows these broad-scan ESCA spectra. Predominant ESCA spectral features are denoted in the usual spectroscopic shorthand: C(1*s*), O(1*s*), N(1*s*), and Na(1*s*) are photoelectrons ejected from the first *s* core-level shell of the elements carbon oxygen, nitrogen, and sodium, respectively; Cl(2*p*) and Na(2*s*) are photoelectrons arising from the *s* and *p* levels of the second electron shell of chlorine and sodium. Broad spectral features labeled KVV are Auger lines for corresponding elements.

Cell (MDCK, epithelioid) attachment rate profiles for the ethylene polymer films are also shown in Fig. 2. Cell attachment measurements were made by simply adding a single-cell suspension of cells in a protein-containing, serum-based medium to culture dishes fabricated with films of interest and incubating. Attachment rate profiles plot the percent of inoculum attached to films at different exposure times (see [82] for details).

TABLE 1 Surface Composition of Ethylene Polymers

	Atom percents excluding hydrogen					
Substrate	C (285.0)	O (532.0)	N (399.9)	Na (65.0)	Cl (199.2)	Zn (1020.9)
PE	98.9	1.1	—	—	—	—
E/MA/Zn	98.4	1.5	—	—	—	—
E/MA/Na	92.3	1.4	—	3.2	3.1	—
E/MA	98.0	2.0	—	—	—	—
E/MA/FBS	78.8	12.7	8.5	—	—	—
PE/FBS	80.5	11.5	8.0	—	—	—

PE = polyethylene, "Alathon" 1645 (with some surface oxidation).
E/MA = ethylene-methacrylic acid copolymer, "Nucrel 0903."
E/MA/Na = Na-neutralized E/MA copolymer, "Surlyn" 1601, 54% Na (with some trace chlorine surface contamination).
E/MA/Zn = Zn-neutralized E/MA copolymer, "Surlyn" 1702, 22% Zn.
"Alathon," "Surlyn," and "Nucrel" are the tradenames of E.I. du Pont de Nemours and Company.

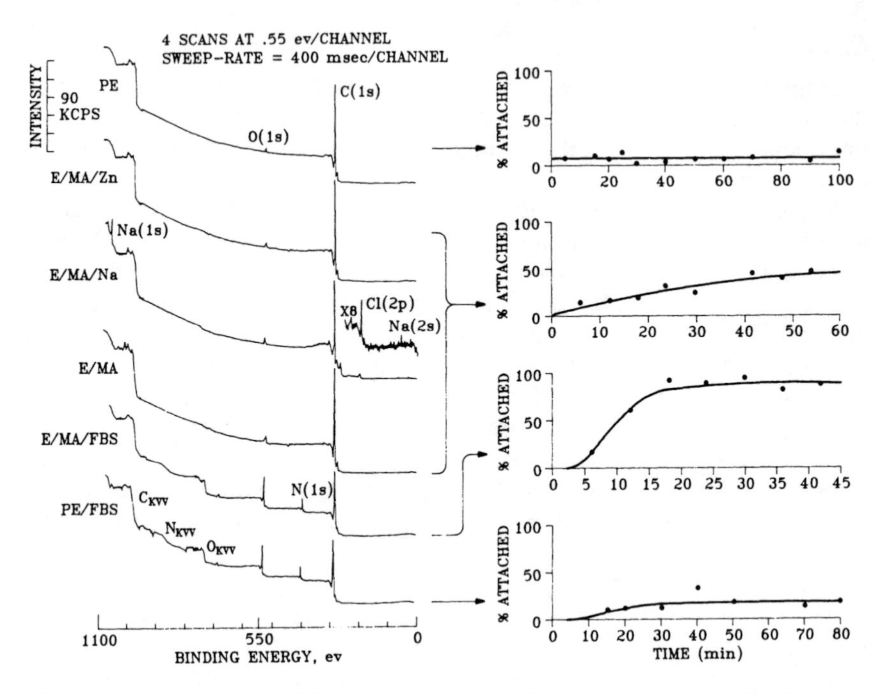

FIG. 2 Comparison of ESCA and cell adhesion for an ethylene-based polymeric film system. Introduction of methacrylic acid (MA, partly neutralized with Na and Zn) into the polyethylene (PE) backbone significantly changes cell (MDCK) adhesion characteristics, but these different polymers are not distinguished by broad-scan ESCA. Preadsorption of fetal bovine serum (FBS) proteins amplifies the cell adhesion difference between PE and E/MA. High-resolution ESCA is required to detect subtle surface chemistry differences that strongly influence interfacial properties that drive cell adhesion.

Films of PE (with some surface oxidation) were found not to be substantially different in surface composition by ESCA from those of E/MA/Na (with minor NaCl contamination), E/MA/Zn, or E/MA.

$$-[CH_2-CH_2]_n- \qquad -[CH_2-CH_2]_n-[CH_2-\underset{CH_3}{\overset{CO_2(H,\ Na^+,\ Zn^{++})}{C}}]_m-$$

Polyethylene — Ethylene/Methacrylate Copolymer
(E/MA = H; E/MA/Na = Na^+; E/MA/Zn = Zn^{++})

Likewise, PE/FBS and E/MA/FBS with preadsorbed fetal bovine serum

(FBS) proteins* could not be clearly distinguished from one another in terms of surface chemical composition by ESCA. By contrast, *cell attachment rates* were very sensitive to differences in surface chemistry between ionomers and plain polyethylene (and among ionomers in detail [82]).

Cells apparently experienced significantly different chemistry at the various film/liquid interfaces that was not observed in the ESCA of film/vacuum surfaces. Clearer correlations between cell attachment and ESCA surface composition begin to emerge only when enhanced sensitivity techniques and peak deconvolution are utilized to more finely resolve surface stoichiometry [26,31]. These results suggest that *subtle differences* in elemental composition or oxidation states are related to *striking differences* in cell adhesion. Apparently, at the highest surface sensitivity available, ESCA just begins to detect compositional differences at the interfacial level. These compositional differences must be somehow connected to interfacial energetics that control cell adhesion. However, this connection can only be a secondary relationship since no information on the interaction of the aqueous phase with surfaces and cells can be deduced from ESCA spectra. That is, the biomedical relevance issue associated with treating dry-state surface chemistry as the primary driver of cell attachment is that the aqueous phase with dissolved proteins is regarded as a simple, neutral carrier of cells that does not interact at film surfaces. As will be shown in the following sections of this chapter, this sort of interpretation is clearly an unwarranted oversimplification.

A more accurate view of the cell attachment process is to recognize that the primary biological response to film surfaces is adsorption (see Sec. IV.D). The secondary event monitored by cell attachment is a response to surface energetics created at the interface by this primary event [40,83]. In recognition of this, a number of predictive theories have been proposed correlating surface energetics with cell attachment [13–22,30, 84–88]. One such theory will be discussed in subsequent sections of this chapter. By contrast, it has yet to be shown that chemistry within the upper 1–10 nm sensed by most spectroscopic methods is primarily linked to biomaterial interfacial properties through a direct cause and effect; at least no widely predictive theory based on spectroscopic measurements has been proposed.

F. Philosophy

Relations between surface properties and the biological response to materials have been the subject of biomaterials research for more than two

* FBS is a product of blood coagulation of fetal calves used in the culture of animal cells in vitro. See Sec. III.E.1 for experimental details.

decades. Despite intensity of effort and application of ever increasingly sophisticated surface analytical techniques, few concepts have emerged that have wide predictive power. Fundamental parameters driving biologic responses to materials are apparently yet to be discovered and interpreted within a predictive theoretical framework. A more reductionistic approach to biomaterials science will be required to elucidate the fundamental relationships between surfaces, interfaces, and the biological response to materials.

From a historical perspective, Baier's proposal that critical surface energy (see the next section) can be directly linked to biocompatibility is perhaps the most penetrating concept among the few generalities offered to explain rules of biocompatibility. This theory, in its most general form, recognizes that surface energetics must control the way biologic fluids interact with materials and that this interaction, in turn, must primarily influence tissue and cell reactions. Although this hypothesis sensibly agrees with scientific intuition, it remains basically untested because, as will be discussed subsequently, critical surface energy does not measure biological-fluid interactions with materials. Consequently, tools appropriate for the task of assessing the merit of this surface-energy concept are still to be developed. It seems evident that these tools cannot be spectroscopically based because surface spectroscopies either have limited surface sensitivity, require high-vacuum sample preparation, or cannot be interpreted in terms of fundamental surface energetics. Tensiometric tools simultaneously meet the requirements of surface sensitivity and yield results that are relevant to the hydrated physiological environment. The challenge is to properly apply tensiometry to biomaterial problems and correctly interpret the results. The rest of this chapter will be dedicated to the discussion of the use and interpretation of tensiometric measurements in biomaterial problems.

III. MEASUREMENT AND INTERPRETATION OF BIOMATERIAL WETTING PROPERTIES

A. Surface Energy, Interfacial Tensions, and Contact Angles

1. Surface Energy

Surface energy is a physical property arising from the placement of molecules at the boundary of a material. These molecules are energetically unique by virtue of being deprived of stabilizing, nearest-neighbor interactions that otherwise would have been experienced in the bulk phase. The loss of these nearest-neighbor interactions leads an excess energy state relative to those in the bulk. This excess energy state is surface energy. In units of energy-per-unit area (ergs/cm^2), surface energy is an intensive thermodynamic property of a material. These units are formally equivalent to a force per unit length (dyne/cm), which physicists call a tension. The terms surface energy and surface tension are synonyms.

2. Interfacial Tension

An interface is formed when two different phases come into contact. Interaction of atoms along the interfacial plane leads to a unique interfacial energy or tension. Interfacial tensions can be measured using contact-angle (tensiometric) techniques. It turns out that the interpretation of liquid/vapor (*lv*) or liquid/liquid (*ll*) interfacial tensions is quite straightforward and unambiguous. Determination of solid/vapor (*sv*) and solid/liquid (*sl*) tensions is, on the other hand, quite ambiguous, requiring application of one of a number of different theoretical approaches [9–12, 89–95]. This ambiguity and the controversy that surrounds it [90,96–100] have been detrimental to the correct and widespread application of interfacial chemistry to biomaterial problems. It is the goal of this section to review some of the basic concepts, with an eye toward separating fact from controversy and identifying that which is practically useful for the biomaterial practitioner.

3. Contact Angles and the Young Equation

The Young equation relates the contact angle θ formed by a droplet of liquid on a substrate to interfacial tensions γ at the (*lv*), (*sv*), and (*sl*) interfaces (Fig. 3c, see [101,102] for an example

$$\gamma_{(sv)} - \gamma_{(sl)} = \gamma_{(lv)} \cos \theta_{(sl)} \equiv \tau \tag{1}$$

The product $\gamma_{(lv)} \cos \theta$ is defined as the wetting or adhesion tension τ, which can be either directly measured or calculated from separate $\gamma_{(lv)}$ and θ determinations. The Young equation is somewhat enigmatic because it relates the measurable quantity τ to the difference between unmeasurable quantities $\gamma_{(sv)}$ and $\gamma_{(sl)}$. Some investigators have sought to "solve" Eq. (1) for the quantity $\gamma_{(sv)}$ and use this value as a single, material-related parameter that characterizes wetting properties. We will see that this is an unnecessary complication for practical biomaterial applications.

4. Critical Surface Tension

Another parameter designed to measure surface energetics that appears in the literature is critical surface energy γ_c [90,101,103,104]. A series of test liquids with different $\gamma_{(lv)}$ values are used in the measurement of θ. Extrapolation of the trend line connecting points on a cos θ plot against $\gamma_{(lv)}$ yields γ_c at the intersection $\cos \theta = 1$, the point of perfect wetting.

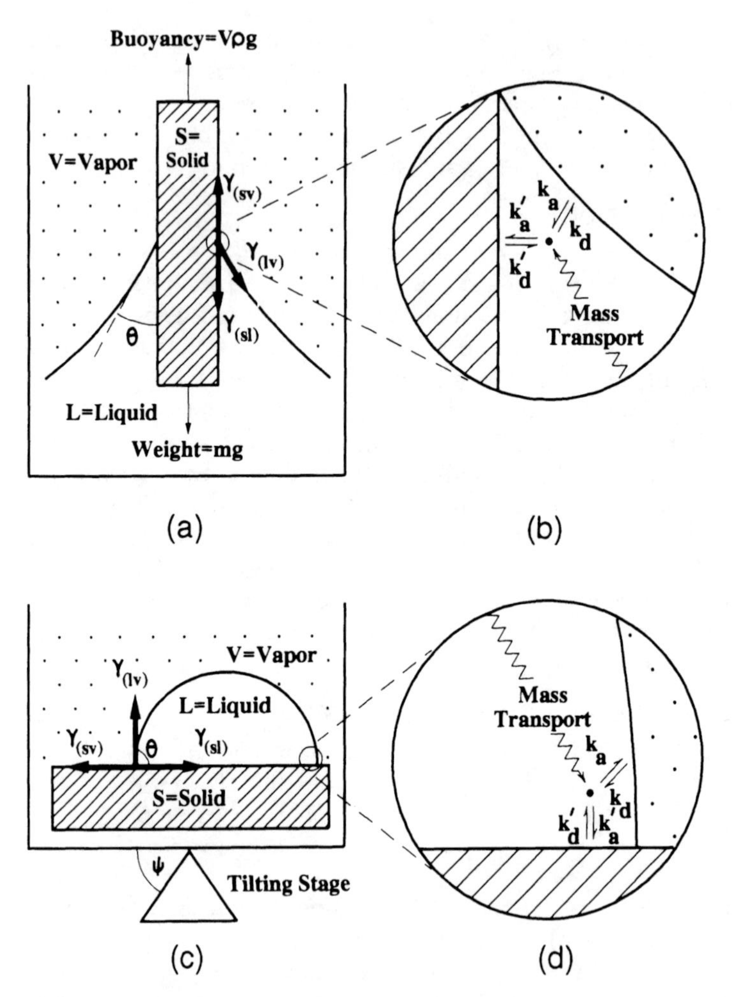

FIG. 3 Illustration of the Wilhelmy balance (a) and contact-angle (c) methods for measuring adhesion tension. (b) and (d) diagram the meniscus regions showing adsorption from bulk to interfacial regions governed by adsorption and desorption rates k_a and k_d, as well as mass transport steps.

Values for γ_c are found to be characteristic of different materials. Historically, the critical surface energy concept stimulated development of theories aimed at solving the Young equation. It is of interest to follow development and application of these theories in colloid and biomaterials surface science.

5. Biomedical Relevance of γ_c and $\gamma_{(sv)}$

The parameters of γ_c and $\gamma_{(sv)}$ fail the test of biomedical relevance because neither measures interaction of the surface with liquid phases containing dissolved proteins. For nearly every biomaterials application imaginable, from in vitro tissue culture to in vivo implants, biomaterials first encounter the biological environment through a proteinaceous liquid phase, setting forth a complex panoply of events ranging from surface reorganization to adsorption of proteins. Any surface energy parameter that effectively disregards these events cannot be, therefore, *primarily* connected to or directly correlated with the biological response to the surface in question. It is for these reasons that γ_c and $\gamma_{(sv)}$ fail as the long-sought, single-material-related parameter measuring biocompatibility.

6. Biomedical Relevance of τ

Examination of the Young equation reveals that $\gamma_{(sl)}$ and τ terms measure solid-/liquid-phase interaction. Either of these terms pass the biomedical relevancy test. However, $\gamma_{(sl)}$ cannot be calculated from the Young equation in a general way because theories allowing separation of the individual (sv) and (sl) terms are restricted in use to liquid phases containing no surface-active solutes such as proteins (and if we ignore the controversy surrounding these theories). By contrast, no special theories or complex relationships are required in the measurement or interpretation of τ, and because τ (an observable) is directly proportional to $\gamma_{(sl)}$ (not observable), it is equally sensitive to interfacial events of concern to biomaterial scientists such as protein adsorption. Furthermore, τ can be measured at any instant in time, whether or not the system is in equilibrium, and correctly interpreted as a force exerted on a substrate by the liquid.

7. Work of Adhesion

One of the more useful thermodynamic relationships involving τ is work of adhesion

$$W = \tau + \gamma_{(lv)} \tag{2}$$

W is the work required to remove liquid from a solid (per unit area of contact), with higher W values reflecting greater interaction of a solid with a liquid. Work of adhesion gives a complete picture of the interplay of interfacial forces that govern liquid, solid, and solute interactions.

B. Concentration-Dependent Interfacial Tensions and Work of Adhesion

1. Surfactant Effect

The surfactant effect is preferential partitioning of a surface-active solute from bulk to interfacial regions. The term partition has been purposely selected here to emphasize the fact that a surface-active solute in aqueous solution is driven from the bulk phase by water/molecule interactions that tend to expel incompatible, hydrophobic domains to a relatively hydrophobic interface. Solute partitioning is a more precise way to think of adsorption [34] and reduction of interfacial tensions is typically the driving force. This partitioning can be either positive, negative, or neutral in direction. Commonly, adsorption is thought of as a *positive* action for solute molecules. In this case, the interphase separating bulk liquid and bulk solid phase becomes enriched in solute and necessarily depleted in solvent. However, interfacial energetics may drive adsorption in a *negative* direction, leading to an interphase depleted in solute relative to bulk concentration but enriched in solvent molecules. Surface energetics can be adjusted so that there is no net partitioning of solute.

2. Biosurfactant Effect

Many biological macromolecules adsorb to interfaces from aqueous solution. As examples, it is common knowledge that blood proteins adsorb to artificial materials in vivo and that serum proteins adsorb from culture media to plastics in vitro. Proteins are in every sense surface-active molecules or, by definition, surfactants. Thus, it can be expected that these molecules should obey many of the same laws that describe and quantify surface activity of ordinary detergents. Stated as a testable hypothesis, biological macromolecules behave as *biosurfactants* and all of protein interfacial chemistry, including adsorption and related phenomena,* can be interpreted in the same surface thermodynamic terminology used to understand and quantify ordinary surfactant effects.

3. Interfacial Tension Curves [149,150]

Adsorption can be measured by monitoring interfacial tensions as a function of surfactant or biosurfactant concentration. A plot of interfacial tension against surfactant concentration on a logarithmic scale is an effective way of presenting data and is here termed an interfacial *tension curve*. Appropriate interfacial tensions are, of course, γ and τ: $\gamma_{(ll)}$ and $\gamma_{(lv)}$ for liquid interfaces, respectively, and τ for solid interfaces. We will be concerned only with (lv) and (sl) interfaces in this chapter, although (ll) interfaces are a unique and somewhat simplified model system to study adsorption fundamentals.

Adsorption to (lv) and (sl) interfaces causes measurable changes in $\gamma_{(lv)}$ and τ over a concentration range that is characteristic of the surfac-

* Excluding, perhaps, specific biological activity such as ligand-receptor interactions and enzymatic activity.

tant, compatibility with solvent (usually buffered physiologic saline for biomaterial applications), and surfactant activity at the interface. It is usually observed that interfacial tensions rise or fall as a function of surfactant concentration, from $\gamma^{\circ}_{(lv)}$ or τ° at infinite dilution to a limiting value $\gamma'_{(lv)}$ or τ' that usually occurs after the critical micelle concentration (CMC) of surfactant is reached. The parameters $\gamma^{\circ}_{(lv)}$ and τ° are inherent material properties measured with pure solvent that can be used to compare the surface energies of different materials. Maximal surfactant effect is measured by $\gamma'_{(lv)}$ and τ'.

Tension curves can be constructed for any surfactant or biosurfactant system under study. Surfactant solutions need not be comprised of a single surface-active component and may actually consist of some unspecified number of molecules because measurements of $\gamma_{(lv)}$ and τ are valid for any composition. This is a critical attribute for biomaterial applications since complex proteinaceous mixtures such as blood, diagnostic samples, or culture media derived from the clinical laboratory can be treated as a single biosurfactant system and characterized by tension curves.

4. Parameterization of Tension Curves

Tension curves are typically sigmoidal in shape (on a logarithmic concentration scale), with a linear region connecting the plateau values $\gamma^{\circ}_{(lv)}$ and $\gamma'_{(lv)}$ or τ° and τ'. The author has found it computationally convenient to fit tension curves to a four-parameter variant of the logistic equation [87,88,105–107], which is a mathematical description of a sigmoidal curve having the form

$$Y = \left\{ \left[\frac{(A - D)}{(1 + (K/X)^{N})} \right] + D \right\} \tag{3}$$

where Y is either $\gamma_{(lv)}$ or τ. $X = \ln C = \log_e$(surfactant dilution), where surfactant dilution can be conveniently expressed so that $X > 0$ by using scales such as parts per trillion, or if the molecular weight is known and the ideal dilute solution approximation applied, the number equivalent to picomol/L (picomolar). The parameter A is the fitted plateau values at infinite surfactant dilution corresponding to either $\gamma^{\circ}_{(lv)}$ or τ°. Similarly, D is the fitted plateau value at the high-concentration limit corresponding to $\gamma'_{(lv)}$ or τ'. The K parameter measures surfactant concentration (in ln C units) at half-maximal $\gamma_{(lv)}$ or τ change, so that surfactants with a low CMC have characteristically low K values. Slopes of tension curves are related to the exponential N, which in this formulation of Eq. (3) is always negative, with higher negative values for steeper curves. Although a logistic equation has no special physical or thermodynamic meaning in application to surfactant effects, the fitting procedure provides a statistical means of utilizing data gathered over the entire ln C range in the extraction of $\gamma^{\circ}_{(lv)}$ and $\gamma'_{(lv)}$ or τ° and τ' parameters with error estimates. These parameters then can be conveniently tabulated, allowing quick comparison of biosurfactant properties.

Tension curves can be used to quantify adsorption in those cases when the composition of the surfactant is known through application of the surface thermodynamics developed by Gibbs in the latter quarter of the 19th century. Gibbs' adsorption isotherm for a single component solution* states that the slope of the linear-like portion of a tension curve is proportional to the amount of surfactant adsorbed at the Gibbs' interface, as measured by the surface excess Γ parameter [34,101,108,109]:

$$\frac{\partial[\gamma_{(lv)}]}{\partial[\ln C]} = -RT[\Gamma_{(lv)}] \tag{4}$$

$$\frac{\partial[\tau]}{\partial[\ln C]} = RT[\Gamma_{(sl)} - \Gamma_{(sv)}] \tag{5}$$

where R is the gas constant and T the Kelvin temperature. Surface excess is a special way of measuring adsorption through concentration-dependent contact angles or tensions, and the reader is encouraged to review details of its theoretical development in any good text on surface thermodynamics. For the purposes of this discussion, however, the somewhat simplified view of surface excess shown in Fig. 4 will provide an adequate physical interpretation of Γ. The *interphase* separating a bulk liquid and bulk solid phase for a unit area of surface is diagrammed in Fig. 4a, showing the accumulation of surfactant due to adsorption. Fig. 4b is the surfactant concentration profile as a function of height within this interphase, with the shaded area referring to the amount adsorbed in excess of bulk/liquid composition, n^{σ}. Gibbs' model of adsorption ascribes n^{σ} to a hypothetical planar area (Fig. 4c) placed somewhere within the interphase that is defined as the *interface* separating liquid and solid phases. Thus, Gibbs' interface is a model of an interphase that replaces continuously varying component densities with constant compositions. The surface excess Γ is an estimate of the amount of surfactant accumulated within the three-dimensional interphase, which is ascribed to a two-dimensional interfacial plane and has, therefore, mol/cm^2 as its units. It is not always intuitively obvious where the Gibbs' interface is positioned and locating three-dimensional molecules at a two-dimensional plane loses a certain amount of physical appeal. However, the Gibbs' model allows measurement of

* Specifically, a solution of a single isomerically pure surfactant in a pure solvent at constant temperature and pressure that exhibits ideal-dilute behavior.

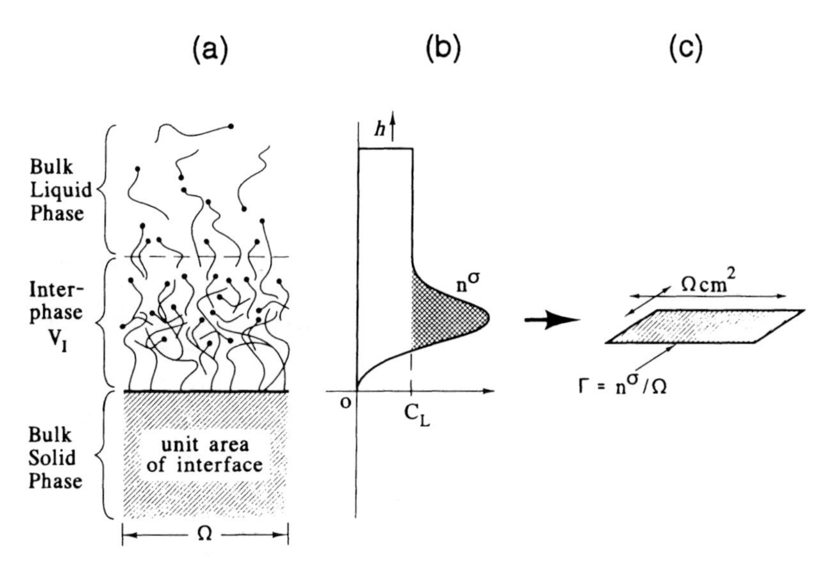

FIG. 4 Depiction of a unit area Ω of surface in contact with a liquid containing a surfactant showing (a) concentration of surfactant within the interphase separating bulk liquid and solid phases by adsorption. The concentration profile as a function of height within the interphase (b) measures the number of adsorbed molecules, with the shaded region referring to the amount *in excess* of bulk concentration C_L. In the Gibbs' model, this excess n^σ is attributed to a planar interface separating bulk phases defining the surface excess parameter $\Gamma \equiv n^\sigma/\Omega$. (From Ref. 150.)

Γ from very sensitive wetting measurements. Surface excess can be a positive or negative number,* corresponding to the conditions of interfacial excess or depletion mentioned in the preceding discussion of surfactant effects.

Adsorption to solids [Eq. (5)] is more complicated than liquids [Eq. (4)] since surfactant can, in principle, accumulate at either or both (*sl*) and (*sv*) interfaces. It can be anticipated that $\Gamma_{(sv)} \approx 0$ for nonvolatile surfactant solutions in contact with nonpolar surfaces under experimental circumstances in which there is no mechanism of adsorption to (*sv*) interfaces [110,111]. Under these highly restrictive conditions, which ap-

---

* Interfacial excesses can be relatively large positive values on the order of hundreds of picomoles per unit area of interface, depending on molecular dimensions in the adsorbed state (see Sec. III.E.2 to III.E.4). By contrast, interfacial depletions must be small negative values because there is only a small number of solute molecules within the interfacial region to be expelled by solvent (water) at hydrophilic surfaces.

parently can be obtained using contact-angle goniometry (see Sec. III.C.2), the slope of a τ curve is directly proportional to $\Gamma_{(sl)}$. This is not necessarily the case for surfaces exhibiting significant wettability and $[\Gamma_{(sl)} - \Gamma_{(sv)}] < 0$ alerts the experimentalist that $\Gamma_{(sv)} > 0$ since $\Gamma_{(sl)}$ can take on only small negative numbers.

5. Adsorption Index [151]

It is more typical than not that liquid phases of practical interest in biomaterials such as blood and bodily fluids are of unknown chemical composition. Thermodynamic concepts such as Gibbs' adsorption equations have, therefore, somewhat limited quantitative utility. Another way to parameterize tension curves that is independent of solution composition is through the differences $[\gamma'_{(lv)} - \gamma^{\circ}_{(lv)}]$ or $[\tau' - \tau^{\circ}]$. These differences are thermodynamic-like indices of laboratory-accessible quantities that measure the net effect of biosurfactant adsorption. Thus, the *adsorption index* has practical value for chemically undefined biosurfactant systems of general interest to the biomaterials community. The utility of the adsorption index will be demonstrated in Sec. V.B.2.

6. Work of Adhesion Curves

W will rise and fall as a function of solute concentration as $\gamma_{(lv)}$ and τ respond to adsorption at the (lv) and (sl) interfaces, respectively. If there is no adsorption to the (sl) interface, τ curves will be flat and W curves will resemble $\gamma_{(lv)}$ curves. If adsorption to the (sl) interface occurs, the appearance of W curves will be dictated by what amounts to constructive or destructive interference of tension curves along the ln C axis. The net effect can be a strong peak in W over a narrow range of solute concentration.

C. Measurement of Interfacial Tensions

1. Wilhelmy Balance

The Wilhelmy balance and contact-angle goniometer are probably the most useful tensiometric tools for the biomaterials practitioner. Both these complementary techniques are generally required for the determination of biomaterial wetting properties: the goniometer for direct observation of θ and the balance for measurement of either or both $\gamma_{(lv)}$ and τ.

The Wilhelmy balance essentially weighs a fluid meniscus adherent to the perimeter of a rigid material as it is immersed (advancing mode) into or withdrawn (receding mode) from a test liquid of interest (Fig. 3a). These force measurements are monitored as a function of depth in the fluid, yielding hysteresis curves from which wetting properties can be determined. Typically, rigid glass or metallic plates, either with or without some coating of interest, are utilized in Wilhelmy balance work, but very good results can be obtained with cylindrical rods or fibers [80,112–116]. The appropriate force balance equation is

$$F = mg = p\gamma_{(lv)} \cos \theta - V\rho g = p\tau - V\rho g \tag{6}$$

where V and ρ are the volume and density of the displaced fluid, respectively, and g is the acceleration due to gravity. Contribution due to τ can be determined from Eq. (6) if the buoyancy $V\rho g$ and sample perimeter p are known. Most frequently, buoyancy is eliminated from Eq. (6) by evaluating force-immersion curves at $V = 0$

$$F\,|_{V=0} = mg\,|_{V=0} = p\gamma_{(lv)} \cos \theta = p\tau \tag{7}$$

Contact angles can usually be driven to zero by making the substrate perfectly wettable by rigorous oxidative treatment, typically by applying flames or oxygen plasmas. Under these special circumstances, interfacial tension $\gamma_{(lv)}$ of the fluid can be determined from mass and perimeter measurements: $\gamma_{(lv)} = mg/p|_{V=0}$. Contact angles can be calculated from τ if $\gamma_{(lv)}$ is known from separate balance measurements. Advancing and receding contact angles correspond to sample immersion and withdrawal cycles, respectively, and are usually different. The difference between advancing and receding contact angles is called contact-angle hysteresis [101,117]. Surface roughness or chemical heterogeneity can give rise to contact-angle hysteresis, and it is not always straightforward to differentiate the two effects. Qualitative observations are that surface roughness increases advancing angles and decreases receding angles. Advancing angles are sensitive to hydrophobic surface domains on a surface and receding angles are characteristic of the relatively hydrophilic domains [117]. Little or no hysteresis is a compelling diagnostic for smooth, chemically homogeneous surfaces and demonstration of high receding angles is the most severe test of hydrophobicity.

Wetting properties, and hence wetting forces measured with the balance, are strongly dependent on adsorption of surface-active solutes (surfactants) to (lv) and (sl) interfaces. In fact, tensiometry is so exquisitely sensitive to adsorption that it is frequently the method of choice to quantify adsorption and measure adsorption kinetics. The meniscus region is depicted in Fig. 3b for a Wilhelmy plate method hysteresis experiment as the solid is immersed into or withdrawn from the liquid phase. Surfactant can adsorb or desorb as liquid moves across the object perimeter and the interfacial region can become depleted in surfactant, triggering movement of solute from the bulk phase to the interfacial region. Thus, a complex set of interrelated events can occur along the three-phase (slv)

line during immersion or withdrawal cycles that can affect the net observed wetting tension [118,119]. The overall rate of wetting tension change is dependent on the interplay between adsorption rates (k_a, k_d, k'_a, and k'_d in Fig. 3b) and mass transport steps. If these rates are fast relative to the time frame of a balance experiment, adsorption effects will be transparent to the investigator and hysteresis curves will be regular. These circumstances can be expected for low-molecular-weight, highly surface-active detergents. Hysteresis curves can be somewhat erratic to uninterpretable, however, when much larger protein molecules or complex mixtures are studied because either or both adsorption kinetics and mass transport affect wetting tension on the same time scale as hysteresis measurements.

2. Contact-Angle Goniometry

A contact-angle goniometer is a horizontal telescope with a reticle eyepiece for observation and measurement of the angle subtended by a small droplet of liquid on a material (Fig. 3c). The goniometer is most useful for planar materials such as plaques or films with one or both surfaces of interest. A tilting stage* allows the drop to advance (leading droplet edge) or recede (trailing droplet edge) as the stage is tilted from horizontal through some angle ψ. A humidified chamber is essential to avoid evaporation, particularly if long equilibration times are required to account for time-dependent process such as adsorption or surface reorganization. Advancing and receding contact angles are usually read at ψ corresponding to incipient rolling of the drop on the substrate. Tilt angle at incipient rolling will be dependent on W, which is controlled by material wetting properties for the test liquid under study [see Eq. (2)]. Generally, $\psi \simeq$ 10–30° for poorly wettable substrates.

As in the Wilhelmy balance case, adsorption to (lv) and (sl) interfaces affects wetting tension and hence the observed contact angle (Fig. 3d). An important difference between the goniometer and balance methods is that the liquid front is more or less stationary on the substrate surface, although vibrations play an important role in determining final advancing and receding contact angles [102,117]. For the most accurate work, therefore, it is important to mount the goniometer on a vibration-isolation table. Adsorption kinetics are observed by monitoring θ as a function of time. Again, as in the Wilhelmy case, these kinetics are controlled by a complicated interplay of adsorption rates (k_a, k_d, k'_a, and k'_d in Fig. 3d) and

* There are a number of alternative methods for measuring hysteresis (see [101] for examples). The tilting stage technique is the author's preference and will be the only method discussed in this work.

mass transport steps. If these rates are fast relative to the time frame of an observation, adsorption effects will be transparent to the investigator and a contact angle will not vary with time. On the other hand, a contact angle will change with time if adsorption or mass transit to interfacial regions is relatively slow.

D. Interpretation of Interfacial Tensions

1. Hypothetical System

Simplicity of wetting measurements belies the surface sensitivity and power of tensiometric techniques. It is all too easy to misuse the information, particularly when chemical interpretations of wetting phenomenon are made. Cautious interpretations bear in mind that wetting measurements probe surface energetics at an essentially two-dimensional interface where adsorption at (sv) and (sl) interfaces control τ. This is particularly important in understanding interfacial phenomena at biomaterial interfaces since biological fluids of interest typically contain many surface-active solutes. It is worthwhile, therefore, to closely scrutinize the interfacial forces that lead to the observables τ and θ and how adsorption can affect experimental values obtained using different protocols.

For this purpose, consider the hypothetical contact angle and Wilhelmy balance experiments outlined in Fig. 5. Substrates for each case are identical smooth, nondeformable plates coated with a uniform surface chemistry of interest, such as a hydrophobic layer, for example. These idealized substrates exhibit a single contact angle with no hysteresis. Liquid phases are likewise identical and contain a dissolved solute that may or may not adsorb to solid interfaces, as dictated by model assumptions. The solute is assumed to be nonvolatile and sufficiently dilute so as not to significantly affect vapor-phase partial pressure or composition.

2. Adsorption to Liquid/Vapor Interfaces

As depicted in Fig. 3, solute can adsorb to (lv) interfaces, as well as (sl) interfaces, during the measurement of interfacial tensions. Adsorption to the (lv) interface can be quantified independently from $\gamma_{(lv)}$ curves as described in previous sections. In the event that the solution under study consists of a single surfactant of known molecular weight, Eq. (4) can be used to calculate $\Gamma_{(lv)}$.

3. Adsorption to Solid Interfaces

There are at least three ways a goniometer and a Wilhelmy balance can be used to measure concentration-dependent τ from which adsorption to the (sl) interfaces can be estimated. To aid in the comparison of these

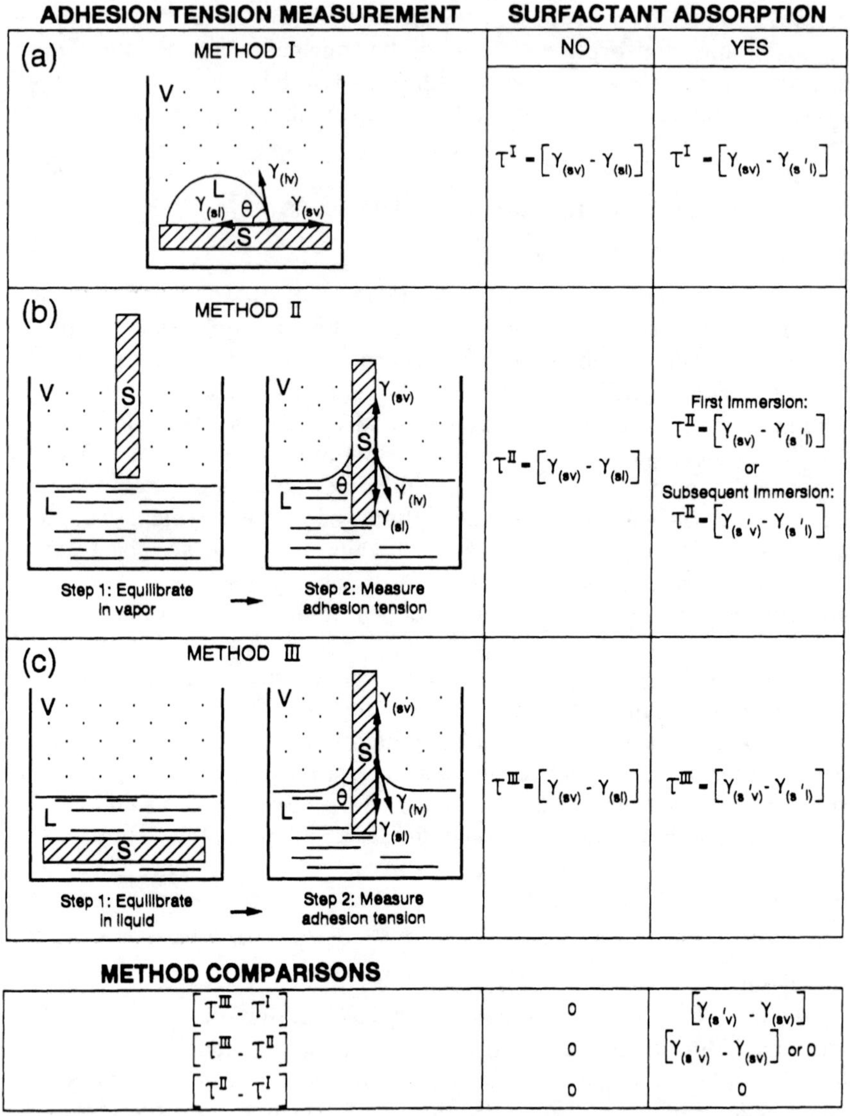

ADHESION TENSION MEASUREMENT	SURFACTANT ADSORPTION: NO	SURFACTANT ADSORPTION: YES
(a) METHOD I V, L, S, $\gamma_{(lv)}$, $\gamma_{(sl)}$, θ, $\gamma_{(sv)}$	$\tau^{I} = [\gamma_{(sv)} - \gamma_{(sl)}]$	$\tau^{I} = [\gamma_{(sv)} - \gamma_{(s'l)}]$
(b) METHOD II V, S, L; V, S, L, θ, $\gamma_{(sv)}$, $\gamma_{(lv)}$, $\gamma_{(sl)}$ Step 1: Equilibrate in vapor → Step 2: Measure adhesion tension	$\tau^{II} = [\gamma_{(sv)} - \gamma_{(sl)}]$	First Immersion: $\tau^{II} = [\gamma_{(sv)} - \gamma_{(s'l)}]$ or Subsequent Immersion: $\tau^{II} = [\gamma_{(s'v)} - \gamma_{(s'l)}]$
(c) METHOD III V, L, S; V, S, L, θ, $\gamma_{(sv)}$, $\gamma_{(lv)}$, $\gamma_{(sl)}$ Step 1: Equilibrate in liquid → Step 2: Measure adhesion tension	$\tau^{III} = [\gamma_{(sv)} - \gamma_{(sl)}]$	$\tau^{III} = [\gamma_{(s'v)} - \gamma_{(s'l)}]$

METHOD COMPARISONS

$[\tau^{III} - \tau^{I}]$	0	$[\gamma_{(s'v)} - \gamma_{(sv)}]$
$[\tau^{III} - \tau^{II}]$	0	$[\gamma_{(s'v)} - \gamma_{(sv)}]$ or 0
$[\tau^{II} - \tau^{I}]$	0	0

FIG. 5 Comparison of three methods of measuring wetting tension. (a) is the contact-angle method (see Fig. 3c). (b) and (c) are Wilhelmy balance methods (see Fig. 3a) with and without an adsorption preequilibration step, respectively. (From Ref. 150.)

three independent methods, solid interfaces with adsorbed solute will be differentiated from the case in which no adsorption occurs by the (s') notation. Thus, $\gamma_{(s'v)}$ and $\gamma_{(s'l)}$ are solid/vapor or solid/liquid interfacial tensions, respectively, for plates with adsorbed solute. It is further assumed that adsorption proceeds to steady state so that it is clear that (s') indicates identical adsorption to solid surfaces between cases. This notation should not be confused with $\gamma'_{(lv)}$ or τ' defined in previous sections.

(a) Contact-Angle Goniometry. Method I (Fig. 5a, also Fig. 3c) is the classical contact-angle measurement in which a dry plate is equilibrated with vapor phase surrounding a small droplet of liquid. Wetting tension for case I is $\tau^{\mathrm{I}} = [\gamma_{(sv)} - \gamma_{(sl)}] = \gamma_{(lv)} \cos\theta$ and must be calculated from separate measurements of $\gamma_{(lv)}$ and θ, as discussed previously. There are two ways this information can be used to estimate values for $[\Gamma_{(sl)} - \Gamma_{(sv)}]$ when the solution composition is known. First, Eq. (5) can be directly applied to τ curves assembled from the individual values calculated from separate θ observations. This has the disadvantages that (1) errors in $\gamma_{(lv)}$ and the θ compound* in calculated τ and (2) the "synthetic" τ is not a direct observable that is usually preferred by experimentalists. These problems can be avoided by calculating surface excess through

$$[\Gamma_{(sl)} - \Gamma_{(sv)}] = -\left\{\frac{[\gamma_{(lv)} \sin\theta \, d\theta/d \ln C]}{RT} + [\Gamma_{(lv)} \cos\theta]\right\} \tag{8}$$

where the term $[d\theta/d \ln C]$ is the slope of the θ vs. $\ln C$ curve. Equation (8) can be derived [102,149] by differentiating Eq. (1) with respect to concentration and allows excess to be calculated from experimental observables.

Adsorption to the (sl) interface will alter the observed wetting tension so that $[\gamma_{(sv)} - \gamma_{(sl)}] \neq [\gamma_{(sv)} - \gamma_{(s'l)}] = \tau^{\mathrm{I}}$. Adsorption does not change $\gamma_{(sv)}$, since by model assumptions, there is no mechanism for solute migration to the (sv) interface on these hypothetical hydrophobic substrates.

(b) Wilhelmy Balance. Methods II and III are Wilhelmy balance experiments that are otherwise identical, except that in method III the slide is first completely immersed in the liquid phase for some length of time. Effect of the immersion pretreatment is important only in the event of adsorption because adsorbed solute will coat the entire slide, including that portion to be exposed to vapor in subsequent measurement steps. For this case with solute adsorption, $\tau^{\mathrm{III}} = [\gamma_{(s'v)} - \gamma_{(s'l)}]$.

---

* Propagation of error in $\gamma_{(lv)}$ and θ into calculated τ values does not lead to a simple root-mean-square of individual error contributions due to the transcendental function.

Method II with adsorption is a special case because it changes modes after the first immersion. If we assume dipping speed is sufficiently slow so that adsorption reaches equilibrium and there is no mechanism of solute deposition at the (sv) interface, then $\tau^{II} = [\gamma_{(sv)} - \gamma_{(s'l)}]$. In the first receding cycle or any subsequent immersions, however, $\tau^{II} = [\gamma_{(s'v)} - \gamma_{(s'l)}]$ because adsorbed solute is exposed at the (sv) interface, as described for method III.

It is very difficult in practice to prevent solute deposition at the (sv) interface using the Wilhelmy balance because of plate movements and vibration-induced ripples in the liquid surface. The net effect is that the three-phase line (slv) rises and falls on a microscopic scale, depositing solute at the (sv) interface by evaporation. A consequence of some or all of these factors is that wetting forces along the perimeter of the specimen can be highly erratic as the solute adsorbs and desorbs (depending on the kinetics of adsorption; see Fig. 3b). Method II force-immersion curves can be particularly time-dependent in nature and quite difficult to interpret.

4. Comparison of Results

Results for the three methods of measuring wetting tension are listed in the comparison table of Fig. 5. In general, identical results will be obtained only for the special case in which no adsorption occurs. When there is adsorption, each method can lead to different results, depending on the condition of solute deposition at the (sv) interface. It is important in comparing results of different methods of measuring interfacial properties of biomaterials to recognize and take into account potential differences in the information obtained.

Under highly controlled conditions of equilibrium adsorption and identically prepared surfaces, information regarding adsorption to the (sv) interface can be obtained by comparing methods III and I since $[\tau^{III} - \tau^{I}] = [\gamma_{(s'v)} - \gamma_{(sv)}]$. This difference is a kind of "tension excess" that effectively measures the additional stress at the vapor interface imparted by adsorbed solute. It is easy to show that this tension excess manifests itself as a difference in work of adhesion: $\Delta W = [W^{III} - W^{I}] = [\tau^{III} - \tau^{I}]$.

E. Example

1. Model System

As an illustration of the effect of surfactant adsorption on the measurement of wetting tension, methods I–III (Fig. 5) were applied to a model surface in construction of tension curves for a low-molecular-weight detergent (tween-80, polyoxyethylene sorbitan monooleate, nominal MW = 1309.68, density (d) = 1.064, used as received from Aldrich) and a relatively higher-molecular-weight protein (human serum albumin, HSA, prepared from globulin-free lyophilized crystals, nominal MW = 69,000, used as received from Sigma Chemical). Fetal bovine serum (FBS, diluted in physiologic saline from whole serum as received from Gibco Laboratories) serves as an example of a chemically undefined mixture of proteins. FBS is a product of blood coagulation that consists mostly of albumin, but at least 500 other proteins can be identified in this complex milieu [120,121].

Model surfaces were smooth glass coverslips with an octadecyltrichlorosilane coating. Plates were prepared batchwise, yielding identically treated specimens with little batch-to-batch variability (saline contact angles varied about 5° within and between batches). However, these surfaces were not perfect as described for the idealized substrate in the preceding section, exhibiting about 20° hysteresis on an advancing contact angle of about 110° (see the headings of Fig. 8).

2. Adsorption to Liquid/Vapor Interfaces

Figure 6 is a $\gamma_{(lv)}$ curve that compares surfactant properties of tween, HSA, and FBS. It is apparent that proteins in solution exhibit surfactant properties that are very similar to the ordinary detergent. The (lv) interfacial behavior of FBS is included in this comparison to illustrate that even though the exact composition of the FBS is unknown, this heterogeneous mixture of biological macromolecules can be phenomenologically treated as a single biosurfactant that partitions between bulk and interfacial regions in a manner very similar to that observed for the purified detergent and protein.

Smooth curves through the data of Fig. 6 result from least-squares fitting to the logistic Eq. (3), as described in the preceding section, yielding characteristic parameters $\gamma^{\circ}_{(lv)}$, $\gamma'_{(lv)}$, $K_{(lv)}$, and $N_{(lv)}$ collected in Table 2. The $\gamma^{\circ}_{(lv)}$ parameter measures liquid interfacial tension at infinite surfactant dilution and is, therefore, equal to the interfacial tension of saline used to prepare the different surfactant solutions. Tween $\gamma'_{(lv)}$ was about 15 dyne/cm lower than that for HSA or FBS. Protein solutions also exhibit a limiting $\gamma'_{(lv)}$ value, probably related to a maximum adsorption at the (lv) interface. Interfacial excess values $\Gamma_{(lv)}$ for tween and HSA (for which the density and MW are known) were calculated from the linear slope region of the $\gamma_{(lv)}$ curves using Eq. (4). For tween and HSA, $\Gamma_{(lv)} = 215.6$ and 97.8 picomol/cm^2, respectively. Greater $\Gamma_{(lv)}$ for tween indicates that adsorbed molecular area is considerably smaller than that of HSA, leading to a higher packing in a unit area of interface, as might be expected on the basis of relative molecular weight.

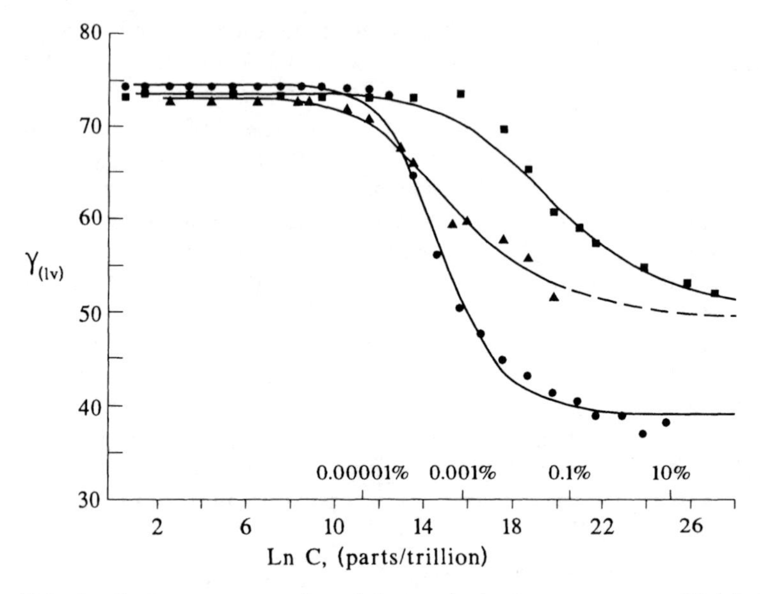

FIG. 6 Surfactant properties of the nonionic detergent tween-80 (circles), fetal bovine serum (FBS, squares), and human serum albumin (HSA, triangles). Smooth curves through the data result from fitting to a logistic equation.

It is of interest to interpret surface excess in terms of molecular configuration in the adsorbed state by calculating the adsorbed area in units of Å^2/molecule from Γ^{-1}. Results are of use on a relative basis, but cannot be taken as quantitative measures of a three-dimensional molecular configuration. This is because the Gibbs' model of a three-dimensional interphase is a two-dimensional interface, as discussed in Sec. III.B.4 and illustrated in Fig. 4. As a consequence, estimates of Å^2/molecule are generally higher than physically possible for known molecular dimensions. Nevertheless, useful insights can be obtained by comparison of adsorbed areas for the same molecule to different interfaces. Results for tween and HSA at the (*lv*) interface of 215.6 and 97.8 picomol/cm^2 correspond to 77 and 170 Å^2/molecule, respectively. No such information can be calculated for FBS since it is a chemically undefined milieu for which no compositional information is readily available to scale interfacial activity.

It is important to stress at this juncture that the correct and quantitative use of Gibbs' relationship is critically dependent on a number of factors. First, the solution must obey the ideal dilute solution approximation, and if Eq. (4) is used, the solution must contain only a single surfactant. Other-

TABLE 2 Surfactant Properties of Tween-80 and Protein Liquid/Vapor Interface

	Fitted parameters[b]				Surface excess[c]
Surfactant system[a]	$\overset{\circ}{\gamma}_{(lv)}$ (dyne/cm)	$\gamma'_{(lv)}$ (dyne/cm)	$K_{(lv)}$ (ln C units)	$N_{(lv)}$	$\Gamma_{(lv)}$ (picomol/cm^2)
Tween-80 (in saline)	73.5 ± 0.6	39.7 ± 0.6	15.9 ± 0.1	−11.4 ± 0.9	215.6 ± 9.6
Human serum albumin (in saline)	73.1 ± 0.5	49.5 ± 2.7	15.8 ± 0.6	−7.3 ± 0.4	97.8 ± 19.8
Fetal bovine serum (in DMEM)	74.6 ± 1.6	54.8 ± 1.4	20.5 ± 0.4	−17.2 ± 3.1	

[a] Physiologic saline (0.9% NaCl, Abbott). DMEM = Dulbecco's modified eagle medium, an isotonic saline solution with added glucose and amino acids (Gibco).
[b] Indicated uncertainties are standard errors of the statistical, least-squares fitting procedure. In all cases, the R-squared goodness-of-fit parameter was 95% or better.
[c] Indicated uncertainty was estimated from error in fitted coefficients.

wise, a more complicated form of Eq. (4) must be applied that accounts for the surface activity of each solution constituent. Second, adsorption must be reversible and have attained equilibrium. Application of Eq. (4) to tween and HSA solutions is an approximation to the extent that these compounds are not isomerically pure, with nominal molecular weights representing the average composition. With respect to reversibility and equilibrium adsorption, kinetic studies [122] suggest that tween adsorption to the (lv) interface is reversible, although adsorption is strongly favored with $k_a/k_d \simeq 10^8$ (see k_a, k_d in Fig. 3b), and steady state is attained well within the time frame of tension measurements reported here. In addition, interfacial tension measurements for tween were observed to be stable for more than an hour. It was concluded that tween is a system that attains steady state to which other systems can be compared. No such specific kinetic information was gathered for HSA or FBS, other than the observation of stable interfacial tensions over time, and the existence of an equilibrium state can be only inferred from behavior similar to that for the tween system. Thus, the accuracy of the calculated interfacial excesses listed in Table 2 is dependent on the validity of purity and reversibility assumptions. These issues should be considered within the framework of the larger picture, which is that the calculation of parameters such as interfacial excess essentially amounts to an interpretation of laboratory measurements (reality) within the context of a model (reversible thermodynamics).

3. Adsorption to Solid Interfaces

The complete characterization of wetting properties of a solid surface exhibiting contact-angle hysteresis involves the measurement of advancing and receding τ. As for the (lv) interface, the correct application of Gibbs' adsorption law for solid interfaces given by Eqs. (5) or (8) is dependent on solute purity, ideal-dilute solution behavior, and attainment of thermodynamic equilibrium. Equilibrium arguments for hysteretic solid surfaces are much more complicated than for liquid interfaces since these surfaces are known to be either or both chemically heterogeneous or rough. Consequently, advancing and receding angles do not represent equilibrium measurements, even if steady-state adsorption can be demonstrated.

A paradigm that can be adopted for interpreting (sl) adsorption treats advancing and receding τ measurements as separate equilibrium values representing a pure "advancing phase" or a pure "receding phase," respectively. Thus, in this perspective, Γ values calculated from advancing and receding θ or τ measurements represent adsorption to advancing and receding phases. Given that advancing and receding contact angles are the largest and smallest stable angles observable, advancing and receding τ measurements will correspond to the smallest and largest tensions, respectively. It follows that calculated Γ values will bracket maximum and minimum adsorption. Again, these calculations represent an interpretation of laboratory measurements (reality) within the context of a model (reversible thermodynamics).

Results obtained by applying methods I and III to the silylated glass slide system are shown in Fig. 7 for tween-80 and HSA (advancing τ = filled circles; receding τ = open circles). Whereas methods I and III yield similar τ curves for tween (Figs. 7a and b), dramatically different results were obtained for HSA (Figs. 7c and d).

Analyzing first the benchmark tween case, method I and III curves were reasonably smooth and sigmoidal, similar in behavior to $\gamma_{(lv)}$ curves for which equilibrium adsorption conditions were thought to prevail. Wetting tension transition from low to high tween concentrations was linear, with a positive slope indicating adsorption to the (sl) interface. Wilhelmy balance hysteresis curves were regular over the entire tween concentration range, exhibiting none of the erratic behavior sometimes associated with strongly adsorbed solutes, even when method II was applied. Example method III hysteresis curves are illustrated in Figs. 8a–c for tween

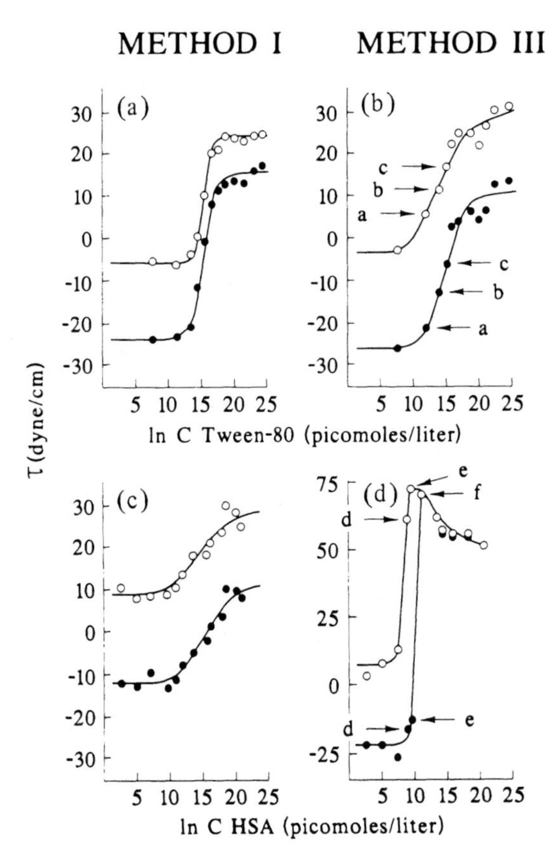

FIG. 7 Advancing (full circles) and receding (open circles) wetting tension curves for tween and HSA measured by methods I and III shown in Figs. 5a and c. Smooth curves through the data of a–c were obtained by fitting to the logistic equation, as described in Sec. III.B.4. Hysteresis curves for concentrations marked a through f are shown in Fig. 8. (From Ref. 150.)

concentrations corresponding to (a), (b), (c) marked on Fig. 7b that extend over the linear slope portion of the τ curve. Note that advancing and receding buoyancy slopes are regular and parallel, indicative of stable advancing/receding contact angles. Method II curves (not shown) were very similar to those of method III. The regularity of hysteresis curves and equivalence of methods II and III are strong evidence that adsorption events are occurring much faster than the time frame of balance measurements and that adsorption rapidly attains steady state. Adsorption

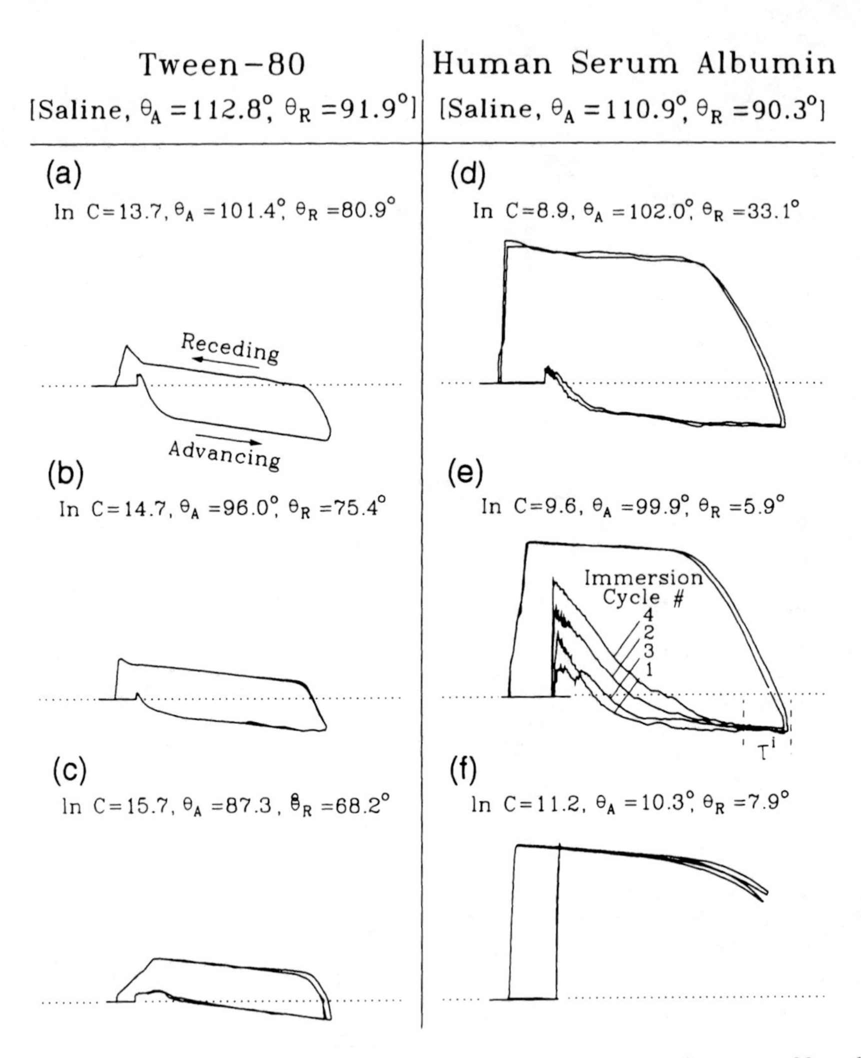

FIG. 8 Comparison of Wilhelmy balance hysteresis curves for tween-80 and HSA for silane-treated glass slides with (a) through (f) corresponding to wetting tension measurements shown in Fig. 7. Erratic wetting behavior exhibited in (e) for four different immersion cycles lead to a stable wetting tension labeled τ^i. (From Ref. 150.)

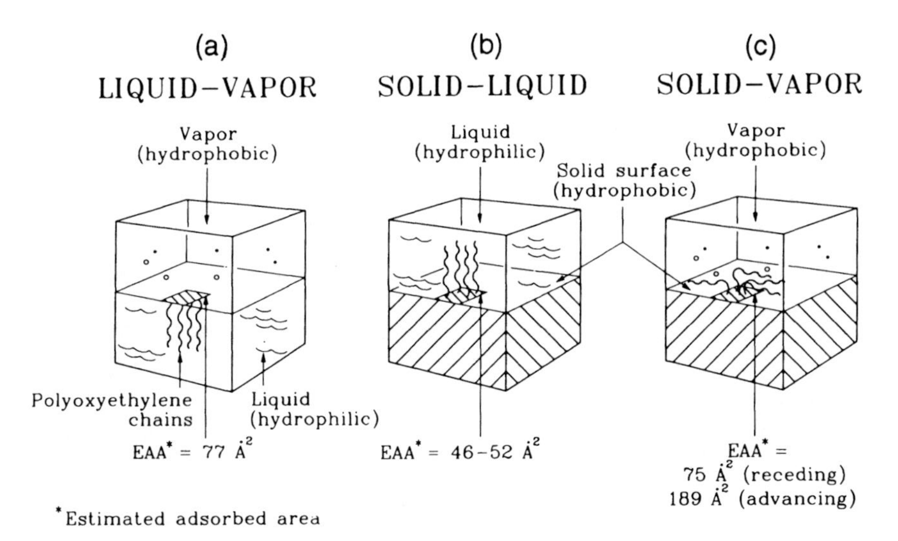

FIG. 9 Illustration of the adsorbed condition of tween-80 at various interfaces inferred from interfacial excess measurements. See discussion of Fig. 4 for detailed interpretation. (From Ref. 150.)

apparently leads to a more hydrophilic interface since τ increases, signaling a reduced contact angle, as might be expected for a surfactant adsorbing with the nonpolar head group down and hydrophilic arms extending into the (sl) interface (see Fig. 9b). It can be concluded from these observations that tween attains adsorption equilibrium at the (sl) interface and may serve as a model system to which protein adsorption behavior can be compared.

Turning now to the results obtained with HSA and reviewing momentarily the τ curves of Figs. 7c and d, we can see again that, in contrast to the tween case, methods I and III yield drastically different measures of HSA wetting. The contact-angle approach yields smooth curves similar to those of the tween case, except that the slope is reduced. Method III (with a 1-h preadsorption step) τ curves have a very sharp transition near $\ln C = 9$, rising a dramatic 70 dyne/cm or more to a nearly completely wetted condition with $\tau' \simeq 73$ dyne/cm ($\theta' \simeq 0$). It is of interest to examine the method III hysteresis curves (Figs. 8d–f) in this sharp transition region occurring over the narrow concentration range labeled (d), (e), and (f) in Fig. 7d. Hysteresis was repeatable and curves were regular up to about $\ln C = 8.9$ (5.5×10^{-5}% HSA, Fig. 8d). Moving to $\ln C = 9.6$ (1×10^{-4}% HSA, Fig. 8e), we note that hysteresis was erratic, signaling adsorption/desorption kinetics, particularly for that portion of the substrate that had been immersed for the longest periods (shallow immersion depth). Wetting tension measurements shown in Fig. 7 corresponding to HSA concentrations were estimated from the portion of the hysteresis curve labeled τ^i in Fig. 8e. Finally, at ln C = 11.2 (5×10^{-4}% HSA, Fig. 8f) and higher concentrations, hysteresis all but vanished and contact angles were less than about 10°.

These concentration-dependent wetting effects can be rationalized in terms of the adsorption phenomena discussed earlier in reference to Figs. 3b and d and effects due to adsorption at the (sv) interface. With low HSA concentrations (ln $C < 8$), there was insufficient HSA adsorbed to either the (lv) or (sl) interfaces to affect $\gamma_{(lv)}$ (Fig. 6) or τ (Figs. 7c and d) curves, respectively. Hysteresis curves were, consequently, regular and very similar to those obtained in pure saline. At higher HSA concentrations, sharply increased τ values signal adsorption during the immersion pretreatment step (Fig. 5c), with increased substrate surface coverage at higher HSA concentrations. As slides were immersed in the advancing mode of the balance experiments, $(s'v) \rightarrow (s'l)$ involving adsorbed HSA rehydration, adsorption, and desorption. Adsorption kinetics are connected with mass transport steps in such a way that solute supply can be limiting at low bulk concentrations. Adsorption kinetics, mass transport, and the speed of experimental measurements can conspire at some critical concentration to yield highly erratic wetting behavior like that observed in Fig. 8e. At HSA concentrations greater than this critical concentration, solute supply meets or exceeds adsorption demands and the substrate surface can be saturated with adsorbed HSA, yielding regular hysteresis curves like that shown in Fig. 8f. This interplay between adsorption kinetics and mass transport, the dynamics of supply and demand, are amplified in method II because "virgin" surfaces (at least during the initial immersion) have more capacity for HSA than the pre-equilibrated method III substrates. As a consequence, method II yields unintelligible results for all but the most dilute or concentrated HSA solutions. By contrast, concentration dependency is not observed in Wilhelmy balance measurements of tween solutions because adsorption and mass transport are never limiting in the time frame of the force measurements.

The only kinetic effects noted in the contact-angle experiments of method I were that angles decay smoothly with time to a steady-state value. It is assumed that the (slv) line was stationary because the goniometer was mounted on a vibration-isolation table and a humidified environmental chamber employed to minimize evaporation. Thus, adsorption is presumed to occur only at the (sl) interface.

4. Comparison of Results

Table 3 collects fitted wetting parameters τ°, τ', $K_{(sl)}$, $N_{(sl)}$ for tween and HSA obtained using methods I–III. It is of use in analyzing this data to first consider some of the generalities and then discuss separately the results for tween and HSA.

Three measurements made with the tilting base technique are listed in Table 3 for method I corresponding to advancing and receding contact angles, as well as contact-angle readings with the base in the horizontal position ($\psi = 0$). Generally speaking, horizontal-position wetting tensions (corresponding τ curves are omitted from Fig. 7 for clarity) were more like advancing than receding tensions. This similarity probably coincides with the unavoidable act of advancing test liquid droplets on the substrate during deposition at the beginning of the experiment. Contact angles tend to remain in an advancing mode once imposed, so long as the base is held in the horizontal position and there is no evaporation around the droplet perimeter [123]. Evaporation, however slight, tends to reduce droplet diameter, which causes the droplet, and hence the contact angle, to recede. Likewise, any droplet vibration tends to reduce advancing angles and increase receding angles [102]. These phenomena are probably responsible for the observation that horizontal-position contact angles were similar but not identical to advancing angles.

In contrast to dynamic measurements made with balance methods II and III, the three-phase (slv) line is essentially motionless in method I. Thus, the mechanisms of transport and deposition of solute at the (sv) interface have been minimized or eliminated.* Thus, for method I, it is useful to assume that $\Gamma_{(sv)} = 0$ and $[\Gamma_{(sl)} - \Gamma_{(sv)}] = \Gamma_{(sl)}$ (see Sec. III.B.4) because information regarding the nature of the adsorbed state at the (sl) interface can be inferred from molecular area calculations. In turn, these estimates can be used to deduce information on the nature of the adsorbed state at (sv) interfaces from method II or III in which $\Gamma_{(sv)} \neq 0$.

Secure wetting parameters could not be obtained for HSA using methods II and III because of the adsorption-kinetics phenomena described in the previous section. Estimated τ values from the hysteresis curves are shown in Fig. 7 for comparison to tween results, but data-fitting procedures (see Sec. III.B.4) could not be applied to these sharp adhesion tension curves.

(a) Tween-80. Methods I–III τ° and τ' parameters were roughly equivalent for tween because τ curves start and end at similar plateau values

---

* Minimized to the extent allowable for smooth glass coverslips with a hydrophobic surface treatment held in an environmental chamber to minimize evaporation and by use of a vibration-isolation table to eliminate droplet movement during angle measurements.

TABLE 3 Surfactant Properties of Tween-80 and Serum Proteins at the Solid/Liquid Interface[a]

	Tween-80					Human serum albumin				
	τ° (dyne/cm)	τ' (dyne/cm)	$K_{(sl)}$ (ln C units)	$N_{(sl)}$	$\Gamma_{(sl)}$-$\Gamma_{(sv)}$ (picomol/cm^2)	τ° (dyne/cm)	τ' (dyne/cm)	$K_{(sl)}$ (ln C units)	$N_{(sl)}$	$\Gamma_{(sl)}$-$\Gamma_{(sv)}$ (picomol/cm^2)
Method I[b,c]										
Advancing	−23.8 ± 0.9	14.2 ± 0.6	15.4 ± 0.1	−18.3 ± 7.9	317.3 ± 13.0	−11.5 ± 0.9	12.4 ± 4.2	16.2 ± 1.0	−7.2 ± 2.2	91.5 ± 23.2
Receding	−5.8 ± 0.5	23.4 ± 0.3	15.6 ± 0.1	−25.1 ± 2.3	358.2 ± 8.2	8.9 ± 1.2	29.3 ± 3.7	15.1 ± 1.2	−6.1 ± 2.6	109.7 ± 21.2
Horizontal	−17.4 ± 1.2	15.8 ± 0.7	15.5 ± 0.1	−23.0 ± 3.9	290.6 ± 23.2	−6.4 ± 1.1	13.7 ± 15.6	16.2 ± 4.8	−6.6 ± 2.0	92.6 ± 16.9
Method II[c]										
Advancing	−24.2 ± 1.0	9.2 ± 0.8	15.1 ± 0.2	−12.8 ± 1.8	251.0 ± 16.8	—	—	—	—	—
Receding	1.3 ± 2.4	32.9 ± 2.7	14.9 ± 0.7	−7.0 ± 2.4	122.8 ± 27.2	—	—	—	—	—
Method III[c]										
Advancing	−26.6 ± 2.5	10.3 ± 1.3	14.3 ± 0.4	−10.0 ± 2.2	229.5 ± 30.9	—	—	—	—	—
Receding	−3.9 ± 2.9	30.0 ± 2.0	13.9 ± 0.6	−6.1 ± 1.6	136.5 ± 25.3	—	—	—	—	—

[a] Tween-80 and human serum albumin in physiologic saline (0.9% NaCl, Abbott). Solid surfaces were octadecyltrichlorosilane-treated glass plates identically treated for each surfactant. Typical advancing and receding contact angles (saline) $\theta_A \approx 113°$, $\theta_R \approx 92°$, respectively.

[b] Advancing and receding contact-angle measurements made using the tilting base technique at 40° from horizontal. Angles read after 1-h equilibration time.

[c] See Fig. 4 and text for discussion of methods.

(Fig. 7). However, the slope connecting plateaus, as measured by the $N_{(sl)}$ parameter, were much steeper for method I and this steeper slope translates into greater surface excess values* (compare rows 1 and 2 for method I under $[\Gamma_{(sl)} - \Gamma_{(sv)}]$, column 6 of Table 3, to rows 4 and 5 and 6 and 7 for methods II and III, respectively). The origin of this steeper slope and larger surface excess values for method I is probably related to the stationary nature of the contact-angle droplet. Thus, from the advancing and receding θ measurements, $317 < \Gamma_{(sl)} < 358$ picomol/cm^2 corresponding to between 46 and 52 Å^2/molecule. This adsorbed area is considerably smaller than the 77 Å^2/molecule estimated from (lv) interfacial tensions. According to this interpretation that assumes $\Gamma_{(sv)} \approx 0$ for contact-angle goniometry on these hydrophobic surfaces, adsorption to a closely packed silane layer at the (sl) interface is more structured than that at the (lv) interface, leading to a higher packing density (see Fig. 9b).

By contrast to the sessile droplet in method I, fluid movement across solid surfaces during immersion and withdrawal modes of methods II and III creates microscopic flow fields that affect solute transport, accumulation, and adsorption at the (sl) and (lv) interfaces [118,119]. In addition to fluid flow, the meniscus can be affected by substrate vibration or oscillation while hanging in the balance, as well as small ripples in the liquid. These factors provide a mechanism for depositing solute at the (sv) interface within the microenvironment of the advancing or receding (sl) line. Solute accumulation at the (sv) interface causes $\Gamma_{(sv)} > 0$ and, consequently, $[\Gamma_{(sl)} - \Gamma_{(sv)}]^{III} < [\Gamma_{(sl)} - \Gamma_{(sv)}]^{I}$ (compare rows 6 and 7 to rows 1 and 2 of column 5). Rapid and reversible adsorption of tween to the (sv) interface also leads to the equivalence of results for methods II and III, despite the fact that method III involves the immersion pretreatment (compare fitted values for τ curves, rows 4 and 6 under columns 1–4 of Table 3 and surface excess values listed in rows 4 and 6 under column 5). As the virgin surface contacts liquid during the first immersion of method II, sufficient tween must accumulate at the (sv) interface so that $[\Gamma_{(sl)} - \Gamma_{(sv)}]^{III} = [\Gamma_{(sl)} - \Gamma_{(sv)}]^{II}$. Of course, it is of no surprise that receding modes of methods II and III were identical since there is no operational difference between methods after the first immersion. It is interesting, however, that $[\Gamma_{(sl)} - \Gamma_{(sv)}]$ values were nearly half those calculated for the advancing mode and that this observation is in sharp contrast to the results obtained by goniometry. Apparently, the amount of solute is deposited at the (sv) interface [per unit area of interface within

* Method I contact-angle data hav been translated into adhesion tension data for graphical comparison to method III results. However, Eq. (8) was utilized for the calculation of surface excess from method I data. See Sec. III.D.3.a.

the (slv) microenvironment] in the receding mode and is about twice that deposited in the advancing mode. Possibly this is due to evaporative effects at the trailing liquid edge as the plate is removed from solution. These (sv)-deposited molecules become redissolved or at least rehydrated as the liquid front reverses direction in the advancing mode.

There is no independent means of estimating $\Gamma_{(sl)}$ values from method III τ curves since only the difference $[\Gamma_{(sl)} - \Gamma_{(sv)}]$ can be calculated. However, an insight into the adsorbed condition of solute at the (sv) interface can be obtained by making the assumption that $\Gamma_{(sl)}$ estimated from the contact-angle case (317–358 picomol/cm^2) is a reasonable expectation value for $\Gamma_{(sl)}$ of method III. This allows a value for $\Gamma_{(sv)}$ to be calculated. First, for the receding mode in which solute deposition by (lv) evaporation is suspected, $\Gamma_{(sv)} = 222$ picomol/cm^2 or 75 Å^2/molecule. This value is quite close to $\Gamma_{(lv)} = 77$ Å^2/molecule obtained from the $\gamma_{(lv)}$ curve. Estimates for the advancing mode yield a much reduced packing density of $\Gamma_{(sv)} = 87.8$ picomol/cm^2 or 189 Å^2/molecule, possibly reflecting partial dissolution and hydration effects at the (slv) line due to the advancing liquid front. Relationships for these adsorbed areas at different interfaces are sketched in Fig. 9.

Advancing and receding W curves shown in Fig. 10a for tween show a substantial peak near $\ln C = 15$ (advancing mode = circles; receding = squares). These peaks in W occur because maxima in τ curves fall earlier on the $\ln C$ axis than the minimum in the $\gamma_{(lv)}$ curve ($K_{(lv)} > K_{(sl)}$; compare column 4, row 1 of Table 2 to column 4, rows 1 and 2 and 6 and 7 of Table 3). Thus, the fall in $\gamma_{(lv)}$ (Fig. 6) with an increasing tween concentration does not fully compensate for the rise in τ (Fig. 7), leading to increased adhesion of the liquid to the substrate. Smooth curves through the data of Fig. 10 were obtained by simply adding fitted logistic functions corresponding to τ and $\gamma_{(lv)}$ tension curves. Methods I and III W curves are qualitatively similar, but there are substantial quantitative differences in the peak magnitude and centroid position along the $\ln C$ axis. These differences can be again attributed to solute adsorption at the (sv) interface, which manifests itself as a "tension excess" related to the difference in work of adhesion $\Delta W = [W^{\mathrm{III}} - W^{\mathrm{I}}] = [\tau^{\mathrm{III}} - \tau^{\mathrm{I}}] = [\gamma_{(s'v)} - \gamma_{(sv)}]$, as discussed in Sec. III.D.1.d. Figure 11a plots ΔW as a function of concentration, with smooth curves obtained by subtracting fitted logistic functions. As the tween concentration increases, there is a steady rise in ΔW up to about $\ln C = 14$ because the liquid meniscus is more strongly adherent to the plate in the Wilhelmy balance method than the droplet in the contact-angle case. Apparently, the $(s'v)$ interface with adsorbed tween is at a higher tension than that of the hydrophobic octadecyl (sv) interface since the difference $[\gamma_{(s'v)} - \gamma_{(sv)}]$ is greater than

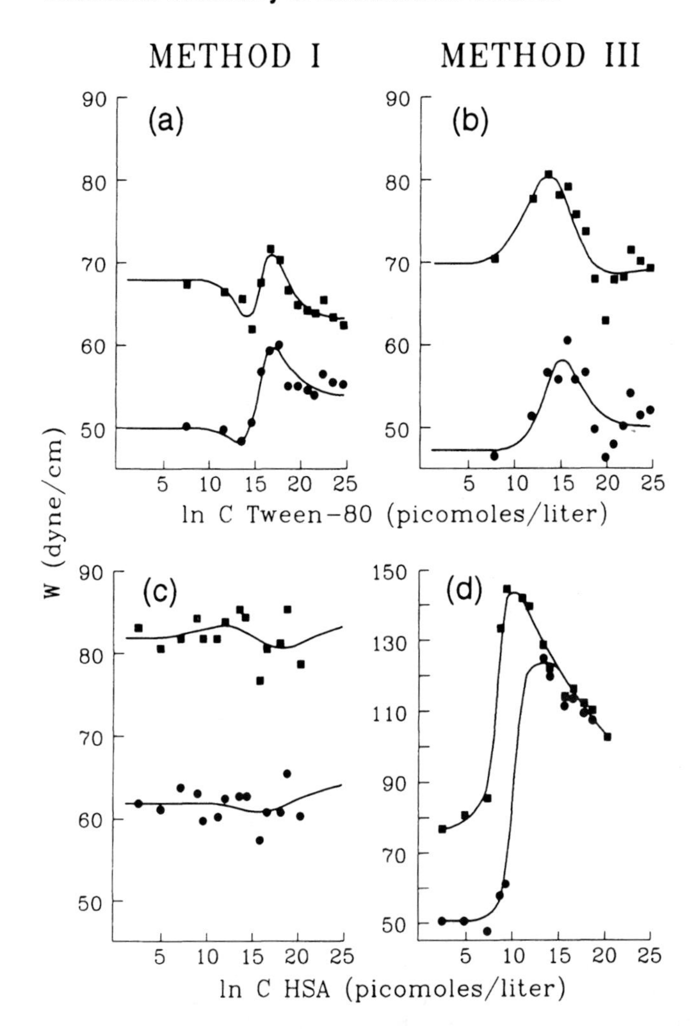

FIG. 10 Work of adhesion curves (advancing = circles, receding = squares) for methods I and III corresponding to Fig. 7. Note strong transition in (a) and (c) associated with surfactant behavior illustrated in Figs. 6 and 7.

zero. Higher $\gamma_{(s'v)}$ at low solute packing density (tween concentrations < ln C = 14 < $K_{(sl)}$) may be related to an incompatibility between polar polyoxyethylene chains and the relatively nonpolar vapor phase. It is interesting that this trend reverses at tween concentrations exceeding ln C = 14. Possibly, higher solute packing density (ln C > 14 > $K_{(sl)}$) leads

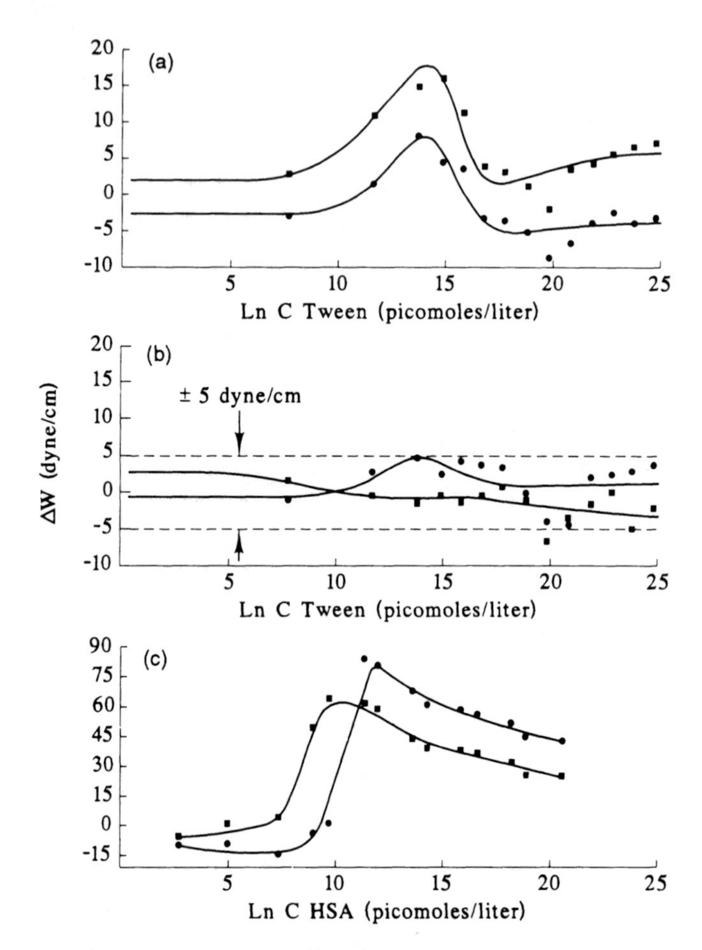

FIG. 11 ΔW curves (advancing = circles, receding = squares) for tween-80 (a and b) and HSA (c). Curves (a) and (c) correspond to the difference in results between methods III and I ($\Delta W = W^{III} - W^{I}$). Curve (b) corresponds to the difference between methods II and I ($\Delta W = W^{II} - W^{I}$). Compare Figs. 5, 7, and 9.

to molecular interactions that tend to shield the polarity of polyoxyethylene chains at the $(s'v)$ interface. Figure 11b plots ΔW for methods III and II and provides a measure of uncertainty in the technique comparison since, by previous discussion, methods III and II should yield identical results for tween ($\Delta W = [W^{III} - W^{II}] = [\tau^{III} - \tau^{II}] = [\gamma_{(s'v)} - \gamma_{(s'v)}] = 0$). Differences less than about ± 5 dyne/cm are probably insignificant.

(b) Human Serum Albumin. As mentioned previously, secure wetting parameters could be obtained for HSA only for method I. Estimated advancing and receding $\Gamma_{(sl)}$ were 92 and 110 picomol/cm^2, respectively, corresponding to 181 and 151 Å^2/molecule. This adsorbed area is quite close to the 170 Å^2/molecule measured at the (lv) interface. According to this interpretation of (sl) interfacial measurements, the configuration of adsorbed HSA is about the same at (sl) and (lv) interfaces. This conclusion is in contrast to the results obtained with tween in which a higher packing density at the (sl) interface was inferred from tension measurements. Possibly, the folded tertiary structure of the larger protein molecule will not allow reconfiguration into a more "packable" unit area.

Figure 10c is the HSA W curve for method I and it is interesting that the curves are quite flat. In contrast to the tween case, the rise in τ (Fig. 7c) nearly completely compensates for the fall in $\gamma_{(lv)}$ (Fig. 6). This is consistent with the above discussion of the tween and HSA adsorbed molecular dimensions. On the one hand, tween appears to adsorb more efficiently to the (sl) interface than to (lv) interfaces, which causes the rate of τ change with respect to surfactant concentration to exceed that of $\gamma_{(lv)}$. Thus, it is observed that $K_{(lv)} > K_{(sl)}$. On the other hand, HSA has about the same packing dimension at (sl) and (lv) interfaces so that adsorption efficiency is roughly the same and $K_{(lv)} \approx K_{(sl)}$ (compare column 4, row 2 of Table 2 to column 9, rows 1 and 2 of Table 3).

Sharp changes in τ measured with method III (Fig. 7d) completely dominate the W curve shown in Fig. 10d and the ΔW curve in Fig. 11c. The ΔW curve rises with increasing concentration, but does not fall to initial values at a high concentration as observed with tween. The interpretation is that the $(s'v)$ interface remains at a high tension excess over that of the (sv) interface due to adsorbed HSA and this trend is not reversed by added HSA to the solution. Possibly related to this excess tension is the result that plates with adsorbed HSA were considerably more fragile to handle than control plates, presumably because the excess tension amplifies surface microcracks and stresses [124].

5. Instrumental Limitations in Tensiometry

A conclusion to be drawn from the model system study is that the utility of the Wilhelmy balance technique in the measurement of biosurfactant properties of biological macromolecules is essentially limited to the measurement of $\gamma_{(lv)}$ curves. Adsorption events at (sl) and (sv) interfaces tend to obscure signals and τ curves are not easily interpreted because $\Gamma_{(sv)} \gg 0$. These limitations might be turned into an advantage, however, if this instrumental approach can be applied in the measurement of adsorption kinetics. Adsorption kinetics are very useful in understanding biosurfactant properties and establish operational regimes in which adsorption kinetics do not obscure force measurements. Toward this end, improved computer control, data acquisition, and data analysis seem essential [80].

Those venturing into the routine application of contact goniometry for the purpose of measuring θ or τ curves for different experimental materials will quickly become aware of the labor intensity and subjectivity in making contact-angle measurements as a function of solute concentration. Here too, computer control and data-acquisition approaches combined with image analysis techniques [125–127] might have significant impact on the routine application of this powerful method.

D. Philosophy

Many biological macromolecules exhibit biosurfactant properties by adsorbing to interfaces from bulk phases. The origin of biosurfactant properties is incompatibility with water, which is simultaneously a good and poor solvent for most proteins [128]. Water molecule interactions drive incompatible, hydrophobic domains of a dissolved macromolecule out of solution to any relatively hydrophobic interface, such as the surface of biomaterials, in order to reduce interfacial tensions by adsorption. Adsorption induces the unfolding of tertiary structure, subsequently exposing many regions that are determinants for biological process such as adhesion, spreading, and recognition. Understanding and controlling adsorption are tantamount to achieving control over material and interfacial properties that, in turn, provides a handle on biomedical applications. Thus, biosurfactant properties are primarily linked to the overall biological response to artificial materials and must be a part of any complete understanding of biocompatibility. Most important, the techniques and tools of surface physical chemistry that have been used for decades in the understanding of surface energetics and wetting phenomena can be adapted in the pursuit of quantifying these biosurfactant properties.

The appropriate energetic property of biomaterial interfaces is the wetting tension τ. Wetting tension includes liquid-phase properties and is sensitive to adsorption events at the (sl) interface. Wetting tension is easily measured, understood, and requires no theory with hidden assumptions to apply. Unfortunately, wetting tension has received little emphasis in the biomaterials literature because of the continued and unnecessary fascination with a separate surface-energy term for materials. This is a curious oversight that ignores the important role liquid-phase composition plays in biocompatibility. The specification of biomaterial wetting properties without also completely specifying liquid-phase composition is at best dangerously incomplete.

IV. COLLOID AND THERMODYNAMIC CONCEPTS IN BIOMATERIALS SCIENCE

A. Interfacial Biophysics

In previous sections of this chapter, the physicochemical nature of biomaterial interfaces was considered, leading to the conclusion that interfacial energetics is a primary determinant of biocompatibility. Various analytical aspects were discussed, and it was shown that the basic tools of surface physical chemistry, particularly tensiometric tools, could be adapted for the characterization of biomaterial interfacial properties. Attention is turned now to conceptual issues associated with the use of these surface analytical measurements in predictive models. The goal of modeling is to factor the biological process under scrutiny into manageable pieces in such a way that the role of interfacial energetics can be identified. This process of "interfacial biophysics" is a quantitative way of thinking about complicated biological process that leads to the formulation of testable hypotheses. The final goal is to understand the structure-function relationships that connect surface chemistry and interfacial properties to the biological response to materials. It is these relationships that provide a rational basis for prospective biomaterials surface preparation.

B. Colloid Chemistry: The Science of Touching

There are a number of interesting parallels between biomaterials and colloid science. Both typically involve large numbers of particles in suspension that can mutually interact or interact with other surfaces: biological cells in the biomaterials case; clay, ceramic, polymer, or emulsion particles in the colloids case. Both typically involve fluid phases containing surfactants that can alter particulate interactions by adsorption: proteins in saline solution in the biomaterials case; aqueous or organic solutions of detergents in the colloids case. Interparticle forces that operate over very small distances are important in both fields because these forces control adsorption and adhesion events that mediate the biological response or colloidal stability, respectively. Understanding interactive forces at close proximity, what one may call the science of touching, is very important in both fields.

A colloid science theory that quantifies interactions at small distances and has biomaterial applications is the so-called DLVO theory (see, e.g.,

[129]). The basis of this theory is that attractive van der Waals potentials and repulsive electrostatic forces are additive. Formulation of these interaction potentials can be quite detailed for each case, but the qualitative predictive aspects for a microscopic cell or particle in close proximity to a macroscopic substrate are quite straightforward.

C. Surface Thermodynamics: A Modeling Tool

An important conceptual tool in modeling is surface thermodynamics. A principal utility of thermodynamics is the ability to handle multicomponent systems in a phenomenological manner. One way that thermodynamics accomplishes this is by lumping a number of microscopic properties into a single, measurable parameter by the judicious assignment of a macroscopic thermodynamic property. A particular example of this that will be discussed subsequently is the assignment of an average wetting property for a biological cell. Quite clearly, biological cells are not hard spheres of uniform material but, in fact, have a nonuniform topography that is patchwise heterogeneous and studded with macromolecules exhibiting very specific biochemical properties. These ultrastructural aspects of cell membranes are assigned an average impact on the macroscopic wetting parameter in the thermodynamic model. Specific cell/substrate interactions that are known to occur between ligands and receptors are included in the overall "affinity" of the cell for substrate, but not separated from nonspecific forces that cells encounter at interfaces.

The ends of the modeling process will justify the means. That is, if the thermodynamic model is predictive of reality, it can be assumed that the biological process behaves in accordance with the rules of the model and that the measurable parameters are the important, controlling factors. Generalities can be extracted from the model and used as a standard against which exceptions can be ordered.

D. Example

A biomaterials example applying DLVO theory and the thermodynamic modeling approach is the study of adhesion of mammalian cells to plastic substrates in vitro. This somewhat simplified biological process serves as an experimental system for which the process of factoring a biological process into component pieces can be applied.

Cell biologists describe cell adhesion to a substrate in terms of four separately identifiable stages. These different phases of adherence are illustrated on a cell attachment rate curve shown in Fig. 12, which plots the percentage of a cell inoculum attached to a substrate (cell number attached/total number of cells × 100) from a sessile fluid phase at various time intervals. The experiment here is the same as that described in the comparison of ESCA and cell adhesion in Sec. II.E. A single-cell suspension in a nutritive medium containing proteins is applied to the substrate of interest, in a culture dish arrangement for simplicity, and cells are allowed to gravitate from suspension to close proximity of the substrate where attachment can occur. Cells that are not removed by gentle buffer washes are defined as adherent and are counted with time [82]. The final stage illustrated in Fig. 12 is cell spreading, which begins well after initial contact and attachment events, continuing during and after the first hour of attachment (note the logarithmic time axis). Cell spreading is quite different from the previous steps that lead to a steady-state adhesion plateau, involving the production of adhesion proteins by cells and the formation of adhesion patches or plaques with progressive adherence to substrate [52,82,87,88,130–132]. These latter steps are more related to the proliferative potential of cells on the substrate than are the simpler, preceding attachment steps. Cell proliferation is undoubtedly dependent on the biological nature of individual cell types, and a detailed study of steps involved in spreading and exponential growth seems less likely to yield generalized information regarding cell/substrate attachment compatibility. The subfunctions to be explored will be, therefore, restricted to *short-term* adhesion steps including protein adsorption, cell contact, and attachment leading to the steady-state plateau labeled $\%I_{max}$ in Fig. 12.

Part of the factoring process is restatement of the biological problem into physical terms. Ruckenstein et al. have viewed short-term cell adhesion as an *attachment kinetics problem* [133–135] and modeled the process by applying DLVO theory [136–138]. In doing so, physicochemical parameters such as cell and substrate surface charge densities have been identified and testable hypotheses assembled. One of these hypotheses is the kinetic saturation of cell adhesion. Briefly stated, kinetic saturation proposes that cell adhesion rates change as the substrate becomes populated with cells because electrostatic interactions become progressively controlled by cell/cell repulsion as surface coverage increases. These effects are illustrated in Fig. 13 for a hypothetical case in which the cells and surface bear a dissimilar net negative charge, with cells at a more negative potential. In this diagram, changes in the free energy (G) of interaction between a cell and surface are plotted as a function of cell/substrate separation. Free energy curves, which originate from the DLVO theory of colloid stability, show that little or no interaction occurs up to about 5–10 nm of cell/substrate separation. Here a weak interactive well (free energy $G < 0$) is encountered. Cells gravitate from suspension and collect at this attractive potential well, usually referred to as a secondary

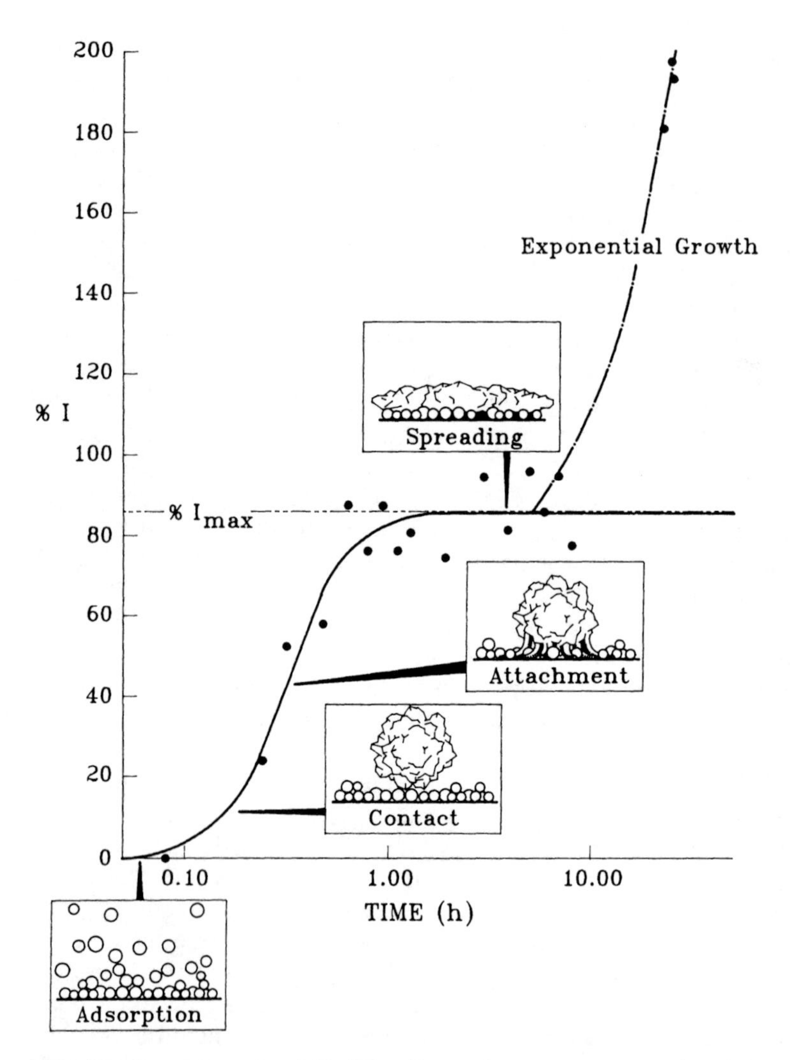

FIG. 12 Representative MDCK cell attachment rate curve for tissue-culture-grade polystyrene plotting percent of inoculum attached (%I) as a function of incubation time. Important attachment events are identified with sketches of cell morphology. Compare contact and attachment stages with the micrograph of Fig. 15a. (Adapted from Ref. 88.)

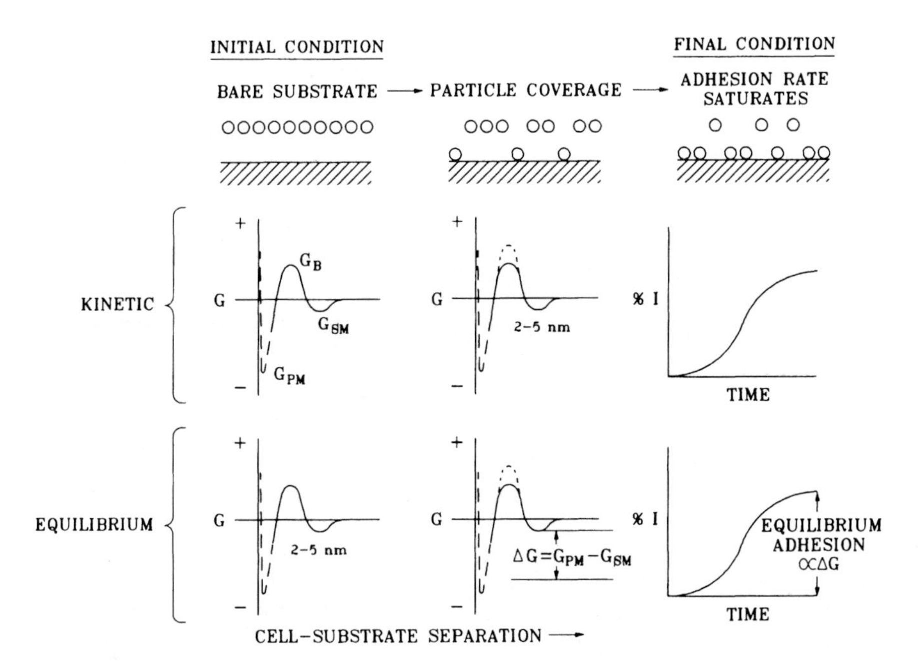

FIG. 13 Free energy curves for cell attachment illustrating the kinetic saturation and equilibrium cell adhesion models. Note that the electrostatic barrier to adhesion G_B increases with increased cell-surface population according to the kinetic saturation model, resulting in a decreasing adhesion rate with time. The equilibrium model purports that steady-state adhesion is controlled ΔG.

minimum G_{SM} in the literature [30,139,140]. These "unattached" cells penetrate an electrostatic barrier (G_B) that occurs because of cell/substrate repulsion and fall into a deeper attractive well very close to the interface. This deeper well is termed a "primary minimum" G_{PM}. Cells within G_{PM} are firmly "attached." The kinetic saturation model purports that changes in the height of G_B as the surface is populated with attached cells lead to the steady-state coverage. In Fig. 13, this increase in G_B (indicated with dashed lines) is concomitant with particle coverage because the total surface charge increases with attached cell density. Vogler and Bussian [82] suggest a somewhat different hypothesis in which it is proposed that the steady-state adhesion plateau is ultimately thermodynamically controlled by the difference in energy states between the attached and unattached cells (ΔG in Fig. 13). A kinetic saturation-like process may connect the starting condition (cells collected at the G_{SM}) and the final, steady-state condition (cells distributed between G_{PM} and

G_{SM}), but this final state is controlled by ΔG. Thus, by using a similar physical model of adhesive forces at close approach (DLVO), two alternative, testable hypotheses have been suggested. The kinetic saturation hypothesis underscores the role of "microscopic" properties such as surface charge densities, whereas the thermodynamic proposal can be stated in terms of "macroscopic" surface wettability properties. This latter proposal will be expanded on here because it is an example of a connection between surface and interfacial chemistry and the biological response of cell adhesion.

The transmission electron micrograph (Fig. 14a, see [88] for details) is a cross section of a dog kidney cell that has just made contact with a polystyrene substrate, which appears phase-dark due to staining. This preparation shows a highly invaginated cell membrane making a few possible attachments to the substrate. Figure 14b is a model of the attachment process showing the formation of a cell/substrate interface (*sc*) with interfacial tension $\gamma_{(sc)}$ and the concomitant destruction of cell/liquid (*cl*) and substrate/liquid (*sl*) interfaces with interfacial tensions $\gamma_{(cl)}$ and $\gamma_{(sl)}$. Again, the model is restricted to short-term adhesion events prior to the spreading of cells on the substrate when a strong biologic component of adhesion can be expected.* Microscopic or molecular aspects of cell membranes are lumped together into a macroscopic, cell-surface energy property. Wetting properties are thus ascribed to individual cells [141]. That is, cells are regarded as nondeformable, spherical particles with characteristic wetting properties that do not significantly alter the microenvironment at cell attachment points through specific metabolic processes.

The reversible work of adhesion W_{adh} of a cell to a substrate is given by the Du Pre' Eq. (9) that is the difference in surface tensions between product $\gamma_{(sc)}$ and reactants $[\gamma_{(cl)} + \gamma_{(sl)}]$. W_{adh} is, therefore, equivalent to the free energy of adhesion ΔG in Fig. 13.

$$W_{\text{adh}} = -\Delta G = \gamma_{(cl)} + \gamma_{(sl)} - \gamma_{(sc)} \tag{9}$$

Physically, W_{adh} is the work to remove a cell from a substrate, leaving all interfaces at equilibrium, particularly with respect to adsorption. Implicit in the Du Pre' equation are the facts that adhesion is reversible and that liquid interfaces are destroyed during adhesion. These latter aspects of the thermodynamic model have elicited considerable debate in the literature [142,143] because the existence of a reversible adhesion equilibrium is difficult to prove. Indeed, the "thermodynamic cell adhesion equi-

* Cells in contact with surfaces for extended periods in time can modify the adhesive environment by the production of proteins, leading to formation of an "extracellular matrix."

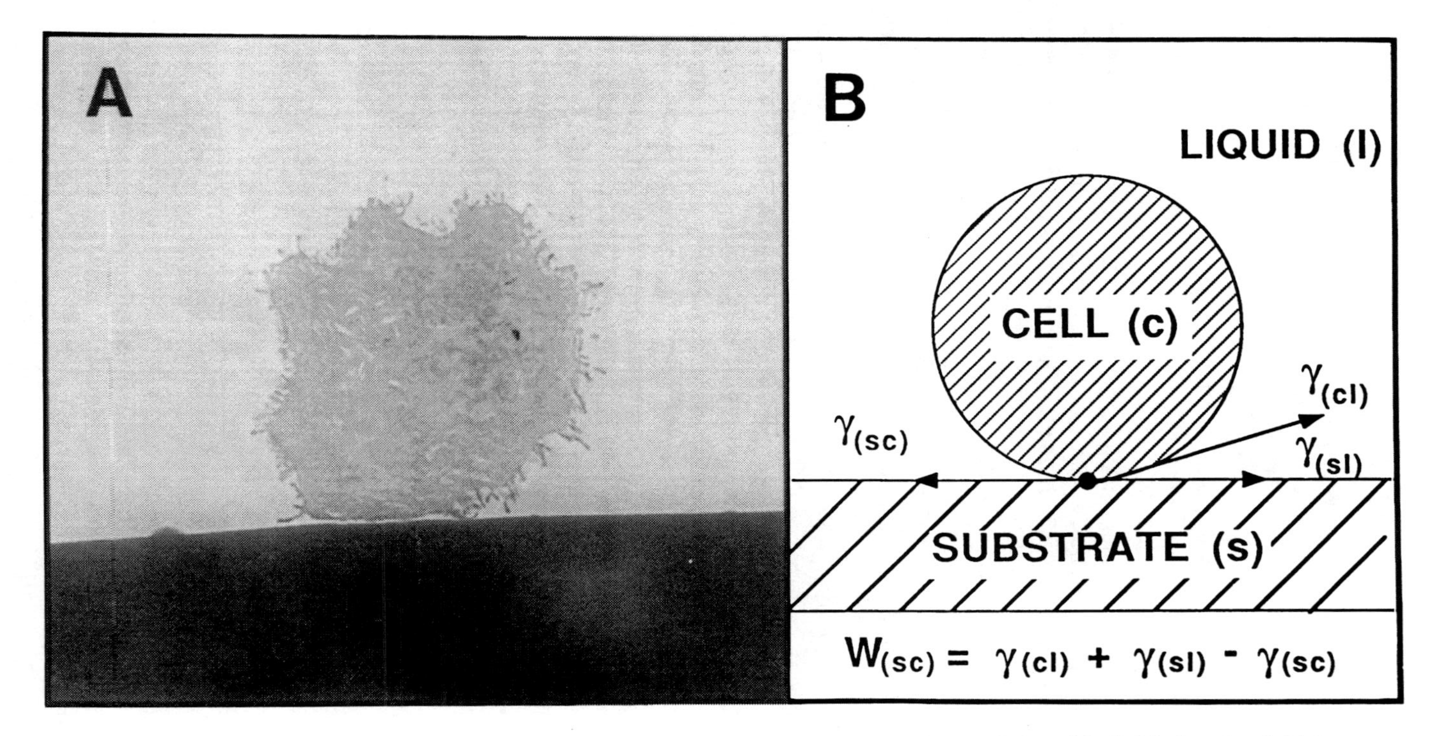

FIG. 14 Transmission electron micrograph (a, 7439×) of a cross section of a cell (MDCK, epithelioid) forming initial contact and attachment to tissue-culture-grade polystyrene from a sessile liquid phase (10% FBS containing DMEM, see text). Cell suspension was in contact with the substrate for 20 min. Physical model (b) of the initial cell adhesion showing important interfacial tensions controlling adhesion. (From Ref. 88.)

librium'' is a somewhat tenuous concept. Detailed ultrastructural analysis of adherent cells show that there are a number of points of close substrate contact along the periphery and underside of an adherent cell [52,88, 130,131]. Some of these contacts may undergo reversible adhesion without the whole cell physically moving from the substrate. Statistically based theories may be more appropriate for interpreting cell adhesion at this microscopic level [144–146].

Perhaps the most pertinent point to be drawn from a consideration of adhesion reversibility issues is that Eq. (9) cannot be expected to apply for all stages of cell adhesion. Indeed, adhesion is probably a misnomer in the current context. Adhesion usually refers to the force necessary to separate adherents and is controlled by short-range forces resulting from the formation of covalent, ionic, hydrogen, and charge-transfer bonds [147]. Perhaps it is more appropriate to use the terms ''attachment'' and ''work of attachment'' to make it clear that the adhesive process under consideration does not include these irreversible adhesive mechanisms.

Methods for assessing the values for solid and cell interfacial tensions in Eq. (9) are very controversial (see Sec. III.A.2), particularly for liquid phases containing proteins, making direct tests of the applicability of the Du Pre' equation to cell adhesion quite difficult. To further simplify the model, the cell adhesion process is abstracted into the two steps illustrated in Fig. 15. The first separates (*cl*) and (*sl*) interfaces into (*c*), (*s*), and (*l*) unit-area components requiring (losing) work $W_{(cl)}$ and $W_{(sl)}$, respectively. This represents an explicit destruction of the (*cl*) and (*sl*) interfaces only implicitly stated in Eq. (9). The second step reassembles (*s*) and (*c*) components to form the (*sc*) interface that defines the cell adhesion (attachment) event. Liquid/liquid (*ll*) interfaces, originally destroyed by the introduction of (*s*) and (*c*) into the fluid system, are likewise reformed. This

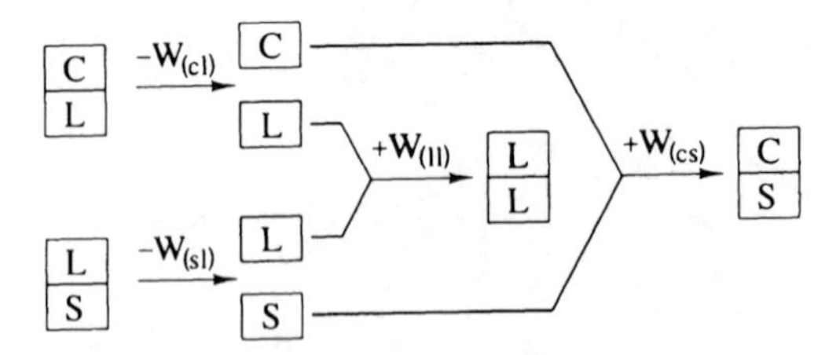

FIG. 15 Abstraction of the cell attachment process showing separation and recombination of unit interfacial areas leading to adhesion (*sc*) of cell (*c*) to substrate (*s*) and recombination of liquid (*ll*) phases.

second step produces work $+W_{(sc)}$ and $+W_{(ll)}$. The overall work of adhesion equivalent to that expressed in Eq. (9) is the sum of these individual work contributions

$$W_{\text{adh}} = W_{(sc)} + W_{(ll)} - W_{(cl)} - W_{(sl)} \tag{10}$$

To many biologically oriented readers, this level of abstraction may seem an even more distant departure from the biological process shown in Fig. 14a than the Du Pre' equation, because the processes outlined in Fig. 15 do not correspond to any of the familiar biological descriptions of adhering cells. In this connection, two points are important. First, none of the γ terms appearing on the right-hand side of Eq. (9) are directly measurable and are, therefore, no less abstract than the W terms in Eq. (10). In fact, some of the terms in Eq. (10) are directly measurable, as will be described subsequently. Second, since we have defined our model as a closed, adiabatic system, we can take advantage of all relevant thermodynamic relationships, including the first law, which allows the total work of a process to be calculated from the sum of any number of intermediate steps leading to the final state. The familiar physical chemistry analog is the calculation of heat of reaction from standard thermodynamic tables.

Some of the W terms in Eq. (10) can be immediately identified with measurable parameters. The work of cohesion $W_{(ll)} = 2\gamma_{(lv)}$. $W_{(sl)}$ follows directly from the definition of work of adhesion given by Eq. (2). By analogy, $W_{(cl)} = \tau_{(cl)} + \gamma_{(lv)}$, where $\tau_{(cl)}$ is the wetting tension of liquid on a unit area of cell membrane. Of course, the accurate measurement of $\tau_{(cl)}$ is experimentally difficult, if not impossible, but for the present purposes it is useful to make this definition since this model ascribes macroscopic thermodynamic properties to microscopic cells. Substituting these values into Eq. (10) leads to the conclusion that $W_{\text{adh}} = W_{(sc)} - \tau_{(cl)} - \tau_{(sl)}$, which is the wetting tension analog of Eq. (9). Here $\tau_{(sl)}$ symbology is used to distinguish cell and substrate wetting tensions. This relationship suggests that cell adhesion for a fixed cell type should vary sensibly as either or both substrate wetting properties or liquid-phase composition is altered. This is indeed found experimentally to be the case by using different substrates with different τ° or by varying the concentration of surfactants and biosurfactants in liquid phases used in cell adhesion experiments [87,88].

It is found that cell adhesion can be driven to zero at some empirically determined surfactant concentration at which $W_{\text{adh}} = 0$. When this occurs, $[\tau^*_{(cl)} - \tau^*_{(sl)}] = W_{(sc)}$, where the asterisk denotes wetting tensions at the surfactant concentration leading to zero adhesion. Using this zero-adhesion point as a reference state allows Eq. (9) to be rewritten entirely in terms of wetting tensions

$$W_{\text{adh}} = \{[\tau^*_{(cl)} - \tau_{(cl)}] + [\tau^*_{(sl)} - \tau_{(sl)}]\} = \{[\text{a constant}] - [\tau_{(cl)} + \tau_{(sl)}]\} \quad (11)$$

In Eq. (11) it has been recognized that $\tau^*_{(cl)}$ and $\tau^*_{(sl)}$ are constants for a given cell and substrate pair. The interesting qualitative prediction of Eq. (11) is that cell adhesion should rise and fall with surfactant concentration as cell and substrate wetting tension curves constructively or destructively combine along the ln *C* axis, just as described for *W* curves and shown in Sec. III.E. This cell adhesion behavior is noted experimentally and a thermodynamic theory further developed from this model predicts short-term cell adhesion in vitro with surprising fidelity [87,88].

E. Philosophy

Thermodynamic principles can be applied to many biomaterial problems involving wetting phenomena by the use of carefully constructed models. The modeling approach is part of the reductionist practice that has yielded many of the familiar laws of nature. It is disappointing that these systematic approaches to biomaterials development have not been more widely adopted by the biomedical community at large and supported by long-term funding so that some of the appropriate "laws of biomaterials" can be discovered [148].

Colloid science is a field with many parallels to biomaterials science that has largely yielded to the reductionist strategy. Colloid stability questions can be addressed in detail using the formalism of the DLVO theory. It is unfortunate that no such formalism is yet available for biomaterial applications. Many of the predictive attributes of colloid science have already had significant impact in biomaterials science and many more will result with a more directed, purposeful application of the thermodynamics of interfacial chemistry.

V. PROSPECTS FOR RATIONAL SURFACE PREPARATION FOR BIOMATERIAL APPLICATIONS

A. Perspective on the Problem

1. Connections Between Biomaterials Function, Interfacial Properties, and Surface Chemistry

A basic tenet of this chapter is that the biological response to artificial materials is primarily driven by interfacial phenomena at biomaterial surfaces and that wetting measurements directly probe these interfacial properties. Wettability, as measured in surface energies or tensions, must be in some way connected to surface chemistry, as described in compositional terms, because it is those functional groups residing at a given surface with specific chemical reactivity that most profoundly distinguish one surface from any other. This line of reasoning leads to a hierarchy of relationships diagrammed in Fig. 16 with some direct connections (solid arrows) and other less direct connections (dashed arrows). Interfacial properties in saline and physiologic media are specifically separated in recognition of the fact that the adsorption of various biosurfactants from a biological fluid phase can effectively mask starting surface chemistry in as yet unpredictable ways. The development of rational, prospective synthetic schemes for biomedical applications is critically dependent on understanding some or all of the connections outlined in Fig. 16.

2. Translating Biomaterial Properties into Physicochemical Terms

Connecting biomaterial and interfacial properties is the most challenging task outlined in Fig. 16. The difficulty arises from the fact that biological or biomedical descriptions of interfacial events are of little practical utility to the materials scientist. Somehow, the biological view of these interfacial events must be *translated* into physicochemical terms usable by a materials scientist in the preparation of biomaterials for various applications. This translation step will require an understanding of interfacial biophysics, involving a systematic subdivision of the overall biological process into discrete components that can be modeled, leading to identification of controlling physicochemical parameters. Testable hypotheses can be assembled from the models and an iterative process of model improvement with a check against experiment leads to a predictive framework against which the biological process can be understood in terms of interfacial properties.

3. Relating Interfacial Properties and Surface Chemistry

The field of surface science has not progressed to the point that interfacial properties, including biosurfactant adsorption, can be directly correlated with the presence of specific chemical functional groups at the interface. This limitation is due, in part, to the surface characterization difficulties discussed in Sec. II of this chapter. Work toward the goal of establishing surface-composition/interfacial-property relations is progressing, but much is to be achieved before this basic knowledge can be widely applied.

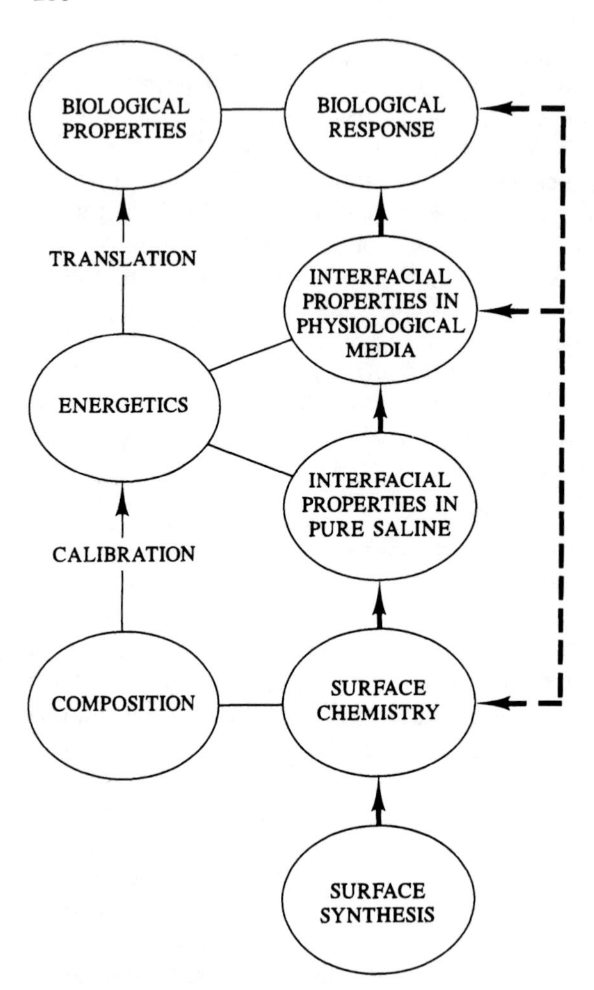

FIG. 16 Connections between material surface chemistry, interfacial properties, and the biological response, with solid lines indicating direct or primary connections and broken lines indirect or secondary connections.

An alternative, pragmatic approach is to correlate synthetic steps used in the preparation of a surface with composition and interfacial properties in a cause-and-effect manner: a calibration of sorts as indicated in Fig. 16. The intermediate step relating surface chemistry to wetting properties in saline simplifies the calibration by unlinking surface composition and protein adsorption.

B. Example

1. Surface-Composition/Interfacial-Property Calibration

The discussion of cell adhesion in vitro of Sec. IV is an illustration of the translation of biomaterial properties into physicochemical terms. The following examples focus on the potential for establishing relationships between surface composition and interfacial properties that are the embodiment of the calibration step shown in Fig. 16.

Any calibration relies on a primary relationship between cause and effect that can be monitored through measurable properties. A useful calibration between surface chemistry and interfacial properties for biomaterial applications must in some way relate surface composition to protein adsorption since adsorption is a ubiquitous event in the physiological environment. One approach to this calibration is to first relate surface chemistry, as determined by ESCA, for example, to material wetting properties measured in pure water or saline and then relate these wetting properties to those obtained in physiologically relevant, protein-

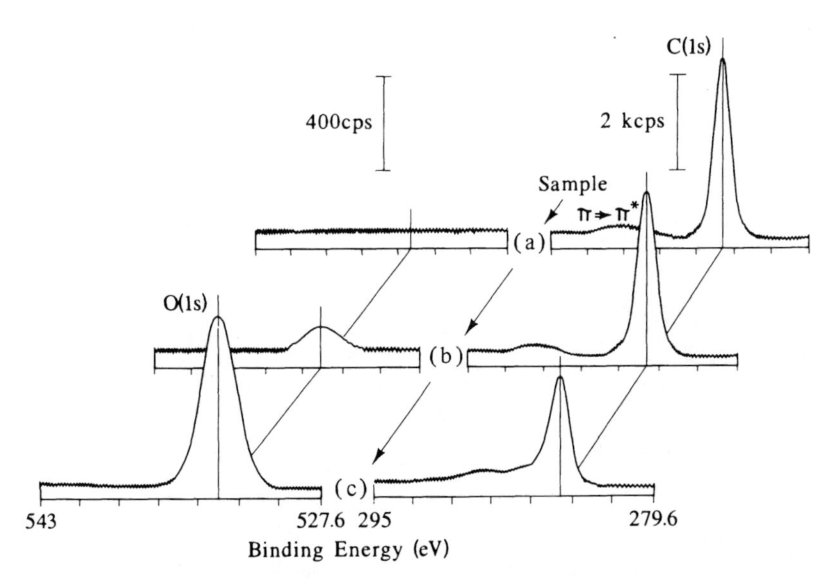

FIG. 17 ESCA spectra of oxygen-plasma-treated polystyrene samples representing the untreated condition (a), intermediate treatment condition (b), and fully treated condition (c) showing evolution in surface oxygen composition and concomitant broadening of the *C*(1*s*) peak. (Adapted from Ref. 151.)

containing solutions. The two-step process is intended to simplify and sharpen causal relationships.

Figure 17 shows ESCA C(1*s*) and O(1*s*) spectra for three samples of polystyrene with different levels of oxidation resulting from oxygen plasma treatment. Sample a is untreated polystyrene exhibiting no surface oxidation, sample b has intermediate oxygen composition at 4.9 atom %, and sample c exhibits maximal surface oxygen achievable using this particular surface treatment method at 21.7 atom %. Surface oxygen is incorporated into the polystyrene backbone of treated specimens in a number of carbon oxidation states indicated by the appearance of an unresolved, high-energy wing on the C(1*s*) spectrum. Samples a–c cover the full range of solvent (saline) wettabilities (see Sec. III.B.3) corresponding to native, untreated polystyrene to fully wettable. Sample a: advancing τ° = 7.5, receding τ° = 25.4 dyne/cm; sample b: advancing τ° = 23.5, receding τ° = 39.8 dyne/cm; sample c: advancing = receding τ° = 72.8 dyne/cm (fully wettable condition exhibiting no contact-angle hysteresis). Different durations of exposure to the oxygen plasma yield surfaces falling variously within this range of wettabilities. Figure 18 col-

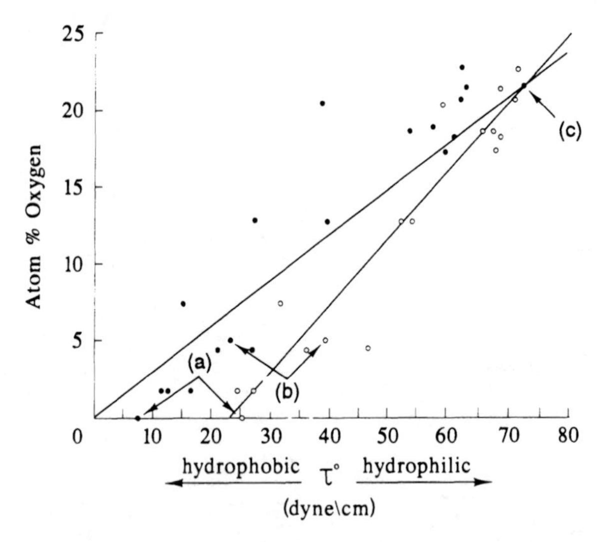

FIG. 18 Monotonic decrease in saline wettability τ° with increasing oxygen atom % and for plasma-treated polystyrene surfaces (advancing = filled circles, receding = open circles). Lines through the data represent a linear best fit to the data, with the constraint that advancing τ° = receding = τ° = 72.8 dyne/cm (no hysteresis) at 21.7 atom % oxygen. Samples (a), (b), and (c) correspond to spectra shown in Fig. 17. (From Ref. 151.)

lects oxygen compositions for polystyrene samples plasma treated in this way. The monotonic increase in wettability with surface oxygen composition* in Fig. 18 suggests a direct linkage between surface chemistry as measured here by ESCA and surface energetics measured by contact angles. That is, for this particular study, there is a primary cause-and-effect relationship between the chemistry of the upper 5-nm surface layers probed by ESCA (the cause) and interfacial properties of the outermost 0.5 nm sensed by contact angles (the effect). It is no surprise that oxidized surfaces are more wettable than nonoxidized polymers, but the linearity of the trends, as suggested by best-fit lines through the data of Fig. 18, is somewhat unexpected given the difference in surface sensitivity between ESCA and contact angles. Possibly, reactive species in the oxygen plasma ostensibly treat only the outermost molecular layers so that, in this special case, ESCA surface sensitivity to oxygen is directly proportional to increased hydrophilicity. Alternatively, the plasma treatment may penetrate quite deeply, but the treatment is very uniform in-depth, so that the outer 0.5-nm composition is the same as at lower strata. However this occurs, Fig. 18 constitutes a rough calibration between surface chemistry and surface energetics.

2. Adsorption Map: A Calibration Tool for Complex Biological Mixtures [151]

The step connecting interfacial properties in pure saline with that in physiologic media shown in Fig. 16 involves the effects of biosurfactant adsorption. Relationships between surface wettability and surfactant adsorption can be quantified using Eqs. (4), (5), and (8) if the chemical composition of the fluid phase is known. Many, if not most, real-world biomaterial problems, however, involve complex biological milieu of unknown chemical composition such as blood, tissue-culture supernate, and diagnostic or bodily fluids. Consequently, surface thermodynamic concepts are not directly applicable to these important cases. The adsorption index introduced in Sec. III.B.5 has been found to be quite useful for these occasions. Basically, the index compares "full-strength" surfactant effect τ' to solvent wettability τ° through the difference $[\tau' - \tau^\circ]$. Full strength is functionally defined and might be, for example, whole blood serum of a detergent mixture at some critical concentration. The chemical

* Some of the "noise" in the data of Fig. 18 may be attributed to position-to-position surface variability of the polystyrene plaques used in this study since samples for ESCA and contact analyses were cut from different portions of the same plaque. Both ESCA and contact angles sample very small portions of a sample surface (>1 mm^2) and do not, therefore, represent a view of the "average" surface composition or wettability over a larger scale. See [151] for more discussion.

composition of the surfactant system need not be known in the measurement or interpretation of these phenomenological, thermodynamic-like adsorption indices. In particular, it is found that the adsorption index is predictive of surfactant adsorption to polymeric solids when plotted against the inherent wettability of the surface τ°. This construction is termed an *adsorption map*. Following the pattern of previous sections, it is useful first to consider results obtained for a simple surfactant like tween, using this as a basis for comparison for more complicated protein solutions.

As shown on the abscissa of the adsorption map of Fig. 19, the practical range of τ° values for polymers against saline is between about -37 (PTFE teflon-like) and 73 (completely wettable) dyne/cm. Practical limits on $[\tau' - \tau^{\circ}]$ are determined by boundary conditions on θ'. That is, full-

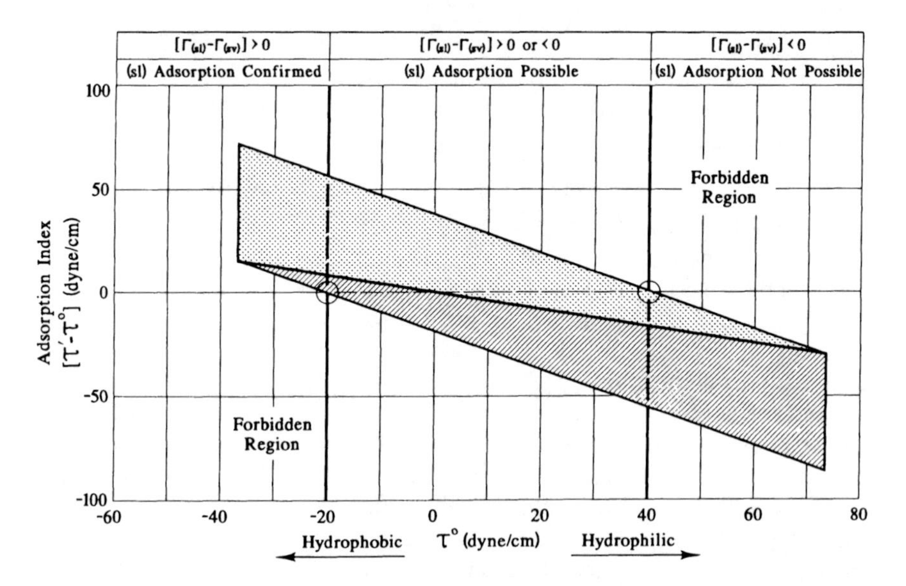

FIG. 19 Adsorption map for tween-80 plotting the adsorption index $[\tau' - \tau^{\circ}]$ against inherent wettability τ°. Upper and lower boundary lines enclose the region of allowed wetting data in a parallelogram, excluding data from the indicated forbidden regions. Points where the wetting parallelogram cross the zero-adsorption line indicated by open circles define wettability boundaries that define adsorption only (left-hand zone), indeterminate region (central zone), and a region where (*sl*) adsorption cannot occur (right-hand zone). The line bisecting the wetting parallelogram into wetting triangles represents the hypothetical case, in which addition of surfactant does not affect the observed contact angle. (Adapted from Ref. 151.)

strength surfactant effect can cause θ' to be no less than 0° (perfect wetting limit, $\cos\theta = 1$) and the practical limit of observable angles is about 120° (nonwettable limit, $\cos\theta = -\frac{1}{2}$). Boundary conditions on τ' are thus $+\gamma'_{(lv)}$ and $-\gamma'_{(lv)}/2$, imposing limits on $[\tau' - \tau°]$ represented by the lines $[\gamma'_{(lv)} - \tau°]$ and $[-\gamma'_{(lv)}/2 - \tau°]$. These limits enclose the set of all real wetting data in the parallelogram shown in Fig. 18. It is physically impossible for wetting results to fall outside the parallelogram. The line bisecting the wetting parallelogram represents the hypothetical condition in which no change in contact angle occurs by the addition of surfactant. In this circumstance, $\cos\theta' = \cos\theta° = [\tau'/\gamma'_{(lv)}] = [\tau°/\gamma°_{(lv)}]$. It follows that $[\tau' - \tau°] = (\tau°\{[\gamma'_{(lv)}/\gamma°_{(lv)}] - 1\})$ is the equation of the θ = constant line. Data for surfactants that adsorb, causing $\theta' < \theta°$, must lie above this line, whereas surfactants that adsorb, causing $\theta' > \theta°$, must lie below the bisecting line. Wetting data for surfactants exhibiting a single mechanism of adsorption to all surfaces are thus further constrained to a wetting triangle represented by either the upper or lower half of the wetting parallelogram. At this juncture, it is important to note that the full range of possible adsorption behaviors for a given surfactant system and any surface with inherent wettability $\tau°$ can be determined by $\gamma°_{(lv)}$ and $\gamma'_{(lv)}$ values alone. This fact is of considerable practical utility since $\gamma_{(lv)}$ is relatively straightforward and can be measured experimentally with good accuracy and precision.

No adsorption occurs when $\tau' = \tau°$ and $[\tau' - \tau°] = 0$. The wetting parallelogram crosses the $[\tau' - \tau°] = 0$ line at discrete points indicated in Fig. 19. The area of the parallelogram falling entirely above this line (left-hand portion of Fig. 19) corresponds to surfaces that must exhibit (*sl*) adsorption. The area of the parallelogram that falls entirely below the $[\tau' - \tau°] = 0$ line (far right-hand portion of Fig. 19) corresponds to surfaces that do not support (*sl*) adsorption. The adsorption condition to surfaces falling within the central portion of Fig. 19 must be ascertained by experiment and cannot be predicted from the confines of the map alone. A particularly informative landmark on the adsorption map is where the *upper boundary* line crosses the zero adsorption at $[\tau' - \tau°] = 0$. Here, at maximal surfactant effect for the system under study, $\tau° = \tau' = \gamma'_{(lv)}$ and no adsorption can occur. It follows that interfaces associated with solid surfaces having inherent wettability $\tau° > \gamma'_{(lv)}$ must exhibit interfacial depletion. This adsorption limit can be read directly from $\gamma'_{(lv)}$ on $\gamma_{(lv)}$ dilution curves and should be of considerable utility to biomaterial scientists who wish to control protein adsorption. Thus, for the specific case of the tween map of Fig. 19, it can be predicted that (*sl*) adsorption must occur for surfaces exhibiting $\tau° < -20$ dyne/cm and cannot occur for surfaces exhibiting $\tau° > 40$ dyne/cm.

Figure 20a plots advancing (full circles) and receding (open circles) adsorption index data for tween-80 on the same family of oxidized polystyrene surfaces discussed in the preceding section. Polystyrene is a surface that is hydrophobic in its native state and has been oxidized to yield surfaces with increasingly hydrophilic character. By contrast, silane-treated glass coverslips (see Sec. III.E.1; filled triangles = advancing, open triangles = receding index data in Fig. 20a) are a surface that is hydrophilic in its native state and has been partially derivatized with silane to yield surfaces with an increasingly hydrophobic character. Data for these distinctly different surface chemistries fall within the wetting parallelogram and generally above the $\theta^\circ = \theta'$ line. Tween data cluster near the $\cos \theta' = 1$ boundary because this surfactant significantly increases the surface wettability of surfaces in the study. Samples (A–D) noted on the adsorption map of Fig. 20a were selected for quantitative adsorption measurements using concentration-dependent contact angles. Figure 20b plots the surface excess difference* $[\Gamma_{(sl)} - \Gamma_{(sv)}]$ against τ° for comparison to the map in Fig. 20a. It is interesting that the monotonic decrease in $[\Gamma_{(sl)} - \Gamma_{(sv)}]$ with τ° mirrors that observed in the adsorption index on the tween adsorption map. More important, $[\Gamma_{(sl)} - \Gamma_{(sv)}] = 0$ and the $[\tau' - \tau^\circ] = 0$ crossing correspond near $\tau^\circ = 40$ dyne/cm, confirming the interpretation that no (sl) adsorption occurs when the index is zero. Thus, the results of the adsorption mapping procedure are corroborated by the more rigorous but labor-intensive method of monitoring adsorption by concentration-dependent interfacial tensions.

HSA and FBS adsorption maps are collected in Figs. 21a and b for the same family of oxidized polystyrene and silane-treated glass slides discussed for the benchmark tween case. Note that there is no ambiguity in comparing the adsorption behaviors of a purified detergent (tween-80) to a purified protein (HSA) to a heterogeneous milieu of proteins (FBS) using the adsorption mapping procedure. It is of interest that the "allowed" areas defined by the wetting triangles for these three surfactant systems are quite different. Adsorption limits for tween, HSA, and FBS are 39.9 ± 0.5, 60.6 ± 0.5, and 56.4 ± 0.8 dyne/cm, respectively, from $\gamma_{(lv)}$ curves for these surfactants. Thus, protein-repellent surfaces must be much more wettable than tween-repellent surfaces, in accordance with the general observation that proteins adsorb to nearly all polymeric materials.

Adsorption maps exhibit linear trends similar to those described in the preceding section. As such, these maps constitute a calibration between the adsorptive capacity of a given surface with a characteristic inherent

* Computed using Eq. (8) of Sec. III.D.

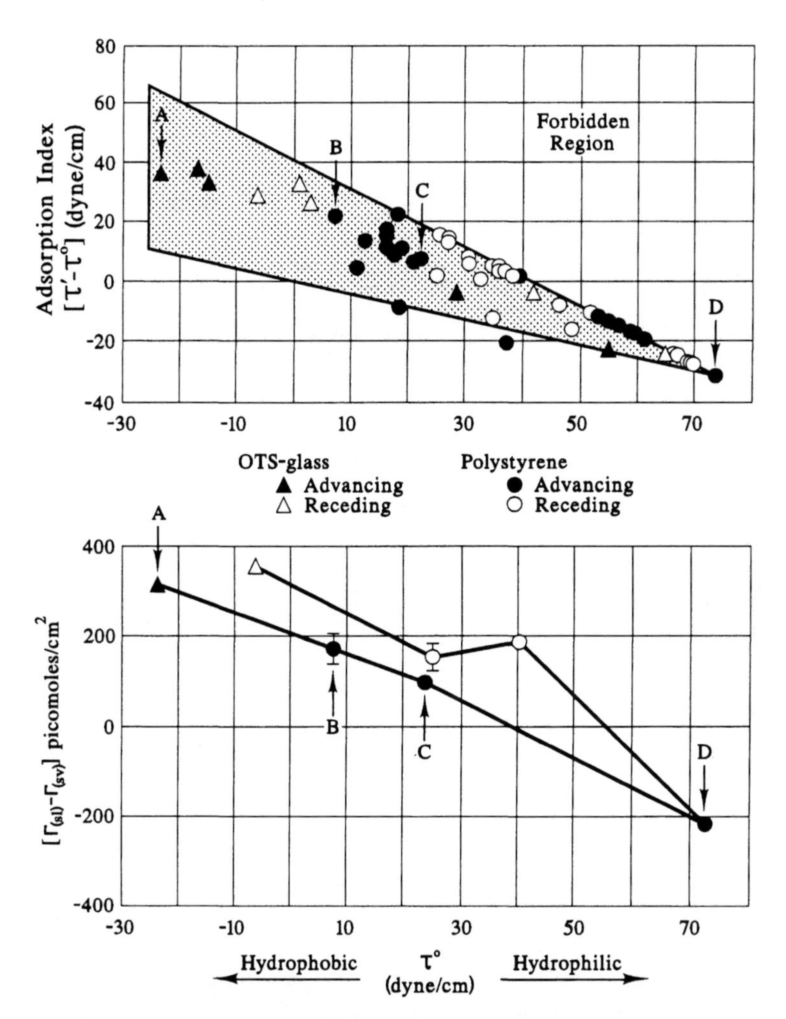

FIG. 20 Tween-80 adsorption map for polystyrene (circles) and octadecylsilane-treated glass slides (triangles) for advancing (solid symbols) and receding (open symbols) adsorption index data. Lower half of the wetting parallelogram is not shown since most data fall within the upper triangle. Points where the wetting triangle crosses the zero adsorption line are indicated with open circles. Calculated surface excess $[\Gamma_{(sl)} - \Gamma_{(sv)}]$ from concentration-dependent contact angles is plotted against τ° for comparison to the adsorption map. (From Ref. 151.)

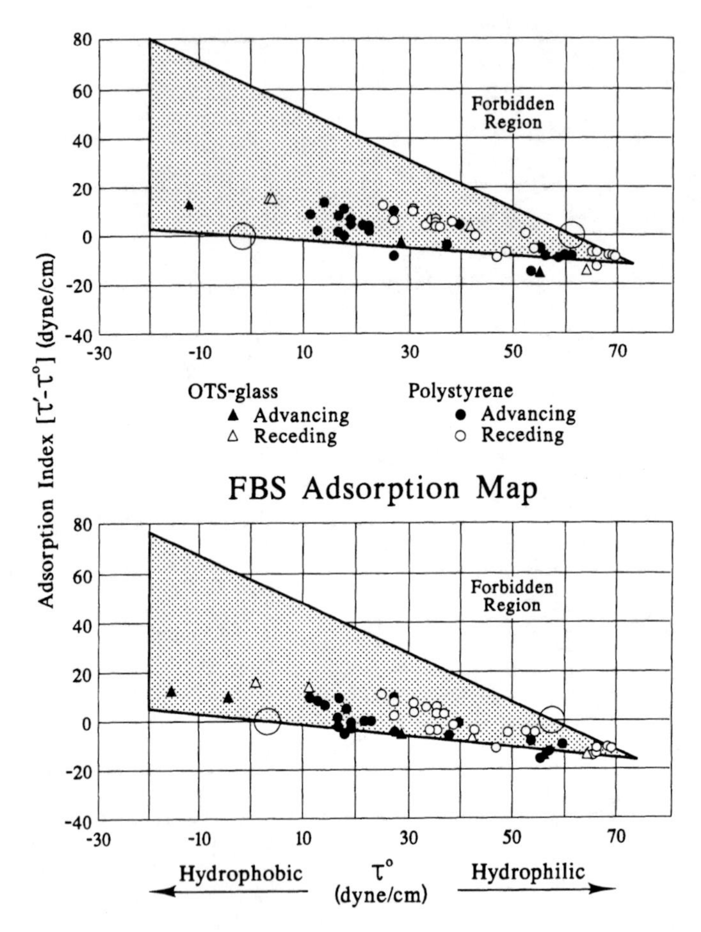

FIG. 21 Adsorption maps for HSA and FBS with polystyrene (circles) and octadecylsilane-treated glass slides (triangles) for advancing (solid symbols) and receding (open symbols) adsorption index data. Note boundaries of the wetting triangles are shifted relative to Figs. 18 and 19 because $\gamma'_{(lv)}$ has increased. Points where the wetting triangles cross the zero-adsorption line are indicated with open circles. (Adapted from Ref. 151.)

wettability, effectively making the connections between surface chemistry and interfacial properties shown in Fig. 16.

C. Philosophy

The rudimentary problem in biomaterials is the translation of the desirable biomedical properties that define biocompatibility for a particular end application into materials properties that lead to the correct selection, design, or synthesis of an appropriate material for that use. This is a particularly difficult problem with respect to surface properties because structure-function relationships connecting biomedical function to surface chemistry and interfacial properties are not yet generally available. Elucidating these structure-function relationships is a fundamental challenge of modern biomaterials surface science.

ACKNOWLEDGMENTS

The author is indebted to Professor P. K. Kilpatrick and Dr. R. E. Johnson for helpful discussion of this work. Collaboration with D. A. Martin and Dr. D. B. Montgomery in the development of the adsorption mapping method is most appreciated. The expert technical assistance of J. Graper and editorial assistance of L. W. Tingen are gratefully acknowledged. R. K. Hanson provided artistic input in the preparation of figures.

REFERENCES

1. R. E. Baier, in *Blood Vessels. Problems Arising at the Borders of Natural and Artificial Blood Vessels* (Proceedings 8th Scientific Conference of Gesellschaft Deutscher Naturforscher und Aerzte) (S. Efferet, and J. D. Meyer-Erkelenz, eds.), Springer-Verlag, Berlin, 1976, pp. 159–162.
2. R. E. Baier, and R. C. Dutton, *J. Biomed. Mater. Res. 3:*191 (1969).
3. R. E. Baier, R. C. Dutton, and V. L. Gott, in *Surface Chemistry of Biological Systems* (M. Blank, ed.), Plenum, New York, 1970, pp. 235–260.
4. R. E. Baier, and V. A. dePalma, in *Management of Arterial Occlusive Disease* (W. A. Dale, ed.), Year Book Medical Pub., Chicago, IL, 1971, pp. 147–163.
5. R. E. Baier, *Bull. N. Y. Acad. Med. 48:*257 (1972).
6. R. E. Baier, in *Adhesion in Biological Systems* (R. S. Manly, ed.), Academic Press, New York, 1970, pp. 15–48.
7. R. E. Baier, and A. E. Meyer, in *Physicochemical Aspects of Polymer Surfaces*, Vol. 2 (K. L. Mittal, ed.), Plenum, New York, 1981, pp. 895–909.

8. R. E. Baier, A. E. Meyer, J. R. Natiella, R. R. Natiella, and J. M. Carter, *J. Biomed. Mater. Res. 18:*337 (1984).
9. C. A. Ward, and A. W. Neumann, *J. Coll. Inter. Sci. 49:*286 (1974).
10. A. W. Neumann, R. J. Good, C. J. Hope, and M. Sejpal, *J. Coll. Inter. Sci. 49:*291 (1974).
11. A. W. Neumann, D. R. Absolom, D. W. Francis, and C. J. van Oss, *Sep. Purif. Methods 9:*69 (1980).
12. J. K. Spelt, D. R. Absolom, and A. W. Neumann, *Langmuir 2:*620 (1986).
13. C. J. van Oss, C. F. Gillman, and A. W. Neumann, in *Phagocytic Engulfment and Cell Adhesiveness*, Marcel Dekker, New York, 1975, pp. 7–152.
14. S. K. Chang, O. S. Hum, M. A. Moscarello, A. W. Neumann, W. Zingg, M. J. Leutheusser, and B. Ruegsegger, *Med. Prog. Technol. 5:*57 (1977).
15. D. R. Absolom, A. W. Neumann, W. Zingg, and C. J. van Oss, *Trans. Am. Soc. Artif. Intern. Organs 25:*152 (1979).
16. A. W. Neumann, D. R. Absolom, C. J. van Oss, and W. Zingg, *Cell Biophys. 1:*79 (1979).
17. D. R. Absolom, C. J. van Oss, R. J. Genco, D. W. Francis, and A. W. Neumann, *Cell Biophys. 2:*113 (1980).
18. A. W. Neumann, O. S. Hum, D. W. Francis, W. Zingg, and C. J. van Oss, *J. Biomed. Mater. Res. 14:*499 (1980).
19. P. J. Facchini, A. W. Neumann, and F. DiCosmo, *Appl. Microbiol. Biotechnol. 29:*346 (1988).
20. J. Steinberg, A. W. Neumann, D. R. Absolom, and W. Zingg, *J. Biomed. Mater. Res. 23:*591 (1989).
21. C. J. van Oss, W. Zingg, O. S. Hum, and A. W. Neumann, *Thrombosis Res. 11:*183 (1977).
22. A. W. Neumann, C. J. Hope, C. A. Ward, M. A. Herbert, G. W. Dunn, and W. Zingg, *J. Biomed. Mater. Res. 9:*127 (1975).
23. J. H. Braybrook, and L. D. Hall, *Prog. Polym. Sci. 15:*715 (1990).
24. H. G. Klemperer, and P. Knox, *Lab. Pract. 26:*179 (1977).
25. A. S. G. Curtis, J. V. Forrester, C. McInnes, and F. Lawrie, *J. Cell Biol. 97:*1500 (1983).
26. W. S. Ramsey, W. Hertl, E. D. Nowlan, and N. J. Binkowski, *In Vitro 20:*802 (1984).
27. M. J. Lydon, T. W. Minett, and B. J. Tighe, *Biomaterials 6:*396 (1985).
28. S. Hattori, J. D. Andrade, J. B. Hibbs, Jr., D. E. Gregonis, and R. N. King, *J. Coll. Inter. Sci. 104:*72 (1985).
29. H. Takayama, T. Tanigawa, A. Takagi, and K. Hatada, *Biomed. Res. 7:* 11 (1986).
30. D. Barngrover, in *Mammalian Cell Technology* (W. G. Thilly, ed.), Butterworths, Boston, MA, 1986, pp. 131–149.
31. B. R. McAuslan, and G. Johnson, *J. Biomed. Mater. Res. 21:*921 (1987).
32. N. F. Owens, D. Gingell, and A. Trommler, *J. Cell Sci. 91:*269 (1988).
33. A. Curtis, and C. Wilkinson, *Stud. Biophys. 127:*75 (1988).

34. R. Aveyard, and D. A. Haydon, in *An Introduction to the Principles of Surface Chemistry*, Cambridge Univ. Press, Cambridge, 1973, pp. 1–30.
35. H. Hantsche, *Scanning 11:*257 (1989).
36. H. W. Werner, and A. Torrisi, *Fresen. J. Anal. Chem. 337:*594 (1990).
37. B. D. Ratner, ed., *Surface Characterization of Biomaterials*, Elsevier, New York, 1988.
38. B. D. Ratner, A. B. Johnston, and T. J. Lenk, *J. Biomed. Mater. Res: Appl. Biomat. 21:*59 (1987).
39. B. D. Ratner, *Prog. Biomed. Eng. 5:*87 (1988).
40. B. D. Ratner, D. G. Castner, T. A. Horbett, and T. J. Lenk, *J. Vac. Sci. Technol. 8:*2306 (1990).
41. M. Ratzsch, H. J. Jacobasch, and K. H. Freitag, *Adv. Coll. Inter. Sci. 31:* 225 (1990).
42. S. Nir, *Prog. Surf. Sci. 8:*1 (1976).
43. F. M. Fowkes, in *Symposium on Physics and Chemistry of Interfaces*, 2nd ed., ACS, Washington, DC, 1971, pp. 154–167.
44. R. J. Good, and E. Elbing, *Ind. Eng. Chem. 62:*54 (1970).
45. F. M. Fowkes, in *Surfaces and Interfaces I* (J. J. Burke, N. L. Reed, and V. Weiss, eds.), Syracuse Univ. Press, Syracuse, NY, 1966, pp. 197–224.
46. W. A. Zisman, *Ind. Eng. Chem. 55:*19 (1963).
47. F. M. Fowkes, *Ind. Eng. Chem. 56:*40 (1964).
48. P. M. Claesson, C. E. Blom, P. C. Herder, and B. W. Ninham, *J. Coll. Inter. Sci. 114:*234 (1986).
49. R. M. Pashley, P. M. McGuiggan, B. W. Ninham, and D. F. Evans, *Science 229:*1088 (1985).
50. J. Marra, *J. Coll. Inter. Sci. 109:*11 (1986).
51. J. Israelachvili, and R. Pashley, *Nature 300:*341 (1982).
52. F. Grinnell, in *International Review of Cytology*, Vol. 53 (G. H. Bourne, and J. F. Danielli, eds.), Academic Press, New York, 1978, pp. 65–144.
53. E. Ruoslahti, M. D. Pierschbacher, A. Oldberg, and E. G. Hayman, *BioTechniques 38*(Jan./Feb. 1984).
54. K. M. Yamada, and D. W. Kennedy, *J. Cell Biol. 99:*29 (1984).
55. D. F. Williams, and R. D. Bagnall, in *Fundamental Aspects of Biocompatibility*, Vol. II (D. F. Williams, ed.), CRC Press, Boca Raton, FL, 1981, pp. 113–127.
56. M. J. Lydon, and C. A. Foulger, *Biomaterials 9:*525 (1988).
57. J. D. Andrade, D. E. Gregonis, and L. M. Smith, in *Physicochemical Aspects of Polymer Surfaces*, Vol. 2 (K. L. Mittal, ed.), Plenum, London, 1981, pp. 911–922.
58. L. Lavielle, and J. Schultz, *J. Coll. Inter. Sci. 106:*438 (1985).
59. J. D. Andrade, and W.-Y. Chen, *Surf. Inter. Anal. 8:*253 (1986).
60. S. H. Lee, and E. Ruckenstein, *J. Coll. Inter. Sci. 120:*529 (1987).
61. J. M. Park, and J. D. Andrade, in *Polymer Surface Dynamics* (J. D. Andrade, ed.), Plenum, New York, 1988, pp. 67–88.
62. J. D. Andrade, in *Polymer Surface Dynamics* (J. D. Andrade, ed.), Plenum, New York, 1988, pp. 1–8.

63. F. Garbassi, M. Morra, E. Occhiello, L. Barino, and R. Scordamaglia, *Surf. Inter. Anal. 14:*585 (1989).
64. M. Morra, E. Occhiello, R. Morola, F. Garbassi, P. Humphrey, and D. Johnson, *J. Coll. Inter. Sci. 137:*11 (1990).
65. M. Morra, E. Occhiello, L. Gila, and F. Garbassi, *J. Adhesion 33:*77 (1990).
66. E. Occhiello, M. Morra, G. Morini, F. Garbassi, and P. Humphrey, *J. Appl. Polym. Sci. 42:*551 (1991).
67. J. L. Brash, and T. A. Horbett, eds., *Proteins at Interfaces*, ACS Symp. Series 343, Washington, DC, 1987.
68. C. D. Bain, and G. M. Whitesides, *J. Am. Chem. Soc. 110:*5897 (1988).
69. G. M. Whitesides, and P. E. Laibinis, *Langmuir 6:*87 (1990).
70. J. W. Bluhm, *Am. Lab. 6* (Nov. 31, 1974).
71. A. W. Neumann, and R. J. Good, *Surf. Coll. Sci. 11:*31 (1979).
72. D. Chatenay, D. Langevin, and J. Meunier, *J. Dispersion Sci. Technol. 3:* 245 (1982).
73. M. D. Lelah, T. G. Grasel, J. A. Pierce, and S. L. Cooper, *J. Biomed. Mat. Res. 19:*1011 (1985).
74. J. L. Valero, A. Prieto, A. Lloris, G. Olivares, and J. Morales, *Microcomput. Appl. 4:*65 (1985).
75. W. G. Pitt, B. R. Young, and S. L. Cooper, *Coll. Surf. 27:*345 (1987).
76. Y. K. Kamath, C. J. Dansizer, S. Hornby, and H.-D. Weigmann, *Text. Res. J. 57:*205 (1987).
77. P. Dryden, J. H. Lee, J. M. Park, and J. D. Andrade, in *Polymer Surface Dynamics* (J. D. Andrade, ed.), Plenum, New York, 1988, pp. 9–24.
78. P. Than, L. Preziosi, D. D. Joseph, and M. Arney, *J. Coll. Inter. Sci. 124:* 552 (1988).
79. T. Tanaka, I. Kon-No, and T. Kadoya, *J. Coll. Inter. Sci. 132:*139 (1989).
80. D. A. Martin, and E. A. Vogler, *Langmuir 7:*422 (1991).
81. E. A. Vogler, *Interfaces Comp. 3:*19 (1985).
82. E. A. Vogler, and R. W. Bussian, *J. Biomed. Mat. Res. 21:*1197 (1987).
83. T. A. Horbett, B. D. Ratner, and A. S. Hoffman, Devices and Technology Branch Contractors Meeting, Program and Abstracts, National Institutes of Health, Bethesda, MD, 1985.
84. D. R. Absolom, F. V. Lamberti, Z. Policova, W. Zingg, C. J. van Oss, and A. W. Neumann, *Appl. Environ. Microbial. 46:*90 (1983).
85. H. J. Busscher, A. H. Weerkamp, H. C. van der Mei, A. W. J. van Pelt, H. P. de Jong, and J. Arends, *Appl. Environ. Microbiol. 48:*980 (1984).
86. M. Uyen, H. J. Busscher, A. H. Weerkamp, and J. Arends, *FEMS Microbiol. Lett. 30:*103 (1985).
87. E. A. Vogler, *Biophys. J. 53:*759 (1988).
88. E. A. Vogler, *Coll. Surf. 42:*233 (1989).
89. A. W. Neumann, *Adv. Coll. Inter. Sci. 4:*105 (1974).
90. M. Morra, E. Occhiello, and F. Garbassi, *Adv. Colloid Inter. Sci. 32:*79 (1990).

91. D. F. Gerson, in *Physicochemical Aspects of Polymer Surfaces*, Vol. 1 (K. L. Mittal, ed.), Plenum, New York, 1981, pp. 229–240.
92. D. K. Owens, and R. C. Wendt, *J. Appl. Polym. Sci. 13:*1741 (1969).
93. S. Wu, *J. Adhes. 5:*39 (1973).
94. B. Janczuk, and T. Bialopiotrowicz, *J. Coll. Inter. Sci. 140:*362 (1990).
95. B. Janczuk, and T. Bialopiotrowicz, *J. Coll. Inter. Sci. 127:*189 (1989).
96. R. E. Johnson, Jr., *J. Phys. Chem. 63:*1655 (1959).
97. J. R. Huntsberger, *J. Adhesion 7:*289 (1976).
98. J. K. Spelt, and A. W. Neumann, *J. Coll. Inter. Sci. 122:*294 (1988).
99. R. E. Johnson, Jr., and R. H. Dettre, *Langmuir 5:*293 (1989).
100. F. M. Fowkes, F. L. Riddle, W. E. Pastore, and A. A. Weber, *Coll. Surf. 43:*367 (1990).
101. J. F. Padday, in *Surface and Colloid Science*, Vol. 1 (E. Matijevic, ed.), Wiley-Interscience, New York, 1969, pp. 40–97.
102. R. E. Johnson, Jr., and R. H. Dettre, in *Surface and Colloid Science*, Vol. 2 (E. Matijevic, ed.), Wiley-Interscience, New York, 1969, pp. 85–153.
103. E. G. Shafrin, and W. A. Zisman, *J. Coll. Sci. 7:*166 (1952).
104. W. A. Zisman, *Adv. Chem. 43:*1 (1964).
105. H. G. Burger, V. W. K. Lee, and G. C. Rennie, *J. Lab. Clin. Med. 80:* 302 (1973).
106. D. J. Finney, in *Statistical Method in Biological Assay*, 3rd ed., Charles Griffin & Co., Ltd., London, 1978, pp. 330–334.
107. D. Rodbard, in *Ligand Assay* (J. Langan, and J. J. Clapp, eds.), Masson Publishing, New York, 1981, pp. 45–102.
108. C. A. Smolders, in *Chemistry of Physical Applications of Surface Active Substrates* (Proceedings of 4th International Congress on Surface Active Agents), Vol. 2 (J. Th. G. Overbeek, ed.), Gordon and Breach Science Publ., New York, 1967, pp. 343–349.
109. E. H. Lucassen-Reynders, *J. Phys. Chem. 67:*969 (1963).
110. C.-S. Gau, and G. Zografi, *J. Coll. Inter. Sci. 140:*1 (1990).
111. R. A. Pyter, G. Zografi, and P. Mukerjee, *J. Coll. Inter. Sci. 89:*144 (1982).
112. G. E. Collins, *J. Text. Inst. 38:*173 (1947).
113. B. Miller, and R. A. Young, *Text. Res. J. 45:*359 (1975).
114. B. Miller, in *Surface Characteristics of Fibers and Textiles*, Part II (M. J. Schick, ed.), Marcel Dekker, New York, 1977, pp. 417–445.
115. B. Miller, L. S. Penn, and S. Hedvat, *Coll. Surf. 6:*49 (1983).
116. B. B. Sauer, and T. E. Carney, *Langmuir 6:*1002 (1990).
117. R. E. Johnson, and R. H. Dettre, in *Contact Angle, Wettability and Adhesion*, Advances in Chemistry, Vol. 43 (F. M. Fowkes, ed.), ACJ, Washington, DC, 1964, pp. 112–135.
118. B. S. Damania, and A. Bose, *J. Coll. Inter. Sci. 113:*321 (1986).
119. S. Padmanabhan and A. Bose, *J. Coll. Inter. Sci. 126:*164 (1988).
120. Y. E. McHugh, B. J. Walthall, and K. S. Steimer, *BioTechniques 1:*72 (1983).
121. D. Barnes, *BioTechniques 5:*534 (1987).

122. E. A. Vogler, *J. Coll. Inter. Sci. 133:*228 (1989).
123. L. S. Penn, and B. Miller, *J. Coll. Inter. Sci. 77:*574 (1980).
124. T. A. Michalske, and B. C. Bunker, *Sci. Am. 257:*122 (1987).
125. S. H. Anastasiadis, J.-K. Chen, J. T. Koberstein, A. F. Siegel, J. E. Sohn, and J. A. Emerson, *J. Coll. Inter. Sci. 119:*55 (1987).
126. C. I. Chiwetelu, V. Hornof, and G. H. Neale, *J. Coll. Inter. Sci. 125:*586 (1988).
127. F. K. Hansen, and G. Rodsrud, *J. Coll. Inter. Sci. 141:*1 (1991).
128. R. H. Pain, in *Biophysics of Water* (Proceedings of Working Conference at Girton College, Cambridge) (F. Franks, ed.), Wiley, New York, 1982, pp. 3–14.
129. P. C. Hiemenz, in *Principles of Colloid and Surface Chemistry*, Marcel Dekker, New York, 1986, pp. 611–736.
130. R. Cornell, *Exper. Cell Res. 58:*289 (1969).
131. M Britch, and T. D. Allen, *Cryo-Lett. 1:*51 (1979).
132. J. M. Schakenraad, Ph.D., Univ. Groningen, The Netherlands, 1987.
133. A. Marmur, W. N. Gill, and E. Ruckenstein, *Bull. Math. Biol. 38:*713 (1976).
134. E. Ruckenstein, A. Marmur, and S. R. Rakower, *Thrombos. Haemostas.* (Stuttg.) *36:*334 (1976).
135. A. Marmur, and E. Ruckenstein, *J. Coll. Inter. Sci. 114:*261 (1986).
136. E. Ruckenstein, and D. C. Prieve, *J. Theor. Biol. 51:*429 (1975).
137. E. Ruckenstein, A. Marmur, and W. N. Gill, *J. Theor. Biol. 58:*439 (1976).
138. R. Srinivasan, and E. Ruckenstein, *J. Coll. Inter. Sci. 79:*390 (1981).
139. P. R. Rutter, in *Cell Adhesion and Motility* (A. S. G. Curtis, and J. D. Pitts, eds.), Cambridge Univ. Press, Cambridge, 1980, pp. 103–135.
140. B. A. Pethica, in *Microbial Adhesion to Surfaces* (R. C. W. Berkeley, J. M. Lynch, J. Melling, P. R. Rutter, and B. Vincent, eds.), Soc. Chemical Industry, London, 1983, pp. 19–45.
141. P. Bongrand, C. Capo, and R. Depieds, *Prog. Surf. Sci. 12:*217 (1982).
142. E. Ruckenstein, and R. Srinivasan, *J. Biomed. Mat. Res. 16:*169 (1982).
143. A. W. Neumann, D. W. Francis, W. Zingg, C. J. van Oss, and D. R. Absolom, *J. Biomed. Mat. Res. 17:*375 (1983).
144. G. I. Bell, M. Dembo, and P. Bongrand, *Biophys. J. 45:*1051 (1984).
145. D. C. Torney, M. Dembo, and G. I. Bell, *Biophys. J. 49:*501 (1986).
146. D. A. Hammer, and D. A. Lauffenburger, *Biophys. J. 52:*475 (1987).
147. Th. F. Tadros, in *Microbial Adhesion to Surfaces* (R. Berkeley, J. Lynch, J. Melling, P. Rutter, and B. Vincent, eds.), Soc. Chemical Industry, London, 1983, pp. 93–116.
148. R. Weiss, *Science 252:*1060 (1991).
149. E. A. Vogler, *Langmuir 8:*2005 (1992).
150. E. A. Vogler, *Langmuir 8:*2013 (1992).
151. E. A. Vogler, *Langmuir* (in press) (1992).

5

Dynamic Contact Angles and Wetting Kinetics

TERENCE D. BLAKE Research Division, Kodak Limited, Harrow, England

I. INTRODUCTION

As we have seen in the preceding chapters, when two immiscible fluids are brought into contact with a solid, one of the fluids may spread spontaneously to form a thin, moreorless uniform film (Fig. 1a). Alternatively, the final state may leave both fluids simultaneously in contact with each other and the solid surface at a line of three-phase contact (Fig. 1b). When one fluid displaces the other, the three-phase line is considered to move across the surface of the solid. Since at least one of the fluids must be a liquid, this phenomenon is commonly referred to by the generic term *wetting* (or dewetting) and the three-phase line is called the *wetting line.*

In many industrial processes, the practical problem is to maximize the speed and uniformity of wetting. If wetting is too slow, then sections of

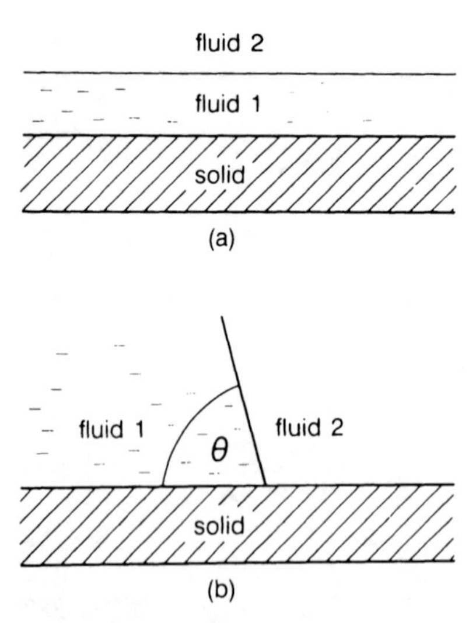

FIG. 1 Possible configurations for two immiscible fluids in contact with a solid. (a) One of the fluids (fluid 1) displaces the other and spreads spontaneously to form a film. (b) Fluid 1 forms a contact angle θ at a three-phase wetting line.

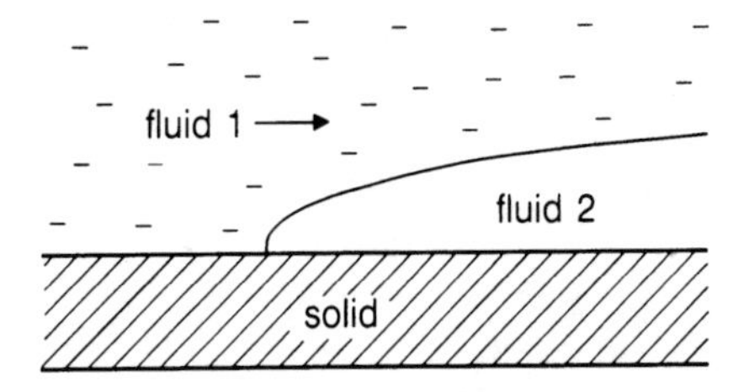

FIG. 2 Entrainment of fluid 2 by fluid 1.

the fluid/fluid interface may advance more rapidly than the wetting line, so that the advancing fluid overruns and entrains the fluid being displaced (Fig. 2). In liquid-coating operations, entrainment of air leads to patchy or uneven deposition and hence to wasted product. In petroleum recovery, entrainment of crude oil by gas or water flood may reduce both the efficiency of recovery and the ultimate yield of the reservoir. In both of these processes, as in many others such as detergency, plant protection, adhesion, and mineral flotation, the dynamics of wetting are of crucial importance.

Although wetting is frequently discussed in purely macroscopic terms, it is important to recognize that at the molecular level, the fluid/fluid interface meets the solid not at a *line* of contact, but within a *zone* of small yet finite dimensions where three interfacial regions merge and the structures and properties associated with one solid/fluid interface give way to those of the other. In noncritical systems, this transition is likely to be abrupt, but even at equilibrium, it will not be static. Just as the local structures and properties of the individual solid/fluid interfaces constantly fluctuate about mean configurations and values, so the structure and location of the three-phase zone must constantly fluctuate about some mean configuration and position with intense molecular activity. The way in which this molecular activity may determine the macroscopic dynamics of wetting is the main theme of this chapter.

II. PHENOMENA AND THEORIES

Before embarking on our exploration of the microscopic world of the three-phase zone, it is helpful first to review the relevant wetting phenomena and examine the problems that beset their theoretical interpretation.

A. Dynamic Contact Angle

Most experimental studies of wetting are concerned with the contact angle θ, that is, the angle between the planes tangent to the fluid/fluid interface and the solid surface at the wetting line (Fig. 1b). Since the significance of the contact angle as a measure of wettability is based on equilibrium thermodynamic arguments, static systems are those most frequently studied. Nevertheless, investigations of *dynamic* contact angles, that is, those associated with moving wetting lines, have become increasingly common. Here, the practical interest centers on using the angle as the geometric boundary condition for the moving fluid/fluid interface at the solid surface.

The earliest, systematic study of dynamic contact angles appears to have been made by Ablett [1]. More recent work includes that of Inverarity [2,3], Schwartz and Tejada [4,5], Hoffman [6], Burley and Kennedy [7,8], Gribanova and Mulchanova [9], Gutoff and Kendrick [10], Petrov and Radoev [11], Berezkin and Churaev [12], Gribanova [13], Burley and Jolly [14], Hopf and Geidel [15], Ström et al. [16,17], and Blake et al. [18]. The wetting velocities investigated range from microns per second to meters per second, although this wide range has seldom been studied for a single system. Figure 3 illustrates some of the geometries used. The angles are usually determined geometrically, but are sometimes obtained by tensiometry or measurement of capillary pressure. A useful account of the early work is given in a theoretical paper by Huh and Scriven [19]. For more up-to-date reviews, see Dussan V. [20], Davis [21], de Gennes [22], and Tilton [23,24].

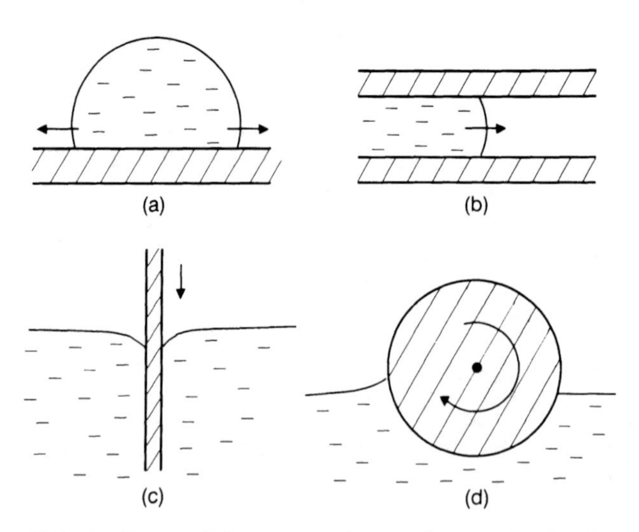

FIG. 3 Some of the geometries used to study the dynamic contact angle: (a) spreading drops; (b) fluid/fluid displacement in capillary tubes or between flat plates; (c) steady immersion or withdrawal of fibers, plates, or tapes from a pool of liquid; (d) rotation of a horizontal cylinder in a pool of liquid.

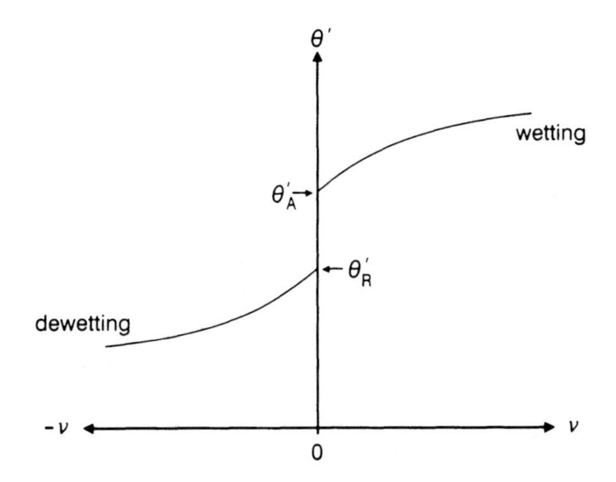

FIG. 4 Schematic representation of the velocity dependence of the experimentally determined contact angle θ', showing static advancing and receding limits θ'_A and θ'_R.

B. Velocity Dependence and Hysteresis

Experiment suggests that dynamic contact angles can deviate significantly from their static values. Although some speed-independent regimes have been reported [25–28], the experimentally observed contact angle θ' is usually found to depend on both the speed and direction of wetting line displacement, that is, θ' is *velocity-dependent*, as illustrated in Fig. 4.

The situation is made more complex by the fact that even the static contact angle is unlikely to be single-valued. In most real systems, the value observed when the wetting line is apparently stationary depends on the history of the system and varies according to whether the wetting line is tending to advance or recede. This phenomenon, which is known as contact-angle hysteresis,* makes it very difficult to determine the true, equilibrium angle θ^0. In general, the static angle has "advancing" and "receding" limits, θ'_A and θ'_R, respectively, such that $\theta'_A \geq \theta^0 \geq \theta'_R$. Moreover, the angle can usually be made to adopt any intermediate value without perceptible movement of the wetting line. If the wetting line is made to advance, then θ' becomes greater than θ'_A and may attain 180° at sufficiently high rates of wetting. Similarly, if the wetting line is made to recede, then θ' becomes less than θ'_R and may attain zero at sufficiently

* See Chap. 1.

high rates of dewetting. Clearly, dynamic advancing angles are found only when $\theta'_A < 180°$, and dynamic receding angles only when $\theta'_R > 0$. At 180° and zero, entrainment of the displaced phase leads inevitably to the loss of three-phase contact.

Contact-angle hysteresis is usually attributed to roughness or heterogeneity of the solid surface, which hinders free movement of the wetting line and so allows the system to adopt metastable configurations of variable geometry [29–32]. However, the precise scale and degree of nonuniformity necessary to cause a detectable effect are unclear. Despite a number of recent theoretical studies, for example, [33–40], there are still major difficulties in relating observed hysteresis to practical measures of surface nonuniformity [41,42]. Furthermore, although most authors agree that the effects of surface irregularities of molecular size will be smoothed out by thermal fluctuations, and even larger-scale barriers could be surmounted by capillary waves on the liquid surface [43], kinetic factors may also allow nonequilibrium angles to persist in apparently static systems if experimental time scales are short compared with the time required for the system to come to equilibrium [44–47]. The observed behavior may then be difficult to distinguish from true hysteresis.

Although fundamentally distinct from one another, hysteresis and velocity dependence of the contact angle are clearly related; for example, both imply thermodynamic irreversibility [48]. In the static case, this is attributed to spontaneous transitions between metastable equilibrium states, whereas in the dynamic case, the system may fail to attain any kind of equilibrium in the time available. Irreversibility then results from spontaneous progress toward equilibrium [45]. In either case, the contact angle changes in such a direction as to oppose movement of the wetting line.

An additional complication is that once the wetting line begins to move, it often does so with a hesitant, stick-slip, shuffling motion [19,26] that is usually accompanied by oscillations in the contact angle. Such unsteadiness may be a further manifestation of the casual nonuniformity of real solid surfaces [26,40,45,49–51], or could be caused by inequalities in the adsorption of surface-active material in the vicinity of the wetting line [52–55]. However, more fundamental explanations should not be ruled out. Perhaps better insight into the underlying mechanisms will follow from a new technique, recently introduced by Stokes et al. [56], in which small-amplitude oscillatory disturbances are used to probe wetting line movement. The response of the system gives an indication of the energy dissipation due to surface heterogeneity and may therefore be useful in testing current theories [49,57–60].

C. Theoretical Challenge

In trying to understand the mechanism by which a liquid wets a solid, the overall theoretical challenge is to interpret observed wetting behavior in terms of the underlying physics and chemistry. However, this has proved difficult, and despite increasing attention, the problem remains only partially resolved. For example, as a result of theoretical hydrodynamic studies, it is now widely recognized that it is necessary to relax the classical boundary condition of no slip at the wall and allow limited slip between fluid and solid in the immediate vicinity of a moving wetting line [19–21,23,24,61–78].† Without slip, hydrodynamics make the unacceptable prediction that the force exerted by the fluid on the solid is locally unbounded. Nevertheless, the physical basis for slip and the appropriate form of the slip condition remain a matter of great speculation [20,23,24, 62–65,67,74,76–84].

Uncertainties also surround the fundamental significance of the dynamic contact angle and how its experimentally determined value can be related to the velocity of wetting and to the material and geometric properties of the system [20,85]. Progress has been hindered by our incomplete understanding of the true nature of the contact angle very close to the wetting line even for ideal systems under equilibrium conditions. According to recent theoretical work [46,47,86–95], intermolecular forces can cause the local angle to deviate significantly from its macroscopic value, with most of the change occurring within, say, 1 nm of the wetting line [88]. To preserve mechanical stability, these deviations are accompanied by corresponding variations in interfacial tension. Ultimately, at the molecular level of the three-phase zone, both the fluid/fluid interface and its contact angle may cease to have any obvious geometric meaning [93].

Other theoretical studies suggest that hydrodynamic forces may also cause the curvature of the fluid/fluid interface to change very rapidly near the solid surface. The magnitude of this effect seems to depend on not only the properties of the fluids, but also the extent to which we are prepared to relax the no-slip boundary condition [23,24,62,65]; the greater the slip, the weaker the effect. Compared with the degree of slip, the precise nature of the slip law is sometimes considered to be relatively unimportant in determining flow, except in the region very close to the wetting line [62,76]. However, a recent numerical study of droplet spread-

† See Chap. 6 for a detailed discussion of the hydrodynamics of wetting.

ing [78] indicates that the rate of spreading depends on the relationship between slip velocity and dynamic contact angle.

An analysis given by Hansen and Toong [96], in which calculations were truncated at a small but arbitrary distance from the wetting line, led them to conclude that some, though not all, observations of velocity-dependent contact angles could be attributed to hydrodynamics. Similar conclusions were drawn by Huh and Mason [65] on the basis of calculations involving model slip conditions. Both these studies had allowed for the possibility that the local contact angle at the solid surface might remain fixed and equal to the equilibrium angle, or might vary with the wetting velocity in its own right. In contrast, many subsequent authors assumed a fixed local angle and ascribed velocity dependence entirely to hydrodynamic bending of the fluid/fluid interface close to the wetting line [68,69,71–73,75,76]. Of all these treatments, perhaps the most complete is that given by Cox [76]. Since practical systems usually show contact-angle hysteresis, several authors [69,71,72,76] set the local angle equal to θ'_A or θ'_R for advancing or receding interfaces, respectively. However, this seems inconsistent with the accepted explanations of contact-angle hysteresis, according to which neither θ'_A nor θ'_R is likely to equal the actual contact angle at the solid surface.

White [88] has proposed that at equilibrium, the local contact angle is always zero (or 180°). This could also be true in the dynamic case, but except when the equilibrium angle already takes these limiting values, it seems highly unlikely that the balance of forces at the wetting line would be so totally indifferent to the motion of the line that the local angle would remain constant.

D. Entrainment

A dynamic contact angle of 180° would appear to simplify the theoretical problem by allowing the liquid to advance with a rolling motion, thus retaining the no-slip boundary condition [97,98]. However, an analysis by Ngan and Dussan V. [99] suggests that serious problems remain even under these conditions. Furthermore, unless the displaced phase were a vacuum (or at least inviscid), the no-slip condition would ensure that this phase was entrained as a film between the advancing fluid and the solid [100,101], in which case the wetting line would effectively disappear. The precise thickness and stability of the entrained film would then depend on the balance of hydrodynamic, capillary, and intermolecular forces [19,102,103].

Scriven and co-workers [104–107] have developed these ideas into a detailed theory. They argue that slip is possible only at relatively low speeds of wetting; at higher speeds, hydrodynamic forces ensure that a submicroscopic film of the displaced phase is *always* entrained even though θ' may appear to be less than 180°. Being unstable, because of negative disjoining pressure,* the film collapses, but the resulting bubbles or droplets are too small to be observed and may rapidly dissolve.

Such a theory is clearly attractive. As well as avoiding problems caused by the no-slip condition, the successive creation and collapse of the film could be responsible for the stick-slip behavior observed at low rates of wetting [23,24]. Unfortunately, confirmation is difficult; a very recent study provides an example of the ingenious experimentation required. Using laser-doppler velocimetry to monitor the flow field very close to the moving wetting line, Mues et al. [109] detected effects that they attributed to a thin, invisible film of the displaced phase (air), extending as far as 120 μm downstream of the apparent wetting line. The effects were detected even when the measured dynamic contact angle was only 135°, that is, considerably less than 180°. Though not conclusive, the results are sufficiently convincing to warrant further investigation.

Mues et al. did not speculate on the fate of the invisible film. In their experiments, wetting rapidly became unstable at speeds corresponding to contact angles above 150°, and further increases in speed led to visible air entrainment. According to Teletzke et al. [104] and Teletzke [106], the entrained film should thicken progressively as the wetting speed is increased, and θ' should attain 180° only when the film becomes sufficiently thick for the effects of disjoining pressure to be negligible. However, this prediction seems to conflict with the common observation that the onset of entrainment is not progressive, but tends to occur abruptly at some critical speed. Any viable theory must also explain why entrainment is frequently preceded by the appearance of a sawtooth wetting line (see Sec. IV, Fig. 10), and why the resulting bubbles and droplets, which tend to have a fairly narrow size distribution, always appear from the trailing vertices of the sawteeth and not at random positions along the wetting line [110].

E. Dynamic Contact Angles: Real or Apparent?

Contact-angle measurements frequently involve either visual or mathematical extrapolations of the macroscopic interfaces. However, in view of the arguments outlined above, it might seem unwise to expect the extrapolated dynamic contact angle to be preserved at very small distances from the solid surface. Many authors stress this by referring to the

* Derjaguin's concept of disjoining pressure [108] is discussed in Chap. 6.

observed angle as the "apparent" dynamic contact angle, for example, [24,65,68,69,96,104], but their caution derives from theoretical considerations, not experiment. Examined through an optical microscope, the fluid/fluid interface appears to meet the solid surface with a well-defined slope. This is true in both the static and dynamic cases. Observations made with a scanning electron microscope [111] have given important insight into the effects of surface roughness and heterogeneity on wetting behavior. Nevertheless, on smooth and homogeneous surfaces, the technique reveals no sudden changes in meniscus curvature down to distances as small as 0.5 μm from a moving wetting line. Unfortunately, for systems in which the static contact angle is greater than zero, we have little direct experimental evidence about the shape of the fluid/fluid interface at distances from the solid surface approaching anything like molecular dimensions.

F. Precursor Films

For systems in which the liquid spreads spontaneously to give a nominally zero contact angle, the situation is somewhat clearer. Interferometry, ellipsometry, electron microscopy, and other techniques have all shown that in such cases, a precursor or primary film moves ahead of the main body of liquid, even though the latter may appear to exhibit a nonzero dynamic contact angle [112–131] (Fig. 5). Precursor films form when the intermolecular forces of attraction between the solid and liquid are sufficiently strong to create positive spreading coefficients and disjoining pressures. These related properties provide the driving force for spreading and are dissipated against viscous drag as the film spreads and thickens [106,114,132–140]. The disjoining pressure is a steep, inverse function of film thickness and therefore creates a correspondingly steep, negative pressure gradient between the bulk liquid and the thin periphery of the film. Hence, liquid is drawn out of the bulk and into the film. Where disjoining pressures are large and the viscosity small, the precursor film

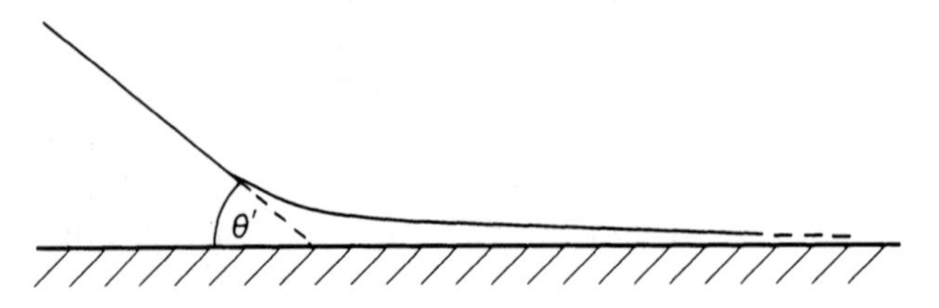

FIG. 5 Formation of a precursor film with apparent contact angle θ'.

may spread quite rapidly for significant distances ahead of the bulk liquid; for water on clean glass, speeds of up to 0.1 m s^{-1} appear possible [106,120,135,140].

In many practical systems, this basic pattern of spreading is modified by thermal gradients, surface-active substances, or volatile components of higher or lower surface tension (e.g., ethanol in water). Each of these can induce Marangoni flow–flow caused by surface tension gradients. This is the origin of the well-known "tears of wine" phenomenon and of related effects in the spreading of liquid mixtures [112,113,141–144]. It is also the probable source of the complex film profiles observed by Bascom et al. [115]. On the other hand, the remarkable molecularly layered profiles reported by Cazabat and co-workers [129–131,145] appear to be due to intrinsic ordering at the solid/liquid interface. Insight into these latter effects has been gained from recent statistical-mechanical models of spreading and wetting [146,147].

A simplifying feature of wetting via a precursor film is that flow within the film is virtually one-dimensional. More significantly, the motion of the bulk liquid is decoupled from that of the wetting line; hence, the liquid may be considered to spread across an already wetted surface [148]. Under these conditions, the hydrodynamic equations can be solved explicitly, and the velocity dependence of the apparent contact angle is predicted to follow simple cubic laws [22,149–151], which echo the equations used to describe the entrainment of liquid films during coating [102,152,153]. Similar relationships are obtained for systems without precursor films, but only if slip is included or the region close to the wetting line is formally excluded from the hydrodynamics [2,73,150,154]. Moreover, because the physics of the region are poorly defined, the slip parameters and cut-off distances have to be treated as adjustable parameters. It has been suggested that the cut-off distances can be defined in terms of long-range intermolecular forces [155,156], but although this approach is instructive, it remains incomplete.

For systems in which a precursor film may be expected, data following cubic laws have been found experimentally [2,3,6,150,154,157–160]. Nevertheless, since the wetting line is removed from the problem, neither these results nor their theoretical interpretation can illuminate the mechanism by which the wetting line advances at the extreme periphery of the film. Proposals that the mechanism might involve evaporation/condensation [112] or surface diffusion [113,115,142,161] derive from entirely independent considerations. Furthermore, there is no evidence to suggest that precursor films are ubiquitous. They are unlikely with systems that exhibit nonzero static contact angles, and their influence on wetting dynamics can be important only if their rate of advance matches or exceeds that of the bulk liquid. In all other cases, the wetting line remains part of the flow field and we are therefore left with the problem of explaining not only how the line moves, but also how its motion influences the experimentally observed contact angle.

G. Viscosity, Surface Tension, and Geometry

Even the most cursory examination of wetting phenomena suggests that both surface and viscous forces have a strong influence on wetting dynamics. In consequence, the experimentally observed dynamic contact angle is usually supposed to depend in some way on the capillary number $Ca = \eta v/\gamma$, where v is the velocity of wetting for a liquid of viscosity η and surface tension γ [6,162]. For example, most of the cubic relationships cited above can be written in the form

$$\theta^3 = f(Ca) \tag{1}$$

This equation appears to be widely applicable for wetting liquids, provided that θ' is small and $Ca \ll 1$ [22]. Along with several related formulas, it has proved especially useful in describing the spreading of liquid drops [73,122,136,150,154,160,163–167]. Various other, sometimes more empirical, correlations have also shown some success in drawing together dynamic contact-angle data obtained for physicochemically similar systems [6–8,10,160,168–173]. In most cases, the advancing contact angle is found to be a monotonically increasing function of Ca, such that the velocity dependence becomes steeper with increasing viscosity and decreasing surface tension. However, the majority of the experimental data are for $Ca < 1$, and there is nothing to indicate that the capillary number is the sole arbiter of the dynamic contact angle, especially in the absence of a precursor film. Thus, for example, Hoffman [6] found it necessary to use shift factors based on the static contact angle in order to collapse data obtained for a range of liquids onto a single, master curve of θ' against Ca.

At high capillary numbers, surface forces are expected to become relatively unimportant and the observed dynamic contact angle should then be determined entirely by the hydrodynamics [174]. Under these conditions, θ' should be influenced by the geometry of the system. Ruschak [175] has highlighted indirect evidence in support of this prediction, namely, the experimental finding that the coating speed at the onset of air entrainment can be changed by altering the flow field [176–178]. As Ruschak also points out, it is commonly observed that the coating speed at air entrainment depends on the coating method. Some of the underlying reasons for this dependency have begun to emerge only recently from flow-visualization studies [179–181] and computer-aided numerical modeling [107,174,182,183].*

As $Ca \rightarrow 0$, the influence of flow geometry should diminish, but the point at which its effect will cease to be detectable is uncertain. Ngan and Dussan V. [184,185] have reported data for silicone oil displacing air between glass slides, which show the measured dynamic contact angle to be dependent on the spacing of the slides (10–120 μm) at capillary numbers as low as 5×10^{-3}. This result, which was anticipated in an earlier theoretical study [85], is supported by the experiments of Legait and Sourieau [186], who used a series of narrow capillaries of diameter 0.1–1.2 mm, and by the computational work of Bach and Hassager [75]. On the other hand, no such effect was found by Zheleznyi [158] in an experimental study of two-phase flow in *dry* capillary tubes of diameter 0.1–0.6 mm at capillary numbers up to about 5×10^{-2}. Tube diameter became important only when the tubes were prewetted (in this case with thin, nonequilibrium films of less than 1 μm). Zheleznyi's results show that the presence of a wetting film causes the apparent dynamic angle to increase with tube diameter. This trend, which also emerges from a theoretical analysis of the effects of equilibrium wetting films given by Starov et al. [187], is the same as that found by Ngan and Dussan V and suggests that their results might have been influenced by the presence of a precursor film of silicone oil. It would be interesting to know whether the capillaries used by Legait and Sourieau were prewetted in any way.

H. Microscopic View

As this brief survey shows, the mechanism of wetting poses many unanswered questions. One reason for this may be that much of the best theoretical work has tended to emphasize only the hydrodynamic aspects of the problem. Surface phenomena and intermolecular forces have been considered, but only in so far as they influence the shape of the fluid/fluid interface near the wetting line. In nearly all cases, the treatments have been based on continuum mechanics. Moreover, although there is widespread recognition of the need for a limited-slip boundary condition for flow in the immediate vicinity of a moving wetting line, the molecular details of this region have frequently been considered not only intractable, but also relatively unimportant, for example, [64,76]. The region has sometimes been neglected; slip formulas and coefficients have been chosen for their mathematical convenience rather than by reference to a physical model; alternatively, the no-slip problem has been avoided by

* See Chap. 6.

postulating flow across a liquid-filled, rough surface [63] or within a porous surface layer [74]. Few studies have been concerned with the precise molecular mechanism by which a wetting line might move across a smooth impermeable solid.

In contrast, the molecular-kinetic approach, first adopted by the present author [103,188] and by Cherry and Holmes [189], not only provides a theoretical basis for slip at the wetting line, but emphasizes the role of molecular events occurring within the three-phase zone as the controlling influence on the wetting process. A similar stance has been taken more recently by Ruckenstein and Dunn [80], Neogi and Miller [82], Hoffman [190], Neogi [142], and Durbin [77]. Parallels may also be drawn with much earlier molecular-kinetic theories concerned with relative motion at interfaces, specifically Tolstoi's theory of slip between liquids and solids [79,191], and Bartenev's theory of dry friction between vulcanized rubber and a solid surface [192].

However, perhaps the most encouraging indication of the importance of microscopic processes has come from molecular dynamics simulations. Early studies showed some success in treating static contact angles [87] and simple dynamic systems [164,193], but realistic simulations of fluid/fluid displacement with a dynamic contact angle have been achieved only very recently [194–196]. Within the limits of statistical resolution, the latest work not only reveals molecular slip at the wetting line, but also demonstrates the velocity dependence of the contact angle, entrainment of nonequilibrium films, and stick-slip wetting. We can expect that as techniques are refined and computing power grows, such simulations will provide an increasingly detailed description of wetting behavior.

The basis for the molecular-kinetic approach to wetting is the statistical mechanics treatment of transport processes developed by Eyring and coworkers [197] and also by Frenkel [198]. Despite the simplicity of this classical framework, the resulting theory of *wetting kinetics* [103,188] has proved successful in accounting for the velocity dependence of the contact angle observed for a wide range of experimental systems [4,5,9,11,13,15–18,83,103,188,199,200]. The extent of this agreement is sufficient to suggest some measure of underlying validity. The bulk of this chapter is therefore devoted to a thorough exploration of the theory and its practical implications.

In the following sections, the basic equations are developed, their predictions are examined, and some of the ways in which the theory may be extended are considered. In particular, separate models of wetting proposed by Blake and Haynes [188] and Cherry and Holmes [189] are combined to take specific account of the influences of both surface and viscous forces on the molecules of the three-phase zone. A rudimentary method for including hydrodynamic effects is also discussed. Finally, we consider the possibility of developing a more complete theory of wetting in which the molecular and hydrodynamic approaches are complementary. At each stage in the analysis, the theoretical predictions are compared with the results of recent experimental studies.

III. WETTING KINETICS

A. Adsorption/Desorption Model: Basic Equations

The principal hypothesis is that the motion of the three-phase *line* is ultimately determined by the statistical kinetics of molecular events occurring within the three-phase *zone*. The overall objective is to find the relationship between the contact angle θ and the wetting velocity v in terms of material properties of the system, that is, we seek some function of the form

$$\theta = f(\theta^0, v, \eta, \gamma, \ldots) \tag{2}$$

The method we adopt is based directly on that used by Eyring and others [197] to interpret, as stress-modified molecular rate processes, various transport phenomena, such as diffusion and viscous flow. Following Blake [103] and Blake and Haynes [188], we make use of a simple model. We suppose that for the wetting line to advance across the solid surface, molecules already adsorbed at localized sites on the initial solid/fluid interface must be displaced by molecules from the advancing fluid. The displacements occur randomly but progressively within the moving three-phase zone and leave a new solid/fluid interface in their wake. Displacement of one fluid by the other need not be complete, and further reorganization or dissolution may occur after the three-phase zone has passed, but any entrained molecules exist initially in the adsorbed state and do not have the properties of the bulk fluid. The origins of this model can be traced to the ideas of Yarnold and Mason [201], who seem to have been the first to suggest that the velocity dependence of the contact angle might be due to the disturbance of adsorption equilibria at the solid/fluid boundaries.

The model is illustrated schematically in Fig. 6. The two fluids, 1 and 2, are taken to be substantially immiscible and may be either two liquids or a liquid and a gas. The contact angle is measured through fluid 1, and the displacement of fluid 2 by fluid 1 is arbitrarily defined as being in the forward direction. There are n adsorption sites per unit area and the average length of each molecular displacement is λ. Provided that the distribution of sites is isotropic, $\lambda \sim n^{-1/2}$.

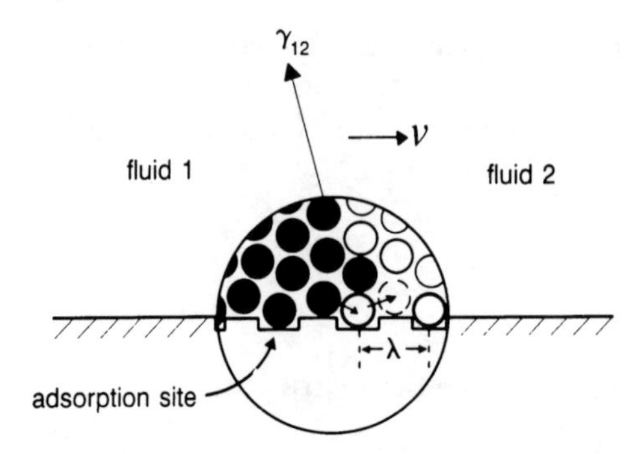

FIG. 6 Adsorption/desorption model of molecular displacement within the three-phase zone. There are n adsorption sites per unit area and the average distance between sites is λ, where $\lambda \sim n^{-1/2}$.

When the system is at equilibrium, the wetting line will appear to be stationary, but at the molecular level, the three-phase zone will usually be in a state of vigourous thermal motion as molecules of one species constantly interchange with those of the other, either by surface migration or via the contiguous bulk phases. Nevertheless, the net rate of displacement will be zero and the local three-phase boundary will merely fluctuate back and forth about its mean position.

If the frequency of molecular displacements in the forward direction is κ_W^+ and that in the backward direction is κ_W^-, then, at equilibrium,

$$\kappa_W^+ = \kappa_W^- = \kappa_W^0 \tag{3}$$

Here, the subscript W is used to denote the wetting process. Application of Eyring's theory of absolute reaction rates [197] allows us to relate the equilibrium frequency κ_W^0 to the molar activation free energy of wetting ΔG_W^*

$$\kappa_W^0 = \left(\frac{kT}{h}\right) \exp\left(\frac{-\Delta G_W^*}{NkT}\right) \tag{4}$$

where k, T, h, and N have their usual significance. In the initial publications [103,188], κ_W^0 was written in terms of the activation *energy* and the partition functions for the activated and initial states. Equation (4) is thermodynamically equivalent.

For the wetting line to advance, molecular displacements in the forward direction must become more frequent than those in the reverse direction, that is, κ_W^+ must become greater than κ_W^-. This can be achieved by applying a forward-directed shear stress to the molecules within the three-phase zone. The stress modifies the profiles of the potential-energy barriers to molecular displacement, lowering barriers in the forward direction and raising those in the backward direction. The greater the stress, the greater the net rate of displacement, and therefore, the greater the velocity of the wetting line. Once a molecular displacement has taken place, the energy used in overcoming the barrier will be dissipated to the remainder of the system and eventually appear as heat. The work done in this way is therefore irreversible. Like all other macroscopic processes, wetting can happen reversibly and isothermally only if it also happens infinitely slowly [202].

According to our model, the locus of each rate-determining molecular displacement is an adsorption site on the solid surface. If the irreversible work done by the shear stress per unit displacement of unit length of the wetting line is w, then the work done on each site is w/n, since there are n sites per unit area. If the energy barriers to molecular displacement are symmetrical, $w/2n$ is used to lower the barriers in the forward direction, whereas the remaining $w/2n$ is used to raise the barriers in the backward direction. Hence, from Eq. (4),

$$\kappa_W^+ = \left(\frac{kT}{h}\right) \exp\left[-\frac{(\Delta G_W^*/N - w/2n)}{kT}\right] \tag{5a}$$

$$\kappa_W^- = \left(\frac{kT}{h}\right) \exp\left[-\frac{(\Delta G_W^*/N + w/2n)}{kT}\right] \tag{5b}$$

and the net frequency of molecular displacement is

$$\begin{aligned} \kappa_W &= \kappa_W^+ - \kappa_W^- \\ &= \kappa_W^0 \exp\left(\frac{w}{2nkT}\right) - \kappa_W^0 \exp\left(\frac{-w}{2nkT}\right) \\ &= 2\kappa_W^0 \sinh\left(\frac{w}{2nkT}\right) \end{aligned} \tag{6}$$

The velocity of wetting is therefore

$$\begin{aligned} v &= \kappa_W \lambda \\ &= 2\kappa_W^0 \lambda \sinh\left(\frac{w}{2nkT}\right) \end{aligned} \tag{7}$$

where v is the velocity *normal* to the wetting line at the point under consideration. For an irregularly shaped wetting line, the overall velocity of wetting V may be quite different.*

The next step, which results in a considerable simplification of the theoretical problem, is to suppose that the shear stress required to drive the wetting line is provided by the out-of-balance interfacial tension forces acting at the wetting line and arising from the change in the contact angle from its equilibrium value θ^0 to some dynamic value $\theta = \theta(v)$. Per unit length of the wetting line, these forces will amount to

$$F = \gamma_{12}(\cos \theta^0 - \cos \theta) \tag{8}$$

where the fluid/fluid interfacial tension γ_{12} is assumed to remain constant. In general, $\theta(v) \neq \theta^0$ and this equation effectively gives a measure of the irreversible work expended in moving the wetting line at a finite velocity. Note that if θ^0 is zero, we may expect F to be increased by the surface pressure of fluid 1 at the solid/fluid 2 interface $\pi_{1,S2}$. However, the effect is likely to be significant only on high-energy surfaces initially devoid of adsorbed films [203].

In Eq. (8), θ and θ^0 are measured through fluid 1, which is the advancing phase when $v > 0$. Since $v > 0$ only when $F > 0$,

$$\theta > \theta^0 \qquad \text{when } v > 0$$
$$\theta < \theta^0 \qquad \text{when } v < 0$$

In the original analysis [103,188], θ and θ^0 were measured through the receding phase, so the RHS of Eq. (8) appeared with a negative sign.

If none of the effort is dissipated elsewhere, the work done by F per unit displacement of unit length of the wetting line can be equated with w in Eq. (7), and we finally arrive at an expression linking θ and v as required

$$v = 2\kappa_W^0\lambda \sinh\left[\frac{\gamma_{12}(\cos \theta^0 - \cos \theta)}{2nkT}\right] \tag{9}$$

The form of this expression is such that in accord with observation, ad-

* As first derived [103,188], Eq. (7) did not contain the factor 2 in the denominator of the argument of sinh. The factor was excluded on the basis that a molecular displacement in the forward direction can occur only at a site initially occupied by a molecule of phase 2, and a backward displacement only at a site initially occupied by a molecule of phase 1. It was, therefore, concluded that the energy barriers were modified by $\pm w/n$ rather than $\pm w/2n$. However, this is incorrect. In unit displacement of unit length of the wetting line, all n sites must be involved, irrespective of the direction of displacement; hence, the factor 2 should be included—as in all other treatments of transport processes based on Eyring theory.

vancing angles increase and receding angles decrease with increasing rate of wetting line displacement.

If the argument of sinh is sufficiently small, that is,

$$\gamma_{12}(\cos \theta^0 - \cos \theta) \ll 2nkT$$

then Eq. (9) simplifies to

$$v = \kappa_W^0 \lambda \gamma_{12} \frac{(\cos \theta^0 - \cos \theta)}{nkT} \tag{10}$$

This linear relationship between $\cos \theta$ and v should always hold as $\theta \rightarrow \theta^0$.

Conversely, if the argument of sinh is sufficiently large, that is,

$$\gamma_{12}(\cos \theta^0 - \cos \theta) \gg 2nkT$$

then κ_W^- is negligible and Eq. (9) becomes

$$\begin{aligned} v &= \kappa_W^+ \lambda \\ &= \kappa_W^0 \lambda \exp\left(\frac{w}{2nkT}\right) \\ &= \kappa_W^0 \lambda \exp\left[\frac{\gamma_{12}(\cos \theta^0 - \cos \theta)}{2nkT}\right] \qquad (v > 0) \end{aligned} \tag{11}$$

In this case, $\cos \theta$ is logarithmically dependent on v, and a plot of $\gamma_{12}(\cos \theta^0 - \cos \theta)$ against $\log v$ should give a straight line of slope $4.606nkT$ and intercept $\log v = \log(\kappa_W^0 \lambda)$. Hence, we can easily obtain the values of n, $\lambda = n^{-1/2}$, and κ_W^0. We can also calculate the effective area per site, $a = 1/n$, and the relaxation time of rate-determining molecular displacements, $\tau_W^0 = 1/\kappa_W^0$. The activation free energy of wetting ΔG_W^* is obtained form κ_W^0 using Eq. (4).

If the experimental data lie in both logarithmic and linear regions, the molecular parameters are most readily determined by curve-fitting, using Eq. (9) directly rather than either limiting form. However, whichever method is chosen, we have little option but to assume that $\theta = \theta'$. In practice, this is unlikely to be a serious limitation since Eq. (8) will always provide a good measure of the irreversible work done by surface tension forces. The only question will be whether this work is expended entirely within the three-phase zone, or whether some of it is dissipated elsewhere by, for example, viscous drag [22,154,156,204]. Because of contact-angle hysteresis, the precise value of θ^0 may also be uncertain. A nonrigorous, but expedient solution is to use θ_A' with advancing angles and θ_R' with receding angles. Alternatively, we can treat θ^0 as an additional unknown.

Although we have defined six molecular parameters, n, λ, a, κ_W^0, τ_W^0, and ΔG_W^*, Eq. (9) contains essentially only two, κ_W^0 and n, from which the others are derived. In consequence, only two should be required to characterize the wetting behavior of a given system. The interrelationships allow us to choose various pairs for this purpose in addition to κ_W^0 and n, for example, ΔG_W^* and n, or κ_W^0 and λ. With some loss of uniqueness, each pair may be combined to give a new quantity that tells us something about the more general wetting characteristics of the system. Thus, we may define the specific activation free energy of wetting: $\Delta g_W^* = n\Delta G_W^*/N$. High or low values of Δg_W^* imply, respectively, strong or weak velocity dependence of the contact angle. Similarly, we may define the "natural" velocity of wetting under quasiequilibrium conditions: $v_W^0 = \kappa_W^0\lambda = \lambda/\tau_W^0$. In this case, high or low values of v_W^0 imply, respectively, weak or strong velocity dependence.

B. Comparison with Experiment

For present purposes, one example is sufficient to demonstrate the potential utility of the theory and to give some feel for the magnitudes of the relevant molecular parameters. Figure 7 shows the advancing contact angle as a function of wetting velocity for a 70% (45.6 mPa s) aqueous solution of glycerol on Mylar polyester tape. The data are taken from Kennedy [205] and Burley and Kennedy [7,8]. The curve through the points was obtained by nonlinear regression using Eq. (9) and assuming that $\theta = \theta'$. The resulting values of n, λ, κ_W^0, etc. are listed in the figure's caption. In this case, virtually identical values are obtained from a simple plot of $\gamma_{12}(\cos\theta^0 - \cos\theta)$ against $\log v$ (see Fig. 7, inset). This is because $\gamma_{12}(\cos\theta^0 - \cos\theta) \gg 2nkT$ for all the data; hence, Eq. (11) is valid throughout.

From Fig. 7, it is evident that the mathematical form of Eq. (9) is well able to account for all the major features of the velocity dependence of the contact angle found with this system. This is true at even the highest velocities, where θ' tends to 180°, and the equation predicts that θ becomes a steep and ultimately infinite function of v. Furthermore, the value of the molecular parameters are reasonable and consistent with our understanding of adsorption and diffusion at solid surfaces [206,207]. Thus, λ is of molecular dimensions and $\Delta G_W^*/N \gg kT$. In addition, the values of κ_W^0 and ΔG_W^* indicate that within the three-phase zone, molecular displacements occur at lower frequencies and require correspondingly higher activation free energies than within the bulk. For the viscous flow of simple liquids, typical values are 10^{11} s^{-1} and 10 kJ mol^{-1}, respectively.

On the other hand, the agreement between Eq. (9) and the experimental data does not *prove* the validity of the molecular-kinetic theory. The

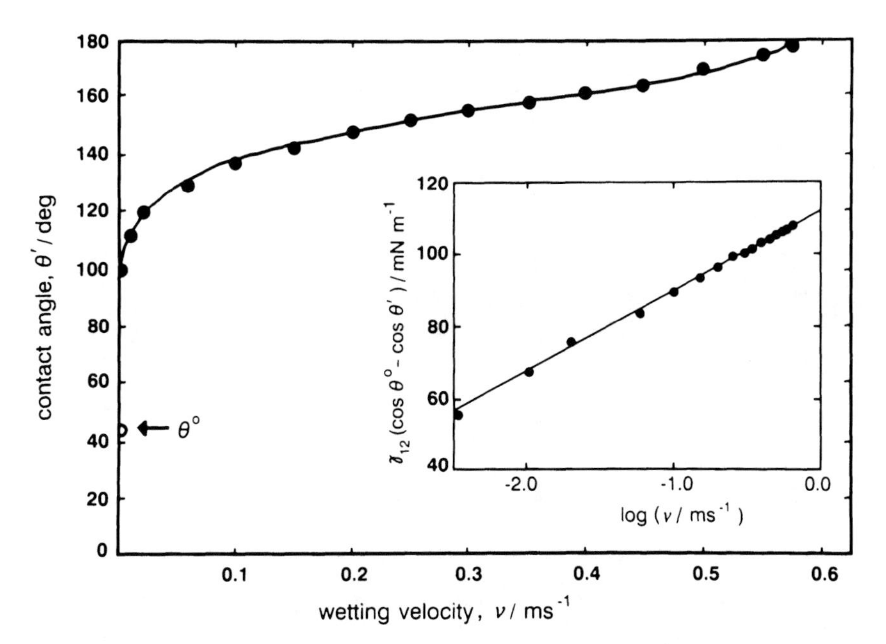

FIG. 7 Dynamic contact angle as a function of wetting velocity for a 70% (45.6 mPa s) aqueous solution of glycerol on Mylar polyester tape. Data of Kennedy [198] and Burley and Kennedy [7,8];—theoretical curve obtained by nonlinear regression using Eq. (9) and assuming that $\theta = \theta'$. Inset shows the same data plotted as $\gamma_{12}(\cos\theta^0 - \cos\theta)$ against $\log v$. Calculated molecular parameters: $n = 1.2 \times 10^{18}\ \text{m}^{-2}$, $\lambda = 0.92$ nm, $\kappa_W^0 = 8.1 \times 10^3\ \text{s}^{-1}$, $\Delta G_W^* = 49$ kJ mol^{-1}, $a = 8.4 \times 10^{-19}\ \text{m}^2$, $\tau_W^0 = 1.2 \times 10^{-4}\ s$, $v_W^0 = 7.4\ \mu\text{m s}^{-1}$, $\Delta g_W^* = 97$ mJ m^{-2}.

purely hydrodynamic treatment given by Cox (which allows for slip, but ascribes the velocity dependence of the contact angle entirely to viscous drag) leads to curves of θ as a function of v that are the same shape as that shown in Fig. 7. (See Cox [76], Fig. 4.) Evidently, either theory can be made to fit the data, and on this criterion alone, there is no simple way by which the comparative merits of the two theories can be established.

Nevertheless, once the molecular parameters have been determined for a given system, Eq. (9) can be used to predict the relationship between θ and v for all θ. A sample curve is shown in Fig. 8. Although theoretical extrapolations outside the experimental range carry some risk, they are one of the main reasons for having a theory. Such extrapolations not only provide practical insight, they also suggest opportunities to evaluate and

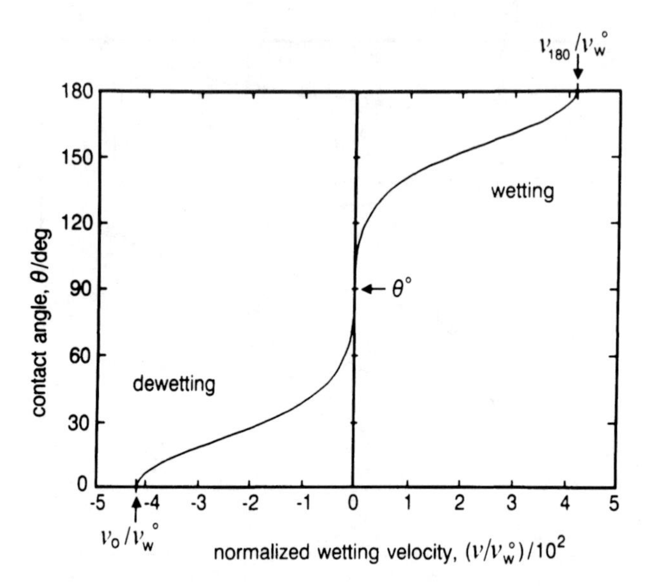

FIG. 8 Relationship between θ and the normalized wetting velocity v/v_W^0 predicted by Eq. (9) for $\theta^0 = 90°$, $\gamma_{12} = 50$ mNm^{-1}, $n = 10^{18}$ m^{-2} ($\lambda = 1$ nm), and $T = 300$K.

extend understanding. Ultimately, the test of any theory is its ability to consistently explain observation.

For example, an examination of the curves generated by Eq. (9) reveals that near θ^0, θ tends to be a steep function of v, with a slope

$$\left(\frac{d\theta}{dv}\right)_{\theta\to\theta^0} = \frac{nkT}{\kappa_W^0 \lambda \gamma_{12} \sin\theta}$$

Figure 9 shows the predicted relationship for the system represented in Fig. 7. The slope at $\theta^0 = 45°$ is 1.45×10^4 rad m^{-1}s, or 8.28×10^5 deg m^{-1}s. This would mean that undetected wetting line motion of only ± 1 mm min^{-1} would cause apparent contact-angle hysteresis of about 28°, a result that supports the view that reported hysteresis may sometimes be of kinetic origin.

At low rates of wetting, it is frequently impossible to sustain the steady movement of the wetting line. Instead, the line becomes pinned at what are presumably topographic, energetic, or adsorptive irregularities at the solid surface. Pinning leads to contact-angle hysteresis and the unsteady, stick-slip motion of the wetting line referred to in Sec. II.B. With some

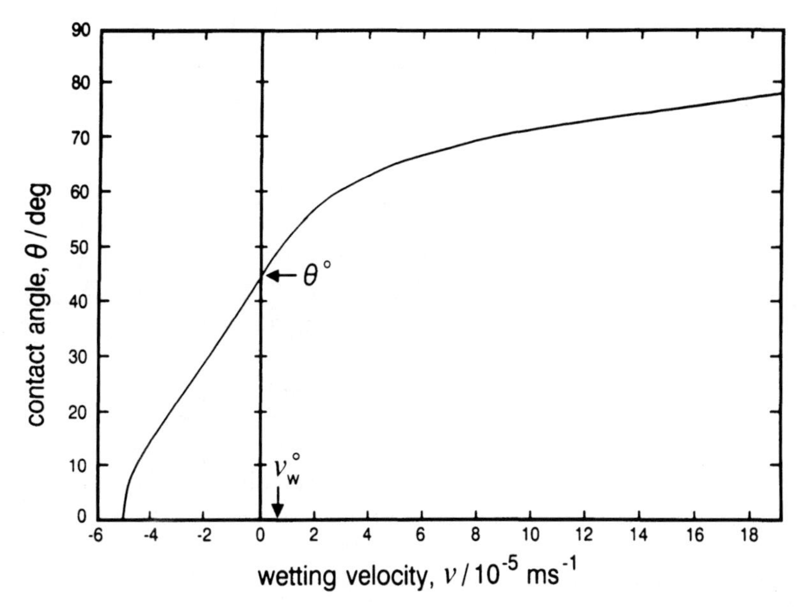

FIG. 9 Predicted relationship between θ and v for the system of Fig. 7 at very low velocities near θ^0. Curve calculated using Eq. (9).

systems, this stick-slip motion may be sufficient to mask any underlying molecular-kinetic effects and could explain those instances in which θ' is reported to be largely independent of the wetting rate, for example, [25–27]. However, in most cases we may expect measured contact angles to reflect both kinetic and hysteresis effects, although it may not always be easy to distinguish unambiguously the two phenomena.

IV. MAXIMUM VELOCITIES OF WETTING AND DEWETTING

A potentially important property of Eq. (9) is that it predicts v to have maximum and minimum values corresponding to dynamic contact angles of 180° and zero, respectively. This is illustrated in Fig. 8 and is readily proved by differentiating Eq. (9) with respect to θ. Accordingly, any attempt to wet a solid at a velocity greater than

$$v_{180} = 2\kappa_W^0\lambda \sinh\left[\frac{\gamma_{12}(\cos\theta^0 + 1)}{2nkT}\right] \tag{12a}$$

is likely to cause entrainment of the displaced fluid. Similarly, any attempt to *dewet* a solid at a velocity greater than

$$-v_0 = 2\kappa_W^0\lambda \sinh\left[\frac{\gamma_{12}(1 - \cos\theta^0)}{2nkT}\right] \tag{12b}$$

will also fail. Computed values of v_{180} and v_0 for the system of Fig. 7 are 0.57 m s^{-1} and -50 μm s^{-1}, respectively. The large disparity in the magnitudes of the two velocities is due to the fairly low value of θ^0 (45°). The symmetry of Fig. 8 results from choosing $\theta^0 = 90°$ and from the fact that the energy barriers to molecular displacements are assumed to be symmetrical; hence, $v_{180} = -v_0$.

Irrespective of such details, nature does appear to place restrictions on the rate at which a wetting line can be made to advance or recede normal to itself. Blake and Ruschak [110] have shown experimentally that when a flat, solid surface (a polymer tape) is drawn into or out of a pool of liquid (an aqueous glycerol solution), limiting velocities are observed that closely approximate v_{180} and v_0. Attempts to exceed these velocities cause the wetting line to lengthen and adopt a sawtooth configuration such that the component of the overall velocity of wetting V, normal to each straight-line segment, remains constant at the limiting value. This is shown in Fig. 10. If ϕ is the angle between the normal to the wetting line and the overall direction of wetting, then

$$\cos\phi = \frac{v_{180}}{V} \tag{13a}$$

Similarly, for dewetting,

$$\cos\phi = \frac{v_0}{V} \tag{13b}$$

At sufficiently high velocities, the displaced phase is entrained from the trailing vertices of the sawteeth. Here, at least a portion of the wetting line must lie normal to the overall direction of wetting and will therefore tend to be drawn into the advancing phase at a rate approximating $(V - v_{180})$ or $(V - v_0)$. The resulting stream, being unstable with respect to capillary forces, will collapse into bubbles or droplets.

Sawtooth wetting lines leading to air entrainment have been widely documented [8,102,168,178], and the existence of maximum velocities of dewetting has been confirmed by Petrov and Sedev [208–210]. Furthermore, the apparent advancing angle at the onset of air entrainment is usually close to 180°, as predicted. Although originally assumed a priori by Derjaguin and Levi [102], this limit has since been confirmed in several

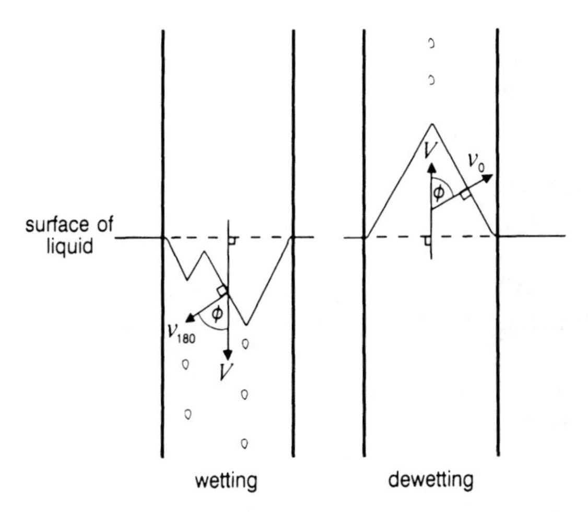

FIG. 10 Formation of sawtooth wetting lines when a flat, solid surface is drawn into or out of a pool of liquid at velocities exceeding v_{180} and $-v_0$, respectively. In order to observe a sawtooth wetting line during dewetting, θ^0 must be greater than zero.

studies [2,3,7,8,10] and is frequently taken as axiomatic. Interestingly, the hydrodynamic theory presented by Cox [76] also predicts limiting velocities at $\theta' = 180°$ and zero, but only if the viscosity of the displaced phase is nonzero. Confirmation that sawtooth wetting lines form when the displaced phase is a vacuum would provide evidence in favor of the present theory.

For systems in which the displaced fluid is more viscous than the displacing fluid, entrainment has been observed at receding angles significantly greater than zero; for example, a limiting angle of about 24° was found with benzene displacing glycerol [103]. Similar behavior may also have been seen during the deposition of Langmuir–Blodgett films, where air displaces water [11] (see Sec. VII). However, in such cases, entrainment seems to result from a hydrodynamic instability, rather than any limit imposed by the moving three-phase line. A theoretical analysis has been given recently by de Gennes [211] and Brochard-Wyart and de Gennes [204]. The phenomenon may also be related to viscous fingering [212,213]. Instability would appear to occur when viscous dissipation outweighs the available driving force for wetting due to surface tension [103]. This leads to expansion of the fluid/fluid interface and the consequent formation of an entrained film. The film decouples the wetting line from the bulk flow and so prevents the receding contact angle from achieving its limiting value of zero. One interpretation is that the instability marks a transition from control by molecular dissipation at the wetting line, to control by hydrodynamic forces [103]. This transition will become progressively more likely as the dynamic receding angle tends to zero, because of increasing viscous drag within the narrowing wedge of liquid, and will be highly favored at the trailing vertices of a sawtooth wetting line.

An analogous situation seems to exist during the dewetting of surfaces that have been coated with metastable films of poorly wetting liquids ($\theta^0 > 0$) [204,214]. A recent experimental study by Redon et al. [215] has shown that upon rupture, the films recede at a constant rate and with a constant, but nonzero receding contact angle (e.g., 23°). Furthermore, the maximum velocity of dewetting follows the general relationship predicted by de Gennes [211]: $v_{max} = -(c\gamma/\eta)\cdot(\theta^0)^3$, where c is a system-dependent numerical constant.

At angles near 180° or zero, premature entrainment may also be triggered by random factors such as roughness and heterogeneity of the solid surface. The maximum velocity of wetting reported by Blake and Ruschak [110], for a 0.1 Pa/s aqueous glycerol solution on polyethylene terephthalate (PET), actually corresponded to an advancing contact angle of 170° rather than 180°. Similar behavior has been noted by Burley and Jolly [14]. Thus, in general, dynamic contact angles of 180° and zero are best regarded as, respectively, upper and lower bounds.

At the practical level, the existence of limiting velocities of wetting and dewetting can pose serious problems in industrial processes such as liquid coating and petroleum recovery; controllable factors that affect these limits are therefore of interest. According to Eq. (9), if κ_W^0 and n remain constant, a reduction in γ_{12} is likely to lead to a *reduction* in v_{180} unless there is a corresponding *increase* in $\cos\theta^0$. Since $\cos\theta^0$ can vary only between ± 1, we can expect any substantial decrease in γ_{12} to increase the risk of entrainment. Thus, the addition of surfactants as "wetting agents" may sometimes be counterproductive. The published experimental evidence, though sparse, seems to support this conclusion [14,178].

Another prediction of practical significance is the effect of temperature. From Eqs. (4) and (9), it is easy to show that the velocity dependence of the contact angle should always decrease with increasing temperature, provided that γ_{12}, θ^0, and ΔG_W^* are only weak functions of temperature and $2\Delta g_W^* > |\gamma_{12}(\cos\theta^0 - \cos\theta)|$. Thus, an increase in temperature should lead to an increase in both v_{180} and $-v_0$. Although very little has been published on this subject, the results of a study by Dyba and Miller

[216] of the temperature dependence of v_{90} (the velocity at which θ' attains 90°) suggest that the predictions are correct. For a variety of simple liquids and surfactant solutions on polymer filaments, v_{90} was found to increase steadily with temperature. Moreover, Arrhenius plots of log v_{90} against $1/T$ were found to be linear, with negative slopes yielding activation energies in the range 10–40 kJ mol^{-1}. The values at the high end of this range substantially exceed those expected for bulk viscous flow; hence, the effect cannot be explained simply by a decrease in viscosity.

V. SOLID/LIQUID INTERACTIONS

According to the Blake and Haynes model, molecular interactions between the liquid and the solid surface are the main hindrance to the motion of the wetting line. We can therefore expect the velocity dependence of the contact angle to be more pronounced if solid/fluid interactions are strong and the number of interaction sites per unit area is large. Strong interactions imply long relaxation times, and if there are also a large number of sites, then the natural velocity of wetting v_W^0 will be small. Conversely, for a weakly interacting system, v_W^0 will be large and θ' should be a relatively weak function of velocity, a result that may explain the very low velocity dependence observed by Johnson et al. [26] for water and hexadecane on low-energy surfaces, and by Morrow and Nguyen for a variety of liquids on polytetrafluoroethylene (PTFE) [27].

The strength of solid/fluid interactions can often be gauged from the static contact angle. Small angles usually indicate strong interactions or at least interactions that are greater than those between the fluid molecules. This means that the kinetics of wetting may actually be *slower* for liquids that show good equilibrium wetting properties than for those that do not. For example, it may prove "easier" to spread a layer of water across a hydrophobic surface than a hydrophilic one, even though, in the former case, the layer would not remain spread, but would subsequently retract and break up into droplets. Since the predicted effects do not follow from simple hydrodynamic arguments, their experimental verification would provide useful support for the molecular-kinetic interpretation of wetting.

Experiments by the author and co-workers [18] have shown that very high rates of wetting are indeed possible for water on the relatively hydrophobic surface of PET (θ'_A = 75–82°). In these experiments, a 35-mm tape of Kodak Estar™ film base was drawn vertically downward into a glass tank filled with distilled water. Dynamic contact angles were determined either directly, with an optical goniometer, or from high-speed video images. Figure 11 shows the resulting data plotted as a function of

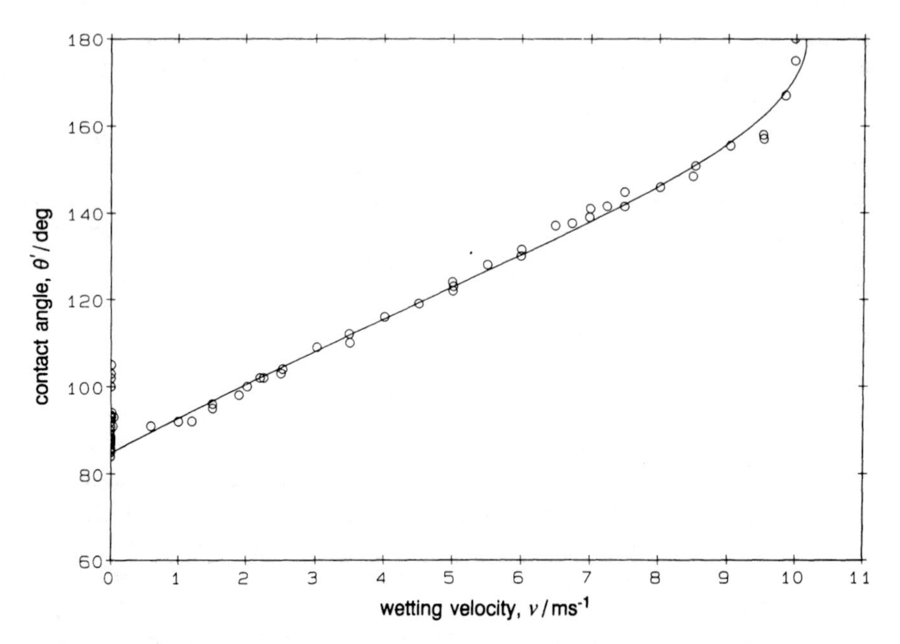

FIG. 11 Advancing contact angle as a function of wetting velocity for water on PET [18]. Theoretical curve obtained by nonlinear regression using Eq. (9).

wetting velocity and reveals a very high value of v_{180} of approximately 10 m s^{-1}. This may be compared with the much lower values, 4.6–6.8 m s^{-1}, reported by Gutoff and Kendrick [10] for water on the more hydrophilic surface of a gelatin-coated polyester tape (static angle, 32°). Although the reduction in θ^0 leads to an increase in the driving force $\gamma_{12}(\cos\theta^0 - \cos\theta)$, this appears to be insufficient to compensate for the reduction in v_W^0; hence, v_{180} is reduced also.

In the author's experiments, θ' proved to be a well-behaved, monotonically increasing function of v over most of its range, showing good agreement with theory (the solid line in Fig. 11). However, at low velocities, the wetting behavior was more complex (Fig. 12). Below about 1 mm s^{-1}, θ' rose steadily on a steep curve, but above this velocity, measurements became difficult as the angle began to alternate between values on the steep curve and lower values on a much shallower curve. No reliable data could be obtained on the steep curve at velocities greater than about 1 cm s^{-1}, and movement of the wetting line was distinctly unsteady and well described as "stick-slip." The unsteadiness died out at about 10 cm s^{-1}, and the data then rose smoothly on the shallower

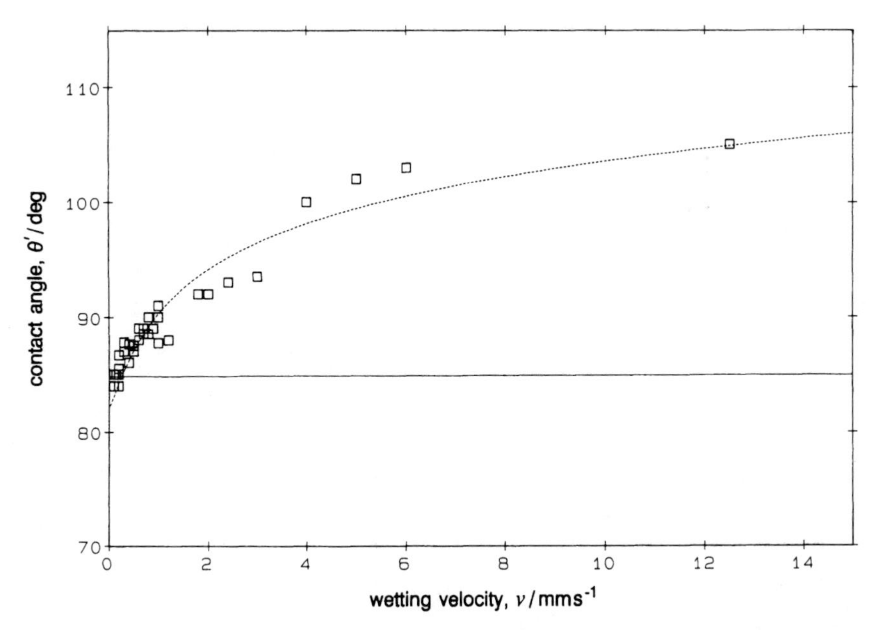

FIG. 12 Low-velocity detail from Fig. 11. Dashed curve obtained by fitting Eq. (9) to the low-velocity data only. The solid line is the start of the curve shown in Fig. 11.

curve up to the maximum wetting velocity. It may be significant that unsteady wetting started as θ' attained 90°, that is, as the free liquid surface became flat and therefore most susceptible to disturbance. Separate high- and low-speed regimes have also been reported by Cain et al. [28] for water advancing on smooth, siliconized glass, and by Buhaenko and Richardson [200] during the deposition of Langmuir–Blodgett monolayers of soaps. In the former case, the two curves were separated by a region in which θ' (~107°) was apparently independent of velocity, but there was no indication of any stick-slip phenomena.

A possible explanation for the complex behavior of water on PET emerges from fitting Eq. (9) separately to the data obtained at high and low velocities. Since the data span more than five decades, it is helpful to plot $\gamma_{12}(\cos\theta^0 - \cos\theta)$ against $\log v$. The result is shown in Fig. 13 and reveals what appear to be two wetting mechanisms: a high-velocity mechanism for which n is large but ΔG^*_W is small, implying many weak interactions between liquid and solid per unit area; and a low-velocity mechanism for which n is smaller but ΔG^*_W is much larger, implying a

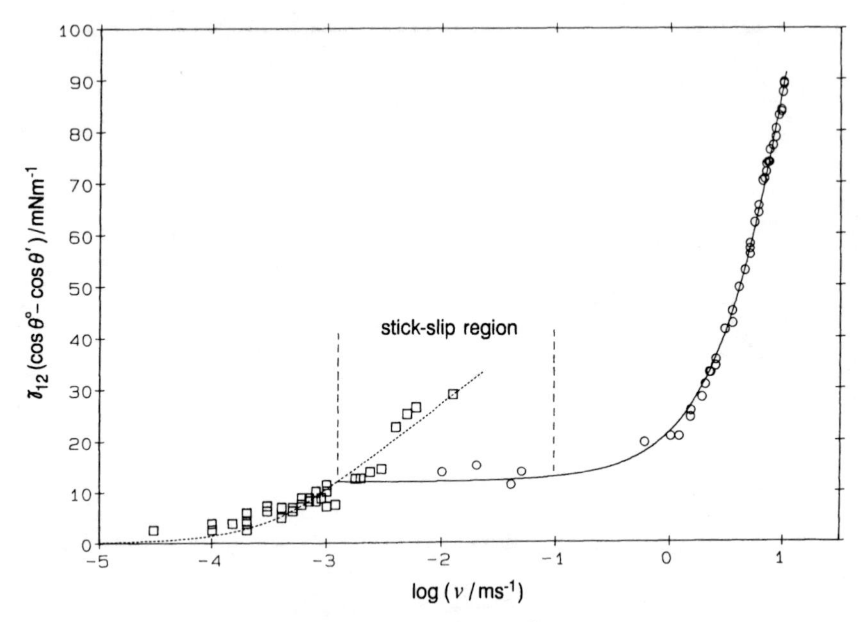

FIG. 13 Data from Fig. 11 plotted as $\gamma_{12}(\cos\theta^0 - \cos\theta)$ against log v. Theoretical curves obtained by fitting Eq. (9) separately to the data obtained at high and low velocities and treating θ^0 as an unknown. High-velocity curve ○: $n = 7.6 \times 10^{18}\ \mathrm{m}^{-2}$, $\lambda = 0.36$ nm, $\kappa_W^0 = 8.6 \times 10^9\ \mathrm{s}^{-1}$, $\Delta G_W^* = 16$ kJ mol^{-1}. Low-velocity curve □: $n = 0.92 \times 10^{18}\ \mathrm{m}^{-2}$, $\lambda = 1.1$ nm, $\kappa_W^0 = 2.5 \times 10^5\ \mathrm{s}^{-1}$, $\Delta G_W^* = 32$ kJ mol^{-1}.

small number of strong interactions per unit area. It may be significant that the values of λ and κ_W^0 for the high-velocity data are comparable with those found for the surface diffusion of mobile water molecules on silica (0.3 nm and $1.7 \times 10^{10}\ \mathrm{s}^{-1}$, respectively) [217].

In the present case, it is tempting to identify the sites of strong interaction with the polar ester groups of the PET chain and the sites of weak interaction with the more numerous CH_2 groups [18]. This would be consistent with the way in which the high-velocity curve in Fig. 13 tends to a limiting, zero-velocity contact angle several degrees higher than the experimental value. Thus, the rapidly moving wetting line appears to experience a solid surface that is more hydrophobic than the real surface. A similar explanation has been adopted by Lowe and Riddiford [218] to account for their observations of dynamic contact angles for water on PTFE.

The process might be envisaged as follows [103]. At sufficiently low rates of wetting, the strong interactions will dominate the kinetics of the three-phase zone, but, despite relatively long relaxation times for each molecular displacement, a more or less equilibrium solid/liquid interface will remain after the passage of the wetting line. At much higher velocities, however, the strongly interacting molecules will not relax within the available time interval, and a nonequilibrium adsorbed layer will be entrained. The adsorbed layer will eventually desorb, but within the three-phase zone, the locus of molecular displacement will effectively shift from the surface of the solid to the plane immediately above the adsorbed layer (Fig. 14). The stick-slip behavior observed at intermediate velocities is simply a macroscopic manifestation of an unsteady transition between alternate planes of slip.

This interpretation of the stick-slip phenomenon is similar to Durbin's explanation of slip in terms of an interfacial yield stress at the boundary between a liquid and its own, strongly adsorbed monolayer [77]. A parallel may also be drawn with the theory of wetting advanced by Scriven and

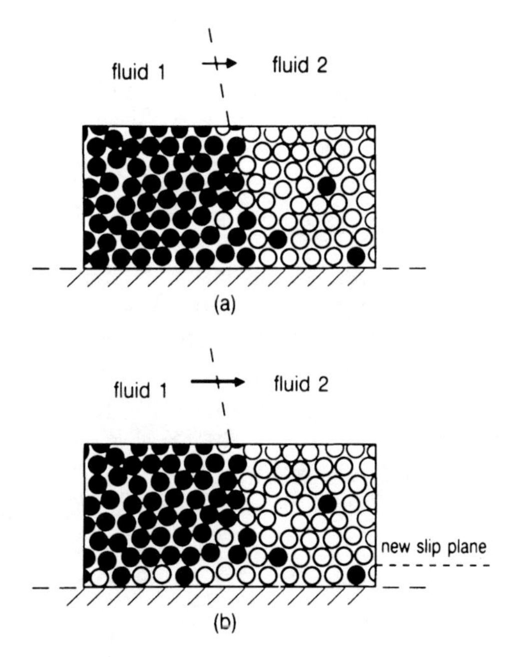

FIG. 14 (a) Quasiequilibrium molecular displacement at low rates of wetting. (b) Incomplete displacement at high rates of wetting.

co-workers [104–107] (see Sec. II.D above), but note that in the present case, the entrained film is considered to be of strictly molecular dimensions and is not expected to engender macroscopic entrainment. The onset of macroscopic entrainment signals a transition from a regime in which fluid/fluid displacement near the wetting line is controlled by surface and capillary forces, to one in which hydrodynamics play the dominant role. Except where θ^0 is zero, the length scales of these regimes are usually quite different. This is why the transition is so abrupt [103].

VI. INFLUENCE OF VISCOSITY

A. Viscous-Flow Model

Although the theory of wetting kinetics has been developed around a simple adsorption/desorption model of the moving three-phase zone, the underlying concepts and mathematical framework do not preclude other, possibly more complex mechanisms [103,188]. One or more of a wide range of elementary processes requiring activation energies could be envisaged as rate-determining steps: For example, surface diffusion [80,82, 113,115,142,161,219]; evaporation and condensation [112,220]; molecular reorientation [44,221]; viscoelastic deformation [222], penetration, swelling, or dissolution of the solid; adsorption or desorption of an additional surface-active component; disruption of electrical double layers; and effects due to microroughness or microheterogeneity of the solid surface [49,56–60]. The only restriction is that the elementary process must involve sufficiently small amounts of material and energy to be susceptible to Maxwell–Boltzmann statistics. Anything larger will cause contact-angle hysteresis.

One mechanism requiring special attention is that proposed by Cherry and Holmes [189]. They were concerned with the spreading of polymers having very high viscosities of the order 10^2 Pa s. They therefore supposed that the rate of spreading was determined simply by viscous flow over arbitrary energy barriers on the solid surface. Invoking the Eyring theory of viscosity [197], they identified the activation free energy of wetting with that for bulk flow, ΔG^*_{vis}. Any contributions arising from interactions with the solid surface were neglected. According to Eyring, the viscosity of a Newtonian liquid is given by

$$\eta = \left(\frac{h}{\text{v}}\right) \exp\left(\frac{\Delta G^*_{\text{vis}}}{NkT}\right) \tag{14}$$

where v is the volume of the "unit of flow." For many simple liquids, the unit of flow is a single molecule and we can set v equal to the molecular volume. For long-chain molecules such as hydrocarbons or polymers, which move in segments, v is the segment volume. For a given polymer, v is approximately constant and independent of the overall chain length.

Cherry and Holmes did not actually derive an equation connecting θ and v, but the appropriate expression can be deduced from their Eqs. (3), (7), and (8). In the present notation,

$$v = \frac{\lambda^2 \delta \gamma_{12}(\cos \theta^0 - \cos \theta)}{\eta \mathrm{v}} \tag{15}$$

where δ is the length of the unit of flow (molecule or polymer segment) in the direction parallel to the wetting line. Equation (15) has the same form as Eq. (10) and likewise predicts $\cos \theta$ to be a linear function of v. More interestingly, it also predicts $\cos \theta$ to vary linearly with the capillary number $Ca = \eta v/\gamma_{12}$. Hence, if θ is small, then $\theta^2 \propto Ca$, a result that stands in contrast to Eq. (1).

B. Combined Theory

The adsorption/desorption and viscous flow models are very useful in developing our picture of the moving wetting line. They are, however, highly simplified. In reality, several mechanisms may be at work. At the very least, the molecules of the three-phase zone will be influenced by both interactions with the solid surface and viscous interactions between the fluid molecules themselves. Any competent treatment of wetting kinetics should therefore take specific account of both surface and viscous effects.

A simple way of formulating a combined theory is to assume that ΔG^*_W has both surface and viscous contributions [83,190,223]. We suppose, as before, that the elementary process for movement of the wetting line takes place at specific sites on the surface of the solid and involves the displacement of an adsorbed molecule of one fluid by a molecule of the other fluid. However, we now recognize that the incoming and outgoing molecules (or molecular segments in the case of polymers) are retarded by not only the attraction of the solid, but also viscous interactions with neighboring molecules in their parent fluids 1 and 2. Formally, we write

$$\Delta G^*_W = \Delta G^*_{\mathrm{vis},1} + \Delta G^*_{\mathrm{vis},2} + \Delta G^*_S \tag{16}$$

where ΔG^*_S is the contribution to ΔG^*_W arising from the retarding influence of the solid alone. In principle, ΔG^*_S could take either positive or negative values (although positive values seem more likely) and could vary with the direction of movement of the wetting line. If we assume it to be independent of direction, then on combining Eqs. (4) and (16), we obtain

$$\kappa_W^0 = \left(\frac{kT}{h}\right) \exp\left[-\frac{(\Delta G^*_{\text{vis},1} + \Delta G^*_{\text{vis},2} + \Delta G^*_S)}{NkT}\right] \tag{17}$$

Furthermore, if we define

$$\kappa_S^0 = \left(\frac{kT}{h}\right) \exp\left(-\frac{\Delta G^*_S}{NkT}\right) \tag{18}$$

then

$$\kappa_W^0 = \kappa_S^0 \exp\left[-\frac{(\Delta G^*_{\text{vis},1} + \Delta G^*_{\text{vis},2})}{NkT}\right] \tag{19}$$

where κ_S^0 is the frequency of molecular displacements at equilibrium when retarded only by surfaces forces, such that $\kappa_W^0 = \kappa_S^0$ when $\Delta G^*_{\text{vis},1} = \Delta G^*_{\text{vis},2} = 0$. Hence, on substituting for the exponential terms in Eq. (19) using Eq. (14),

$$\kappa_W^0 = \kappa_S^0 \left(\frac{h^2}{\eta_1 v_1 \eta_2 v_2}\right) \tag{20}$$

and using this expression to substitute for κ_W^0 in Eq. (9), we finally arrive at a new expression relating v and θ, containing both surface- and viscosity-related terms as required

$$v = 2\kappa_S^0 \lambda \left(\frac{h^2}{\eta_1 v_1 \eta_2 v_2}\right) \sinh\left[\frac{\gamma_{12}(\cos\theta^0 - \cos\theta)}{2nkT}\right] \tag{21}$$

Various simplifications of this expression are possible. If one of the fluids is a gas, then for that fluid, $\Delta G^*_{\text{vis}} \sim 0$; hence,

$$v = 2\kappa_S^0 \lambda \left(\frac{h}{\eta v}\right) \sinh\left[\frac{\gamma_{12}(\cos\theta^0 - \cos\theta)}{2nkT}\right] \tag{22}$$

where η and v refer to the remaining liquid phase. Near θ^0, this equation simplifies further to the linear form

$$v = \frac{\kappa_S^0 \lambda h \gamma_{12}(\cos\theta^0 - \cos\theta)}{nkT\eta v} \tag{23}$$

so that cos θ is once again proportional to *Ca* [compare Eq. (15)]. Alternatively, away from θ^0, Eq. (22) reduces to the exponential form

$$v = \kappa_S^0 \lambda \left(\frac{h}{\eta v}\right) \exp\left[\frac{\gamma_{12}(\cos\theta^0 - \cos\theta)}{2nkT}\right] \qquad (v > 0) \tag{24}$$

and a logarithmic relationship exists between $\cos \theta$ and the product ηv.

C. Comparison with Experiment

Provided that the molecular flow volumes are known or can be determined from bulk flow characteristics, the new equations linking θ and v still contain essentially only two unknown molecular parameters. However, because of the increased scope of the underlying theoretical model, the equations now take explicit account of viscosity. They predict that if all other factors are held constant, an increase in viscosity will lead to an increase in the velocity dependence of the contact angle and, hence, to a decrease in the maximum velocities of wetting and dewetting. A reduction in viscosity will have the opposite effect. In either case, the quantities $(\eta_1 \mathrm{v}_1 \eta_2 \mathrm{v}_2) v_{180}$ and $(\eta_1 \mathrm{v}_1 \eta_2 \mathrm{v}_2) v_0$ should remain fixed.

Support for this latter prediction is to be found in a very recent paper by Ström et al. [17]. Their results, for water (phase 1) displacing paraffin oil and various alkanes (phase 2) from polystyrene surfaces at low capillary numbers are plotted in Fig. 15 as v_{180} against $h/\eta \mathrm{v}$. The molecular flow volumes were estimated from data reported by Glasstone et al. [197]

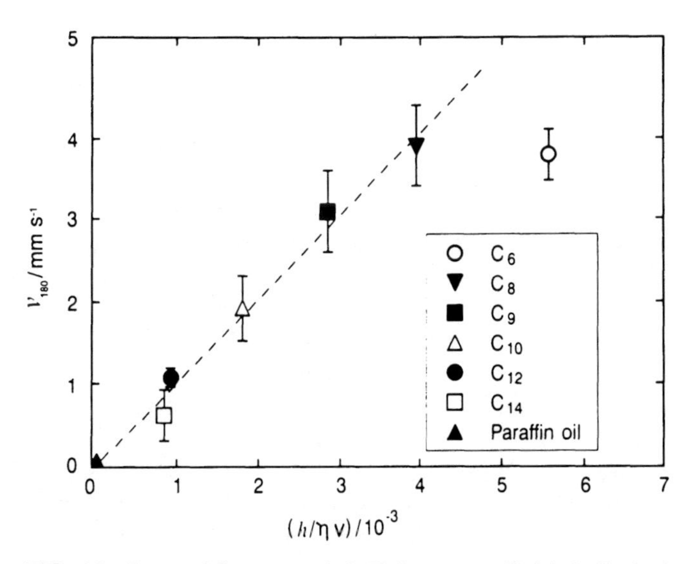

FIG. 15 Data of Ström et al. [17] for water (fluid 1) displacing various alkanes (fluid 2) from polystyrene surfaces at low capillary numbers.

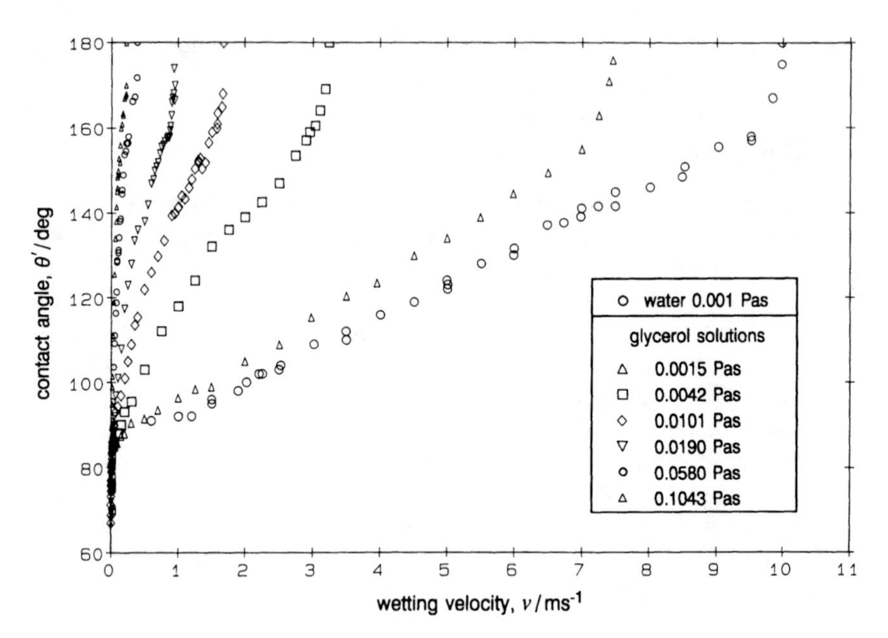

FIG. 16 Dynamic contact angle as a function of wetting velocity for water and aqueous glycerol solutions on PET [18].

(pp. 497–500). Evidently, the expected linear dependence holds well for the higher alkanes, but not for n-octane or n-hexane (which falls outside the figure). As shown below, broad agreement between Eq. (21) and experiment can also be demonstrated for liquids of widely ranging viscosity displacing air from surfaces as varied as PET and glass. However, with the more viscous liquids, discrepancies arise as $\theta' \rightarrow 180°$ and $Ca \rightarrow 1$.

Figure 16 shows data obtained in the author's laboratory for water and aqueous glycerol solutions of differing viscosities on PET tape [18]. The data for water are the same as in Fig. 11, with low-velocity points omitted for clarity. The effect of increasing viscosity is clearly substantial and follows the predicted trends. With systems such as these, comprising a series of liquids of similar chemical composition and a single solid, κ_S^0, n and λ will be moreorless constant. If we also assume v to be constant, then according to Eq. (22), a plot of $\gamma_{12}(\cos\theta^0 - \cos\theta)$ against $\log(\eta v)$ should give a single master curve for all the liquids. At sufficiently high velocities, the plot will be linear. Figure 17 depicts the result for the data of Fig. 16. As with water on PET, the high- and low-velocity data for the glycerol solutions appear to fall on separate curves, although the distinc-

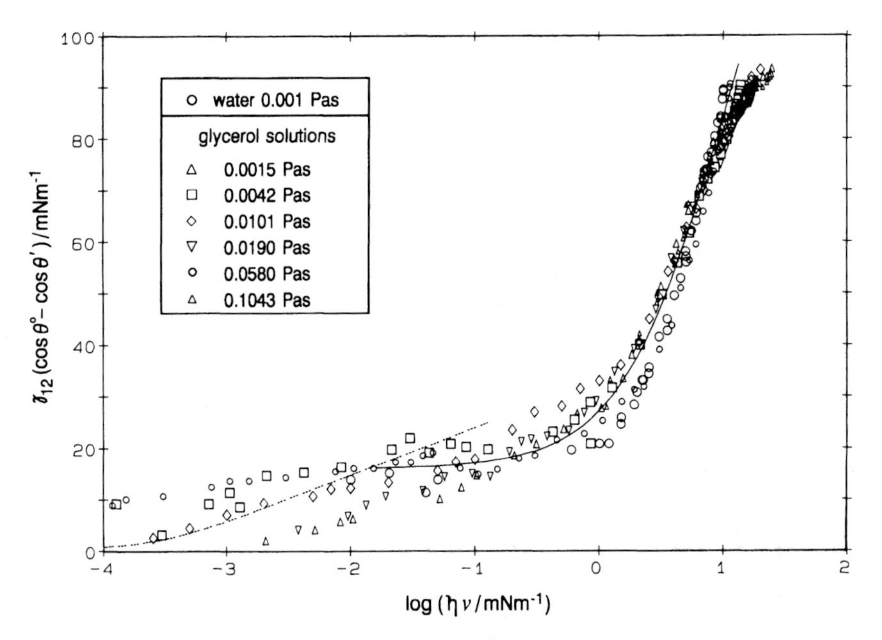

FIG. 17 Data of Fig. 16 plotted as $\gamma_{12}(\cos \theta^0 - \cos \theta)$ against log v. Theoretical curves obtained using Eq. (22) with data assigned to each curve by inspection and treating θ^0 as an unknown. High-velocity curve: $n = 4.8 \times 10^{18}$ m^{-2}, $\lambda = 0.46$ nm, $\kappa_W^0 = 3.6 \times 10^9$ s^{-1} to 3.5×10^7 s^{-1}, $\kappa_S^0 = 6.6 \times 10^{11}$ s^{-1}, $\Delta G_S^\ddagger = 5.6$ kJ mol^{-1}, $\Delta g_S^\ddagger = 45$ mJ m^{-2}. Low-velocity curve: $n = 0.49 \times 10^{18}$ m^{-2}, $\lambda = 1.4$ nm, $\kappa_W^0 = 1.8 \times 10^5$ s^{-1} to 1.7×10^3 s^{-1}, $\kappa_S^0 = 3.2 \times 10^7$ s^{-1}, $\Delta G_S^\ddagger = 30$ kJ mol^{-1}, $\Delta g_S^\ddagger = 24$ mJ m^{-2}.

tion is less clearcut. Stick-slip behavior was observed at low viscosities, but it damped out as the viscosity increased and shifted to correspondingly lower velocities. Nevertheless, it seemed appropriate to assume two wetting mechanisms when fitting the theory (solid lines) to determine the characteristic molecular parameters listed in the caption. Since aqueous glycerol solutions are highly associated, v is unknown and may well vary over the concentration range investigated (0–86% glycerol). In view of this uncertainty, the spread of the data is reasonable, and we may conclude that Eq. (22) provides a fairly good description of the velocity dependence of the contact angle observed with these systems. The quoted values of κ_S^0, $\Delta G_S^\ddagger$ and the component of $\Delta g_W^\ddagger$ attributable to interactions with the solid surface, $\Delta g_S^\ddagger = n\Delta G_S^\ddagger/N$, were calculated by assuming v to equal the molecular volume of glycerol.

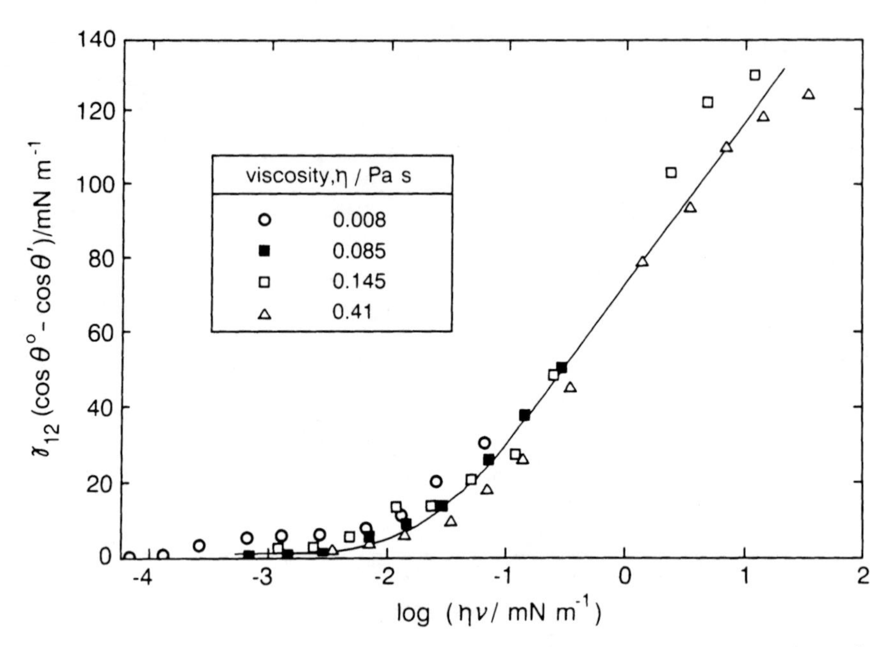

FIG. 18 Dynamic contact-angle data of Inverarity [2,3] for aqueous glycerol solutions on E-glass filaments.—Equation (22) with $n = 2.3 \times 10^{18}\ m^{-2}$, $\lambda = 0.66$ nm, $\kappa_W^0 = 3.8 \times 10^6\ s^{-1}$ to $7.4 \times 10^4\ s^{-1}$, $\kappa_S^0 = 5.5 \times 10^9\ s^{-1}$, $\Delta G_S^* = 17$ kJ mol^{-1}, and $\Delta g_S^* = 66$ mJ m^{-2}.

Further evidence supporting the utility of Eq. (22) is provided by Figs. 18–21. The first of these shows experimental results reported by Inverarity [2,3] for 59–95% aqueous glycerol solutions on E-glass filaments. The agreement with theory is good and, in this instance, the data show a smoothly asymptotic approach to the log(ηv) axis at $\theta' = \theta^0$ as v tends to zero. Thus, there is no indication of more than one wetting mechanism. The quoted values of κ_S^0, ΔG_S^*, and Δg_S^* are again based on the molecular volume of glycerol.

Figure 19 shows Inverarity's results for 5–25% solutions of polystyrene in xylene on E-glass filaments. In these systems, the polymer is likely to adsorb at the glass/solution interface and the rate-controlling step for wetting will probably involve the displacement of individual polymer segments. The length of each segment will depend on the molecular structure of the polymer, but should be largely independent of concentration. Hence, v should be roughly constant and we can expect the separate sets of data to fall on a single curve as shown. Since the actual value of v is

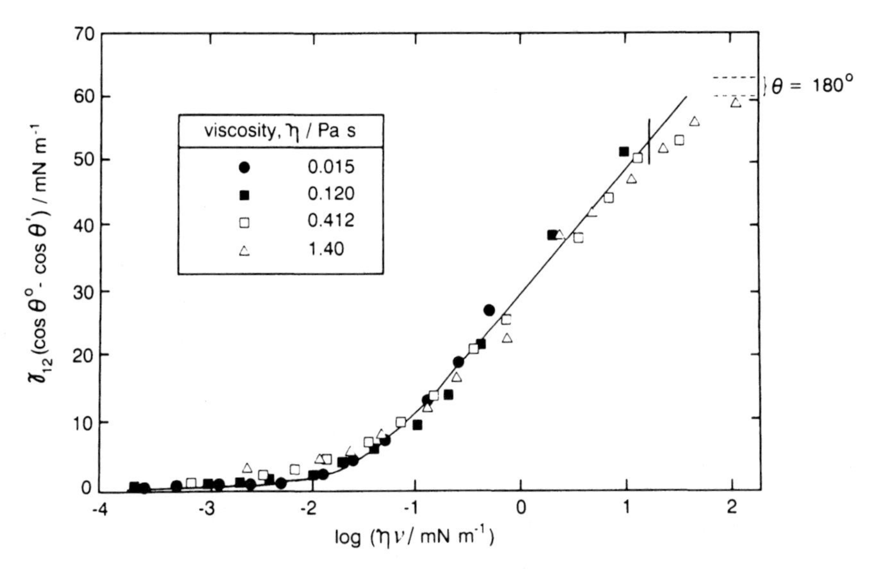

FIG. 19 Dynamic contact-angle data of Inverarity [2,3] for solutions of polystyrene in xylene on E-glass filaments.—Equation (22) with $n = 0.99 \times 10^{18}$ m^{-2}, $\lambda = 1.0$ nm, $\kappa_W^0 = 1.7 \times 10^6$ s^{-1} to 1.8×10^4 s^{-1}, $\kappa_S^0 = 3.9 \times 10^{10}$ s^{-1}, $\Delta G_S^* = 13$ kJ mol^{-1}, and $\Delta g_S^* = 21$ mJ m^{-2}.

unknown, the molecular parameters were calculated by assuming that v $= \lambda^3$. This is equivalent to assuming a segment molecular weight of about 550.

At high values of ηv, the data of Fig. 19 begin to deviate from theory as θ' tends to 180°. Because of this, points to the right of the short vertical line were arbitrarily excluded from the curve-fitting routine in order to improve agreement at smaller angles. The same procedure was used for Figs. 20 and 21 where the deviations are more marked. Comparison with Figs. 17 to 18 suggests that the effect becomes significant only with relatively viscous liquids; even then, Eq. (22) appears to work well for dynamic contact angles up to about 150°, and the choice of variables yields a good correlation at all angles.

The data in Fig. 20 were obtained by Inverarity for solutions of polyester resin in styrene on E-glass filaments. The graphs were plotted by assuming a surface tension of 44 mN m^{-1} [224]. With these liquids, v should be almost exactly constant and independent of viscosity. This should also be true for the two silicone oils investigated by Hoffman [6] (Fig. 21) and may explain why the correlation is particularly good for both systems despite the wide range in viscosity.

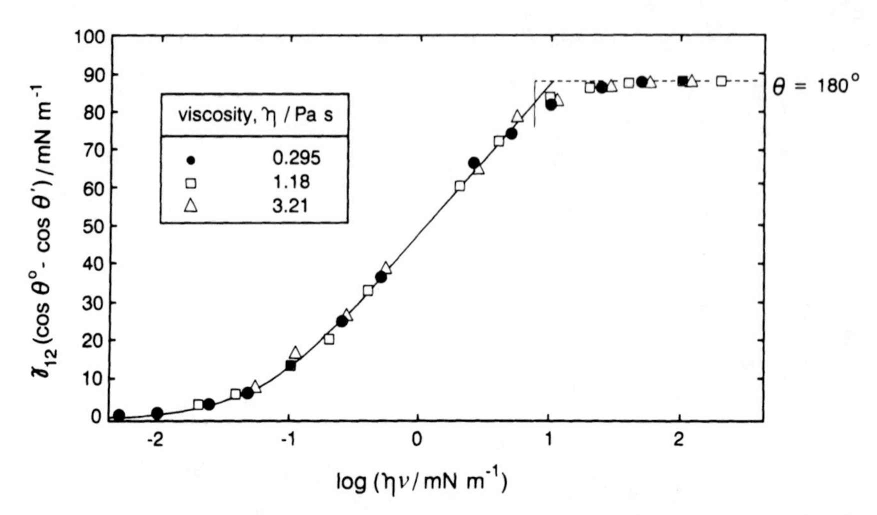

FIG. 20 Dynamic contact-angle data of Inverarity [2,3] for polyester resins in styrene on E-glass filaments.—Equation (22) with $n = 2.1 \times 10^{18}$ m^{-2}, $\lambda = 0.69$ nm, $\kappa_W^0 = 2.9 \times 10^5$ s^{-1} to 2.6×10^4 s^{-1}, $\kappa_S^0 = 1.1 \times 10^{11}$ s^{-1}, $\Delta G_S^* = 10$ kJ mol^{-1}, and $\Delta g_S^* = 35$ mJ m^{-2}.

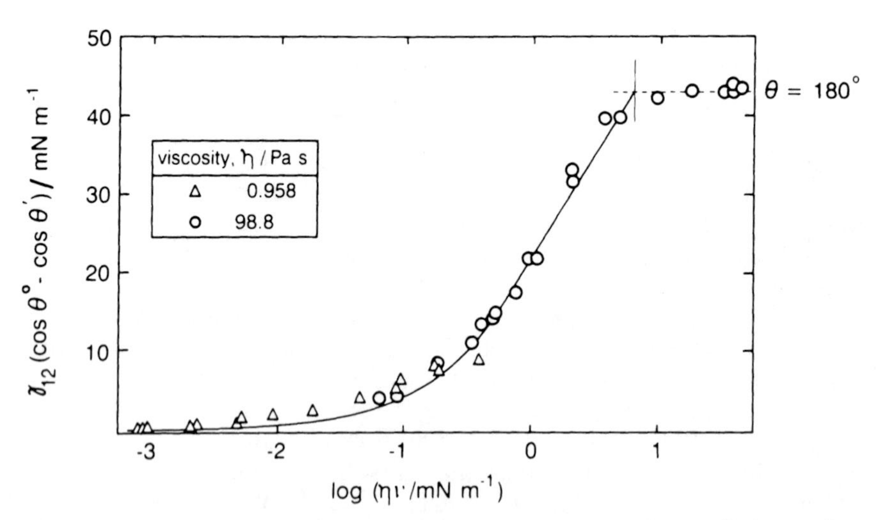

FIG. 21 Dynamic contact-angle data of Hoffman [6] for silicone oils in a cylindrical glass capillary.—Equation (22) with $n = 1.6 \times 10^{18}$ m^{-2}, $\lambda = 0.8$ nm, $\kappa_W^0 = 2.3 \times 10^5$ s^{-1} and 2.3×10^3 s^{-1}, $\kappa_S^0 = 1.7 \times 10^{11}$ s^{-1}, $\Delta G_S^* = 8.8$ kJ mol^{-1}, and $\Delta g_S^* = 23$ mJ m^{-2}.

The molecular parameters for Fig. 20 were calculated by assuming v to be the same as that for linear polyesters (segment molecular weight approximately 500; see [197], pp. 500–503). Those for Fig. 21 were calculated for $v = \lambda^3$ (equivalent to a segment molecular weight of approximately 300). The resulting values, together with those obtained in the preceding examples, appear to be sensible despite the uncertainty in the molecular flow volumes. The magnitude of ΔG_S^* is of special significance. For systems where the liquid is likely to interact strongly with the solid, for example, aqueous glycerol on glass, ΔG_S^* is substantial and approaches the energy of a hydrogen bond. However, where the solid/liquid interactions are comparatively weak, for example, silicone fluids on glass, then ΔG_S^* falls to much lower values. In the former case, ΔG_{vis}^* and ΔG_S^* are of similar magnitude (18–28 kJ mol^{-1} and 17 kJ mol^{-1}, respectively), and we may conclude that viscous and surface interactions are of equal importance. In the latter case, ΔG_{vis}^* is up to five times greater than ΔG_S^* (33 or 45 kJ mol^{-1}, compared with 8.8 kJ mol^{-1}), so viscous interactions clearly dominate.

The magnitude of Δg_S^* is also significant, since it gives a direct indication of the surface forces opposing the movement of the wetting line. For the systems considered here, the values are similar to the surface tensions of the corresponding liquids. Once again, this is a reasonable result and one that may shed light on the detailed mechanism of molecular displacement within the three-phase zone.

VII. INFLUENCE OF FLOW

A. Deviations from Theory

The examples given in the preceding section demonstrate that Eq. (22) can provide a consistent description of the velocity dependence of the contact angle for a wide range of systems over some five orders of magnitude in ηv and four orders of magnitude in viscosity alone. This success lends support to the underlying theoretical assumptions, including the basic molecular-kinetic mechanism, the utility of Eq. (8), and the identification of θ' with θ.

Nevertheless, the comparison with experiment also reveals a progressive deviation from theory as $\theta' \rightarrow 180°$. The deviation is just discernible in Figs. 17 and 18, but is very evident in Figs. 19–21, and appears to grow with viscosity. The trend is such that θ' is smaller and approaches its limiting value at wetting velocities higher than predicted. The onset of entrainment is therefore delayed. In particular, the polyester resins and silicone oils appear to show a nearly asymptotic approach to 180°, not observed with liquids of lower viscosity such as aqueous glycerol solutions (Fig. 16).

Near 180°, accurate measurements of θ' become increasingly difficult, and it is not always possible to identify the precise velocity at which the limiting value is attained (or if, in fact, it is attained at all). Some of the reported values may therefore be suspect. Nevertheless, it would be unreasonable to suppose that the deviations from theory were due entirely to experimental error, especially as the same trends are found with data reported for widely differing experimental conditions. Other explanations must therefore be sought. One possibility that seems more consistent with experiment is that near 180°, wetting is driven by forces in addition to the out-of-balance interfacial tension force $\gamma_{12}(\cos\theta^0 - \cos\theta)$.

Hoffman [190] has proposed the existence of a secondary, "tank-tread" wetting mechanism that comes into play only when the advancing contact angle exceeds 90°. He assumed that for $\theta < 90°$, the wetting line advances solely by diffusion along the solid surface, and using arguments similar to those outlined above, he derived an expression analogous to Eq. (22). He then went on to suppose that once θ equals or exceeds 90°, molecules at the liquid/vapor interface near the wetting line begin to diffuse toward the adjacent solid driven by intermolecular forces of attraction. This extra flux of molecules causes the wetting line to advance more rapidly than by surface diffusion alone and eventually becomes the dominant mechanism as θ tends to 180°. The final equations include an additional Eyring expression to account for the extra flux and a simple combining law of the form

$$v = v_X + v_Z \cot(180 - \theta) \qquad (\theta \geq 90°) \tag{25}$$

were v_X is the velocity component due to diffusion along the solid surface and v_Z the tank-tread component normal to the solid surface. The attraction of the solid is approximated by integrating the dispersion forces arising from surface molecules only.

Hoffman showed that with reasonable choices for the various molecular parameters involved, his equations generate a curve that follows closely his earlier experimental results for high-viscosity oils [6]. However, notwithstanding this excellent agreement, some of his underlying theoretical arguments seem questionable. For example, it is not obvious why the tank-tread mechanism should operate only when $\theta \geq 90°$; the intermolecular forces of attraction that are supposed to provide the driving force exist at all angles. In any case, these forces are already accounted for in Hoffman's theory by his use of an expression identical to Eq. (8) to calculate v_X. Furthermore, the trigonometrical logic of Eq. (25) predicts that the velocity of wetting will become unbounded as $\theta \to 180°$ and the supplementary angle closes to zero. Although this is not necessarily impossible (for a wetting process akin to lamination), it is manifestly not true for all systems.

B. Hydrodynamic Effects

A detailed examination of Figs. 17–21 reveals that the deviation from the present theory begins when Ca exceeds about 0.2, that is, when viscous forces begin to be comparable with surface tension forces. This could be rationalized if one were to suppose that the effects were of hydrodynamic origin. Such an explanation would be welcome, because it would offer a link between continuum hydrodynamics and the molecular world of the three-phase zone. In more practical terms, it might explain the known effects of hydrodynamics on air entrainment in liquid-coating processes. Since the deviation from Eq. (22) seems to occur only with liquids of moderate to high viscosities, viscous forces appear to be the prime suspect. Within the microscopic confines of the three-phase zone, inertial effects are unlikely to be significant. It is worth noting that Cox [76] reports similar deviations between his hydrodynamic predictions and Hoffman's data as θ' approaches 180°, a result that he also suggests may be due to the fact that Ca is no longer small compared with unity (as was assumed in his analysis).

A simple way of linking hydrodynamics with the molecular kinetics of wetting has been advanced by Petrov and Radoev [11] to explain the behavior they observed during Langmuir–Blodgett deposition of Y-type multilayers of methyl arachidate and arachidic acid. In their experiments, they measured the velocity dependence of both the receding contact angle and dynamic capillary height L_C, that is, the height of the moving wetting line above the free liquid surface.

With methyl arachidate, a plot of $\cos\theta'$ against the logarithm of the velocity of dewetting ($-v$ in the present notation) gave a straight line in agreement with Eq. (11) written for a receding wetting line:*

$$\begin{aligned} v &= -\kappa_{\bar{W}}\lambda \\ &= -\kappa_W^0\lambda \exp\left(\frac{-w}{2nkT}\right) \qquad (26) \\ &= -\kappa_W^0\lambda \exp\left[\frac{\gamma_{12}(\cos\theta - \cos\theta^0)}{2nkT}\right] \qquad (v < 0) \end{aligned}$$

* Note that Petrov and Radoev used the equation as originally derived, that is, without the factor 2 in the denominator of the exponent.

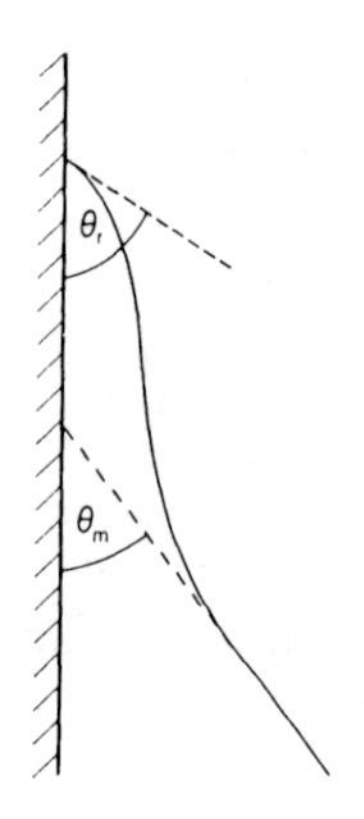

FIG. 22 Schematic representation of the hydrodynamic deformation of a receding meniscus according to Petrov and Radoev [11]. Here, θ_r is the actual macroscopic dynamic contact angle at the tip of the deformed region, and θ_m the angle measured by extrapolation of the undeformed portion of the meniscus.

At the highest dewetting velocity investigated, which was about 25 mm s^{-1}, the receding contact angle was approximately 30° and there was no tendency to entrain the liquid subphase. Furthermore, the measured value of L_C was the same as that calculated by assuming the meniscus to have a quasistatic profile, that is, a profile determined entirely by the balance of surface tension and gravitational forces.

During the deposition of arachidic acid, rather different behavior was observed. The velocity dependence of the contact angle was greater, and the angle was found to fall to zero, giving liquid entrainment at a limiting dewetting velocity of about 12.5 mm s^{-1}. For receding contact angles down to about 32°, the data again followed Eq. (26), but beyond this point, there was a progressive deviation from theory, such that the measured contact angle was smaller and approached zero at a substantially smaller velocity of dewetting than predicted. At the same time, L_C became larger than before and eventually exceeded L_0, the value for a static meniscus with $\theta = 0$. It may also be significant that the wetting line became increasingly irregular with the appearance along its length of separate regions of localized entrainment, an effect which suggests that the maximum velocity of dewetting had been reached already, even though the receding contact angle was apparently greater than zero (see Sec. IV).

Because of the behavior of L_C, Petrov and Radoev attributed the departure from theory to the viscous deformation of the liquid/vapor interface at small distances from the moving solid substrate (Fig. 22). Within this region, viscous drag reduces the force transmitted to the wetting line, so that the actual, macroscopic, dynamic contact angle θ_r is larger than that measured by extrapolation of the undeformed (i.e., quasistatic) portion of the meniscus θ_m. Using lubrication theory [152,153], Petrov and Radoev deduced that the viscous drag per unit length of the wetting line varies as $Ca^{2/3}$ and leads to a modified version of Eq. (26), which may be written as

$$v = -\kappa_W^0 \lambda \exp\left[\frac{\gamma_{12}(\cos\theta_m - \cos\theta^0 - b(-v)^{2/3})}{2nkT}\right] \qquad (v < 0) \qquad (27)$$

where b is a coefficient that depends on the detailed profile of the liquid/vapor interface. Petrov and Radoev showed that with $b = 0.6$, the modified equation is in better agreement with their results than Eq. (26). A later study confirmed these findings [225].

C. Modified Theory

In the experiments of Petrov and Radoev, the capillary number was always less than 10^{-3}. This is at least two orders of magnitude smaller than that at which deviations from Eq. (22) are seen with an advancing wetting line and would seem to indicate a much greater sensitivity to hydrodynamic effects. Presumably, the difference is attributable to the very different flow geometries for receding contact angles near zero and advancing angles near 180°. Viscous drag effects are clearly much greater for small angles than large ones. However, this cannot be the whole story. Although the Petrov and Radoev model is able to account for dynamic receding contact angles that appear to approach zero at dewetting velocities *smaller* than predicted by the basic equations of wetting kinetics, the model does not explain why advancing contact angles should sometimes approach 180° at wetting velocities *higher* than expected. Indeed, the two phenomena would appear to be mutually exclusive. Nevertheless, in view of the experimental facts and strong suspicion that hydrodynamic factors are implicated in both cases, some attempt to rationalize the situation seems worthwhile.

An alternative way of viewing the problem is to suppose that the viscous drag of the liquid is able to act directly on the molecules of the three-phase zone, promoting or retarding the motion of the wetting line and thereby decreasing or increasing the velocity dependence of the contact angle [83]. To a very rough approximation, the viscous stress close to the wetting line will be proportional to ηu, where u is some velocity characteristic of the local flow field. For the advancing fluid, the direction of flow will be inward, *toward* the wetting line with $u \sim v$. For the displaced fluid, flow will be outward, *away* from the wetting line with $u \sim -v$. In both cases, the work done by viscous forces will be dissipative and will therefore vary as $\eta|v|$, but flow inward will tend to promote wetting, whereas flow outward will tend to retard wetting.

Since we have defined the displacement of fluid 2 by fluid 1 as being in the forward direction, the net addition to the driving force for wetting will be equal to $(\alpha_1\eta_1 - \alpha_2\eta_2)|v|$, where α_1 and α_2 are numerical coefficients. The values of these coefficients will be determined by the flow fields within the respective phases immediately adjacent to the wetting line. In particular, the values will depend on the ratios of molecular and hydrodynamic length scales within the three-phase zone. If hydrodynamic forces are important, these length scales are likely to be similar; hence, to a first approximation, we can expect $\alpha \sim 1$. The total driving force for wetting is obtained by combining the new expression with Eq. (8)

$$F = \gamma_{12}(\cos\theta^0 - \cos\theta) + (\alpha_1\eta_1 - \alpha_2\eta_2)|v| \tag{28}$$

The principal equation linking θ and v then becomes

$$v = 2\kappa_S^0\lambda\left(\frac{h^2}{\eta_1 v_1 \eta_2 v_2}\right) \times \sinh\left[\frac{\gamma_{12}(\cos\theta^0 - \cos\theta) + (\alpha_1\eta_1 - \alpha_2\eta_2)|v|}{2nkT}\right] \tag{29}$$

Once again, various simplifications are possible. If one of the fluids is a gas, we may neglect all the viscous terms for that phase. Note, however, that the sign of $\alpha\eta|v|$ for the remaining liquid will depend on whether we choose that liquid to be fluid 1 or fluid 2. If we choose fluid 1, then

$$v = 2\kappa_S^0\lambda\left(\frac{h}{\eta_1 v_1}\right)\sinh\left[\frac{\gamma_{12}(\cos\theta^0 - \cos\theta) + \alpha_1\eta_1|v|}{2nkT}\right] \tag{30}$$

According to this equation, the flow within an advancing liquid will favor wetting and cause θ to approach 180° at higher wetting velocities than predicted by Eq. (22); hence, entrainment of the displaced phase will be postponed. Conversely, the viscous drag of a receding liquid will hinder dewetting and cause θ to approach zero at a smaller dewetting velocity than predicted; hence, liquid entrainment will occur earlier. Both effects should become increasingly important as the viscosity of the liquid increases. As we have seen, these trends are in accord with experiment.

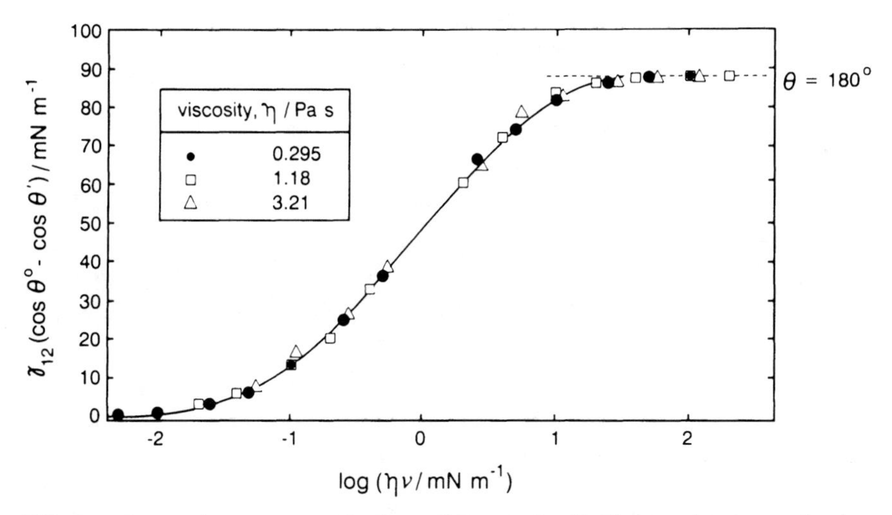

FIG. 23 Dynamic contact-angle data of Inverarity [2,3] for polyester resins in styrene on E-glass filaments.—Equation (30) with $\alpha = 0.65$. Other parameters as in Fig. 20.

For a receding wetting line, if $\gamma_{12}(\cos\theta - \cos\theta^0) \gg 2nkT$, then Eq. (30) reduces to

$$\begin{aligned} v &= -\kappa_W^- \lambda \\ &= -\kappa_W^0 \lambda \exp\left(\frac{-w}{2nkT}\right) \\ &= -\kappa_S^0 \lambda \left(\frac{h}{\eta_1 v_1}\right) \exp\left[\frac{\gamma_{12}(\cos\theta - \cos\theta^0) - \alpha_1 \eta_1 |v|}{2nkT}\right] \end{aligned} \qquad (v < 0) \qquad (31)$$

Except for the absence of the ⅔ power in v, this is analogous to the Petrov and Radoev equation, Eq. (27).

A direct test of Eq. (30) for an advancing wetting line is provided by Fig. 23. This shows Inverarity's data from Fig. 20, together with the theoretical curve obtained using the original molecular parameters and choosing $\alpha = 0.65$ by inspection. Reference to the earlier figure shows that, near 180°, Eq. (30) provides a much better fit than Eq. (22), with no loss of agreement at smaller angles. Good agreement is also found if Eq. (30) is applied to Hoffman's data from Fig. 21. The new plot is shown in Fig.

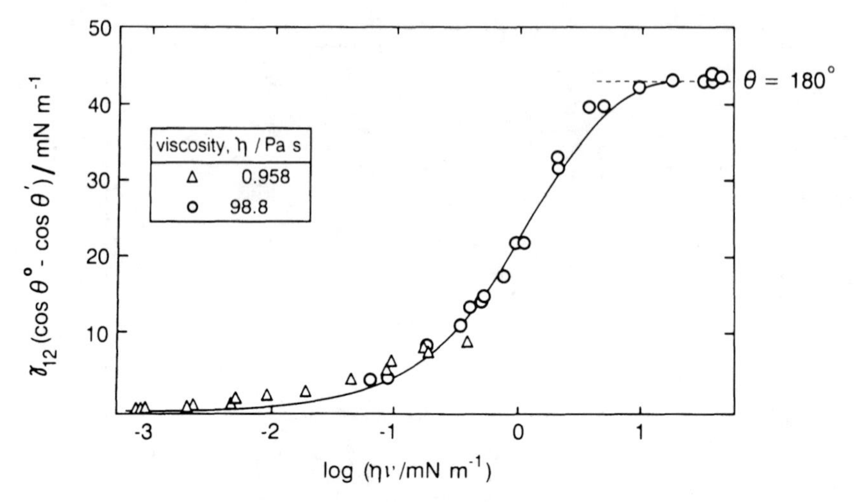

FIG. 24 Dynamic contact-angle data of Hoffman [6] for silicone oils in a cylindrical glass capillary.—Equation (30) with $\alpha = 0.92$. Other parameters as in Fig. 21.

24. In this case, the curve was calculated with $\alpha = 0.92$. Despite its schematic nature, Eq. (30) appears to be effective.

VIII. FUTURE DIRECTIONS

In the foregoing pages, wetting has been modeled as a stress-modified molecular rate process analogous to other transport processes such as viscous flow and diffusion. The theory presented has been based on three principal components: a molecular model of the moving three-phase zone; a surface tension driving force; and a hydrodynamic contribution that can be positive or negative according to the details of the local flow field near the wetting line. In formulating each component, substantial simplifications have been made to keep the approach tractable. Nevertheless, useful equations have been derived linking the dynamic contact angle to the velocity of wetting. These equations are in surprisingly good agreement with experiment; moreover, they successfully account for the influence of surface tension and viscosity, and, to a limited extent, for the effects of fluid flow. They also predict other experimentally verified phenomena, such as the existence of maximum velocities for wetting and dewetting, and offer practical explanations for the influence of substrate wettability and temperature.

Although the theory has obvious limitations, fundamental improvements are unlikely without a significant increase in complexity. For example, major difficulties attend efforts to develop a rigorous theory of transport phenomena in liquids that goes beyond the simple Eyring and Frenkel models, yet retains their practical utility [226]. Corresponding difficulties also confront attempts either to replace the surface tension driving force, Eq. (8), by a proper intermolecular force description of the three-phase region, or to improve the integration with hydrodynamics.

Some progress in these directions has been achieved by Ruckenstein and Dunn [80]. They employed a diffusional model of the moving three-phase zone similar to that used by Blake and Haynes, but based on Frenkel's kinetic theory rather than that of Eyring. Following Miller and Ruckenstein [133], they assumed a planar fluid/fluid interface and described the intermolecular forces responsible for wetting by simple Hamaker theory. The intermolecular forces are considered to give rise to a gradient in chemical potential that drives the wetting process. The principal result is a linear slip equation that can be shown to have the same form as Eq. (10), but with the term $\gamma_{12}(\cos \theta^0 - \cos \theta)$ replaced by a Hamaker expression. Neogi and Miller [82] later extended this approach by using lubrication theory to estimate hydrodynamic effects at small contact angles. Again, the mechanism for wetting line movement was assumed to be surface diffusion and the resulting slip law was linear.

Despite differing explanations for slip, a linear slip law has also been assumed in most of the other hydrodynamic treatments of wetting. Formally, we may write

$$v_{\text{slip}} = \beta\sigma \tag{32}$$

where v_{slip} is the slip velocity induced by wall shear stress σ and β the coefficient of slip. In this case, the no-slip condition is recovered when $\beta = 0$. In order to retain the no-slip condition away from the wetting line, the value of β is usually made to vary according to some specified distribution.

However, if Eqs. (9), (21), or (29) are stripped to their elements, what remains is a *nonlinear* slip condition [83] of the form

$$v_{\text{slip}} = A \sinh(B\sigma) \tag{33}$$

Although the former equations were developed to account for molecular displacements within the three-phase zone, the underlying theory is not restricted to that region. Thus, Eq. (33) should be valid at all points on the solid/liquid interface, although the coefficients A and B will obviously vary according to the local environment.

Away from the wetting line, σ will usually be too small for v_{slip} to become significant; hence, the argument of sinh will be small and Eq. (33) will revert to a linear condition similar to Eq. (32) with $\beta \sim AB$. However, note that whereas Eq. (32) describes slip at a mathematical plane, Eq. (33) refers to the motion of molecules of small but nonzero thickness.

At a wetting line, the hydrodynamic studies tell us that much larger (though nevertheless finite) values of σ are likely, so Eq. (33) should eventually become logarithmic. Furthermore, the first approximation of setting $\sigma = \gamma_{12}(\cos\theta^0 - \cos\theta)/\lambda$ will allow us to estimate values for the coefficients A and B by experiment. Thus, in principle, we have a formula that could be combined with intermolecular force theory and sound hydrodynamics to determine the practical consequences of slip in the macroscopic domain. The resulting equations are bound to be nonlinear, so numerical methods are probably essential. The most promising approaches would appear to be those of Lowndes [68], Christodoulou and Scriven [107,183], and Tilton [23,24], who all used finite-element algorithms to simulate the dynamics of wetting. As usual, these studies employed a linear slip condition and ad hoc estimates of the slip coefficient (or the slip length $\epsilon = \beta\eta$). Perhaps the main contribution of the work reviewed here is to reemphasize the importance of having a realistic model for the moving three-phase zone and to suggest that the correct slip condition is essentially nonlinear and can successfully be described in terms of simple molecular kinetics.

ACKNOWLEDGMENTS

I wish to thank the Eastman Kodak Company and Kodak Limited for supporting this work. I am also indebted to colleagues past and present, especially J. M. Haynes, K. J. Ruschak, and J. G. Petrov, for stimulating discussions of the ideas presented here.

SYMBOLS

Roman Alphabet

A	coefficient in Eq. (33)
a	effective area per adsorption site at the solid/fluid interface
B	coefficient in Eq. (33)
b	coefficient in Eq. (27)
c	system-dependent numerical constant
Ca	capillary number, $Ca = \eta v/\gamma$
ΔG_S^*	component of ΔG_W^* due to interactions between liquid and solid

$\Delta G^*_{\rm vis}$	activation free energy of bulk viscous flow
ΔG^*_W	molar activation free energy of wetting
Δg^*_s	component of Δg^*_W due to interactions between liquid and solid, $\Delta g^*_s = n\Delta G^*_s/N$
Δg^*_W	specific activation free energy of wetting, $\Delta g^*_W = n\Delta G^*_W/N$
F	force per unit length of the wetting line
h	Planck's constant
k	Boltzmann's constant
L_C	dynamic capillary height
L_0	capillary height for a static meniscus with $\theta = 0$
N	Avogadro's number
n	number of adsorption sites per unit area of solid/fluid interface
u	characteristic flow velocity close to the wetting line
V	overall velocity of wetting
v	molecular flow volume
v	velocity of wetting; note that the symbol v is used in the figures
$v_{\rm max}$	maximum velocity of wetting or dewetting
v^0_W	natural velocity of wetting, $v^0_W = \kappa^0_W\lambda$
$v_{\rm slip}$	velocity of slip
v_X	in Eq. (25), component of v due to diffusion along the solid surface
v_Z	in Eq. (25), component of v due to diffusion normal to the solid surface
T	temperature
w	work done per unit displacement of unit length of the wetting line; also, subscript used to indicate variables associated with the wetting process

Greek Alphabet

α	numerical coefficient in Eqs. (28–31)
β	coefficient of slip
γ	surface tension
δ	length of the molecular unit of flow in the direction parallel to the wetting line
ϵ	slip length, $\epsilon = \beta\eta$
η	coefficient of viscosity
θ	contact angle
θ'	experimentally observed contact angle
θ^0	equilibrium contact angle
θ'_A	static advancing contact angle

θ'_R static receding contact angle
θ_m in Fig. 22, contact angle measured by extrapolation of the undeformed portion of the meniscus
θ_r in Fig. 22, actual, macroscopic, dynamic contact angle
κ_W net frequency of molecular displacements at the wetting line
κ_W^+ frequency of molecular displacements in the forward direction
κ_W^- frequency of molecular displacements in the backward direction
κ_W^0 frequency of molecular displacements at equilibrium
κ_S^0 frequency of molecular displacements at equilibrium when retarded only by surface forces
λ average length of an individual molecular displacement in the three-phase zone.
π surface pressure
σ wall shear stress
τ_W^0 relaxation time of rate-determining molecular displacements
ϕ angle between the normal to the wetting line and the overall direction of wetting

REFERENCES

1. R. Ablett, *Phil. Mag. 46:*244 (1923).
2. G. Inverarity, The Wetting of Glass Fibres by Liquids and Polymers, Ph.D. thesis, Univ. Manchester, 1969.
3. G. Inverarity, *Brit. Polym. J. 1:*254 (1969).
4. A. M. Schwartz, and S. B. Tejada, NASA Contract Rept. CR-72728, 1970.
5. A. M. Schwartz, and S. B. Tejada, *J. Coll. Inter. Sci. 38:*359 (1972).
6. R. L. Hoffman, *J. Coll. Inter. Sci. 50:*228 (1975).
7. R. Burley, and B. S. Kennedy, *Brit. Polym. J. 8:*140 (1976).
8. R. Burley, and B. S. Kennedy, *Chem. Eng. Sci. 31:*901 (1976).
9. E. V. Gribanova, and L. I. Mulchanova, *Kolloidn. Zh. 40:*30, 217 (1978).
10. E. B. Gutoff, and C. E. Kendrick, *AIChE J. 28:*459 (1982).
11. J. G. Petrov, and B. P. Radoev, *Coll. Polym. Sci. 259:*753 (1981).
12. V. V. Berezkin, and N. V. Churaev, *Kolloidn. Zh. 44:*417 (1982).
13. E. V. Gribanova, *Kolloidn. Zh. 45:*422 (1983).
14. R. Burley, and R. P. S. Jolly, *Chem. Eng. Sci. 39:*1357 (1984).
15. W. Hopf, and Th. Geidel, *Coll. Polym. Sci. 265:*1075 (1987).
16. G. Ström, M. Fredriksson, P. Stenius, and B. Radoev, *J. Coll. Inter. Sci. 134:*107 (1990).
17. G. Ström, M. Fredriksson, and P. Stenius, *J. Coll. Inter. Sci. 134:*117 (1990).
18. T. D. Blake, G. N. Batts, and R. A. Dobson (to be published).
19. C. Huh, and L. E. Scriven, *J. Coll. Inter. Sci. 35:*85 (1971).

20. E. B. Dussan V., *Ann. Rev. Fluid Mech. 11:*371 (1979).
21. S. H. Davis, *J. Appl. Mech. 50:*977 (1983).
22. P. G. de Gennes, *Rev. Mod. Phys. 57:*827 (1985).
23. J. N. Tilton, Visocapillary Slip Flows with Special Application to Microdisplacement in Porous Media, Ph.D. diss., Univ. Houston, 1985.
24. J. N. Tilton, *Chem. Eng. Sci. 43:*1371 (1988).
25. G. E. P. Elliott, and A. C. Riddiford, *J. Coll. Inter. Sci. 23:*389 (1967).
26. R. E. Johnson, Jr., R. H. Dettre, and D. A. Brandreth, *J. Coll. Inter. Sci. 62:*205 (1977).
27. N. R. Morrow, and M. D. Nguyen, *J. Coll. Inter. Sci. 89:*523 (1982).
28. J. B. Cain, D. W. Francis, R. D. Venter, and A. W. Newman, *J. Coll. Inter. Sci. 94:*123 (1983).
29. B. V. Derjaguin, *C. R. Acad. Sci. USSR 51:*361 (1946).
30. R. J. Good, *J. Amer. Chem. Soc. 75:*5041 (1952).
31. R. E. Johnson, Jr., and R. H. Dettre, in *Contact Angle, Wettability, and Adhesion*, Advances in Chemistry Series 43 (R. F. Gould, ed.), ACS, Washington, DC, 1964, p. 112; *J. Phys. Chem. 68:*1744 (1964).
32. R. E. Johnson, Jr., and R. H. Dettre, in *Surface and Colloid Science*, Vol. 2 (E. Matijevic, ed.), Wiley-Interscience, New York, 1969, pp. 85–153.
33. A. W. Neumann, and R. J. Good, *J. Coll. Inter. Sci. 38:*341 (1972).
34. C. Huh, and S. G. Mason, *J. Coll. Inter. Sci. 60:*11 (1977).
35. L. Boruvka, and A. W. Neumann, *J. Coll. Inter. Sci. 65:*315 (1978).
36. R. G. Cox, *J. Fluid Mech. 131:*1 (1983).
37. J. F. Joanny, and P. G. de Gennes, *J. Chem. Phys. 81:*552 (1984).
38. Y. Pomeau, and J. Vannimenus, *J. Coll. Inter. Sci. 104:*477 (1985).
39. L. W. Schwartz, and S. Garoff, *J. Coll. Inter. Sci. 106:*422 (1985).
40. L. W. Schwartz, and S. Garoff, *Langmuir 1:*219 (1985).
41. J. F. Oliver, C. Huh, and S. G. Mason, *Coll. Surf. 1:*79 (1980).
42. S. J. Hitchcock, N. T. Carroll, and M. G. Nicholas, *J. Mater. Sci. 16:*714 (1981).
43. E. Bayramli, T. G. M. van de Ven, and S. G. Mason, *Coll. Surf. 3:*279 (1981).
44. R. S. Hansen, and M. Miotto, *J. Am. Chem. Soc. 79:*1765 (1957).
45. T. D. Blake and J. M. Haynes, in *Progress in Surface and Membrane Science*, Vol. 6 (J. F. Danielli, M. D. Rosenberg, and D. A. Cadenhead, eds.), Academic Press, New York, 1973, pp. 125–138.
46. G. A. Martynov, V. M. Starov, and N. V. Churaev, *Kolloidn. Zh. 39:*472 (1977).
47. V. M. Starov, *Kolloidn. Zh. 45:*699 (1983).
48. D. H. Everett, *Pure Appl. Chem. 52:*1279 (1980).
49. K. M. Jansons, *J. Fluid Mech. 154:*1 (1985).
50. A. Engel, and W. Ebeling, *Phys. Lett. A 122:*20 (1987).
51. J. G. E. M. Fraaije, M. Cazabat, X. Hua, and A. M. Cazabat, *Coll. Surf. 41:*77 (1989).
52. E. Bayramli, T. G. M. van de Ven, and S. G. Mason, *Coll. Surf. 3:*131 (1981).

53. M. A. Cohen Stuart, and A. M. Cazabat, *Prog. Coll. Polym. Sci. 74:*64 (1987).
54. H. M. Princen, A. M. Cazabat, M. A. Cohen Stuart, F. Heslot, and S. Nicolet, *J. Coll. Inter. Sci. 126:*84 (1988).
55. B. B. Sauer, and T. E. Carney, *Langmuir 6:*1002 (1990).
56. J. P. Stokes, M. J. Higgins, A. P. Kuschnik, S. Bhattacharya, and M. O. Robbins, *Phys. Rev. Lett. 65:*1885 (1990).
57. E. Raphaël, and P. G. de Gennes, *J. Phys. Chem. 90:*7577 (1989).
58. J.-M. di Meglio, and D. Quéré, *Europhys. Lett. 11:*163 (1990).
59. J. F. Joanny, and M. O. Robbins, *J. Chem. Phys. 92:*3206 (1990).
60. F. Brochard, and P. G. de Gennes, *Langmuir 7:*3216 (1991).
61. E. B. Dussan V., and S. H. Davis, *J. Fluid Mech. 65:*71 (1974).
62. E. B. Dussan V., *J. Fluid Mech. 77:*665 (1976).
63. L. M. Hocking, *J. Fluid Mech. 76:*801 (1976).
64. L. M. Hocking, *J. Fluid Mech. 79:*209 (1977).
65. C. Huh, and S. G. Mason, *J. Fluid Mech. 81:*401 (1977).
66. H. P. Greenspan, *J. Fluid Mech. 84:*125 (1978).
67. S. Levine, J. Lowndes, E. J. Watson, and G. Neale, *J. Coll. Inter. Sci. 73:*136 (1980).
68. J. Lowndes, *J. Fluid Mech. 101:*631 (1980).
69. L. M. Hocking, *Quart. J. Mech. Appl. Math. 34:*37 (1981).
70. H. P. Greenspan, and B. M. McCay, *Stud. Appl. Math. 64:*95 (1981).
71. L. M. Hocking, *Proceedings of 2nd International Colloquium on Drops and Bubbles*, JPL Publ. 82-7, NASA-JPL, Pasadena, CA, 1982, pp. 315–321.
72. L. M. Hocking, and A. D. Rivers, *J. Fluid Mech. 121:*425 (1982).
73. A. A. Lacey, *Stud. Appl. Math. 67:*217 (1982).
74. P. Neogi, and C. A. Miller, *J. Coll. Inter. Sci. 92:*338 (1983).
75. P. Bach, and O. Hassager, *J. Fluid Mech. 152:*173 (1985).
76. R. G. Cox, *J. Fluid Mech. 168:*169, 195 (1986).
77. P. A. Durbin, *J. Fluid Mech. 197:*157 (1988).
78. P. J. Haley, and M. J. Miksis, *J. Fluid Mech. 223:*57 (1991).
79. D. M. Tolstoi, *Dokl. Acad. Nauk SSSR 85:*1089 (1952).
80. E. Ruckenstein, and C. S. Dunn, *J. Coll. Inter. Sci. 59:*135 (1977).
81. E. Ruckenstein, and P. Rajora, *J. Coll. Inter. Sci. 96:*488 (1983).
82. P. Neogi, and C. A. Miller, *J. Coll. Inter. Sci. 86:*525 (1982).
83. T. D. Blake, AIChE Spring Meeting, New Orleans, LA, 1988, paper 1a.
84. B. V. Paranjape, and R. E. Robson, *Phys. Chem. Liq. 21:*147 (1990).
85. F. Y. Kafka, and E. B. Dussan V., *J. Fluid Mech. 95:*539 (1979).
86. M. V. Berry, *J. Phys. A: Math. Nucl. Gen. 7:*231 (1974).
87. G. Saville, *J. Chem. Soc. Faraday II 73:*1122 (1977).
88. L. R. White, *J. Chem. Soc. Faraday I 73:*390 (1977).
89. B. V. Zheleznyi, *Kolloidn. Zh. 40:*239 (1978).
90. I. B. Ivanov, B. V. Toshev, and B. P. Radoev, in *Wetting, Spreading and Adhesion* (J. F. Padday, ed.), Academic Press, New York/London, 1978, pp. 37–60.

91. K. K. Mohanty, H. T. Davis, and L. E. Scriven, American Physics Society 31st Meeting on Fluid Dynamics, Los Angeles, CA, 1978; *Bull. Amer. Phys. Soc. 23:*996 (1978).
92. P. C. Wayner, Jr., *J. Coll. Inter. Sci. 77:*495 (1980).
93. R. E. Benner, Jr., L. E. Scriven, and H. T. Davis, *Faraday Symp. Chem. Soc. 16:*169 (1981).
94. N. V. Churaev, V. M. Starov, and B. V. Derjaguin, *J. Coll. Inter. Sci. 89:* 16 (1982).
95. J. S. Rowlinson, and B. Widom, *Molecular Theory of Capillarity*, Clarendon Press, London, 1982.
96. R. J. Hansen, and T. Y. Toong, *J. Coll. Inter. Sci. 37:*196 (1971).
97. D. J. Benney, and W. J. Timson, *Stud. Appl. Math. 63:*98 (1980).
98. L. M. Pismen, and A. Nir, *Phys. Fluids 25:*3 (1982).
99. C. G. Ngan, and E. B. Dussan V., *Phys. Fluids 27:*2785 (1984).
100. F. P. Bretherton, *J. Fluid Mech. 10:*166 (1961).
101. J. Bataille, *C. R. Acad. Sci. 262A:*843 (1966).
102. B. V. Derjaguin, and S. M. Levi, *Film Coating Theory*, Focal Press, London, 1964.
103. T. D. Blake, The Contact Angle and Two-Phase Flow, Ph.D. thesis, Univ. Bristol, 1968.
104. G. F. Teletzke, H. T. Davis, and L. E. Scriven, AIChE Annual Meeting, Chicago, IL, 1980.
105. K. Miyamoto, and L. E. Scriven, AIChE Annual Meeting, Los Angeles, CA, 1982, paper 101g.
106. G. F. Teletzke, Thin Liquid Films: Molecular Theory and Hydrodynamic Implications, Ph.D. thesis, Univ. Minnesota, 1983.
107. K. N. Christodoulou, and L. E. Scriven, AIChE Meeting, San Francisco, CA, 1984.
108. B. V. Derjaguin, and M. Kusakov, *Acta Physicochim USSR 10:*25, 153 (1939).
109. W. Mues, J. Hens, and L. Boiy, *AIChE J. 35:*1521 (1989).
110. T. D. Blake, and K. J. Ruschak, *Nature 282:*489 (1979).
111. J. F. Oliver, and S. G. Mason, *J. Coll. Inter. Sci. 60:*480 (1977).
112. W. B. Hardy, *Phil. Mag. 38:*49 (1919).
113. D. H. Bangham, and Z. Saweris, *Trans. Faraday Soc. 34:*554 (1938).
114. E. D. Shchukin, Yu. V. Goryunov, G. I. Den'shchikova, N. V. Pertsov, and S. D. Summ, *Kolloidn. Zh. 25:*108 (1963).
115. W. D. Bascom, R. L. Cottington, and C. R. Singleterry, in *Contact Angle, Wettability, and Adhesion*, Advances in Chemistry Series 43 (R. F. Gould, ed.), ACS, Washington, DC, 1964, pp. 355–379.
116. W. Radigan, H. Ghiradella, H. L. Frisch, H. Schonhorn, and T. K. Kwei, *J. Coll. Inter. Sci. 49:*241 (1974).
117. H. Ghiradella, W. Radigan, and H. L. Frish, *J. Coll. Inter. Sci. 51:*522 (1975).
118. R. Williams, *Nature 266:*153 (1977).

119. G. C. Sawicki, in *Wetting, Spreading and Adhesion* (J. F. Padday, ed.), Academic Press, New York/London, 1978, pp. 361–375.
120. A. Marmur, and M. D. Lelah, *J. Coll. Inter. Sci. 78:*262 (1980).
121. M. D. Lelah, and A. Marmur, *J. Coll. Inter. Sci. 82:*518 (1981).
122. G. C. Smith, and C. Lea, *Surf. Inter. Anal. 9:*145 (1986).
123. O. Teschke, and M. A. Tenan, *Langmuir 4:*234 (1988).
124. D. Quéré, J.-M. di Meglio, and F. Brochard-Wyart, *Revue Phys. Appl. 23:* 1023 (1988).
125. L. Léger, M. Erman, A. M. Guinet-Picart, D. Ausserre, C. Strazielle, J. J. Benattar, F. Rieutord, J. Daillant, and L. Bosio, *Revue Phys. Appl. 23:* 1047 (1988).
126. J. Daillant, J. J. Benattar, L. Bosio, and L. Leger, *Europhys. Lett. 6:*431 (1988).
127. R. L. Kao, D. T. Wasan, A. D. Nikolov, and D. A. Edwards, *Coll. Surf. 34:*389 (1988/89).
128. O. Teschke, M. A. Tenan, and F. Galembeck, *J. Coll. Inter. Sci. 127:*88 (1989).
129. F. Heslot, A. M. Cazabat, and P. Levinson, *Phys. Rev. Lett. 62:*1286 (1989).
130. F. Heslot, N. Fraysse, and A. M. Cazabat, *Nature 338:*640 (1989).
131. F. Heslot, A. M. Cazabat, P. Levinson, and N. Fraysse, *Phys. Rev. Lett. 65:*599 (1990).
132. B. V. Derjaguin, *Kolloidn. Zh. 17:*191 (1955).
133. C. A. Miller, and E. Ruckenstein, *J. Coll. Inter. Sci. 48:*368 (1974).
134. J. Lopez, C. A. Miller, and E. Ruckenstein, *J. Coll. Inter. Sci. 56:*460 (1976).
135. G. F. Teletzke, H. T. Davis, and L. E. Scriven, AIChE Annual Meeting, New Orleans, LA, 1981.
136. V. M. Starov, *Kolloidn. Zh. 45:*1154 (1983).
137. P. G. de Gennes, *C. R. Acad. Sci. 298 II:*111 (1984).
138. F. Brochard, and P. G. de Gennes, *J. Physique Lett. 45:*L-597 (1984).
139. H. Hervet, and P. G. de Gennes, *C. R. Acad. Sci. 299 II:*499 (1984).
140. G. F. Teletzke, H. T. Davis, and L. E. Scriven, *Chem. Eng. Comm. 55:* 41 (1987).
141. J. Thomson, *Phil. Mag. 10:*330 (1855).
142. P. Neogi, *J. Coll. Inter. Sci. 105:*94 (1985).
143. D. Pesach, and A. Marmur, *Langmuir 3:*519 (1987).
144. S. M. Troian, X. L. Wu, and S. A. Safran, *Phys. Rev. Lett. 62:*1496 (1989).
145. A. M. Cazabat, *Advan. Coll. Inter. Sci. 34:*73 (1991).
146. A. B. Abraham, P. Collet, J. de Coninck, and F. Dunlop, *Phys. Rev. Lett. 65:*195 (1990).
147. A. B. Abraham, P. Collet, J. de Coninck, and F. Dunlop, *J. Stat. Phys. 61:*509 (1990).
148. V. Ludviksson, and E. N. Lightfoot, *AIChE J. 14:*674 (1968).
149. G. Friz, *Z. Agnew. Phys. 19:*374 (1965).

150. L. H. Tanner, *J. Phys. D: Appl. Phys. 12:*1473 (1979).
151. V. V. Kalinin, and V. M. Starov, *Kolloidn. Zh. 48:*907 (1986).
152. L. Landau, and V. G. Levich, *Acta Physicochim. USSR 17:*42 (1942) (see also V. G. Levitch, *Physicochemical Hydrodynamics*, Prentice Hall, Engelwood Cliffs, NJ, 1962, Chap. 12).
153. B. V. Derjaguin, *Zh. Eksp. Teoret. Fiz. USSR 15:*9 (1945).
154. O. V. Voinov, *Fluid Dyna. 11:*714 (1976); *Isvest. Akad. Nauk SSSR, Mekh. Zhid. Gaz. 5:*76 (1976).
155. C. del Cerro, and G. J. Jameson, AIChE Annual Meeting, New York, 1977.
156. P. G. de Gennes, X. Hua, and P. Levinson, *J. Fluid Mech. 212:*55 (1990).
157. T. A. Coney, and W. J. Masica, Tech. Note D-5115, NASA, Washington, DC, 1969.
158. B. V. Zheleznyi, *Dokl. Akad. Nauk SSSR 207:*647 (1972).
159. B. V. Zheleznyi, and T. V. Korneva, *Dokl. Akad. Nauk SSSR 249:*569 (1979).
160. J. D. Chen, *J. Coll. Inter. Sci. 122:*60 (1988).
161. A. M. Cazabat, F. Heslot, and N. Fraysse, *Prog. Coll. Polym. Sci. 83:*52 (1990).
162. W. Rose, and R. W. Heins, *J. Coll. Sci. 17:*39 (1962).
163. A. M. Cazabat, and M. A. Cohen Stuart, *J. Phys. Chem. 90:*5845 (1986).
164. B. D. Summ, V. S. Yushchenko, and E. D. Shchukin, *Coll. Surf. 27:*43 (1987).
165. F. T. Dodge, *J. Coll. Inter. Sci. 121:*154 (1988).
166. L. Leger, in *Physicochemical Hydrodynamics—Interfacial Phenomena* (M. G. Verlade, ed.), Plenum, New York, 1988, pp. 721–740.
167. F. Brochard-Wyart, H. Hervet, C. Redon, and F. Rondelez, *J. Coll. Inter. Sci. 142:*518 (1991).
168. W. L. Wilkinson, *Chem. Eng. Sci. 30:*1227 (1975).
169. T. S. Jiang, S. G. Oh, and J. C. Slattery, *J. Coll. Inter. Sci. 69:*74 (1979).
170. E. Rillaerts, and P. Joos, *Chem. Eng. Sci. 35:*883 (1980).
171. M. Bracke, F. de Voeght, and P. Joos, *Prog. Coll. Polym. Sci. 79:*142 (1989).
172. M. N. Esmail, and M. T. Ghannam, *Canad. J. Chem. Eng. 68:*197 (1990).
173. R. T. Foister, *J. Coll. Inter. Sci. 136:*266 (1990).
174. S. F. Kissler, and L. E. Scriven, AIChE Winter Meeting, Orlando, FL, 1982, paper 45d.
175. K. J. Ruschak, *Ann. Rev. Fluid Mech. 17:*65 (1985).
176. S. M. Levi, and V. I. Akulov, *Zh. Nauchn. Prikl. Photogr. Kinematogr. 9:*124 (1964).
177. S. M. Levi, *Zh. Nauchn. Prikl. Photogr. Kinematogr. 11:*401 (1966).
178. R. T. Perry, Fluid Mechanics of Entrainment through Liquid–Liquid and Liquid–Solid Junctures, Ph.D. thesis, Univ. Minnesota, 1967.
179. J. Hens, and L. Boiy, *Chem. Eng. Sci. 41:*1827 (1986).
180. P. M. Schweizer, *J. Fluid Mech. 193:*285 (1988).
181. L. E. Scriven, and W. J. Suszynski, *Chem. Eng. Prog. 24* (Sept.:24) (1990).

182. S. F. Kistler, and L. E. Scriven, in *Computational Analysis of Polymer Processing* (J. R. Pearson, and S. M. Richardson, eds.), Applied Science, Barking, 1983, Chap. 8, pp. 243–299.
183. K. N. Christodoulou, and L. E. Scriven, *J. Fluid Mech. 208:*321 (1989).
184. C. G. Ngan, and E. B. Dussan V., *J. Fluid Mech. 118:*27 (1982).
185. C. G. Ngan, and E. B. Dussan V., *J. Fluid Mech. 209:*191 (1989).
186. B. Legait, and P. Sourieau, *J. Coll. Inter. Sci. 107:*14 (1985).
187. V. M. Starov, N. V. Churaev, and A. G. Khvorostyanov, *Kolloidn. Zh. 39:*201 (1977).
188. T. D. Blake, and J. M. Haynes, *J. Coll. Inter. Sci. 30:*421 (1969).
189. B. W. Cherry, and C. M. Holmes, *J. Coll. Inter. Sci. 29:*174 (1969).
190. R. L. Hoffman, *J. Coll. Inter. Sci. 94:*470 (1983).
191. T. D. Blake, *Coll. Surf. 47:*135 (1990).
192. G. M. Bartenev, *Dokl. Acad. Nauk SSSR 96:*1161 (1954).
193. E. D. Shchukin, and V. S. Yushchenko, *Kolloidn. Zh. 39:*331 (1977).
194. J. Koplik, J. R. Banavar, and J. F. Willemsen, *Phys. Rev. Lett. 60:*1282 (1988).
195. J. Koplik, J. R. Banavar, and J. F. Willemsen, *Phys. Fluids A1:*781 (1989).
196. P. A. Thompson, and M. O. Robbins, *Phys. Rev. Lett. 63:*766 (1989).
197. S. Glasstone, K. J. Laidler, and H. J. Eyring, *The Theory of Rate Processes*, McGraw-Hill, New York, 1941, and papers cited there, especially, R. E. Powell, W. E. Roseveare, and H. Eyring, *Ind. Eng. Chem. 33:*430 (1941).
198. J. I. Frenkel, *Kinetic Theory of Liquids*, Oxford Univ. Press, Oxford, 1946.
199. T. D. Blake, *Ver. Deut. Ing. Berlin 182:*117 (1973).
200. M. R. Buhaenko, and R. M. Richardson, *Thin Solid Films 159:*231 (1988).
201. G. D. Yarnold, and B. J. Mason, *Proc. Phys. Soc. London B62:*121, 125 (1949).
202. B. D. McLaughlin, and P. L. de Bruyn, *J. Coll. Inter. Sci. 30:*21 (1969).
203. R. J. Good, and N. J. Lin, *J. Coll. Inter. Sci. 54:*52 (1976).
204. F. Brochard-Wyart, and P. G. de Gennes, *Advan. Coll. Inter. Sci. 39:*1 (1992).
205. B. S. Kennedy, Dynamic Wetting and Air-Entrainment at a Liquid/Solid/Gas Junction, Ph.D. thesis, Herriot-Watt Univ., Edinburgh, 1975.
206. J. H. de Boer, *The Dynamical Character of Adsorption*, Clarenden Press, Oxford, 1953.
207. G. Ehrlich, and K. Stolt, *Ann. Rev. Phys. Chem. 31:*603 (1980).
208. J. G. Petrov, and R. V. Sedev, *Coll. Surf. 13:*313 (1985).
209. J. G. Petrov, *Coll. Surf. 17:*283 (1986).
210. R. V. Sedev, and J. G. Petrov, *Coll. Surf. 34:*197 (1988/89).
211. P. G. de Gennes, *Coll. Polym. Sci. 264:*463 (1986).
212. P. G. Saffman, and G. I. Taylor, *Proc. R. Soc. Lond. A245:*312 (1958).
213. J.-F. Joanny, and D. Andelman, *J. Coll. Inter. Sci. 119:*451 (1987).
214. F. Brochard-Wyart, J.-M. di Meglio, and D. Quéré, *C. R. Acad. Sci. 304 II:*553 (1987).

215. C. Redon, F. Brochard-Wyart, and F. Rondelez, *Phys. Rev. Lett. 66:*715 (1991).
216. R. V. Dyba, and B. Miller, *Textile Res. J. 40:*884 (1970).
217. J. W. Clark, P. G. Hall, A. J. Pidduck, and C. J. Wright, *J. Chem. Soc., Faraday Trans. I 81:*2067 (1985).
218. A. C. Lowe, and A. C. Riddiford, *J. Chem. Soc. D: Chem. Comm. No. 6:*387 (1970).
219. A. M. Cazabat, N. Fraysse, and F. Heslot, *Coll. Surf. 52:*1 (1991).
220. P. C. Wayner, Jr., *Coll. Surf. 52:*71 (1991).
221. H. S. van Damme, A. H. Hogt, and J. Feijen, *J. Coll. Inter. Sci. 114:*167.
222. M. E. R. Shanahan, *J. Phys. D: Appl. Phys. 21:*981 (1988).
223. T. D. Blake, Plymouth, New Hampshire, 1980, unpublished lecture.
224. B. Janczuk, and T. Bialopiotrowicz, *J. Coll. Inter. Sci. 127:*189 (1989) and papers cited there.
225. J. G. Petrov, *Z. Phys. Chemie, Lepzig 266:*706 (1985).
226. S. A. Rice, J. P. Boon, and H. T. Davis, in *Simple Dense Fluids* (H. L. Frisch, and Z. W. Salsburg, eds.), Academic Press, New York/London, 1968, pp. 251–402.

6
Hydrodynamics of Wetting

STEPHAN F. KISTLER Magnetic Media Technology Center, 3M Company, St. Paul, Minnesota

I. INTRODUCTION

In *dynamic wetting*, a liquid displaces another fluid, often air, from a solid surface. In *forced wetting*, externally imposed hydrodynamic or mechanical forces cause the interfacial area between the liquid and solid to increase beyond conditions of static equilibrium. Typically, either the liquid front or the solid is driven at a constant speed U. Forced wetting is an essential mechanism in industrial coating processes where a thin layer of liquid is continually deposited onto a moving solid substrate. The commonly encountered failure of the liquid to displace sufficient air from the substrate can severely limit the maximum speed at which coating can proceed. Forced wetting also plays an important role in many other technologies, such as polymer processing (mold filling, composite manufacture, lamination, etc.) and enhanced oil recovery. *Spontaneous spreading*, on the other hand, is the unsteady migration of a liquid over a solid toward thermodynamic equilibrium. The driving force is not imposed externally, but stems from liquid/solid interactions that ''sense'' a reduction in free energy with an increase in solid area covered by liquid. Spontaneous spreading is of practical relevance in the application of paints and adhesives, migration of printing inks on paper, agricultural plant treatment, imbibition of rocks and other porous media, flotation, and numerous other technological and biological systems.

On macroscopic length scales observable by naked eye, or even on the scale of a few microns accessible with optical microscopes, the most prominent feature of dynamic wetting is the *wetting line* where the advancing fluid/fluid interface seems to intersect the solid surface at a measurable *dynamic contact angle* θ_D. Figures 1a–c illustrate the appearance of the dynamic wetting region in forced liquid/air displacement for a liquid/solid combination with a static contact angle well below 90°. At low speeds, the advancing angle θ_D remains near its static value (Fig. 1a). Typically, though, small increments in speed suffice to rapidly increase θ_D (Fig. 1b; θ_D is measured through the liquid). At high speeds of the sort encountered in industrial coating processes, contact angles often become less sensitive to speed and typically fall in the range $120° < \theta_D < 170°$. Eventually, at a critical speed, the dynamic contact angle approaches 180°. Beyond that limit, visible amounts of air are entrained between the liquid and solid (Fig. 1c). The wetting line may become serrated and unsteady before the onset of visible *air entrainment*.

This set of phenomena and many related observations of forced wetting and spontaneous spreading are the subject of the present review. Earlier summaries of the literature [1–10], as well as a rapidly expanding body of newer research results, testify that dynamic wetting has been studied extensively in both experiment and theory. For many systems of practical interest, nevertheless, the hydrodynamic and physico-chemical mechanisms that control how, and whether, a liquid displaces an immiscible fluid from a solid surface remain unresolved.

The crux is the disparity between the submicroscopic length scales at which the wetting process takes place and the macroscopic scales at which

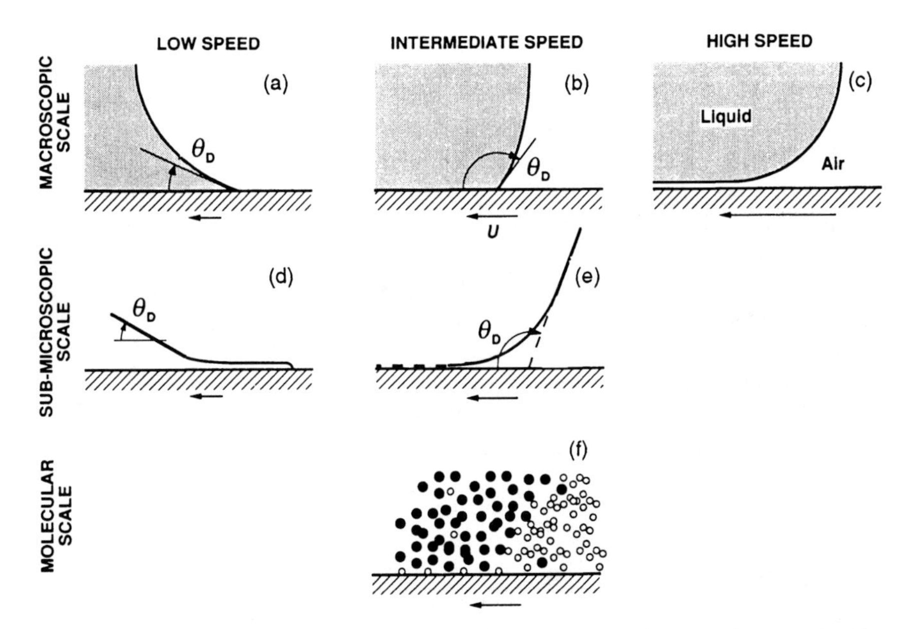

FIG. 1 Schematic of three-phase juncture in dynamic wetting: (a–c) macroscopic length scale of at least several tens of microns; (d) and (e) submicroscopic scale between 30 and 1000 Å; (f) molecular scale. All sketches are drawn in a frame of reference moving with the advancing liquid front.

experimental observations are typically made. From a fundamental scientific point of view, the key challenge is to identify, measure, and model the local mechanisms by which a liquid displaces another fluid from a solid surface. From a more pragmatic point of view, on the other hand, the main challenges are to sort out the material properties and flow parameters that dominate dynamic wetting as it is observed on a macroscopic scale; to calculate macroscopic flow fields under conditions under which the advancing liquid appears to completely displace the receding fluid; and to predict the critical conditions beyond which successful displacement fails catastrophically. Unfortunately, submicroscopic wetting physics and macroscopic wetting dynamics are intricately linked. Progress toward identifying the linkage—and thereby resolving many controversies in the published literature—has been slow for several reasons.

For one, experimental observations on length scales accessible by conventional optical techniques yield only limited insights into the local wetting physics. Sophisticated techniques have been advanced to probe the submicroscopic composition and structure of solid/vacuum interfaces, but most cannot be applied in the presence of liquids. In many regimes of dynamic wetting, the nature of the local displacement process remains therefore subject to conjecture and speculation. On perfectly smooth surfaces, dynamic wetting is probably the result of molecular events (see Fig. 1f) that are subject to significant statistical fluctuations and dominated by short-range forces acting on a scale of a few molecular diameters (i.e., a few Å). On larger, but still submicroscopic length scales (i.e., 0.1 μm or less), dynamic wetting can also be affected by long-range fluid/solid interactions such as van der Waals forces and, for ionic species, electric double-layer forces. At low wetting rates, microstructural forces can drive the surface migration of *primary precursor films* of near-molecular thickness [7,11,12] and the advance of somewhat thicker *secondary films* [6,7,13,14] (see Fig. 1d). Primary layers of volatile species may also form by condensation out of the vapor phase [15]. At higher wetting rates, in contrast, intermolecular forces may leave one or a few molecular layers of the receding fluid adsorbed to the solid [16]. At higher rates yet, especially in forced wetting, long-range fluid/solid interaction forces have to compete with hydrodynamic forces in the inescapable flow of the second fluid. Together, both types of forces control whether the receding fluid is entrained as a thin film that at first may collapse and remain invisible [8,10,17] (see Fig. 1e). At sufficiently high displacement speeds, hydrodynamic forces in the receding phase will cause catastrophic failure of the advancing liquid to successfully wet the solid (see Fig. 1c). On rough or porous substrates, capillary imbibition is likely to facilitate primary wetting at low speeds of the macroscopic front. At higher speeds, on the other hand, the receding fluid may become trapped and entrained in the hollows of the microtexture [17,18]. Evidently, the concept of a moving wetting line at which a fluid/fluid interface "slides" over a solid surface is merely a macroscopic substitute for much more intricate submicroscopic wetting dynamics that can proceed in many different regimes. Therefore, dynamic wetting lines are sometimes referred to as *apparent*, a qualifier that is adopted in this review.

Dynamic contact angles must also be considered apparent when measured with conventional optical means, which typically resolve the shape of the fluid/fluid interface at best within 20 μm from the three-phase juncture. At smaller distances, the interface may be sharply curved. Measured contact angles might therefore depend on the length scale that the experiment can resolve, and there is no unique definition for an apparent dynamic contact angle. On near-molecular scales, geometric concepts such as interfaces and contact angles lose their meaning altogether, even under static equilibrium conditions [19,20]. The dynamic interface deformation at submicroscopic scales might not only be controlled by microstructural forces [6,8,13,21], but might also arise from hydrodynamic forces induced by the flow of which dynamic wetting is inevitably part [22–25]. The macroscopic flow field and geometry, and other externally imposed forces such as those arising from electrostatic fields, may also affect the apparent dynamic wetting behavior. Identifying the essential mechanisms from data published in the literature is difficult because few investigators covered the entire range from low speeds where θ_D remains near its static value to high speeds where θ_D approaches 180°. Further difficulties in sorting out the unifying concepts stem from the potential influence of various nonideal effects, such as surfactants or other traces of impurities, volatile components, microroughness and porosity of the solid, chemical heterogeneities and soluble contaminants on the solid surface, electrical charges carried by the solid, and internal polarization of the solid surface. Isolating these effects requires rigor in cleanliness and material characterization that is standard practice in surface chemistry and static wetting studies, but is all too often ignored in experimental studies of dynamic wetting.

The challenges for theoretical analysis of dynamic wetting are no less formidable. Purely hydrodynamic treatments, albeit quite difficult in themselves, cannot elucidate the local wetting mechanisms. A well-known, unphysical force singularity arises when conventional continuum theory is applied right down to the apparent three-phase region [2]. In addition, a boundary condition is needed for the angle at which the interface intersects the solid surface [3]. When surface tension effects are important, the imposed angle can have a significant effect on the macroscopic flow field, yet in predictive theories, apparent dynamic contact angles ought to be dependent variables. Most mathematical treatments of dynamic wetting have been preoccupied with finding ad hoc remedies for the inadequacies of purely hydrodynamic models, rather than refining the models to account for the physics on the small length scales that matter. Even for refined models, however, agreement between theoretical predictions and experimental measurements at macroscopic and even microscopic length scales is insufficient evidence from which to conclude that a particular model correctly captures the submicroscopic wetting physics. Different local models can yield the same dynamics at macroscopic length scales—not only simple models that rely on ad hoc boundary conditions to produce well-posed boundary-value problems, but also refined models that seek to describe the wetting physics. Unequivocal verification of refined models requires therefore measurements on the small length scales that the theories attempt to resolve.

Despite formidable difficulties inherent in dynamic wetting studies, considerable progress has been made, especially over the last 10–15 years. The main objective of this chapter is to provide a comprehensive guide to these developments, and to synthesize a unified view of dynamic wetting that heretofore has not been available in a single reference. The chapter focuses on steady-state forced displacement of a gas by a liquid, with a bias toward moderate and high speeds of the sort encountered in industrial coating operations. Forced liquid/liquid displacement and spontaneous spreading, which typically proceed at much lower speeds, are also addressed, but only to the extent that they shed light on the mechanisms at work where a liquid front advances over a solid. The chapter is written primarily from a fluid mechanical point of view, with an emphasis on macroscopic experimental observations and hydrodynamic theories that attempt to explain these observations. Fundamental attempts to probe and predict the local wetting physics are contrasted with macroscopic measurements and flow models, but are not the main theme of the chapter.* Reflecting the emphasis in the published literature, the present chapter concentrates on ideal liquid/solid systems in which electrostatic surface charges, surface-active agents, volatility of the liquid, and non-Newtonian behavior are presumed to be of no consequence.† Nonideal effects, however, may have played a role in some of the experiments reviewed here, even if the original authors did not explicitly acknowledge their influence.

Section II takes a fresh look at the experimental evidence available to date. The main aim is to establish to what extent hydrodynamic forces rather than specific material effects control apparent dynamic contact angles (Sec. II.A) and the critical limit of visible air entrainment (Sec. II.B). In particular, Sec. II attempts to separate the influence of surface wettability, liquid/solid interactions, and peculiar low-speed behavior from the effects of viscous stresses, surface tension, gravity, inertia, macroscopic flow geometry, and other flow parameters. Section II.D.1 assesses whether spontaneous spreading experiments corroborate the main findings from forced wetting studies. Section II also emphasizes the important role that the receding gas phase can play, especially the viscous stresses it generates. In addition, Sec. II seeks to identify useful empirical correlations for apparent dynamic contact angles (Sec. II.A.5), macroscopic spreading rates (Sec. II.D.1), and air entrainment velocities (Sec. II.B).

Section II also gives a digest of what is known from experiments about the details of interfacial shape and flow field close to the three-phase

* For an extensive discussion of how molecular events might influence wetting dynamics, the reader is referred to Chap. 5 by T. D. Blake in this volume.

† Chapter 3 by A. Bose in this volume describes how wetting by solutions differs from wetting by pure liquids.

juncture. Section II.C reviews rare attempts to detect and measure the kinematics that give rise to apparent wetting line motion. Section II.D.2 summarizes more extensive data on the shape and expansion of precursor films. Although not ubiquitous, such films play a central role in the field of dynamic wetting because they allow detailed measurements of interfacial profiles at molecular distances from a solid surface.

Section III.A assesses how well conventional hydrodynamic theories predict experimental observations on macroscopic length scales. Section III.A.1 recounts why such theories are unable to capture the physics of the local displacement process. Sections III.A.2 and III.A.3 examine to what extent ad hoc boundary conditions at a putative wetting line can remedy the shortcomings of conventional continuum theory. These sections also identify the best strategies for calculating flows with moving wetting lines when resolving the submicroscopic wetting physics is not of primary interest. Section III.A.4 focuses on a particular set of ad hoc boundary conditions that attribute dynamic variations in an apparent contact angle exclusively to hydrodynamic bending of the interface close to the wetting line, yet capture essential aspects of macroscopic wetting dynamics observed in experiments remarkably well. Section III.A.5 discusses whether these boundary conditions have any chance of mimicking the physics of the local wetting dynamics. Section III.A.6 investigates to what extent conventional hydrodynamic continuum theory predicts the influence of the size, geometry, and other features of the macroscopic flow. Section III.A.7 summarizes attempts to account for specific material effects, especially surface roughness and heterogeneity. Section III.A.8 shows that a preexisting liquid film can significantly enhance dynamic wetting rates. The flow in the receding phase can also be very important, as Sec. III.A.9 shows for a gas and Sec. III.A.11 for an immiscible liquid. Finally, Sec. III.A.10 points out that two-phase flow theories with ad hoc boundary conditions at the wetting line can approximately predict the critical parameters at the onset of visible air entrainment. From a practical point of view, predicting the maximum speed at which dynamic wetting is successful is often most important.

Section III.B reviews what can be learned about the physics of the local displacement process from augmented hydrodynamic continuum theories that account for microstructural forces. Such theories rely on the concept of disjoining pressure and are particularly well suited to describe the formation and fate of films thinner than 0.1 μm. In dynamic wetting, the role of thin films is most prominent in two limiting regimes: at low rates of spontaneous spreading, at which completely wetting liquids form precursor films that migrate ahead of the macroscopic front (Sec. III.B.1), and at high rates of forced wetting, at which a film of the receding phase may be entrained before the onset of catastrophic wetting failure (Sec. III.B.2).

Section III.C finally summarizes the current understanding of how dynamic molecular events may combine to advance a fluid/fluid interface across a solid surface. At present, such an understanding must be gleaned from idealized models that treat the dynamics in the immediate neighborhood of the three-phase juncture as a molecular-rate process (Sec. III.C.1), or confine a limited number of molecular particles in a small box and perform a computer simulation of their individual motions (Sec. III.C.2).

II. EXPERIMENTAL EVIDENCE

A. Apparent Dynamic Contact Angles in Forced Wetting

1. Experimental Methods

Most attempts to understand and quantify forced wetting have focused on measuring apparent dynamic contact angles. In a commonly used configuration, a liquid front is pumped through a capillary with a circular [26–34] (Fig. 2a) or square cross section [35]. In a variant, the meniscus advances between two narrowly spaced plates in either parallel [25,36,37] or radial flow [38]. These arrangements, referred to as *capillary displacement*, are ideally suited to probe the fundamental mechanisms of dynamic wetting because gravity and inertial effects are insignificant [6]. On the other hand, matching solid materials, viscosity ranges, and speeds with those of practical interest is often difficult and sometimes impossible. Many practical applications are mimicked more easily in the *plunge-tank* configuration in which a solid substrate is immersed into an open bath of liquid. The solid can be a Wilhelmy plate [39–42] (Fig. 2b), a fiber or rod [40,43–45], or a continuous strand of plastic film [16,46–51] (Fig. 2c). The last arrangement, often called a *plunging-tape* configuration, has become a "workhorse" among coating engineers trying to assess how measured values of θ_D relate to "coatability" and maximum speed of wetting. Other configurations include rotating cylinders partially immersed in a pool of liquid [52–54] and even actual coating devices [46,55–57]. Zvan et al. [58] recently demonstrated that coating from the tip of a syringe needle (Fig. 2d) is particularly advantageous because it yields a crisp silhouette of the meniscus shape and requires only small samples of coating solutions, yet replicates essential aspects of the dynamics in industrial coating processes.

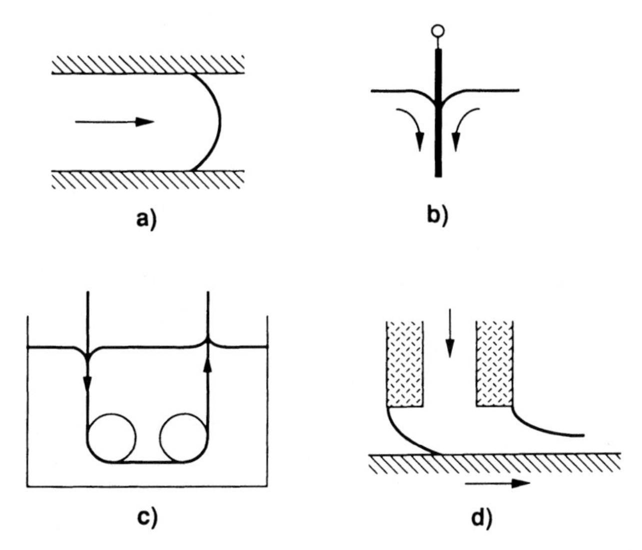

FIG. 2 Selected configurations for measuring apparent dynamic contact angles in forced wetting: (a) capillary displacement; (b) Wilhelmy plate; (c) plunge tank; (d) syringe-needle extrusion coating.

Many studies of forced wetting have relied on viewing the three-phase region through a low-power microscope. At typical magnifications of 20× to 50×, the advancing fluid/fluid interface appears to intersect the solid surface at a well-defined angle—akin to the static contact angle of a sessile drop on a partially wetted solid. Because the interface may be distorted close to the solid, each technique for measuring dynamic contact angles defines a particular version of an apparent angle. Most authors have shied away from specifying a distance from the solid at which the angles reported represent the slope of the interface. To emphasize the distinction between different apparent angles, the present review ascribes a second subscript to the symbol θ_D. Dynamic contact angles are often measured with a protractor [26,43,46,47,49–51,57,59] or a digitizer-microcomputer system [53,59], either from photos or more recently also video images [16,42,59] (angles measured in this manner are denoted as $\theta_{D,T}$). Apparent contact angles have also been read directly through a microscope with a goniometer eyepiece [16,59] ($\theta_{D,R}$). In capillary displacement, an apparent contact angle is usually calculated from the apex height of the meniscus, which is assumed to remain quasistatic [25,27–29,31,34,36,37] ($\theta_{D,M}$). Other techniques include force measurements [39,51,60–62], which are converted into angles via the assumption that the contact angle right at the solid varies with speed and viscous stresses are negligible ($\theta_{D,F}$); pressure drop measurements [26,30,32,33], which are translated into angles with West's formula ($\theta_{D,\Delta P}$); adjustment of the depth of immersion of a rotating cylinder [52] or the angle of immersion of a tape [49] until the interface appears flat ($\theta_{D,\alpha}$); and direct measurement of the meniscus profile with laser-Doppler velocimetry [56] ($\theta_{D,\mathrm{LDV}}$). Standard deviations as small as $\pm 1°$ or $\pm 2°$ have been claimed for θ_D [16,37,43,44,49]. Such repeatability appears overstated since, even for static contact angles, it would require experimental techniques more elaborate [63] than are commonly used for dynamic measurements.

2. Relevant Parameters

Even though a plethora of experimental data has been published, a casual survey of the literature reveals a limited consensus as to which material properties, flow parameters, and geometric variables influence apparent dynamic contact angles. The only observation universally accepted is that θ_D increases monotonically with rising displacement speed U, and that θ_D increases more rapidly the more viscous the liquid is [43,64]. Without a consolidated point of view, apparent dynamic contact angles must be presumed to depend on a long list of variables

$$\theta_D = f(\theta_0; Ca, We, Bo; \lambda_\eta, \lambda_\rho; L_i/L, \ldots; \varphi, \psi, \chi, \ldots; N_A, \ldots) \tag{1}$$

Here, θ_0 is the static contact angle that is typically reported along with θ_D data, $Ca \equiv \eta_1 U/\gamma_{12}$ is the capillary number, $Bo \equiv \rho_1 g L^2/\gamma_{12}$ the Bond number, and $We \equiv \rho_1 U^2 L/\gamma_{12}$ the Weber number, where γ_{12} is the interfacial tension, η_1 the liquid viscosity, ρ_1 the liquid density, g the gravitational acceleration, and L a characteristic macroscopic length scale of the flow (sometimes, the so-called *capillary length* $L_\gamma \equiv \sqrt{\gamma_{12}/\rho_1 g}$ is appropriate). The dimensionless groups Ca, Bo, and We compare, respectively, the importance of viscous stresses, gravity, and inertial effects to that of surface tension. The viscosity ratio $\lambda_\eta \equiv \eta_2/\eta_1$ and density ratio $\lambda_\rho \equiv \rho_2/\rho_1$ compare the fluid properties of the displaced phase (2) to those of the advancing liquid (1). The ratio L_i/L is a short-hand notation for various geometry effects, where $L_i (i = 1, 2, \ldots)$ represents relevant length scales other than L. The symbols φ, ψ, and χ denote dimensionless measures for surface roughness, porosity, and electrostatic surface charge. Finally, N_A is a generic representative for various dimensionless groups that compare fluid/solid interaction forces with hydrodynamic forces.

Experimental systems studied in the literature were all too often nonideal systems chosen primarily for their industrial significance. To determine the predominant parameters among those listed in Eq. (1)—and thereby identify the key mechanisms controlling the dynamic increase of θ_D with U—a judicious distinction between liquid/solid systems with *complete wetting* ($S \geq 0$ and $\theta_E = 0$) and those with *partial wetting* ($S < 0$ or $\theta_E > 0$) is extremely helpful [6]. Here, $S \equiv \gamma_{S2} - \gamma_{S1} - \gamma_{12}$ is the so-called *spreading coefficient*, γ_{S1} and γ_{S2} are the solid/fluid interfacial tensions, and θ_E is the *equilibrium contact angle* in Young's equation, that is, $\gamma_{S2} - \gamma_{S1} - \gamma_{12} \cos \theta_E = 0$. For liquids that completely wet a given solid, the static angle θ_0 is for all practical purposes equal to the equilibrium angle θ_E, that is, $\theta_0 \approx \theta_E = 0$. For liquid/solid pairs with partial static wetting, the appropriate choice of θ_0 is more intricate and is addressed in Sec. II.A.4 below.

3. Liquid/Solid Systems with Complete Equilibrium Wetting

For nonvolatile liquids that completely wet the solid when permitted to spread spontaneously, the dynamics of forced wetting follow a nearly universal behavior. Hoffman [29] was the first to postulate that for $\theta_E \approx 0$, the apparent dynamic contact angle depends solely on capillary number $Ca \equiv \eta_1 U/\gamma_{12}$

$$\theta_D = f_{\text{Hoff}}(Ca) \tag{2}$$

Hoffman deduced the universal function (2) from a systematic study of silicone oils displacing air in glass capillaries. Earlier, Inverarity [43] noticed that data from experiments with fibers plunging into different organic liquids would condense onto a single master curve when plotted as cos $\theta_{D,T}$ versus log $\eta_1 U$. Hoffman's [29] experiments covered a wide range of capillary numbers ($4 \times 10^{-5} < Ca < 36$), allowing the contact angle to vary from a few degrees to $\theta_{D,M} \approx 180°$. Inertia effects were negligible and the influence of gravity was small ($We < 1.6 \times 10^{-5}$, $Bo \approx 0.5$). Figure 3 shows that Hoffman's data, as well as more recent data by others [31,37], fall close to a universal master curve.

Hoffman [29] did not provide an explicit mathematical form for the function $f_{\text{Hoff}}(Ca)$. The solid line in Fig. 3 is calculated from

$$\theta_D = f_{\text{Hoff}}(Ca) = \cos^{-1}\left\{1 - 2\tanh\left[5.16\left(\frac{Ca}{1 + 1.31\,Ca^{0.99}}\right)^{0.706}\right]\right\} \tag{3}$$

Equation (3) is similar to a correlation proposed by Jiang et al. [65] [see Eq. (9) below], but is based exclusively on data points for completely wetting systems with high liquid viscosities and modified to improve the fit at large Ca for such systems. When $Ca < O(0.1)$, Eq. (3) reduces to

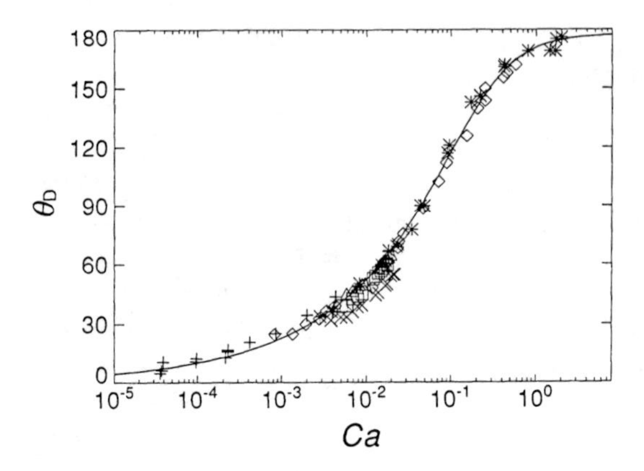

FIG. 3 Apparent dynamic contact angles for silicone fluids displacing air in capillary tubes (+ η_l = 0.958 Pa·s, ∅ = 1.955 mm [29]; * 98.8 Pa·s, ∅ = 1.955 mm [29]; ◇ 5.72 Pa·s, ∅ = 2 mm [31]) and between glass plates (△ 0.97 Pa·s, gap = 1.2 mm; □ 0.97 Pa·s, gap = 0.7 mm; × 0.97 Pa·s, gap = 0.1 mm [37]). The solid line is calculated from Eq. (3).

$$\theta_D = 4.54\, Ca^{0.353} \quad (\theta_D \text{ in radians}) \tag{4}$$

The empirical form (4) is nearly equivalent to the well-known correlation

$$\theta_D^3 \approx c_T Ca \tag{5}$$

Equation (5) is commonly attributed to Tanner [66] because he is considered to have been the first to derive the $\theta_D \sim Ca^{1/3}$ power law from hydrodynamic theory without presuming a prewet surface and to verify the result with experiments [6]. Voinov [67], however, had obtained the same result earlier from a hydrodynamic analysis that, like Tanner's [66], excluded the immediate vicinity of the wetting line (see Sec. III.A.4). Voinov [68] also recognized that the form (5) fits Hoffman's data. For that reason, the present review refers to Eq. (5) as the *Hoffman–Voinov–Tanner law*. Figure 4 shows that Eq. (5) applies not only when $\theta_D < 10°$, or $Ca \ll 1$, as is often suggested in the literature [6,9], but competently describes the apparent dynamic contact angles of completely wetting liquid/solid systems up to $\theta_D \leq 135°$, or equivalently $Ca \leq O(0.1)$.

At low and moderate capillary numbers, that is, $Ca \leq O(0.1)$, apparent dynamic contact angles are independent of the flow configuration in which they are measured. Figure 5, as well as Fig. 4, show that dipping Wilhelmy plates into silicone oils [42,61], plunging polyester tapes into a bath of silicone oil [59], and also extrusion coating of PET films with several

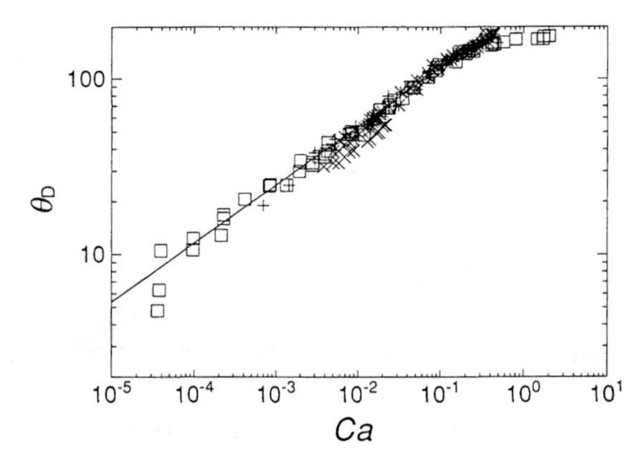

FIG. 4 Validation of Hoffman–Voinov–Tanner law (5) for completely wetting liquid/solid systems (the data are the same as in Figs. 3 and 5): □ capillary displacement [29,31]; × displacement between parallel glass plates [37]; + Wilhelmy plates [42,61]; ◇ plunging polyester tape [59]; * extrusion coating out of syringe needle onto polyester tape [59].

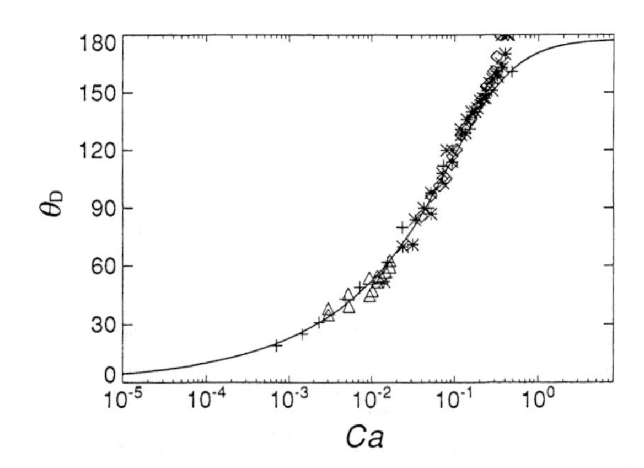

FIG. 5 Apparent dynamic contact angles in presence of gravity and inertia effects for liquid/solid systems with complete wetting: (a) Wilhelmy plate dipping in silicone fluids (+ untreated polystyrene and oxidized polystyrene [42], △ high- and low-density polyethylene [61]); (b) polyester tape plunging into silicone fluid (◇ [59]); (c) extrusion coating out of syringe needle onto polyester tape (* oleic acid, isobutyl alcohol, and silicone fluids [59]). The solid line is calculated from Eq. (3).

liquids by the syringe-needle method [59] yield the same curve for $\theta_D = f_{\mathrm{Hoff}}(Ca)$ as pushing a front of silicone oil through a glass capillary. Further evidence that the macroscopic flow geometry is insignificant comes from the insensitivity of $\theta_{D,T}$ data to the angle of entry of plunging tapes [49,50,54].

The agreement of dynamic contact angle data for different flow geometries, which has been asserted explicitly or stipulated implicitly by other investigators before [51,61,65,69–71], suggests that macroscopic flow effects are of minor consequence. Despite suggestions to the contrary [3,6,41,65], gravity has no detectable effect on the relationship between θ_D and Ca even when it controls the meniscus shape away from the wetting line, as in Wilhelmy-plate and plunging-tape systems. Likewise, inertia effects can be ignored, at least up to $Ca \leq O(0.1)$. Inertia is negligible in Wilhelmy-plate systems because immersion speeds are low and high viscosities are needed to reach large capillary numbers. In plunge tanks operating with continuous tapes and low-viscosity liquids, in contrast, inertia can build substantially as the immersion speed U rises. Figures 4 and 5 include a data set for a PET tape plunging into a 5 mPa·s silicone oil [59]. The data fall right on the universal curve up to $Ca \leq 0.2$, even though the Reynolds number $Re_\gamma \equiv \rho_1 U L_\gamma / \eta_1$ based on the capillary length $L_\gamma \equiv \sqrt{\gamma_{12}/\rho_1 g}$ reaches about 250. In the syringe-needle configuration, Reynolds numbers based on the coating gap are lower, but readily exceed $Re > O(1)$ for low-viscosity liquids with no apparent effect on θ_D.

Apparent dynamic contact angles are also remarkably insensitive to the technique that is used to measure θ_D. Figures 3 and 5 include contact angles inferred from a quasistatic meniscus shape ($\theta_{D,M}$), the tangent to the interface on photo-micrographs ($\theta_{D,T}$), direct visual observation through a goniometer eyepiece ($\theta_{D,R}$), and force measurements ($\theta_{D,F}$), yet no systematic differences are discernible. Attempts to directly compare different measurement techniques in the same flow configuration [26,37,62,70,72,73] also indicate that a lack of resolution of the meniscus shape close to the solid accounts only for small differences in apparent dynamic contact-angle data. Encouraged by this finding, some early investigators [26,72] jumped to the conclusion that the angles measured are the same as those at submicroscopic length scales. Evidence is mounting, however, that the slope of the interface close to the solid can be substantially smaller than the slope visible to the naked eye or measured with a low-power microscope. Most extensive evidence comes from systems that form precursor films at very low spreading rates (see Sec. II.D.2). However, concave meniscus bending has also been documented at elevated capillary numbers ($10^{-4} < Ca < 10^{-2}$), at which the imposed wetting speed most likely exceeds the rate of spontaneous precursor spreading [45] (see also Sec. III.A.5). At high capillary numbers when θ_D is not too far from 180°, the interface might approach the solid with an angle of inclination close to 180° [8,17,56,74], but direct experimental evidence for convex meniscus bending in the immediate vicinity of the three-phase juncture has not been produced to date. Mues et al. [56] measured free-surface profiles close to a moving wetting line by means of laser-Doppler-velocimetry (LDV). They argued that within a distance of 10 μm from the solid the free surface bends sharply in the direction of substrate motion to meet the solid at an angle considerably larger than the apparent angle θ_D. The LDV setup, however, did not resolve the local bending, for the theoretical size of the measuring volume was 15.3 μm and the practical value probably larger. On the length scale that was resolved, the LDV data indicate slight concave bending in the direction opposite to the substrate motion, as measured by Dussan V. et al. [45].

The data in Figs. 3 and 5 suggest furthermore that universal dynamic wetting behavior on completely wettable surfaces holds, regardless of the particular liquid/solid pair being investigated. For silicone fluids, measurements on glass yield the same θ_D versus Ca data as on lower-energy polymeric surfaces [42,59,61]. Switching the solid from an untreated polystyrene to an oxidized polystyrene [42] or a high-density polyethylene to low-density polyethylene [61] does not alter the apparent contact angle either. Changes in surface roughness of a glass capillary tube are inconsequential as well [34]. In addition, not only silicone fluids, but also oleic acid and isobutyl-alcohol follow Hoffman's [29] universal behavior $\theta_D = f_{\mathrm{Hoff}}(Ca)$ when coated on PET [59]. Altogether, forced-wetting data available for systems with $S > 0$ suggest that dynamic contact angles at optically accessible length scales are rather insensitive to changes in spreading coefficient S or other parameters that quantify specific fluid/solid interactions.

The preeminent role of capillary number, $Ca \equiv \eta_1 U/\gamma_{12}$, in correlating apparent dynamic contact angles for completely wetting liquid/solid systems supports the hypothesis that the primary mechanism for the rise of θ_D with U is a purely hydrodynamic competition between viscous stresses and capillary pressure. This hypothesis seems to contradict the notion advocated above that θ_D is insensitive to the macroscopic flow. A possible explanation for the apparent paradox is given in Sec. III.A.4 below. It supposes that the pertinent length scale at which the visco-capillary stress balance sets the apparent contact angle is much smaller than the macroscopic length scale where the particular flow geometry is important and inertia and gravity may influence the hydrodynamics, yet larger than the submicroscopic region where microstructural forces and possibly even molecular events dominate the local wetting dynamics. The explanation is consistent with the observation that liquid/air interfaces often remain quasistatic at macroscopic length scales [29,47]. On the other hand, the explanation also rationalizes why specific material effects often play a secondary role.

Even though most data points in Figs. 3, 4, and 5 fall close to the universal curve $\theta_D = f_{\text{Hoff}}(Ca)$, several systematic deviations can be identified that, thus far, have been brushed aside. A first deviation may arise at very low capillary numbers ($Ca \leq 10^{-4}$) at which, unfortunately, accurate θ_D data for forced wetting are rare. The data available exhibit significant scatter [29,61] and appear to follow a power law slightly different from $\theta_D \sim Ca^{1/3}$ [61]. A hypothetical explanation might be the influence of specific material interactions that, at low displacement speeds, would be most able to successfully compete with hydrodynamic forces. Schwartz and Tejada [44] reported a peculiar low-speed regime for fibers plunging into completely wetting liquids. Their observations are reminiscent of low-speed anomalies commonly seen for systems with partial wetting (see Sec. II.A.4) and may be related to the substantial roughening of the fiber surfaces in Schwartz and Tejada's experiments. Other material effects, such as those arising from surfactants or contaminants, volatile components or impurities, non-Newtonian rheology and soluble layers on the solid, might also give rise to measurable deviations from Hoffman's [29] universal θ_D versus Ca curve, but need to be documented more systematically for forced wetting with well-characterized liquid/solid systems and $\theta_0 = 0°$.

A second systematic deviation might arise from a weak influence of the characteristic size L of the global flow. It is quite distinct from the negligible influence of the geometric configuration of the global flow at low and moderate capillary numbers ($Ca \leq 0.1$). An influence of L has been documented most unequivocally for capillary displacement between narrowly spaced glass plates: $\theta_{D,M}$ data of Ngan and Dussan V. [37] agree with Hoffman's [29] data at the widest gap between the plates (1.2 mm), but are 10–15° lower at the narrowest gap (0.1 mm). A more recent study by Ngan and Dussan [25] reveals a similar *size effect* but, for reasons that are yet unclear, $\theta_{D,M}$ is less sensitive to U than in the original study [37]. A size effect has also been reported for capillary displacement in tubes [27,30]. Other data for capillary displacement, however, reveal no clear-cut size effect [28]. Changing the diameter L_D of a plunging fiber also seems to have a negligible influence on apparent dynamic contact angles [43], but it is not clear a priori whether L_D is the most important length scale. Most plunge tanks measure at least several centimeters between the plunging tape or fiber and the confining walls, but the pertinent length scale is probably the capillary length $L_\gamma \equiv \sqrt{\gamma_{12}/\rho_1 g}$. For liquid/air interfaces, the capillary length is $L_\gamma = O(2\text{ mm})$, which is comparable to the characteristic diameters in most capillary displacement experiments. A size effect may indeed have gone unnoticed in most dynamic wetting studies because the characteristic size of laboratory configurations is typically $L = O(1 \text{ mm})$. In industrial coating configurations, on the other hand, the meniscus is often forced to curve over distances much smaller than 1 mm. A size effect might therefore account for some of the differences between the apparent dynamic wetting behavior observed in industrial coating practice and that seen in idealized laboratory configurations [75].

If a size effect were generic for most liquid/solid systems and flow configurations—which needs be established with further experiments—it would have significant implications. In particular, a size effect would establish that apparent dynamic contact angles are *not* material functions that depend solely on fluid properties, specific fluid/solid interactions, and displacement speed, but are also influenced by the macroscopic flow [25]. If a local contact angle could be identified that is a material function—which remains a conjecture—a size effect alone would be sufficient proof that measured, apparent angles θ_D are different from contact angles close to the solid [25,45]. A generic size effect would also disaffirm the notion of a "universal" dynamic wetting behavior. Size effects reported to date, however, are only minor aberrations from Hoffman's [29] curve $\theta_D = f_{\text{Hoff}}(Ca)$. Even in the study of Ngan and Dussan [37], an order-of-magnitude reduction in the characteristic size L shifts the data by less than a factor of 2 in Ca. By comparison, more than four orders of magnitude in Ca are required to raise the apparent dynamic contact angle from values near $\theta_0 = 0°$ to $\theta_D \approx 180°$ (see Fig. 3).

A third systematic deviation from the universal curve $\theta_D = f_{\text{Hoff}}(Ca)$ sets in above $Ca \geq O(0.1)$, or equivalently $\theta_D \geq 135°$. It is most pronounced for low-viscosity liquids for which θ_D rapidly increases toward 180° (see Fig. 5). For high-viscosity liquids, in contrast, θ_D data approach 180° asymptotically, as in Hoffman's [29] original report (see Fig. 3). Two-phase-flow theory [24,76] suggests that the difference in dynamic wetting behavior between low- and high-viscosity liquids arises primarily from viscous effects in the receding gas phase (see Sec. III.A.9). Supporting evidence comes from experiments in which the displaced fluid is a liquid rather than a gas. In such experiments, θ_D rises much more rapidly with increasing U than in experiments with liquid against air [31,33,34,49,71]. The influence of the receding phase is commonly overlooked when that phase is air, but can affect the flow in the advancing liquid quite strongly when θ_D is near 180° and the liquid viscosity is low. If airflow effects are important, air escape into rough or porous surfaces may measurably alter the apparent dynamic wetting behavior [17]. In addition, once a narrow air wedge forms between the liquid and solid close to $\theta_D \to 180°$, long-range interactions forces, electrostatic attraction between the liquid and solid, and even macroscopic hydrodynamic forces may help exclude the airflow and thereby assist dynamic wetting. Volatility of the advancing liquid may also play an important role [77], but its impact has not been quantified systematically.

Above $Ca \geq O(0.1)$ when the advancing meniscus can deviate substantially from a quasistatic shape, the macroscopic flow may also have a stronger influence than is evident in Figs. 3 and 5. Global flow effects are suspected to be particularly important in industrial coating processes that often operate near $150° < \theta_D \leq 180°$. Practical experience suggests that the macroscopic wetting dynamics depend on the coating configuration used [75] and can be influenced by changes in flow parameters other than U, η_1, and γ_{12}. However, no systematic data have been published that unequivocally establish differences in the θ_D versus Ca curve for the same liquid/solid combination from one coating configuration to another, especially not for well characterized, completely wetting liquid/solid systems. The delayed onset of catastrophic air entrainment in curtain coating as compared to most other industrial coating configurations (see Sec. II.B) provides only indirect evidence that inertia and other characteristics of the macroscopic flow field might also affect θ_D during successful coating [78]. Recent measurements with a syringe-needle probe indicate that widening the coating gap or starving the premetered flow can cause a sudden, very dramatic rise of θ_D toward 180° [58]. The rapid rise of θ_D, however, appears to be related to a hydrodynamic instability that breaks up the coating bead, rather than a dynamic wetting phenomenon. At higher flow rates and smaller gaps at which the liquid successfully bridges the gap between the needle tip and moving substrate, the apparent contact angle becomes insensitive to flow rate. A weak influence of the gap remains and may be related to the size effect discussed above.

4. Liquid/Solid Systems with Partial Equilibrium Wetting

The dynamics of liquids advancing over partially wettable solid surfaces are not as "universal" as those of completely wetting systems. A major complication arises in choosing an appropriate value of the static contact angle θ_0. The equilibrium angle θ_E is rarely seen in experiments because most surfaces are rough and inhomogeneous in chemical composition, at least on submicroscopic scales. Technical surfaces may also be contaminated with impurities and carry nonuniform electrostatic charges and surface polarization. Nonideal surfaces are susceptible to *contact-angle hysteresis* between a recently advanced angle θ_A and recently receded angle θ_R, that is, $\theta_A \geq \theta_E \geq \theta_R$ (see Chap. 1 in this volume). Hysteresis arises primarily from the many metastable positions the wetting line can assume [6,79], but may in part also be a dynamic phenomenon because, over the time scale allowed for practical experiments, equilibrium might not be reached [16]. For some systems, the appropriate value of θ_0 may be close to the recently advanced angle θ_A. However, the time scale of static contact-angle measurements is rarely specified in published studies of dynamic wetting, and the angle θ_0 reported along with dynamic data may, in fact, be larger than θ_A. In addition, many systems with partial wetting exhibit peculiar low-speed behavior, and a *pseudostatic* angle $\theta_0^* > \theta_0$ that results from extrapolating θ_D data at moderate and high displacement speeds to $U \to 0$ may be more successful than $\theta_0 \approx \theta_A$ in correlating dynamic contact-angle data, as discussed in more detail below (see Fig. 7b).

Disregarding contact-angle hysteresis and other low-speed effects, Hoffman [29] suggested that the universal behavior (2) remains valid for partially wetting systems if the nonzero static contact angle θ_0 is absorbed in a shift factor $f_{\text{Hoff}}^{-1}(\theta_0)$, that is,

$$\theta_D = f_{\text{Hoff}}[Ca + f_{\text{Hoff}}^{-1}(\theta_0)] \tag{6a}$$

or

$$g_{\text{Hoff}}(\theta_D) - g_{\text{Hoff}}(\theta_0) = Ca \tag{6b}$$

Here, $g_{\text{Hoff}}(\theta) \equiv f_{\text{Hoff}}^{-1}(\theta)$ is the inverse function of f_{Hoff}. For $\theta_D \leq 135°$, $g_{\text{Hoff}}(\theta)$ is closely approximated by θ^3/c_T [see Eq. (5)], and the universal form (6b) simplifies to

$$\theta_D^3 - \theta_0^3 \cong c_T Ca \tag{7}$$

Hoffman's postulate implies that the apparent dynamic contact angle θ_D remains close to the static angle θ_0 at low Ca, but rapidly loses its sensitivity to θ_0 when the capillary number exceeds $Ca_0 \equiv g_{\text{Hoff}}(\theta_0)$. For instance, if $\theta_0 = 90°$, then Ca_0 is about 0.04; if $\theta_0 = 45°$, on the other hand, Ca_0 is only 0.005. Figure 6a confirms that Eq. (6) satisfactorily captures the dynamic wetting behavior of selected liquid/solid systems with $\theta_0 > 0°$. Included in the figure are Hoffman's original data for capillary displacement [29], as well as data by others for configurations in which gravity and inertia affect the macroscopic flow away from the wetting line, namely, Wilhelmy plates [42,61], plunging tapes [49], and syringe-needle extrusion coaters [59]. Liquid viscosities vary from 2 mPa·s to 4.88 Pa·s, and static contact angles fall in the range $12° \leq \theta_0 \leq 69°$.

Even though the data for $\theta_0 > 0°$ in Fig. 6a gather near the universal curve (6), they exhibit considerably more scatter than the data for $\theta_0 = 0°$ in Figs. 3 and 5. Individual data points deviate from Eq. (6) by as much as 20°. The scatter makes it difficult to ascertain whether the deviations are systematic or random. At small capillary numbers, Eq. (6) is even

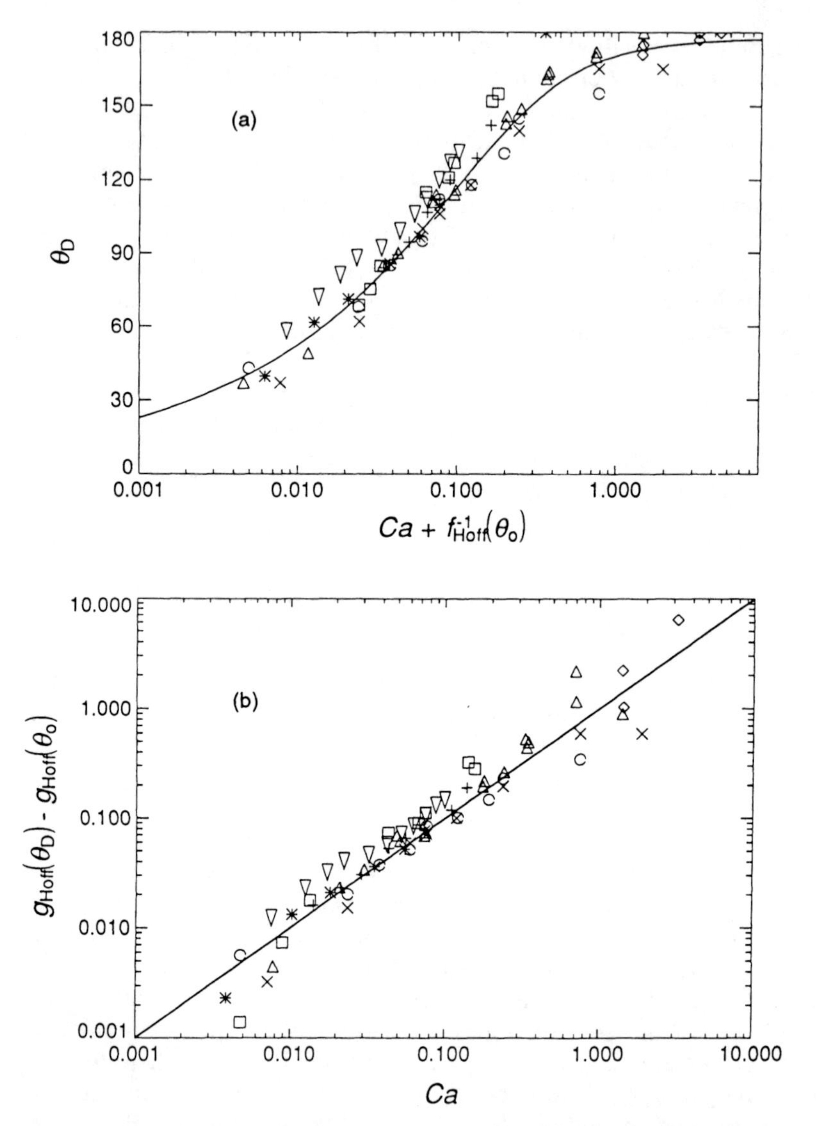

FIG. 6 Apparent dynamic contact angles for liquid/solid combinations with partial equilibrium wetting: capillary displacement [29] (□ Santicizer 404, 11.2 Pa·s, $\theta_0 = 67°$; △ Ashland Chemical Admex 760, 109.3 Pa·s, $\theta_0 = 69°$; ◇ silicone fluid, Pa·s, $\theta_0 = 12°$); Wilhelmy plates in silicone fluids [42] (△ 0.995 Pa·s, PTFE with $\theta_0 = 31°$; ○ 4.88 Pa·s, untreated polystyrene with $\theta_0 = 12°$; × 4.88 Pa·s, PTFE with $\theta_0 = 19°$); polyester tapes in plunge tank [49] (+ corn oil, 59 mPa·s, $\theta_0 = 68°$; * isopropanol, 2.3 mPa·s, $\theta_0 = 31°$); and extrusion coating with syringe needle [59] (▽ di-iodomethane, 2.5 mPa·s, $\theta_0 = 22°$).

more difficult to validate from Fig. 6a because the data collapse near $Ca_0 \equiv f^{-1}_{\mathrm{Hoff}}(\theta_0)$. Figure 6b expands the low-speed range in a plot of $g_{\mathrm{Hoff}}(\theta_D) - g_{\mathrm{Hoff}}(\theta_0)$ vs. Ca that should yield a straight line if Eq. (6) were satisfied [the function g_{Hoff} is calculated from Eq. (3)]. Even though most data points do not fall right on the line, the generalized correlation (6) adequately approximates the data sets selected in Fig. 6, especially over the range $0.01 < Ca < 0.3$.

Many liquid/solid systems with partial static wetting fail altogether to follow the generalized correlation (6). Figure 7a, for example, reveals vast discrepancies among θ_D data by different investigators who all measured dynamic contact angles on polyester tapes in plunge tanks filled with water/glycerine mixtures of comparable viscosity ($0.042 < \eta_l < 0.058$ Pa · s). Because of the similarity of the experiments, hydrodynamic effects such as inertia and gravity cannot account for the differences. The data by Burley and Kennedy [47] are particularly noteworthy, for they indicate that θ_D rises dramatically at extremely low displacement speeds ($Ca < 5 \cdot 10^{-3}$).

Evidently, for liquid/solid systems with incomplete equilibrium wetting, the capillary number Ca and static contact angle θ_0 are often insufficient to account for all the mechanisms that give rise to dynamic variations of the visible contact angle. The systems selected in Fig. 6 consist mostly of simple, nonpolar liquids whose interactions with the solids are probably dominated by dispersive van der Waals forces. On the other hand, the water/glycerine mixtures reported in Fig. 7a, like other ionic or polar species, are susceptible to induced dipolar interactions and electrostatic charge effects. The poor correlation between the data for water/glycerine and Eq. (6) suggests that specific liquid/solid interactions may alter the apparent dynamic wetting behavior and account for the discrepancies in Fig. 7a. However, Seebergh and Berg [61] recently investigated the dynamic wetting of numerous liquid/solid combinations at low capillary numbers and were unable to identify statistically significant differences that set apart systems with only dispersive interactions from those with acid–base interactions. They concluded that the static contact angle alone is enough to account for the impact of specific liquid/solid interactions on the apparent dynamic wetting behavior. The low-speed data, however, exhibited significant scatter, and Seebergh and Berg's [61] conclusion ought to be substantiated further for liquid/solid systems with smooth surfaces for which the interaction parameters can be quantified.

Another possible explanation for the failure of Eq. (6) to successfully describe θ_D versus Ca data for all liquid/solid systems is the impact of peculiar low-speed behavior, which is quite common for systems with partial static wetting. For water on polyethylene and siliconed glass sur-

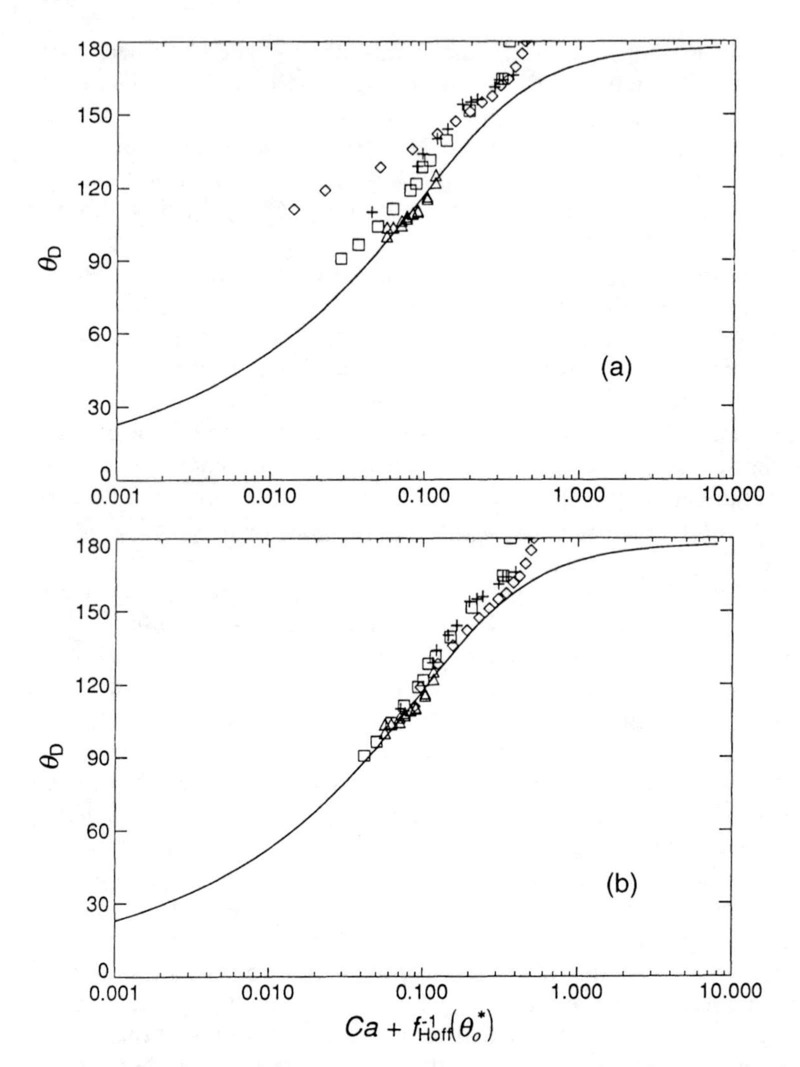

FIG. 7 Apparent dynamic contact angles ($\theta_{D,T}$) for water/glycerine mixtures in plunge-tank experiments: ◇ 45.6 mPa·s, Mylar® polyester tape with $\theta_0 = 45°$, $\theta_0^* = 110°$ [47]; + 42 mPa·s, polyethylene theraphtalate with $\theta_0 = 55°$, $\theta_0^* = 85°$ [59]; □ 58 mPa·s, polyethylene theraphtalate with $\theta_0 \approx 66°$, $\theta_0^* = 80°$ [16]; △ 49.2 mPa·s, polyethylene with $\theta_0 = 93°$, $\theta_0^* = 93°$ [51].

faces, for instance, a finite albeit very small displacement speed $U_{\min} > 0$ has been reported before the apparent contact angle was seen to exceed its static value [38]. For several systems, a transition has been noted between distinct low- and high-speed wetting regimes, suggesting different mechanisms by which the wetting line advances [16,32,34,44,61]. In the low-speed regime, the observed values of θ_D are typically well above θ_0, but the rate of increase of θ_D with U is measurably slower than in the high-speed regime [32,44,61]. For other systems, especially polar liquids such as water, glycerine, or polyethylene glycol displacing air from low-energy surfaces, θ_D rises initially with increasing displacement speed at extremely low capillary numbers, but then remains nearly constant over a finite speed interval before increasing again at higher, yet still small values of Ca [16,38,42,43,52,80,81]. Similar observations have also been reported for the displacement of organic liquids by water [33]. The transition from an initial rise of θ_D at very low Ca to an intermediate plateau of constant θ_D may be accompanied by intense fluctuations in instantaneous apparent contact angle that arise from spasmodic jumps of the advancing meniscus [16]. Unsteady, jerky advance of the apparent wetting line, often referred to as *stick-slip* [2], is very typical for systems with incomplete static wetting. It causes significant scatter and poor reproducibility of θ_D data for $\theta_0 > 0$ and seems to be most severe on rough and heterogeneous surfaces [16,38,60,61,82–85]. Roughness, heterogeneity, contaminants, electrical surface charges, internal polarization of the surface zone of nonconductive substrates, and possibly other surface characteristics are notably unmeasured and uncontrolled in most published studies of dynamic wetting, yet are probably a major cause for stick-slip jumps at low speeds of forced wetting due to temporary pinning of the wetting line to surface irregularities [85]. Stick-slip has also been attributed to the adsorption of surface-active molecules ahead of the wetting line [60,84,86] and to an unsteady transition of molecule displacement at alternate molecular planes [16]. Toward higher speeds, stick-slip is often perceived to diminish and disappear [16], but might merely become noise that is no longer detectable. Some authors report that the wetting line undergoes fitful motion even at high coating speeds [53].

Even though the discrepancies noted in Fig. 7a appear to persist at moderately high speeds ($Ca < 0.1$), they could arise primarily from peculiar low-speed phenomena rather than a complete failure of the universal relationship (6) over the entire capillary number range. Figure 7b supports the notion that a distinct low-speed wetting mechanism accounts for much of the discrepancies between Eq. (6) and experimental data for systems with partial static wetting. In Fig. 7b, the shift factor $Ca_0^* \equiv f_{\mathrm{Hoff}}^{-1}(\theta_0^*)$ is based on a modified, *pseudostatic* angle θ_0^*, rather than the static contact angle θ_0 reported in the original publications. Upon empirical adjustment of the angle θ_0^* for each data set, the data points collapse close to the universal curve. For systems that exhibit a plateau with a nearly constant value of θ_D^* over an intermediate range of Ca [16], θ_0^* turns out to be close to θ_D^*. For other systems that exhibit a continuous increase of θ_D at low speeds [32,61], experimental data may agree better with the modified universal form $g_{\mathrm{Hoff}}(\theta_D) - g_{\mathrm{Hoff}}(\theta_0^*) = Ca$ if θ_0^* were itself allowed to depend on speed. The function $\theta_0^* = f(U)$ could, in fact, be inferred from experimental data [33,34]. The procedure for identifying the values of θ_0^* in Fig. 7b is strictly ad hoc. Nevertheless, the apparent success of the pseudostatic angle θ_0^* in absorbing dramatic discrepancies between different data sets supports the conjecture that above $Ca > O(10^{-3})$, a high-speed regime prevails that is controlled by the same viscocapillary mechanisms that presumably dominate the dynamic increase of θ_D with U in the simpler case of complete wetting. Superimposed on this nearly universal dynamic wetting regime, one may hypothesize, is a second regime in which temporary contact-line pinning sometimes distorts θ_D away from its static value θ_0 at extremely low speeds, and related stick-slip motion increases viscous dissipation and thereby raises θ_D to higher values at lower capillary numbers than Eq. (6) suggests. Modern methods of surface analysis, however, need to be brought to bear in carefully controlled experiments to further substantiate the conjecture that drastic deviations from the universal curve (6) arise mostly from anomalous low-speed wetting phenomena rather than other, poorly understood material effects that might persist at high displacement speeds. Such experiments ought to become a major focus of future research into dynamic wetting under conditions of partial equilibrium wetting.

Apart from discrepancies that are peculiar to partially wetting liquid/solid systems, systematic deviations from the universal curve (6) may also arise from the mechanisms that cause deviations from $\theta_D = f_{\mathrm{Hoff}}(Ca)$ for systems with complete wetting (see Sec. II.A.3). Because of the scatter in θ_D data for $\theta_0 > 0°$, however, such deviations are more difficult to identify than for systems with $\theta_0 \approx 0°$. In particular, a size effect has not been documented for systems with partial wetting. At high capillary numbers ($Ca > 0.1$), the data in Figs. 6 and 7 nevertheless substantiate a systematic difference between high-viscosity liquids that approach the limit $\theta_D \rightarrow 180°$ asymptotically and low-viscosity liquids that experience an accelerated rise of θ_D toward 180°. Other careful measurements with water/glycerine mixtures of low and moderate viscosity ($\eta_1 \leq 0.1$ Pa·s) also reveal a steep slope of θ_D vs. U near 180° [16].

At capillary numbers above $Ca \geq O(0.1)$, macroscopic flow effects probably cause further deviations from Hoffman's generalized correlation

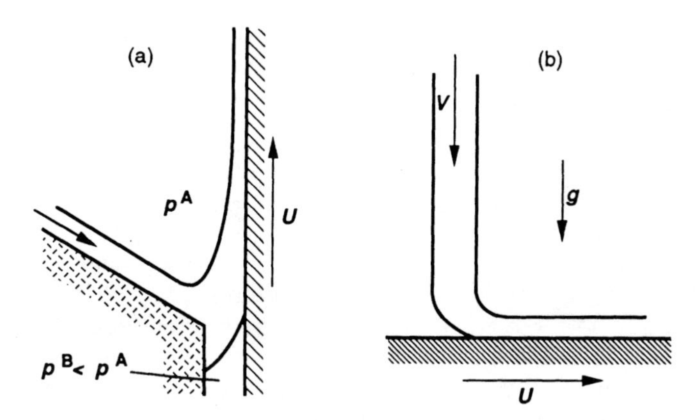

FIG. 8 Schematic of common precision coating configurations: (a) slide coating and (b) curtain coating.

(6). For instance, apparent dynamic contact angles for water/glycerine mixtures extruded out of a syringe needle onto partially wettable PET tapes exhibit similar effects of coating gap and premetered flow rate as for completely wetting silicone oils [58] (see Sec. II.A.3). Chen [57] measured apparent dynamic contact angles in a slide coater and found that Eq. (9) in Sec. II.A.5 below systematically overestimates θ_D. In slide coating (Fig. 8a), the liquid flows as a film down an inclined plate and forms a bridge, commonly referred to as a *coating bead*, where it is transferred across a narrow gap onto a moving substrate. Chen [57] advocated his finding as unequivocal proof that the dynamic wetting behavior in actual coating devices cannot be inferred from plunge-tank experiments or other idealized laboratory configurations. Unfortunately, Chen [57] measured θ_D on a prewetted surface of plated steel. Dynamic contact angles on previously wet surfaces are prone to be lower than those on "dry" substrates [28,87], even when the excess liquid is removed with a scraper blade [53] (see also Sec. III.A.8). Chen [57] furthermore found that, above $Ca > 0.1$, θ_D can be reduced by lowering the subambient pressure, $p^B < p^A$, which is often applied behind coating beads to stabilize the flow (see Fig. 8a; p^A is the ambient pressure). Another study of slide coating at $Ca = 0.16$, in contrast, suggests that the pressure difference $p^A - p^B$ has a negligible influence on θ_D [56].

5. Empirical Correlations

Numerous empirical correlations have been proposed in attempts to provide specific versions of Eq. (1). Many result from standard procedures of dimensionless analysis and express the apparent dynamic contact angle as a simple power relationship of the pertinent dimensionless groups [40,41,46,47,49,50,53]. A representative example is the empirical form

$$\theta_D = 6.24\ Ca^{0.22} N_\gamma^{0.099} \lambda_\eta^{-0.36} \tag{8}$$

by Gutoff and Kendrick [49] in which $N_\gamma \equiv We^2 Bo^{-1} Ca^{-4} = \rho_1 \gamma_{12}^3/(\eta_1^4 g)$ is the so-called property parameter. Correlations like Eq. (8), as well as their dimensional counterparts that typically yield higher correlation coefficients [49], have had some success in unifying data that appear widely scattered in unprocessed form, but fail to shed light on the key mechanisms that control dynamic wetting. The reasons are apparent now (see Sec. II.A.3 and II.A.4): The correlations typically emphasize hydrodynamic forces such as gravity and inertia that appear to be of minor consequence, at least in plunge tanks, yet ignore essential dimensionless groups that account for flow in the displaced phase and specific material effects; most correlations even neglect the static contact angle and completely ignore stick-slip and related low-speed phenomena; the assumed power-law form is inadequate to describe the influence of key parameters, that is, Ca, θ_0 or θ_0^*, and λ_η; the correlations are often based on data that cover only narrow speed ranges; and much of the data stems from plunge-tank experiments with water/glycerine mixtures and polyester substrates, a precarious choice that is particularly susceptible to anomalous low-speed behavior (see Fig. 7).

Other correlations have been aimed at capturing the universal behavior $\theta_D = f(Ca, \theta_0)$ discussed above. For liquid/solid systems with complete wetting, Equation (3) provides a good fit to experimental data for high-viscosity liquids. For low-viscosity liquids, Eq. (3) works well below $Ca \leq O(0.1)$, but should be modified at higher capillary numbers to account for the influence of the ratio $\lambda_\eta \equiv \eta_2/\eta_1$ between the viscosities of the receding gas and the advancing liquid (Foister [71] proposed such a modification for liquid/liquid displacement). Notable correlations for liquid/solid pairs with partial wetting are, apart from Eq. (6), those by Jiang et al. [65]

$$H \equiv \frac{\cos\theta_0 - \cos\theta_D}{\cos\theta_0 + 1} = \tanh(4.96\ Ca^{0.702}) \tag{9}$$

and Bracke et al. [51]

$$H \equiv \frac{\cos\theta_0 - \cos\theta_D}{\cos\theta_0 + 1} = 2\sqrt{Ca} \tag{10}$$

The term $[\cos\theta_0 - \cos\theta_D]$ can be interpreted as an "out-of-balance Young's force" and is often considered a driving force for dynamic wetting [88–95]. Such a force is an essential ingredient in several theoretical developments (see Sec. III.A.4 and III.C.1). It appears also in numerous empirical [32,46,61] as well as semiempirical models for the dynamic variations of apparent contact angles [42,87,96,97]. Many of these models yield results that are special cases of

$$\Delta \cos \theta \equiv \cos \theta_0 - \cos \theta_D = b\varphi(\theta_0)Ca^a \tag{11}$$

Equation (10) is an example of the form (11) with $a = \frac{1}{2}$. For $Ca \leq O(0.1)$, Eq. (9) also specializes to (11) with an exponent of $a = 0.702$. When $(\theta_D - \theta_0)/\theta_0 \ll 1$, the generalized Hoffman–Voinov–Tanner law (7) can be written in the form (11) as well

$$\Delta \cos \theta \equiv \cos \theta_0 - \cos \theta_D \sim Ca \tag{12}$$

A linear increase of $\Delta \cos \theta$ with Ca, however, approximates (7) only for finite static contact angles, that is, $\theta_0 > 0$. For completely wetting systems with $\theta_D \ll \pi/2$, the Hoffman-Voinov-Tanner law (5) is equivalent to

$$\Delta \cos \theta = 1 - \cos \theta_D \sim Ca^{2/3} \tag{13}$$

For $\theta_0 = 0$ and $Ca \ll 1$, Eq. (9) closely approximates (13) and hence agrees with the data in Fig. 4. On the other hand, the empirical correlation (10) reduces to $\theta_D \sim Ca^{1/4}$ and does not fit low-speed data for completely wetting liquid/solid combinations.

The functional form of Eqs. (9) and (10) is attractive because it normalizes dynamic contact-angle data between 0 and 1, no matter what the static angle θ_0 is. The fundamental significance of $H \equiv [\cos \theta_0 - \cos \theta_D]/[\cos \theta_0 + 1]$, however, is not clear. Because of the denominator $[\cos \theta_0 + 1]$, the empirical correlations (9) and (10) are inconsistent with the mathematical form (6) postulated by Hoffman [29] for systems with partial wetting. Above $Ca \geq O(0.01)$, nevertheless, the forms $g_{\text{Hoff}}(\theta_D) - g_{\text{Hoff}}(\theta_0)$ and $H(\theta_D, \theta_0)$ yield correlations that are quite close, as Fig. 9 shows, and so Eqs. (9) and (10) adequately approximate the data selected in Fig. 6. Semiempirical models of the form (11) typically suggest that $\frac{1}{2} < a < \frac{2}{3}$ and also fit experimental data at moderate and high Ca quite well [42,96]. Because of scatter in data for $\theta_0 > 0$ (see Figs. 6 and 7), none of the correlations for partially wetting systems emerges as being clearly superior. Apparent dynamic contact angles calculated from Eqs. (6) and (9) approach the limit $\theta_D \rightarrow 180°$ asymptotically, as data for high-viscosity liquids suggest. Equation (10), in contrast, is based on data from plunge-tank experiments with low-viscosity water/glycerine mixtures [51] and predicts that θ_D increases rapidly and reaches 180° at $Ca = 0.25$.

At low capillary numbers, the differences among various correlations for $\theta_0 > 0$ become significant (see inset in Fig. 9). Moreover, empirical

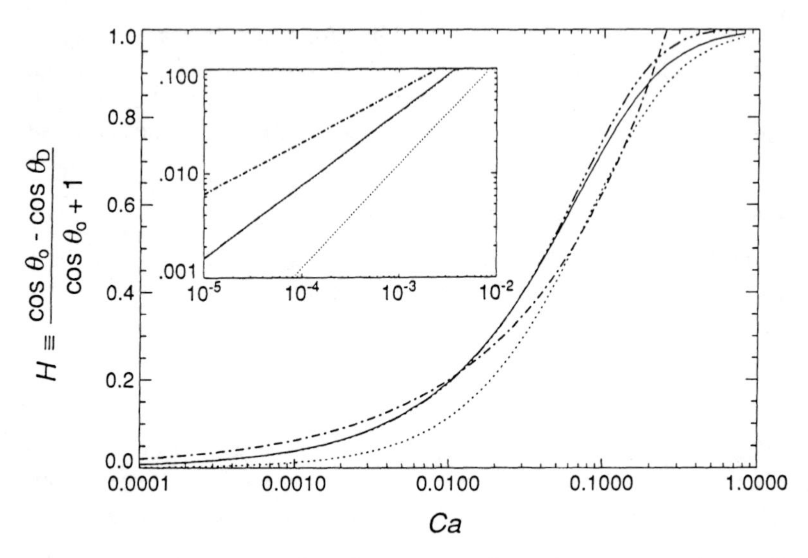

FIG. 9 Empirical correlations for apparent dynamic contact angle as a function of capillary number: —···— Eq. (9); —·— Eq. (10); —— Eq. (3) for $\theta_0 = 0°$; ------ Eqs. (6a) and (3) with $\theta_0 = 90°$. The inset shows an exploded view of H versus Ca at low displacement speeds.

correlations of the form (11) with $a > 0.5$ derived from data at moderate and high Ca overestimate the rate of increase of $\Delta \cos \theta$ with Ca, yet underestimate the observed θ_D values [61]. The universal form (6b), which corresponds to $a = 1$ [see Eq. (12)], performs particularly poorly in describing low-speed data for partially wetting systems. Seebergh and Berg [61] identified a low-speed regime in which a correlation of the form (11) still captures the overall trend of their extensive data for $Ca \ll 1$, but with an exponent of $a = 0.42$ that is distinctly lower than $a = 0.54$, which provided the best fit to their own data at higher speeds. The value $a = 0.42$ is remarkably close to $a = 0.4 \pm 0.05$ that Stokes et al. [32] identified for the low-speed displacement of a glycerol/methanol mixture by a mineral oil in a capillary tube. Stokes et al. [32] obtained accurate readings of the apparent contact angle ($\theta_{D,\Delta P}$) at low speeds by analyzing the harmonic content of the response of the system to a small-amplitude oscillatory flow superimposed on an otherwise steady flow, thereby measuring the pressure drop as well as its derivatives with respect to U. Near $Ca = 3 \cdot 10^{-4}$, the data of Stokes et al. [32] reveal a distinct crossover to a more conventional wetting regime with $a \approx 1$, as in Eq. (12).

B. Critical Limit of Visible Air Entrainment due to Forced Wetting

In successful dynamic wetting, the advancing liquid displaces all but invisible traces of the air previously in contact with the solid. When a liquid is forced to advance at too high a speed, however, visible amounts of air become trapped between the liquid and solid, regardless of the wettability of the surface. The failure of an advancing liquid to displace sufficient air is referred to as *air entrainment*. In precision coating, air entrainment often restricts the maximum speed at which a uniform liquid film can be laid down. The excess air breaks up into patches and bubbles, causing visual flaws and functional defects in the coated film. When the receding phase is an immiscible liquid, it becomes entrained in visible amounts at much lower speeds than air [31,34,46,49,98]. The entrained liquid may form a uniform film, but often breaks up into drops or more irregular globules because of various instabilities [27,31,99].

Visible air entrainment is a threshold phenomenon. Above a distinct speed U_{AE}, the coated layer, or the pool when the liquid is contained in a plunge tank, becomes inundated with a large number of small but visible bubbles. At speeds just below U_{AE}, apparent dynamic contact angles are typically seen to come close to 180° [16,43,46,48,100]. An accurate reading of θ_D near 180°, though, is difficult [42,53] and insufficient data have been published to substantiate that θ_D reaches 180° exactly when $U = U_{AE}$, as was first conjectured by Derjaguin and Levi [100]. Wetting lines can be remarkably straight when θ_D is well below 180°, but often assume a serrated, sawtoothlike shape near the onset of visible air entrainment [48,50,53,64,100–102]. V-shaped air pockets that penetrate as far as 20 mm into the advancing liquid front have been well documented, most extensively for water/glycerine mixtures with viscosities below $\eta_l \leq 200$ mPa·s [48,50,53,64,101,102]. However, they are not ubiquitous. Other, more chaotic modes of catastrophic air entrainment can also limit the maximum speed of dynamic wetting [46,64,78]. In addition, the apparent wetting line may become unsteady at speeds well below U_{AE} [46,50,54, 64,102], and the flow nearby is susceptible to numerous hydrodynamic instabilities.

Failure of successful liquid/air displacement in a purely two-dimensional flow to three-dimensional instability modes is a possible explanation for the formation of the V-shaped air pockets, but remains a conjecture that awaits theoretical verification (see Sec. III.A.10). Scriven [77] interpreted serrated wetting lines in terms of nucleation, growth, and termination of wetting. He drew an analogy between the V-shapes that accompany visible air entrainment and similar V-shapes that form at the leading edge of a liquid layer during startup of a coating process. Earlier, Blake and Ruschak [48] postulated that sawtooth-shaped wetting lines arise because there is a maximum speed at which dynamic wetting can proceed, and that the velocity component normal to the wetting line is rate-determining. To substantiate their postulate, Blake and Ruschak [48] measured the angle ϕ subtended between the straight-line segments of a sawtooth-shaped wetting line past the onset of air entrainment and the undeformed, straight wetting line at lower speeds. The measurements confirmed that ϕ is related to U by the simple formula $\cos \phi = U_V/U$, where U_V is the speed at which the sawtooth shape first forms and $U > U_V$ [48]. The data available, however, do not undisputably establish that the speed at which visible air bubbles are first entrained, U_{AE}, is the same as U_V. The postulate by Blake and Ruschak [48] nevertheless rationalizes why entrained air bubbles preferentially form at the trailing tips of the V-shapes. Still, air bubbles can also become trapped in the advancing liquid by repeated collapse and rebuilding of air pockets [102,103]. The postulate by Blake and Ruschak [48] implies also that the onset of air entrainment should be delayed to greater speeds when the direction of substrate motion is not normal to the wetting line. In misconstrued attempts to verify this, some authors [50,54] changed the angle of entry of a solid into a pool in the wrong direction, with the result that the velocity of the solid surface remained normal to the wetting line and U_{AE} changed little.

Measurements of the critical speed at the onset of visible air entrainment, most often performed in the plunge-tank configuration [16,43,46, 49,50,64], suggest that the liquid viscosity η_1 is the predominant parameter affecting U_{AE}. Figure 10a confirms that a simple power law in viscosity alone

$$U_{AE}(\text{m/s}) = 0.048[\eta_1(\text{Pa}\cdot\text{s})]^{-0.74} \tag{14}$$

correlates U_{AE} data for various liquid/air/solid systems remarkably well. Similar correlations have been proposed in the literature [46,49,50, 64,104]. They often involve additional parameters such as γ_{12}, but nonetheless yield exponents for the viscosity dependence of U_{AE} not much different from -0.74 in Eq. (14). Even though successful in capturing the basic trend of experimental data, power-law correlations of dimensional variables are unsatisfactory because they lack a physical basis. A simple dimensionless correlation for the onset of air entrainment derives from Eq. (10) when θ_D is set to 180° and the corresponding speed is taken to be U_{AE} [51]

$$Ca_{AE} \equiv \eta_1 U_{AE}/\gamma_{12} = 0.25 \tag{15}$$

Equation (15) suggests that, apart from liquid viscosity, surface tension

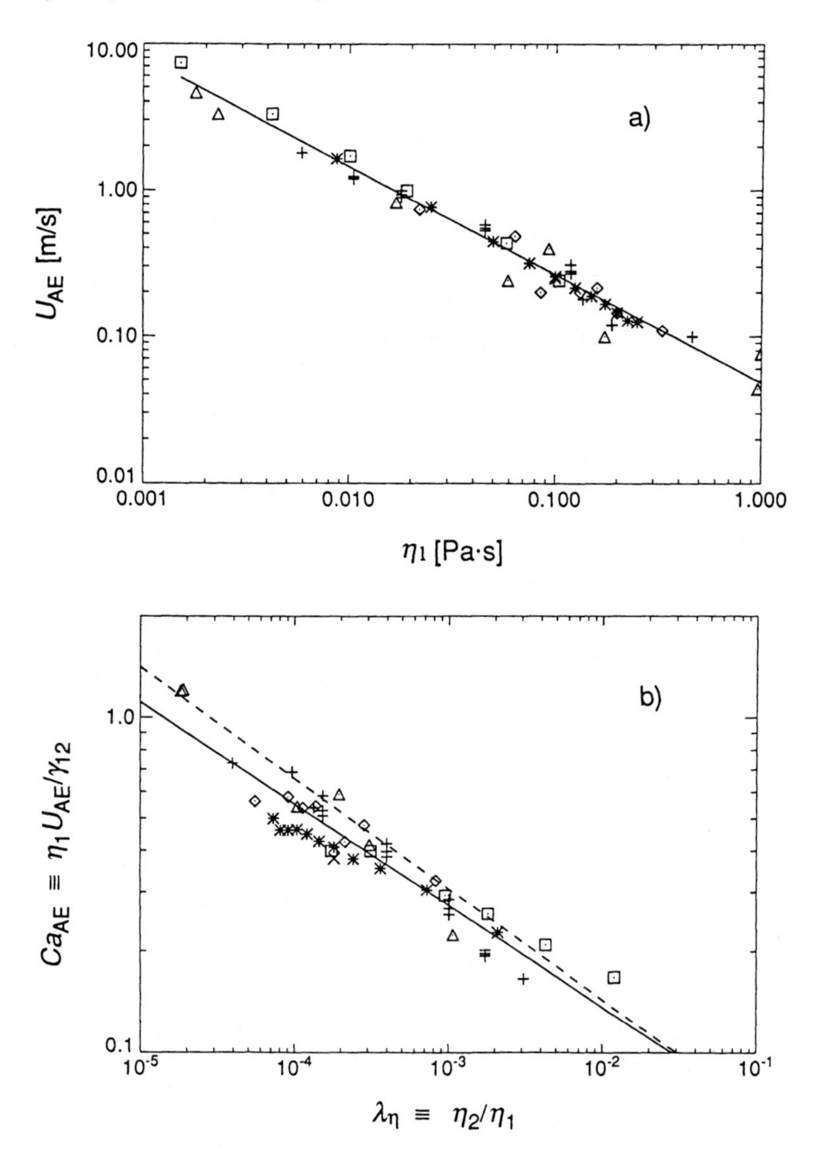

FIG. 10 Onset of visible air entrainment in plunge-tank experiments: (a) air entrainment velocity U_{AE} vs. liquid viscosity η_1 (—— Eq. (14); + Burley and Kennedy [64]; × Blake and Ruschak [48]; △ Gutoff and Kendrick [49]; ◇ Burley and Jolly [50]; □ Blake [16]; * Bracke et al. [51]); (b) critical capillary number $Ca_{AE} \equiv \eta_1 U_{AE}/\gamma_{12}$ vs. viscosity ratio $\lambda_\eta \equiv \eta_2/\eta_1$, where $\eta_2 = 1.83^{-5}$ Pa·s is the air viscosity (—— Eq. (17); – – – – Eq. (16) with $\lambda_\rho = 10^{-3}$).

is the only parameter that influences U_{AE}, even when $\theta_0 > 0°$. In accord with experiments [46,64,104], Eq. (15) predicts that higher surface tensions postpone air entrainment to higher speeds. Published data from plunge-tank experiments, however, indicate that U_{AE} must depend on parameters other than η_1 and γ_{12}, since critical capillary numbers cover the range $0.2 < Ca_{AE} < 1.3$ [49,50,64,101].

The present review contends that, for a given flow configuration, the viscosity ratio $\lambda_\eta \equiv \eta_2/\eta_1$ is a key parameter correlating the onset of visible air entrainment—second in importance only to Ca_{AE}. Figure 10b shows that the critical capillary number at the onset of air entrainment in plunge tanks rises as the viscosity η_1 of the advancing liquid increases in comparison to the viscosity η_2 of the air to be displaced. Indeed, experimental data fall close to a straight line when log Ca_{AE} is plotted as a function of log λ_η. One may argue that the ratio λ_η is a trivial dimensionless group because η_2 is a constant for air. The failure of successful displacement of a liquid by another liquid, however, follows the same trend: As the viscosity of the receding liquid increases in comparison to that of the advancing liquid, the critical speed U_{LE} at the onset of visible entrainment drops dramatically. Figure 11 summarizes this trend for the few data sets that have been published to date [31,34,49,97]. Most empirical correlations [46,49,50,53,64] ignore the influence of the receding phase, in particular its viscosity, and hence fail to capture an essential aspect of the physics of air entrainment. A noteworthy exception is

$$Ca_{AE} = 0.00019\, \lambda_\eta^{-0.33}\, \lambda_\rho^{-0.74} \tag{16}$$

which Gutoff and Kendrick [49] derived from data for visible air entrainment, as well as failed liquid/liquid displacement. When λ_ρ is set to 10^{-3}, Eq. (16) adequately describes the data in Figure 10b. For air entrainment alone, the correlation

$$Ca_{AE} = 0.034\, \lambda_\eta^{-0.30} \tag{17}$$

fits the data slightly better. The scatter of the data points in Figs. 10b and 11 indicates that other mechanisms that are not accounted for by λ_η and Ca_{AE} or $Ca_{LE} \equiv \eta_1 U_{LE}/\gamma_{12}$ must play a role. Nonetheless, the apparent importance of these dimensionless groups suggests that the onset of massive entrainment of either air or another liquid is strongly influenced by a competition among purely hydrodynamic forces, in particular, viscous stresses in both phases and capillary pressure.

Other hydrodynamic forces may also affect the onset of visible air entrainment. Curtain coating epitomizes industrial high-speed coating technology in which macroscopic hydrodynamics are used to great advantage to delay the onset of deleterious air entrainment. In curtain coat-

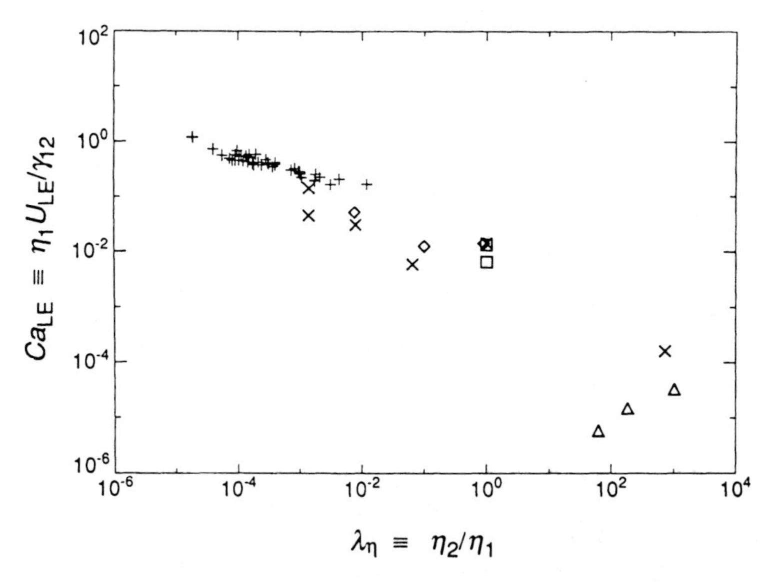

FIG. 11 Critical capillary number at onset of excessive liquid and air entrainment as a function of viscosity ratio, $\lambda_\eta \equiv \eta_2/\eta_1$: + air entrainment data from Fig. 10; △ liquid/liquid displacement in plunge tank [49]; ◇ liquid/liquid displacement in capillary [31]; × liquid/liquid displacement in capillary [34]; □ liquid/liquid displacement in capillary (data by Stokes et al., published in Ref. 97.).

ing, a thin liquid sheet falls freely over a distance of 20–500 mm before it impinges onto a moving substrate (Figure 8b). Curtain coating speeds can exceed those shown in Fig. 10 by an order of magnitude or even more [46,78,105–107]. It is presumably the inertia gained during the free fall that causes significant stagnation pressures at the point of impact on the moving substrate and thereby prevents excessive amounts of air from getting trapped between the liquid and substrate [77,78]. Support for this conjecture comes from the observation that the increase in U_{AE} as compared to plunge-tank data is most dramatic when the wetting line is placed right underneath the falling sheet [107]. While the impact of the falling curtain allows very high coating speeds, the air boundary layer that builds along the moving substrate may deflect the liquid sheet and disrupt successful dynamic wetting [78,105]. When the free-fall velocity at the point of impact or the flow rate are too high for a given coating speed, a portion of the liquid wends its way against the direction of substrate motion and forms what looks like a "heel" [78]. In the presence of a heel, air entrainment velocities in curtain coating are more comparable to those in plunge tanks [78,107], but may become susceptible to a bistable hysteresis between increasing and decreasing coating speeds [78,107].

Other industrial coating configurations may also make judicious use of hydrodynamic effects to control the onset of catastrophic air entrainment, but evidence in the published literature is ambiguous. Gutoff and Kendrick [108] reported that air entrainment velocities in slide coating (see Fig. 8a; see Sec. II.A.4 for a description of slide coating) are close to those measured in plunge-tanks and insensitive to changes in gap clearance or subambient pressure applied behind the coating bead, provided that the coated film is sufficiently thick to avoid loss of stability of the coating bead. Other investigators, in contrast, were able to apply gelatine/water solutions with a slide coater at speeds five times as high as those suggested by Eq. (14) [104] and found that reductions in the coating gap [109] and subambient pressure behind coating beads [104,110] significantly increased the maximum speed of successful air displacement in slide and extrusion coating. The high coating speeds reported by Ishizaki et al. [104] might, in part, be due to the shear-thinning of the gelatine solutions [107], which would reduce the effective viscosity near the wetting line. Improvements in maximum coating speeds from reduced gaps and subambient pressures may in part be related to the mechanisms that cause a size effect for the apparent dynamic contact angle below the onset of visible air entrainment (see Sec. II.A.3). However, more data than those available in the open literature are needed to elucidate all the mechanisms by which macroscopic flow fields affect the onset of visible air entrainment in different coating devices.

The industrial experience that the maximum speed of successful wetting depends on the coating method employed [75] and can be influenced by small changes in coating parameters other than η_1 and γ_{12} may also be related to macroscopic flow instabilities that could be mistaken as catastrophic air entrainment. For example, an advancing liquid front may repeatedly attach to the solid ahead of the instantaneous wetting line and thereby leave patches of trapped air behind [111]. Such a mode of *air entrapment* is well documented for reverse roll coating where it is sometimes referred to as *seashore instability* [112]. Another example is the common failure of a liquid to successfully bridge the gap between a stationary coating device and a moving substrate. The data of Gutoff and Kendrick [108] for slide coating reveal that U_{AE} drops significantly below a critical flow rate or above a critical gap. Likewise, when a narrow strand of liquid is coated from the tip of a syringe needle, apparent dynamic contact angles jump to 180° prematurely and a clearly visible air film is entrained at lower speeds than usual when the premetered flow rate is insufficient or the gap is too wide to maintain a stable liquid bridge between the nozzle and moving solid [58]. Apparent contact angles in slide coating have also been noted to not exceed 150° or 160° before the onset of massive air entrainment [56]. Even if a sudden jump of θ_D to 180° is initiated by macroscopic flow instabilities, it is probably accelerated by viscous drag in the displaced airflow (see Sec. II.A.3 and III.A.9).

The critical speed at the onset of visible air entrainment is also influenced by the properties of the solid surface. Increasing roughness, for instance, appears to be an effective means to raise U_{AE} [113,114]. Scriven [17] gave a lucid description of how rough or porous surface textures provide escape routes for air, thereby delaying the onset of catastrophic air entrainment. Electrostatic surface charges may affect U_{AE} not only when uncontrolled [50], but can also be used to great advantage in so-called *electrostatic-assist coating*. However, no scientific studies have been published to date on charge effects. The impact of surface wettability on U_{AE} is not fully resolved either. Several authors [43,47,49–51,64,113] attest that various materials with a wide range of θ_0 yield very similar air entrainment velocities in plunge-tank experiments (see also Fig. 10). Perry [46], on the other hand, found substantial differences in U_{AE} between two sides of a magnetic tape and attributed them to variations in static contact angle (20° and 70°). Much of Perry's observation, however, might have been related to differences in roughness of the two sides. Ishizaki et al. [104] attempted to correlate U_{AE} with the dispersion component of the solid surface energy. Their data, however, revealed only minor differences for gelatine/water solutions coated on polyethylene terephtalate, triacetyl cellulose, and polyethylene. Increases in U_{AE} were more substantial when a so-called "subbing layer" of dried gelatine was applied to the plastic film before coating the test solutions [104]. Further evidence for surface effects on U_{AE} comes from systems with prewetted surfaces. On partially immersed cylinders equipped with scraper blades to remove excess liquid, critical capillary numbers Ca_{AE} are consistently, albeit only slightly, higher than on dry surfaces [53,101]. Without a scraper blade, a thick liquid film is left on the cylinder surface, and critical capillary numbers can be an order of magnitude higher, that is, $Ca_{AE} \leq O(10)$ [115].

The influence of surface roughness on the onset of visible air entrainment raises a central question: Under what conditions does a liquid advancing over a solid completely displace the fluid previously in contact with the solid surface? Following earlier suggestions [100,116], Scriven and co-workers [8,13,17,117] advanced the hypothesis that an invisibly thin layer of air is entrained even when θ_D is well below 180°. The hypothesis suggests that, below the onset of visible air entrainment and on smooth surfaces, the thin layer collapses into bubbles small enough to rapidly dissolve into the liquid. On rough or porous surfaces, the hypothesis contends, relief of air into the surface texture and bubble collapse dynamics jointly control the fate of the small amounts of entrained air invisible to the naked eye. Using differential interference microscopy, Miyamoto [10,118] detected small "craters" in dried films of gelatine that were coated on PET at speeds below, but not too far from, catastrophic air entrainment. He deduced that these craters were remains of collapsed bubbles and thus validate the hypothesis. Miyamoto [10,119] produced further evidence in support of the hypothesis from experiments in which the dynamic wetting region was surrounded with a gas that dissolves in the liquid more easily than air, such as CO_2: The coating speed above which minute craters were first detectable by microscopy increased as much as threefold. Unfortunately, Miyamoto's evidence is based on curtain coating that exhibits rather anomalous dynamic wetting behavior [78,105–107]. In addition, Miyamoto did not quantify the roughness of the PET substrates he used. Measurements on paper surfaces [114] indicate that the entrained air volume increases with surface roughness even below the onset of catastrophic air entrainment.

C. Kinematics of Liquid Advance

Attempts to visualize the kinematics near an advancing wetting line suggest that the liquid is *deposited* on the solid surface, in accord with the basic notion of a coating process. Yarnold [80] coined the term *rolling motion* to describe his observation that dust particles sprinkled on top of a mercury drop sliding down an inclined plane advance at a velocity that is considerably greater than that of the drop as a whole. Liquid motion along the interface toward the apparent contact line has since been confirmed by dye marks placed into liquids [120–122], particles sprinkled on liquid surfaces [43,46], hydrogen bubbles generated in liquids [47,55], and direct measurement with laser-Doppler velocimetry [56]. When forced wetting is induced by a moving substrate, the liquid accelerates along the free surface as it approaches the apparent wetting line [47,56] and appears to make a sharp turn at the line [56]. Downstream of the apparent wetting line, the liquid distant from the solid accelerates toward the speed of the substrate in a boundary layer [47,55]. Recent experiments [107], as well as earlier finite-element calculations [78], indicate that the boundary layer is of Sakiadis' type [123]. Photos of the motion of dye spots placed at the interface between two highly viscous liquids [121] reveal that particles of one fluid are transported from the fluid/fluid interface onto the fluid/solid interface, whereas the other fluid is ejected from the three-phrase juncture along a separating stream surface. In capillary displacement, the ejection leads to a toroidal eddy as first proposed by Prutow and Ostrach [124] and visualized later by Dussan [122]. In high-speed coating processes, air must be expulsed from the dynamic wetting region at considerable velocity.

The only quantitative measurements of velocity fields close to apparent wetting lines are those by Mues et al. [56] who devised a laser-Doppler velocimeter (LDV) with a small measuring volume to probe liquid flow in slide coating (see also Sec. II.A.3). Mues et al. deduced that for $Ca = O(0.1)$, the liquid moves at about 50% of substrate speed U at the apparent wetting line and attains full substrate speed only a considerable distance downstream (≤ 125 μm). Unfortunately, Mues et al. did not include raw data in the published paper. Unprocessed velocity data apparently had a tendency to jump to the substrate speed when the measuring volume came within 20 μm or so from the solid surface. Mues et al. [56] employed electronic filtering to suppress readings near the substrate speed and only then measured liquid velocities close to the solid that were considerably less than substrate speed. They offered entrainment and collapse of a thin layer of air as an explanation, in support of the hypothesis put forth by Scriven and co-workers [8,10,17,117]. However, dissolution of the subbing layer applied to the substrate prior to coating might also have affected the laser-Doppler readings. In addition, laser-Doppler velocimetry with a measuring volume of 10 μm or more is not able to resolve the small length scales near the wetting line where the flow field undergoes dramatic changes. The air film may break up into bubbles much closer to the apparent wetting line than Mues et al. [56] suggest, and the curious laser-Doppler readings near the solid could be a consequence of dissolving bubbles moving already at substrate speeds rather than apparent slip over macroscopic distances induced by an air film.

D. Spontaneous Spreading

1. Macroscopic Spreading Dynamics

Conventional experimental studies of spontaneous spreading have typically been motivated by practical applications, where the rate at which a liquid covers a solid surface is of primary interest. In recent years, however, sophisticated studies of spontaneous spreading have also been aimed at identifying and probing the physical and physico-chemical mechanisms that control the local wetting dynamics. In the most common laboratory setup, a small sessile drop of liquid is placed on a flat horizontal solid surface, as illustrated in Fig. 12, and its evolution monitored as a function of time [11,66,70,73,91–93,125–132]. Alternative configurations

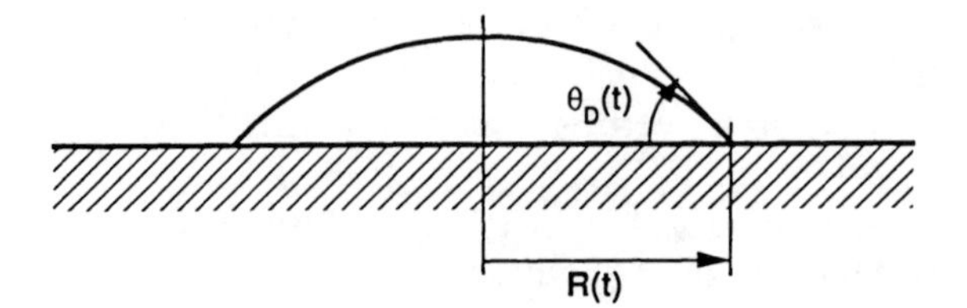

FIG. 12 Schematic of liquid drop spreading on flat, horizontal solid surface.

include capillary rise in a glass tube [81,133,134] or along a solid wall [11,135], and also capillary imbibition [91]. Experiments with spreading drops have several advantages over forced-wetting studies: A wider variety of solid materials may be used; the solid surfaces can be carefully prepared; the entire setup can be placed in an environmental chamber that provides a controlled atmosphere and ensures cleanliness; very low wetting rates can be achieved; and optical techniques much more sophisticated than silhouette photo-micrography can be devised to measure apparent contact angles [66,73,129,131,132,136]. The main drawback of spontaneous spreading experiments is that the wetting process is inherently transient. The interpretation of spreading rates requires some care [3] that was not fully appreciated by early investigators [91–93,126,127]. As a consequence, the early literature does not present a clear picture of the parameters most relevant to spontaneous spreading.

The most fundamental macroscopic observable is the variation of the apparent dynamic contact angle $\theta_D(t)$ as a function of the dimensionless spreading rate $Ca(t) \equiv \eta_1[dR(t)/dt]/\gamma_{12}$. Many early investigators reported only the radius of the drop as a function of time $R(t)$. Some even failed to treat liquids that only partially wet the solid separately from liquids which completely spread out, the case most commonly studied. Others used fairly large drops whose spreading behavior may have been influenced by gravity. Also, poorly characterized test liquids such as highly viscous glass or polymer melts were often employed. Many of these materials were probably non-Newtonian and might have been affected by thermal gradients. Essential characteristics of the solid surface, such as roughness, may also have been of consequence but were rarely quantified. Roughness can significantly enhance spreading rates [11,137] and even allow partially wetting liquids to spread spontaneously [137]. Volatile components and impurities may have been additional uncontrolled factors. They can give rise to surface tension gradients and are known to significantly alter macroscopic spreading behavior [11,130,138–141]. If surface tension gradients are sufficiently large and impose surface stresses directed toward the "dry" substrate, even partially wetting liquids can "spread out" [139,142]. On the other hand, if the surface stresses point in the adverse direction, even completely wetting liquids may retract [139–141]. In either case, spreading under the influence of surface-active materials can lead to highly irregular drop configurations.

Marmur and Lelah [5,130] attempted to bring some order to earlier spreading-drop data. They correlated their own data for completely wetting systems, as well as data by others [66,91–93,125–127], by the simple power law

$$A \equiv \pi R^2(t) = kt^{2n} \tag{18}$$

Most of the data fall in the range $0.1 \leq n \leq 0.14$. Some data sets, though, reflect very slow spreading with n as low as 0.033, and others imply rapid spreading with n as high as 0.314. Lelah and Marmur [130] found that n depends on temperature for water spreading on glass, and that humidity and pH have an effect as well. Certain alcohols also exhibited anomalous spreading behavior [130]. Lelah and Marmur's observations exemplify that specific material effects can be very important, yet that empirical correlations such as (18) have some use in comparing and classifying spreading data for different liquid/solid systems. However, much research remains to be done to fully quantify the consequences of surface roughness [11,137], volatile components and mixtures [11,130,138–141], thermal gradients resulting from evaporation [143], surfactants added on purpose, other contaminants present by accident, non-Newtonian rheology, electrolytes, electrostatic surface charges [130], and other nonideal material effects that all might cause peculiar macroscopic spreading behavior.

Recent, careful measurements with silicone oils spreading on glass and silicone wafers are consistent with the main findings from forced wetting studies (see Sec. II.A.3). The exponent n in Eq. (18) is near $\frac{1}{10}$ for small drops whose shape is immune to gravity ($R \leq L_\gamma \equiv \sqrt{\gamma_{12}/\rho_1 g}$), and hence the drop radius increases as $R \sim t^{1/10}$ [66,70,71,73,93,129,131,137,144]. For larger drops whose spreading dynamics are influenced or dominated by gravity ($R > L_\gamma$), on the other hand, the power-law index is $n \approx \frac{1}{8}$ and the drop expansion follows $R \sim t^{1/8}$ [11,93,137,144,145]. For small, thin droplets that retain the shape of a spherical cap and meet the solid at a small apparent contact angle ($\theta_D \ll \pi/2$), Eq. (18) can be rewritten as [6,66]

$$U(t) = \frac{dR(t)}{dt} \sim \theta_D^{(1-n)/3n} \tag{19}$$

For $n = \frac{1}{10}$, Eq. (19) is consistent with the Hoffman–Voinov–Tanner law

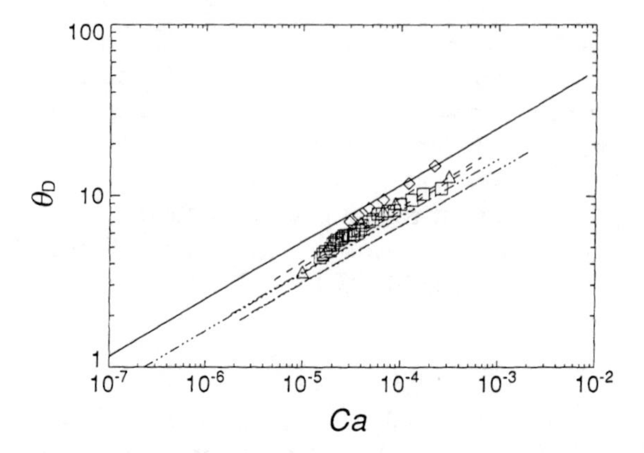

FIG. 13 Apparent dynamic contact angles for the spreading of silicone fluids on glass (— — Tanner [66]; □ △ ◇ Chen's [70] data sets 14, 9, and 5 for drop volumes of $0.34 \cdot 10^{-4}$, $1.70 \cdot 10^{-4}$, and $3.05 \cdot 10^{-4}$ cm^3) and silicone wafers (—···— Ausserré et al. [131]). The solid line is the universal function (3) for comparison.

(5). Figure 13 shows that θ_D versus Ca data for spreading silicone drops [66,70,73,131] not only corroborate the functional form $\theta_D^3 \sim Ca$, but fall close to the universal curve (2) that Hoffman [29] first identified for capillary displacement (see Fig. 3) and that since has been confirmed by other forced-wetting experiments (see Figs. 4 and 5). For $n = 1/10$, Eq. (19) together with Eq. (18) yields $\theta_D \sim t^{-0.3}$. This power law has been verified independently by direct measurements of apparent contact angles as a function of time [66,70,128,129,131]. Typically, however, experimental data correlate better with $R \sim t^{1/10}$ and $\theta_D \sim Ca^{1/3}$ than with $\theta_D \sim t^{-0.3}$ [70].

The agreement of spreading-drop data with the universal correlation (2) suggests that the dynamic variations of apparent contact angles in spontaneous spreading arise from the same mechanisms as in forced wetting. In particular, the agreement reinforces the conjecture put forth in Sec. II.A.3 that the primary mechanism represents a competition between capillary pressure and viscous stresses close to but away from the immediate vicinity of the wetting line, and that only systematic deviations from the universal wetting behavior arise from specific material effects. The agreement between spontaneous spreading and forced wetting also furnishes additional evidence that the θ_D versus Ca relationship is independent of the flow configuration in which measurements are made and, for nonvolatile liquid/solid systems with complete equilibrium wetting, is rather insensitive to the nature of the liquid/solid interactions as measured by the spreading coefficient S [131]. The accelerated expansion of large drops that are affected by gravity may be misconstrued as proof that gravity enhances dynamic wettability. However, even very large drops that follow the $R \sim t^{1/8}$ power law and also medium-sized drops that may fall between $R \sim t^{1/8}$ and $R \sim t^{1/10}$ [146] obey the universal relationship $\theta_D^3 \sim Ca$ between the instantaneous spreading rate and apparent contact angle [145].

As in the case of forced wetting (see Figs. 3 and 5), systematic deviations from the universal spreading behavior may be observed, even for simple, nonvolatile liquids that completely wet the solid. The deviations in spontaneous spreading might be particularly noticeable because wetting rates are typically very low. At low rates—one may conjecture—specific material interactions are most likely to successfully compete with the visco-capillary effects that presumably prevail at higher Ca. Beaglehole [147], for instance, observed systematic deviations from $n = 1/10$ for silicone oils spreading on glass, fused silica, and freshly cleaved mica surfaces and thought that differences in chemical composition between substrates are a probable cause. The spreading of high-molecular-weight polydimethylsiloxane can deviate even more drastically from the universal behavior: On some surfaces, macroscopic spreading stops altogether after a finite time [148]; on other surfaces, spreading may accelerate [93,129]. Even in Fig. 13, some of the data points follow power laws slightly different from $\theta_D \sim Ca^{1/3}$. In addition, most data points gather below the universal correlation (3). Chen's [70] data suggest that the latter discrepancy might be related to a size effect, for the angle θ_D diminishes slightly as the size of the drop is reduced. However, the effect is not as clearcut as the one measured by Ngan and Dussan V. [25,37] for forced wetting. Other data sets for different drop sizes show even less of a size effect [131]. Dussan V. [3], on the other hand, identified a strong size effect upon transforming the data of Schonhorn et al. [126] into a plot of θ_D versus Ca. Foister [71] also documented a size effect for spreading drops surrounded by an immiscible liquid. Evidently, the different causes for systematic deviations from the universal spreading behavior need to be identified further—even for simple, nonvolatile liquids that completely spread on structureless surfaces.

2. Submicroscopic Precursor Films

When a completely wetting liquid spreads sufficiently slowly ($S > 0$, $\theta_D \ll \pi/2$), an invisibly thin *precursor film* may propagate ahead of what the naked eye or an eye aided by a conventional microscope perceives to be a wetting line (see Fig. 1d). Hardy's [15,149] pioneering observations of reduced static friction on solid surfaces near spreading droplets furnished the first experimental evidence for such films. The existence of precursor films has since been confirmed by interference patterns [11,138,150], impeded condensation of humid air on a portion of the solid surface surrounding spreading drops [11,138], anomalous wetting behavior of small droplets of a different liquid placed close to spreading drops [11], ellipsometry [11,147,151–156], electron microscopy of quenched droplets of molten glass [127], electrical conductivity measurements [157], polarized reflection microscopy [131,153], and x-ray reflectivity [153,158].

Precursor films can expand by several mechanisms. Hardy [15] thought that so-called *primary films* form only by evaporation, transport through the vapor phase and condensation on the solid. Later experiments, however, established that primary spreading can also arise from migration along the solid [11,138]. In the latter case, the first molecular layers advance probably by a surface-diffusion mechanism [11]. *Secondary films* ranging from a few molecular layers to 0.1 μm in thickness are driven primarily by microstructural forces between the fluids and solid [6,7]; much thicker secondary films can also arise from surface tension gradients [11,130,138]. The term precursor film is not used consistently throughout the literature, for it may refer to primary films, secondary films, or both. For some liquid/solid systems, especially those without volatile components, a distinction between primary and secondary films may not even be appropriate. Further caution is in order because the older literature [15] used the term "secondary spreading" to refer to what is better called *bulk spreading* [7] of the macroscopic front.

Spreading liquids that form precursor films furnish unequivocal proof that apparent dynamic contact angles measured with conventional optical techniques can be substantially larger than the angle of inclination of the liquid/gas interface very close to the solid [11,127]. Detailed measurements of the interfacial profile confirm that apparent wetting lines are merely regions of localized meniscus curvature [11,147,152,154]. Ellipsometry and x-ray reflectivity measurements reveal that precursor films are usually much thinner than 1 μm at the transition to the macroscopic front [11,147,152,154] and often decay to one or a few molecular layers at the leading edge [147,152,154]. The extremity of precursors may advance as a stack of distinct molecular layers [135,155,159]. The velocity of the leading edge can be as high as 10 cm/s for water spreading on glass [160], but is well below 1 μm/s for other liquids with higher viscosities [11]. In contrast to the macroscopic spreading dynamics—which for some simple, nonvolatile liquids follow nearly universal laws that are rather insensitive to the chemical composition of the solid surface (see Sec. II.D.1)—the submicroscopic structure and dynamics of precursor films depend on the spreading coefficient S [131] and are affected by the local inhomogeneities in surface energy [147,153]. In addition, roughness enhances the expansion of precursor films, especially when the surface texture has a preferred direction, as may result from polishing [11,151,161]. The extent and thickness of precursor films are also related to the state of the macroscopic drop, in particular its instantaneous apparent contact angle [131,147,152]. Vice versa, in special situations precursor films can have a profound effect on the macroscopic spreading dynamics. For example, two spreading drops may migrate toward each other when placed not too far apart [15], and water droplets spread faster on small rather than large glass slides [160].

III. THEORIES OF DYNAMIC WETTING

A. Conventional Hydrodynamics of Liquids Advancing over Solids

1. Breakdown of Hydrodynamic Continuum Theory

A strictly hydrodynamic analysis of dynamic wetting faces serious difficulties. Most prominent is the failure of conventional continuum theory to describe how a liquid can advance over a solid while displacing another fluid [2,3,22,121]. When steady, incompressible, and Newtonian flow is enforced right down to the apparent wetting line—together with boundary conditions that impose no slip along a smooth, homogeneous, and impervious solid surface, continuity of traction and no mass transfer at the fluid/fluid interface, and a local contact angle* in the range of $0° \leq \vartheta_w < 180°$—the calculated velocity field turns out to be multivalued at the wetting line. As a result, viscous stresses and pressure are predicted to increase as $1/r$, where r is the distance from the line, and the force that would be required to advance the liquid becomes logarithmically infinite. Likewise, the total viscous dissipation near the wetting line becomes unbounded. This calculated singularity disagrees, of course, with experimental evidence. If the singularity was real, it would give rise to drastic changes in temperature and material properties. These, in turn, would contradict the assumption of uniform viscosity and density, a fundamental premise in the Navier–Stokes system of equations.

The unbounded force singularity was already evident in single-phase analyses of corner flow with a fixed shear-free boundary [116,162,163].

* The symbol ϑ is used to denote contact angles in theoretical models. It is different from the symbol θ for experimentally observed angles in order to emphasize that theoretical values, in particular those imposed as boundary condition right at the solid, do not necessarily coincide with physical reality.

Huh and Scriven [2] were the first to systematically lay out the consequences of the singularity for fluid/fluid displacement at a moving three-phase line, speculate on its physical origins, and suggest potential remedies. Dussan V. and Davis [121] emphasized that the unbounded force is a dynamic consequence of the multivalued velocity field at the wetting line and *not* the result of a kinematic incompatibility between no slip on the solid and impenetrability of the moving interface to either fluid. Indeed, the multivalued flow fields that Dussan V. and Davis [121] deduced from purely kinematic arguments—and which Huh and Scriven [2] calculated earlier from the full set of dynamic equations under the more restrictive assumption of a flat interface—seem to be consistent with macroscopic observations of liquid accelerating toward the three-phase region and abruptly changing flow direction very close to the solid (see Sec. II.C).

In the immediate vicinity of the wetting line, the mathematical singularity that results from the multivalued velocity is so severe that the posed continuum problem cannot be solved. In local analyses of the complete equation set, the interface deforms so much that the contact angle ϑ_w at the solid surface cannot be prescribed [2,3,164]. A boundary condition for ϑ_w is needed to complete the mathematical description of the displacement flow when surface tension is significant. In the lubrication-flow approximation, the singularity at the three-phase juncture can sometimes be ignored in an outer solution in which surface tension effects are neglected [165–170]. The calculated free-surface profiles, however, exhibit either a shock-type discontinuity at the advancing edge or meet the solid at a 90° contact angle. Both features indicate a breakdown of the small-slope assumption that is essential in rigorous lubrication theory. Nonetheless, for reasons explained by Hocking [171], analyses ignoring capillarity can be useful and correctly predict the spreading behavior $R(t) \sim t^{1/8}$ of large drops that are dominated by gravity [165–167,171] (see Sec. II.D.1). When capillary pressure is included in the lubrication theory and the singularity is not removed, the slope of the interface remains infinite at the contact line [172]. When an angle smaller than 90° is imposed, film profile equations that account for capillary pressure exhibit a singular limit of no solution at small film thicknesses [66,173].

The difficulties that stem from the singularity are not an artifact of failed mathematical approximations [121], but signal a breakdown of one or several of the premises underlying traditional hydrodynamic theory [2,22]. Within a small *cutoff length* L_δ from the wetting line, the surface microtexture of the solid, nonuniform liquid properties, volatility, intermolecular forces, statistical fluctuations due to molecular events, or other still poorly understood *cutoff mechanisms* must alter the physics suggested by macroscopic continuum theory and allow liquids to advance over solids without the resistance of inordinately high forces. A lower bound for L_δ is the size of a few molecules, as is often conjectured for dynamic wetting on smooth surfaces [164,174,175]. The cutoff length may approach macroscopic dimensions when the surface is rough [95] or upon entrainment of a thin layer of air [56]. Resolving the dynamics within the cutoff region requires refined theories that account for the physical and physico-chemical mechanisms at the length scales at which the displacement process takes place (see Sec. III.B and III.C). Any contrivance that makes the velocity field single-valued, however, also alleviates the unbounded force singularity [121] and permits, in an ad hoc manner, a purely hydrodynamic analysis of free surface flows with dynamic wetting lines.

2. Ad Hoc Removal of Unbounded Force Singularity

The simplest approach to treating dynamic wetting in conventional hydrodynamic continuum theory is to exclude the cutoff region altogether from the analysis and thereby avoid identifying a specific cutoff mechanism [27,66,67,76,94,177]. Theories that ignore viscous effects at distances smaller than L_δ from the putative wetting line predict finite values of the net force F_v and total dissipation Φ_v in an advancing liquid front. For instance, in a flat liquid wedge with a small apparent angle $\vartheta_a \ll \pi/2$ as illustrated in Fig. 14, the force and dissipation become [2,6]

$$F_v = \frac{3\eta U}{\vartheta_a} \ln \frac{1}{\epsilon_\delta}, \qquad \Phi_v = \frac{3\eta U^2}{\vartheta_a} \ln \frac{1}{\epsilon_\delta} \tag{20a,b}$$

Here, $\epsilon_\delta \equiv L_\delta/L$, and L is the characteristic length scale of the macroscopic flow.

By far the most common procedure for coping with the unbounded force singularity is to insist that the Navier–Stokes equations remain valid

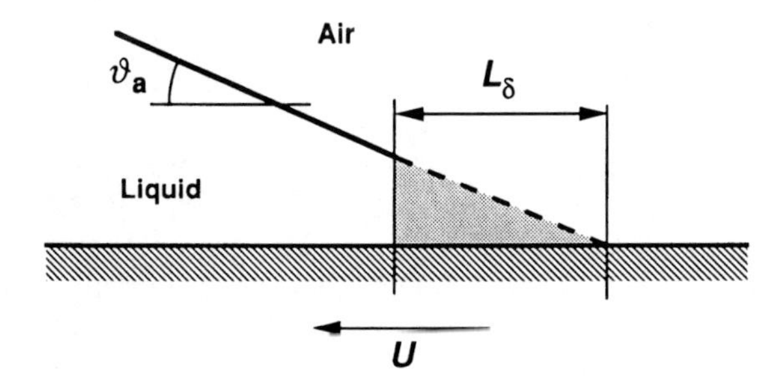

FIG. 14 Schematic of cutoff region in simple wedge flow. (After Ref. 6.)

within the cutoff region, but to stipulate a specific cutoff mechanism, namely, slip of the fluids over the solid. The most popular model for slippage is Navier's [178] classical boundary condition that makes the flux of momentum tangential to the wall proportional to the velocity discontinuity [18,164,179,180]. For two-dimensional flow, this condition is

$$\mathbf{tn}:\mathbf{T} = \frac{1}{\epsilon_\beta}\,\mathbf{t}\cdot(\mathbf{u}_s - \mathbf{t}) \tag{21}$$

Here, **t** and **n** are the unit tangent and normal vectors to the solid surface, **T** is the stress tenor measured in units of $\eta_1 U/L$, $\mathbf{u}_s$ is the liquid velocity on the solid surface measured in units of U, and $\epsilon_\beta \equiv \beta\eta_1/L$ is a dimensionless slip coefficient. Numerous other slip boundary conditions have also been imposed near moving wetting lines. Some generalize Navier's model (21) so that the slip coefficient becomes a function of the thickness of the liquid film advancing over the solid [2,181–183]. Others prescribe a slip velocity profile as a function of distance from the contact line [97,180,184,185], allow free slip ($\mathbf{tn}:\mathbf{T} = 0$) over a preset distance [164], impose a yield criterion for free slip if the wall shear stress exceeds a critical value [186], or relate the effective velocity of the liquid on a porous substrate to the local film thickness and permeability of the solid [187,188]. In numerical solution schemes, local slip can enter through the discretization even when it is not specified as a boundary condition [172,189]. The unbounded force singularity has also been successfully removed by means of one-sided variational constraints [190].

The length $L_\beta = \beta\eta_1$ in (21) is often referred to as the slip length, but the term *extrapolation length* [6] is preferred in this chapter because it is more descriptive: L_β measures the distance from the solid surface to a fictitious point within the solid at which the extrapolated liquid velocity would be the same as that of the solid. The term *slip length* is reserved for the length of the region over which slip affects the kinematics and is therefore a particular version of a cutoff length L_δ. The slip length is denoted by the symbol L_s that, in the present review, is used independently of a particular slip model. Even though L_β is related to L_s, the two are not necessarily the same. For the simple wedge flow discussed above [see Eq. (20)], for instance, $L_s = L_\beta/\vartheta_a$ when $\vartheta_a \ll \pi/2$ [6,191].

Slip boundary conditions are effective for formulating well-posed free-surface flow problems with moving wetting lines [192]. However, slip has been used mostly for its mathematical convenience, and its physical relevance is not obvious a priori. In standard fluid mechanics, the confidence in the no-slip assumption for liquids flowing over solids is based on direct experimental evidence and agreement with macroscopic theories that rely on the assumption [193]. The confidence is reinforced by calculations of flow over rough surfaces which show that, even when perfect slip is assumed at submicroscopic length scales, the flow appears to obey the no-slip boundary condition at length scales larger than the asperities [194,195]. Molecular dynamics simulations further corroborate the no-slip hypothesis for liquids [196]. For sufficiently strong fluid/wall interactions, such simulations even predict several adsorbed layers of molecules immobilized at the wall [197,198]. On the other hand, apparent wall slip has been observed in experiments for the single-phase flow of Newtonian liquids in very fine capillaries, especially when the solid is poorly wetted by the liquid (see Blake [199] and references therein). Adhesive failure between liquids and solids is also well documented for polymer melts and polymer solutions [200–202] for which it is often thought to be responsible for processing instabilities of practical importance [203,204].

Slip is probably an essential mechanism in dynamic wetting—at least at molecular length scales [191] at which dynamic motions of individual molecules can combine to advance a liquid front over a dry solid or a solid covered by a few adsorbed layers [16,69,88,89,205]. Specific molecular processes that have been invoked to rationalize slippage include a time delay required to complete liquid bonding to the surface [164], cohesive failure of liquid/liquid bonds between molecular layers parallel to the solid [186], and surface diffusion induced by a gradient of the fluid/solid interaction potential [175,206]. The last model for slip resembles the mechanistic picture of an activated-rate process of molecules advancing laterally across energy barriers from one adsorption site to another [16,69, 88,89] (see Sec. III.C.1, as well as Chap. 5 in this volume). Recent molecular dynamics simulations of immiscible liquid/liquid displacement on smooth surfaces corroborate that the no-slip hypothesis breaks down within a few molecular diameters from a moving wetting line because of large stresses nearby [196,207]. Other justifications for the use of slip boundary conditions near dynamic wetting lines suggest that the relative motion between the liquid and solid is not physical, but merely apparent. The liquid may slide over a rough surface filled with either the liquid itself or a displaced fluid [18], skip over a thin, entrained film of the receding phase [8,10,17,56,117], advance in an unsteady motion because of microscopic roughness [95], or spread on top of a porous medium filled with liquid [187]. Apparent slippage may arise at length scales considerably larger than molecular dimensions [56,95]. de Gennes [208] submitted that entangled polymers flowing over smooth, passive surfaces should also experience anomalously high extrapolation lengths that, in spontaneous spreading, cause a peculiar macroscopic "foot" near advancing wetting lines [6,209]. Such foot structures, up to 50 μm in size and distinct from much thinner precursor films, have indeed been observed for spreading droplets of polymer melts [127,210], including polydimethylsiloxane (PDMS) [93]. More recent experiments with high-molecular-weight PDMS on smooth silicone wafers, however, revealed much smaller "bumps" that are comparable in height to the radius of gyration of the polymer chains [148,153]. Such bumps have been attributed to changes in monomer mobility on different surfaces that may give rise to a reptation rather than a slip regime [148,211].

Although slip near advancing wetting lines is plausible for several reasons, hydrodynamic models that rely on slip as an ad hoc boundary condition predict several unphysical characteristics of the flow within the cutoff region. One such characteristic is the pressure field that remains singular at the wetting line even when the liquid is allowed to slip over the solid [164,186,212]. The singularity, which now is integrable, stems from the discontinuity in shear stress where the fluid/fluid interface intersects the solid. It is analogous to the singularity encountered in flow calculations at static contact lines [213]. If pressures were negative and infinite as suggested by continuum theory, local evaporation and recondensation, and even cohesive failure due to cavitation [2,24], would have to be essential mechanisms when a liquid advances over a solid. Accounting for evaporation and condensation relaxes the kinematic constraint of impenetrability of the fluid/fluid interface [143,214] and can alter the local dynamics of the displacement process so as to alleviate the unbounded force singularity [215]. Inordinately low pressures might also give rise to local deformation of the solid and thereby affect the wetting dynamics [216]. Like pressure, extension rates remain singular at the wetting line [164] and shear rates, even though finite for most slip models, can become very high, in particular when the slip length is taken to be of molecular dimensions. For Navier's slip model (21), for instance, the maximum shear rate is proportional to $1/\epsilon_\beta \equiv L/L_\beta$ [164,179]. The response to extremely high deformation rates might be non-Newtonian [2], even for liquids that appear to be Newtonian when probed in standard rheological equipment. The impact of non-Newtonian rheology on dynamic wetting remains largely unexplored [217,218].

Another disconcerting feature predicted by hydrodynamic theories of dynamic wetting relates to the kinematics at the wetting line. Removal of the force singularity requires perfect slip, that is, material points that at a given instant are located at the wetting line must have the same velocity as the line itself [121,179]. Therefore, in a frame of reference that moves with the apparent wetting line, the liquid experiences dramatic deceleration immediately upstream of the apparent wetting line and comes to an abrupt stop at the line [164]. For some slip models [164,185], fluid parcels on the fluid/fluid interface never even arrive at the putative wetting line [3]. The notion that the wetting line always consists of the same material points contradicts the basic concept of coating a fluid on a solid while displacing another [121,185]. The consequences of specifying perfect slip at the wetting line are probably most questionable for systems containing surface-active materials. Calculated flow fields can exhibit significant differences, depending on whether the surface active material is forced to accumulate upstream of the wetting line [219] or is allowed to pass freely through the line [220,300].

The Galerkin finite-element method allows computing macroscopic flow solutions without enforcing perfect slip at dynamic wetting lines. Because the method solves a weak form of the governing equations, the discretization can be managed so as to satisfy at the wetting line both the kinematic condition of no mass transfer through the interface and the condition of an impervious solid surface [78,220]. Computed flow fields exhibit much reduced deceleration upstream of the wetting line and, in steady-state solutions, nonzero liquid speeds at the wetting line. The need to impose equal velocities of the liquid at the wetting line and the line itself can also be avoided by setting the local contact angle to $\vartheta_w = 180°$ [221–225]. The resulting *rolling motion* relinquishes the need for slip of the advancing liquid over the solid. However, slip is still required in the receding phase, unless that phase is allowed to be entrained [8] or is neglected altogether, as is most often the case. If the receding fluid is a gas, rarified gas flow might give rise to an additional, physical slip mechanism [46,117,222].

The unphysical features of flow fields calculated with the help of ad hoc slip-flow models arise because hydrodynamic analyses of dynamic wetting are refined at inordinately small length scales. In matched asymptotic expansions, the size of the inner region is the same as the size L_s of the cutoff region [18,179,185], which is often taken to be less than $L_s < 0.1$ μm and sometimes even of molecular dimensions (see Sec. III.A.5). The unphysical pressure and stress singularities discussed above develop on length scales significantly smaller than L_s. Similarly, in numerical discretizations, the size of the smallest elements or cells neighboring the wetting line must be refined well below the slip length L_s to achieve convergence of the solutions away from the wetting line [174,226,227]. At length scales below 0.1 μm, conventional continuum theory ought be augmented to account for microstructural forces—and at molecular scales it breaks down altogether.

The predominant influence of microstructural forces on the local wetting mechanics is best understood for the slow spreading of nonvolatile liquids that completely wet a smooth solid and form a precursor film (see

Sec. III.B.1). de Gennes et al. [191] pointed out that long-range fluid/solid interactions can dominate the local wetting dynamics of systems with partial wetting as well, provided that the solid is smooth and the static contact angle small. For a simple wedge flow with $0 < \theta_E \ll \pi/2$ and $Ca \ll 1$ [see Eq. (20)], van der Waals forces truncate the unbounded viscous force singularity at a distance $L_\delta = O(L_a/\theta_E^2)$ from the wetting line, where L_a is a molecular size (i.e., a few Å) [191]. Local slippage, by comparison, yields a cutoff length $L_\delta = O(L_\beta/\theta_E)$ that is significantly smaller than L_a/θ_E^2, at least when L_β is also of the order of a few molecular diameters. Even if van der Waals interactions dominate slippage as a cutoff mechanism for the apparent viscous force singularity, however, augmented continuum theories cannot resolve the innermost structure of the moving three-phase juncture where short-range forces and dynamic molecular events prevail. Indeed, conservative interaction forces alone cannot alleviate the singularity at the leading edge of an advancing liquid [121]. Augmented continuum theories are even less useful when the static contact angle does not satisfy $\theta_E \ll \pi/2$. In this case, which is often encountered in practical applications, long-range fluid/solid interactions influence the three-phase region only on length scales smaller than $O(L_a)$ [6,191]. To complete a mathematical analysis of dynamic wetting on smooth surfaces for $\theta_E = O(\pi/2)$, macroscopic hydrodynamic theories may have to be coupled directly with stochastic theories that describe the dynamic molecular events of the displacement process (see Sec. III.C). The same type of analysis would presumably also be needed for liquid/solid systems with $\theta_E \ll \pi/2$, or even $\theta_E = 0$, when the apparent dynamic contact angle no longer satisfies $\theta_D \ll \pi/2$. Ultimately, when the apparent dynamic contact angle comes close to 180° in rapid forced wetting, the entrainment and subsequent collapse of a thin film of the receding phase become amenable again to analysis by augmented continuum theories that account for long-range interaction forces (see Sec. III.B.2).

In spite of serious shortcomings in the cutoff region, purely hydrodynamic theories with ad hoc slip flow assumptions are remarkably successful in predicting the macroscopic wetting kinematics of advancing liquid fronts. Matched asymptotic expansion procedures for $\epsilon_s \equiv L_s/L \ll 1$ [24,164,179,185] yield outer solutions that replicate the seemingly discontinuous kinematics observed in experiments at macroscopic length scales [47,56,121] (see also Sec. II.C). For slide coating, finite-element simulations with extensive mesh refinement near the wetting line [227] predict accelerations of the liquid along the free surface toward the wetting line that closely agree with LDV measurements [56]—even when the velocity at the wetting line is set to zero.

The apparent success of purely hydrodynamic theories is possible because the details of the presupposed wetting dynamics within the cutoff region have only a minor influence on calculated flow fields and free surface shapes away from the putative wetting line. The feature that affects the macroscopic flow most is the slip length L_s, but matched asymptotic expansions for $\epsilon_s \equiv L_s/L \ll 1$ and $Ca \rightarrow 0$ indicate that the influence of L_s is logarithmically weak [24,164,179,185] [see also Eqs. (22) and (26) below]. The particular mathematical form of slip models with fixed L_s does not even enter the leading-order solution in the outer region [24,164, 179,185]. When a model allows L_s or other measures for the amount of local slippage to depend on the calculated flow field [182,186,187], the predicted macroscopic wetting dynamics may become sensitive to the details of the slip model at leading order, but the influence remains logarithmically weak. At leading order, slip need not even be invoked as a cutoff mechanism when $\epsilon_\delta \equiv L_\delta/L \ll 1$. Theories that account for the spreading of a precursor film (Sec. III.B.1), assume a preexisting film on the solid (Sec. III.A.8), or ignore the cutoff region altogether (Sec. III.A.4) all predict the same asymptotic structure for the outer, macroscopic flow as hydrodynamic theories that rely on ad hoc slip flow models. This central result of wetting hydrodynamics suggests that the unbounded force singularity that arises in conventional hydrodynamic theory, and the way the singularity is removed, are of less fundamental scientific significance than a large portion of the literature on dynamic wetting insinuates.

Without refined theories for the local wetting dynamics, however, the slip length L_s, or some other cutoff length L_δ, remains an empirical input parameter. It cannot be inferred unequivocally from comparisons between experiments and purely hydrodynamic models because such models contain a second empirical input parameter, namely, the local contact angle ϑ_w at which the mathematical interface intersects the solid at the putative contact line [3,25,37].

3. Dynamic Contact-Angle Boundary Conditions

The need to specify a local contact angle ϑ_w arises because conventional continuum theory cannot describe the mechanisms that set its equilibrium value on ideal surfaces ($\vartheta_w = \theta_E$), influence its hysteresis on rough or heterogeneous surfaces ($\theta_A > \vartheta_w > \theta_R$), and control its dynamic variations when $U > 0$ ($\vartheta_w = f[U, \ldots] > \theta_A$). Unlike the particular cutoff model chosen to remove the viscous force singularity, the imposed value of ϑ_w can govern the shape of the advancing interface over its entire extent, especially at low capillary numbers when surface tension effects are important. Therefore, identifying appropriate boundary conditions for ϑ_w becomes the outstanding issue in conventional wetting hydrodynamics.

Many hydrodynamic analyses of flows with moving wetting lines forfeit a priori any endeavor to predict the dynamic wetting behavior, even on an apparent macroscopic scale. Instead, they rely on ad hoc boundary conditions to prescribe the dynamic variations of the apparent contact angle $\vartheta_w = \theta_D$ as observed in conventional experiments at nearly macroscopic length scales, treating $\vartheta_w = f(U, \ldots)$ as a constitutive equation. Some boundary conditions merely account for contact-angle hysteresis at very low displacement speeds [186]. More complicated models attempt to replicate both hysteresis and dynamic variations of θ_D over wider parameter ranges [181–183,228–235]. The functional form of $\vartheta_w = f(Ca; \theta_A, \theta_R)$, however, is often chosen for its mathematical convenience rather than physical relevance. In finite-element calculations with coarse meshes that focus on the outer flow away from the three-phase region, imposing an arbitrary fixed value of $\vartheta_w = \theta_D$ has been particularly popular [112,180, 218,226,236]. In simulations of polymer processing flows with highly viscous liquids, $\vartheta_w = 180°$ is a common choice [224,225,237]. For industrial coating processes, prescribed contact angles in the range $120° < \vartheta_w < 160°$ are typical [112,218,236]. These values may be acceptably close to physical reality for $Ca \approx O(0.1)$ (see Figs. 3 through 7), but the exact choice for ϑ_w may still influence essential features of the computed solutions, in particular, the sensitivity and stability of the coating flows to external perturbations. Prescribing ϑ_w according to an empirical correlation for $\theta_D = f(Ca; \theta_0)$ can yield more accurate predictions [57].

Even though useful for analyzing macroscopic flows of practical interest, ad hoc boundary conditions for $\vartheta_w \approx \theta_D$ defeat a major purpose of the analysis, for they presuppose the most essential aspect of the dynamics. Agreement between theory and experiments [182,235] provides at best a self-consistency test for the empirical ϑ_w versus U law, but yields no insights into the wetting dynamics. Besides, $\vartheta_w = \theta_D(U)$ is not well suited as a general-purpose boundary condition. Experiments suggests that θ_D may depend on the size and possibly other features of the macroscopic flow. Hence, $\theta_D = f(U)$ would have to be measured separately for each flow geometry to be analyzed [45]. Imposing dynamic contact angles obtained from experiments is complicated further because angles measured by low-resolution optical techniques may be merely apparent (see Sec. II.A.3). In refined solutions of the Navier–Stokes system that faithfully satisfy the ad hoc boundary-value problem near the wetting line, computed angles of inclination of the interface at small, but finite distances away from the three-phase line ϑ_D, can deviate substantially from the value imposed at the solid ϑ_w, even when $Ca \ll 1$. When $\epsilon_\delta \ll 1$, the calculated interface can become highly curved at length scales that would be invisible to the naked eye or a low-power microscope, as was noted a long time ago by Hansen and Toong [22] and van Quy [184]. In purely hydrodynamic theories, the local bending of the meniscus is governed by a balance between normal viscous stresses and capillary pressure. It can be sizable even when the cutoff length is chosen to be well outside the range of influence of microstructural forces and when the imposed value ϑ_w is intended to capture the dynamic state of a macroscopically observable contact angle θ_D. Tilton's [226] finite-element analysis of liquid/liquid displacement in a capillary provides a case in point.

Building on earlier suggestions [3,22,25,185,238], Dussan V. et al. [45] advanced a procedure for imposing dynamic contact angles in macroscopic flow calculations that is consistent with the possibility of local interface deformation. The procedure relies on the insight that the details of the ad hoc model in the cutoff region—that is, the mathematical form of the slip boundary condition, slip length ϵ_s itself, and local contact angle ϑ_w, all of which may be velocity-dependent—affect the macroscopic dynamics of the outer free-surface flow only through the particular combination [45]

$$\vartheta_R \equiv G^{-1}\left\{G(\vartheta_w) + Ca \ln \frac{R}{L_s} + Ca\left[1 + \frac{Q_i(\vartheta_w)(\vartheta_w - \sin\vartheta_w \cos\vartheta_w)}{2 \sin\vartheta_w}\right]\right\} \tag{22}$$

The so-called *intermediate angle* ϑ_R defined by Eq. (22) measures the slope of the interface at a distance R from the wetting line where the meniscus exhibits measurable amounts of local curvature. Equation (22) is based on results from matched asymptotic analyses of creeping flow with three regions of expansion that are illustrated schematically in Fig. 15 [23,24]: an inner region of size $O(\epsilon_s \equiv L_s/L)$; an outer region of size $O(1)$; and a region in between whose characteristic width ϵ_i falls in the intermediate range $\epsilon_s \ll \epsilon_i < O(1)$ (here, length is made dimensionless with L). Equation (22) is valid to first order in capillary number for $\epsilon_s \ll 1$ and $Ca \to 0$ while $Ca \ln (\epsilon_s^{-1}) = O(1)$ (see Sec. III.A.4). The function $G(\vartheta)$ in (22) is defined by

$$G(\vartheta) \equiv \frac{1}{2}\int_0^{\vartheta} \frac{\varphi - \sin\varphi \cos\varphi}{\sin\varphi}\, d\varphi \tag{23}$$

G^{-1} is its inverse function, that is, $\vartheta = G^{-1}[G(\vartheta)]$, and $Q_i(\vartheta_w)$ depends

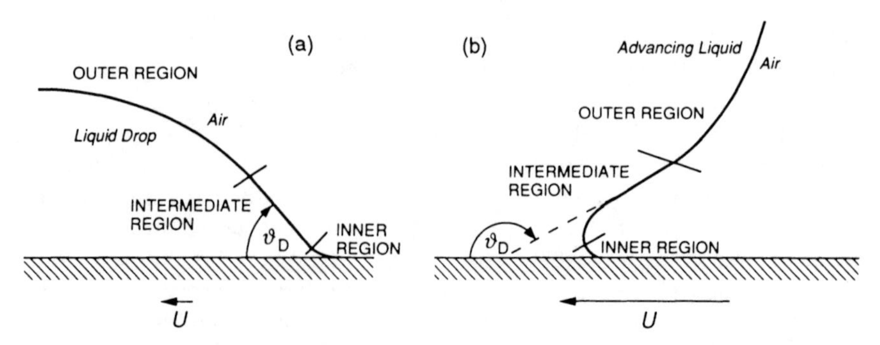

FIG. 15 Schematic of three regions of expansion in asymptotic analyses of Stokes' equation with ad hoc slip boundary condition in cutoff region: (a) slow spontaneous spreading and (b) rapid forced wetting (the three regions are not drawn to scale).

on the slip boundary condition chosen. For $\vartheta \leq 135°$, $G(\vartheta)$ is closely approximated by [67]

$$G(\vartheta) \approx \frac{\vartheta^3}{9} \tag{23a}$$

Equation (22) implies that the angle ϑ_R is independent of the macroscopic flow. In contrast to θ_D, ϑ_R should therefore be a measurable material property of a particular liquid/solid system that could be determined from a careful comparison between theory and measurements of the interfacial shape near the wetting line in one particular flow geometry. Measured values of ϑ_R could then be used as a boundary condition in analyses of other flow geometries without the need to resolve the local wetting dynamics when the outer flow is of main interest, or to specify an ad hoc model for the cutoff mechanism and local contact angle when a refined theory is not available [45]. In particular, the angle ϑ_R could serve as an empirical parameter in the expression [23,24]

$$\vartheta(r) \equiv G^{-1}\left\{G(\vartheta_R) + Ca \ln \frac{r}{R}\right\} \tag{24}$$

which describes the slope of the interface in the intermediate region (here, r is the distance from the wetting line).

To validate the usefulness of the intermediate angle ϑ_R as a boundary condition in flow calculations away from the immediate vicinity of the wetting line, Ngan and Dussan V. [25] sought to predict the size effect that they had observed earlier in experiments [37] (see also Sec. II.A.3).

Unable to directly measure the meniscus shape of a front of silicone oil displacing air between two glass slides, Ngan and Dussan V. [25] deduced families of $\{(R, \vartheta_R)\}$ indirectly from apparent dynamic contact angles $\theta_{D,M}$ at different spacings between the slides L. Results based on Eq. (24) replicated the correct trend, that is, a reduction of the apparent angles with diminishing L, but showed a weaker size effect than was measured. Dussan V. et al. [45] produced more direct experimental evidence validating Eq. (24) by immersing a glass rod into a bath of silicone oil at constant speed and measuring the meniscus profiles near the wetting line through a long working distance microscope. The measurements were restricted to $R > 20$ μm and thus covered mostly the outer region. As a consequence, Dussan V. et al. had to use $\vartheta_{m0} = G^{-1}[G(\vartheta_w) + Ca \ln(\epsilon_s^{-1})]$ rather than ϑ_R as an empirical parameter in a composite solution for the interface profile in both the intermediate and outer region. Nonetheless, calculated profiles agreed well with measured ones.

The procedure advocated by Dussan V. and co-workers [25,45] is potentially quite useful, but is only applicable when the asymptotic theory upon which Eq. (24) is based is valid (see Sec. III.A.4). In addition, the procedure is not genuinely predictive because it involves an empirical parameter (i.e., ϑ_R) that has to be established from difficult measurements very close to the wetting line. The measurements must be repeated for each liquid/solid system, making the experimental task even more formidable. By identifying the intermediate angle ϑ_R as a material function, Dussan V. and her collaborators [25,45] imply that ϑ_R is a lump sum of specific material effects in the immediate vicinity of the wetting line. They therefore share the view with others [8,16,17,69,88,89,236,239] that apparent dynamic contact angles cannot be predicted from first principles without resolving the submicroscopic wetting physics.

In sharp contrast, the nearly universal behavior $\theta_D = f(Ca, \theta_0)$ that is described in Sec. II.A suggests that the increase of θ_D with rising viscosity η_l or displacement speed U is mostly of hydrodynamic origin and fairly immune to specific physico-chemical processes very close to the solid. Conventional continuum theory should therefore be able to predict the universal curve.

Two main approaches have been pursued in attempts to predict apparent dynamic contact angles from strictly hydrodynamic models. One approach makes use of the singular limit $Ca \rightarrow \infty$ in which the capillary pressure term in the traction condition along the free surface becomes negligible, except perhaps in a small inner region at the wetting line. In this limit, the interface cannot transmit the influence of the local wetting physics upstream. Calculated apparent dynamic contact angles ϑ_D are dominated by the flow field away from the wetting line and become dependent variables that can be computed without specifying a boundary condition for the local angle ϑ_w [78]. Predicted values of ϑ_D, however, are sensitive to the distance from the wetting line at which the angle reading is taken from the calculated meniscus profile [240]. Typical angles fall in the range $150° < \vartheta_D < 180°$ [78]. When the no-slip condition is enforced right up to the wetting line and the analysis refined toward the line, predicted angles at which the interface meets the solid surface approach 180° [240,241], and the limit of a rolling motion is recovered [221–223]. Certain time-stepping algorithms used in computational analyses of transient viscous fronts implicitly enforce a rolling motion without slip and also predict dynamic contact angles near 180° [242]. Hoffman [29] was the first to suggest that apparent contact angles should be 180° when viscous forces alone control the shape of the interface. Even in the limiting regime of a rolling motion, however, apparent contact angles calculated at a finite distance away from the wetting line may be substantially less than 180° [240].

An alternative approach, relevant when surface tension effects remain significant, ascribes all the dynamic variations in the apparent contact angle to visco-capillary bending of the interface close to the three-phase juncture. Commonly, a fixed, small value for the slip length is chosen (i.e., ϵ_s = const. $\ll 1$) and the local contact angle is set equal to the static contact angle (i.e., $\vartheta_w = \theta_0$) [23,24,97,164,171,174,184,243–245]. However, ϑ_w set equal to 180° [221,222] and dynamic models for $\vartheta_w = f(U)$ have also been used [97]. The particular choice $\vartheta_w = \theta_0$ has led to auspicious agreement between theory and experiment, especially for liquids displacing air, and deserves more detailed discussion.

4. Predicted Apparent Dynamic Contact Angles

Among the first successful predictions of apparent dynamic contact angles from hydrodynamic theories were those based on computational techniques. Lowndes [174] used a finite-element technique to solve the complete Stokes equations for single-phase capillary displacement with local slip. He assumed that $\vartheta_w = \theta_0$ and $L_\beta = 1$ nm and deployed elements significantly smaller than O(1 nm) near the wetting line to ensure that local bending of the meniscus was accurately resolved. Apparent contact angles calculated from the apex height of computed meniscus profiles agreed to within $\pm 13°$ with the experimental θ_D data of several investigators [26,27,29]. Earlier, Hansen and Toong [22] also calculated apparent dynamic contact angles that corroborated experimental data [27]. Their results, however, were restricted to $Ca \leq 3 \times 10^{-3}$ and so the deviations of the calculated angles ϑ_D from θ_0 remained moderate. Hansen and Toong [22] combined numerical and conventional methods of mathematical anal-

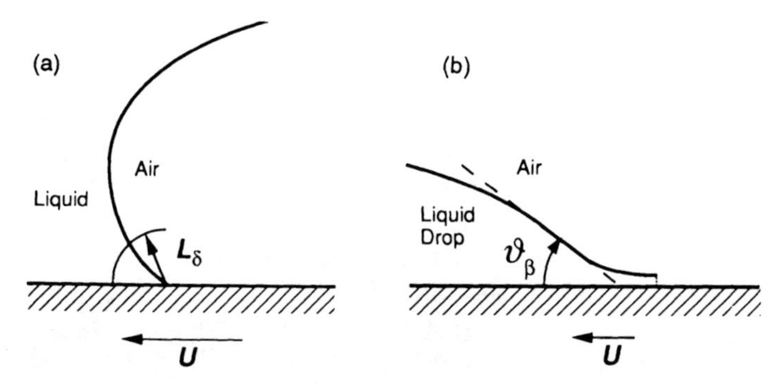

FIG. 16 Schematic of analyses that avoid the vicinity of wetting line: (a) excluded circular region [22,67] and (b) truncated film [66].

ysis. They excluded a circular region of radius 10 nm $\leq L_\delta \leq 0.1$ μm around the contact line, as illustrated in Fig. 16a, and set the slope of the interface to θ_0 at the edge of that region.

Early attempts to analyze the local meniscus deformation near dynamic wetting lines by means of matched perturbation expansions fell short in predicting realistic variations of apparent contact angles because analyses relied on asymptotic limits outside the range covered by most experiments. When wall slip permeates the entire flow, that is, $\epsilon_s \equiv L_s/L = O(1)$, perturbation solutions are uniformly valid for $Ca \rightarrow 0$, but corrections to the nearly static meniscus shape remain small, that is, $O(Ca \mid \ln \epsilon_s \mid)$ [181]. Hence, calculated apparent angles ϑ_m remain close to the value ϑ_w imposed at the solid surface. When slip is confined to the neighborhood of the wetting line but the slip length is larger than submicroscopic dimensions ($0.001 < \epsilon_s < 0.1$), numerical solutions of lubrication-flow equations for spreading drops [182,228,243] yield meniscus profiles that deviate substantially from a quasistatic, spherical-cap shape. In particular, the calculated profiles exhibit foot structures of the sort suggested by Brochard and de Gennes [6,209] and observed in some experiments (see Sec. III.A.2). When slip is confined to submicroscopic distances from the wetting line ($\epsilon_s \rightarrow 0$), perturbation analyses become singular and require matching between an inner region of size $O(\epsilon_s \equiv L_s/L)$ and an outer region of size O(1) [25,164,179,185,238] (here, length is made dimensionless with L). In the limit of $Ca \rightarrow 0$, however, the inner and outer regions overlap only if $Ca \ln(\epsilon_s^{-1})$ also tends to zero [23,24]. The limit $Ca \ln(\epsilon_s^{-1}) \rightarrow 0$ is unattractive because ϑ_m again cannot deviate much from ϑ_w. Huh and Mason [164] nonetheless attempted to compare, with modest success, their asymptotic result $\vartheta_m - \vartheta_w = 4\ Ca/\pi\ (\ln \epsilon_\beta^{-1} + 1.188)$ for capillary displacement with experimental data. Huh and Mason [164] relied on domain perturbation techniques and, thus, their analysis was further restricted to $\vartheta_w \approx 90°$ and $\vartheta_m \approx 90°$. Similar restrictions apply also to the uniformly valid series expansion by which van Quy [184] earlier approximated the meniscus shape of a liquid advancing in a capillary.

When calculated apparent dynamic contact angles deviate substantially from the angle imposed at the solid, a third region is needed to link the inner region, where the slope of the meniscus is dominated by ϑ_w, to the outer region, where up to $Ca < O(0.1)$ the interface closely resembles a static meniscus that appears to intersect the solid at a dynamic contact angle ϑ_m. Voinov [67] was the first to identify a region where the slope of the interface changes rapidly as the solid is approached to meet a local angle dictated by molecular forces, but varies slowly further away from the solid to blend with the macroscopic meniscus. Like Hansen and Toong [22], Voinov [67] excluded the cutoff region from his analysis and prescribed an angle ϑ_δ at a distance L_δ from the putative wetting line. He derived a profile equation for the meniscus close to but not right at the wetting line by balancing the local capillary pressure with the normal stresses acting on a straight wedge with the local slope $\vartheta(r)$ and by calculating the normal stresses from Moffat's [162] solution. The result is

$$G[\vartheta(r)] - G(\vartheta_\delta) = Ca \ln \frac{r}{L_\delta} \tag{25}$$

where $G(\vartheta)$ is defined by Eq. (23). The interface profile (25) is, in essence, the same as that described by Eq. (24). As suggested in Sec. III.A.2, the shape of the meniscus outside the cutoff region is insensitive to the specific cutoff mechanism that is invoked to alleviate the unbounded force singularity. Undeservedly, Voinov's [67] key contribution has not been widely recognized. Boender et al. [177], apparently unaware of Voinov's work, integrated an interfacial equation similar to (25) numerically and found excellent agreement with finite-element solutions of the complete equation set with local slip [174,226].

Hocking [23,244], who was evidently also unfamiliar with Voinov's [67] paper, was the first to deduce from formal expansion procedures that the interface profile near the wetting line has a double structure (see Fig. 15). In the limit of $Ca \ln(\epsilon_s^{-1}) = O(1)$ while $Ca \to 0$ and $\epsilon_s \to 0$, which is much more interesting than the limit $Ca \ln(\epsilon_s^{-1}) \to 0$ studied before [25,164,179,185,238], an *intermediate region* is required to properly match the inner and outer regions. The intermediate region is, of course, the same as the one identified and analyzed by Voinov [67]. Its characteristic width, $\epsilon_i \equiv 1/|\ln \epsilon_s| \gg \epsilon_s$, is much larger than the size of the cutoff region where conventional continuum physics break down, slippage dominates when imposed as an ad hoc boundary condition to alleviate the force singularity, and microstructural forces are most influential when incorporated in refined theories. On the other hand, the intermediate region is much smaller than the macroscopic flow, of which the wetting line is part. Lacey [245] independently identified the importance of the parameter ϵ_i, but employed a multiscale expansion technique rather than matching procedures. Pismen and Nir [222] also investigated a region of size ϵ_i.

To date, Cox [24] has performed the most general analysis of the consequences of the three-region structure near dynamic wetting lines. At leading order in Ca, an apparent dynamic contact angle ϑ_{m0} can be calculated without solving for a specific flow field and interfacial shape in the outer region or specifying a cutoff mechanism in the inner region. Cox [24] adopted the notion that slip is necessary to remove the force singularity, but other cutoff mechanisms would yield to the same leading-order result, which is remarkably simple

$$g(\vartheta_{m0}, \lambda_\eta) - g(\vartheta_w, \lambda_\eta) = Ca \ln(\epsilon_s^{-1}) + O(Ca) \tag{26}$$

At order Ca^{+0}, the outer meniscus profile remains quasistatic, and ϑ_{m0} is merely the angle at which the extrapolated macroscopic profile intersects the solid. The function $g(\vartheta, \lambda)$ is

$$g(\vartheta, \lambda) \equiv \int_0^{\vartheta} \frac{d\beta}{f(\beta, \lambda)} \tag{27}$$

where

$$f(\vartheta, \lambda) \equiv \frac{2 \sin\vartheta\{\lambda^2(\vartheta^2 - \sin^2\vartheta) + 2\lambda[\vartheta(\pi - \vartheta) + \sin^2\vartheta] + [(\pi - \vartheta)^2 - \sin^2\vartheta]\}}{\lambda(\vartheta^2 - \sin^2\vartheta)[(\pi - \vartheta) + \sin\vartheta\cos\vartheta] + [(\pi - \vartheta)^2 - \sin^2\vartheta](\vartheta - \sin\vartheta\cos\vartheta)} \tag{28}$$

At the next higher order Ca^{+1}, Cox's generic result is [24]

$$g(\vartheta_m, \lambda_\eta) - g(\vartheta_w, \lambda_\eta) = Ca \ln(\epsilon_s^{-1}) + Ca\left[\frac{q_i}{f(\vartheta_w, \lambda_\eta)} - \frac{q_0}{f(\vartheta_m, \lambda_\eta)}\right] + O(Ca^2) \tag{29}$$

Here, q_i is an integration constant that depends on the details of the inner solution, that is, λ_η, ϑ_w, and the specific slip model used. Similarly, q_0 must be determined from a particular outer solution and is a function of

λ_η, ϑ_m, $d\vartheta_m/dt$, and the global flow field. Earlier, Hocking and Rivers [23] obtained

$$G(\vartheta_m) - G(\vartheta_w) = Ca \ln\left(\frac{R_0}{L_\beta}\right) + Ca \ln\left[\frac{R(t)}{R_0}\right] + Ca[Q_i(\vartheta_w) - Q_0(\vartheta_m)] \quad (30)$$

for the particular problem of a drop spreading by capillarity on a flat surface. Hocking and Rivers' result (30) has a structure very similar to Cox's Eq. (29). In Eq. (30), $R(t)$ is the instantaneous radius of the drop, R_0 is the equilibrium radius at the static contact angle, and $Ca \equiv \eta_1 U/\gamma_{12}$ is formed with $U = dR/dt$. The functions $Q_i(\vartheta_w)$ and $Q_0(\vartheta_m)$ are analogous to $q_i/f(\vartheta_w, \lambda_\eta)$ and $q_0/f(\vartheta_m, \lambda_\eta)$ in Eq. (29); $G(\vartheta)$ is the same as defined in Eq. (23). At leading order, Hocking and Rivers' result (30) is the same as Voinov's [67] main result

$$\frac{1}{2}\int_{\vartheta_\delta}^{\vartheta_{m0}} \frac{\varphi - \sin\varphi\cos\varphi}{\sin\varphi}\, d\varphi = G(\vartheta_{m0}) - G(\vartheta_\delta) = Ca \ln(\epsilon_\delta^{-1}) \quad (31)$$

Voinov showed that Eq. (31) is well approximated by

$$\vartheta_{m0}^3 - \vartheta_\delta^3 = 9\, Ca \ln(\epsilon_\delta^{-1}) \quad (\vartheta_{m0} \leq 135°) \quad (32a)$$

$$\frac{9\pi}{4} \ln \frac{1 - \cos\vartheta_{m0}}{1 + \cos\vartheta_{m0}} + (\pi - \vartheta_{m0})^3 - \vartheta_\delta^3 = 9\, Ca \ln(\epsilon_\delta^{-1}) \quad (\vartheta_{m0} \geq 135°) \quad (32b)$$

If the viscosity of the receding phase is negligible, $g(\vartheta, \lambda = 0) = G(\vartheta)$ and Cox's leading order result (26) is the same as (31) and hence is also closely replicated by Eqs. (32a) and (32b).

Even at leading order, hydrodynamic theories predict apparent dynamic contact angles that agree remarkably well with experimental data when ϑ_w or ϑ_δ is set to θ_0, and ϵ_s is taken to be independent of U and adjusted empirically to optimize the fit between theory and experiment. For liquid/solid systems with complete wetting ($\vartheta_w = 0$) and negligible air viscosity ($\lambda_\eta \ll 1$), Eq. (26), as well as its predecessors, Eqs. (30) and (31), corroborate the experimental finding that the apparent contact angle ϑ_{m0} is primarily a function of capillary number in both forced wetting (see Figs. 3 and 5) and spontaneous spreading (see Fig. 13). For $\lambda_\eta = 0$, in fact, the inverse functions $\vartheta_{m0} = g^{-1}[Ca \ln(\epsilon_s^{-1})]$ or $\vartheta_{m0} = G^{-1}[Ca \ln(\epsilon_\delta^{-1})]$ provide good mathematical descriptions of Hoffman's universal function $\theta_D = f_{\mathrm{Hoff}}(Ca)$, as Fig. 17 shows. Equations (26) and (31) not only replicate the shape of the experimental θ_D versus Ca curve,

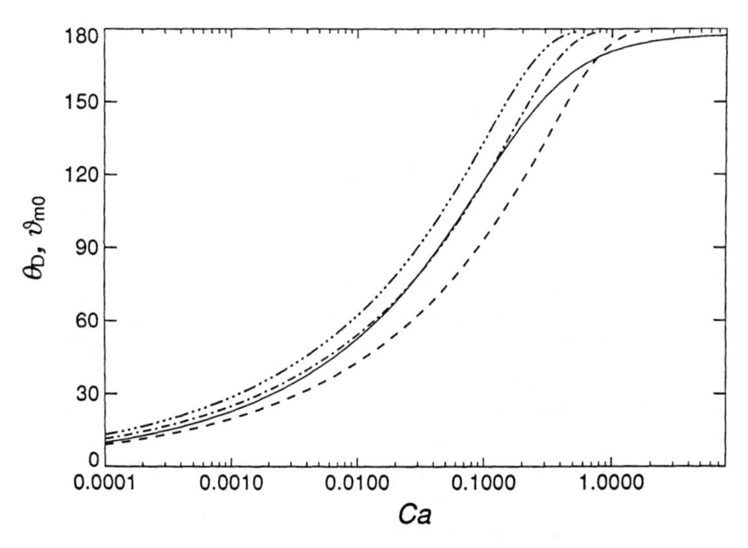

FIG. 17 Comparison between Hoffman's [29] universal function $\theta_D = f_{\mathrm{Hoff}}(Ca)$ [—— Eq. (3)] and Cox's [24] leading-order result (26) (------ $\epsilon_s = 10^{-2}$, —·— $\epsilon_s = 10^{-4}$, —···— $\epsilon_s = 10^{-6}$; $\lambda_\eta = 0$, $\vartheta_w = 0°$).

for which the theory leaves no adjustable parameters, but also properly anticipate the range of Ca over which θ_D rises from very small values to 180°, for which the empirical parameter ϵ_s allows only a logarithmically small shift in Ca. The only systematic deviation occurs above $Ca \geq 0.2$, or equivalently $\theta_D \geq 135°$, where calculated values of ϑ_{m0} rise more rapidly with increasing Ca than measured θ_D data (see Fig. 17). Nonetheless, predicted apparent dynamic contact angles approach 180° asymptotically when $\lambda_\eta = 0$, as is observed in experiments with highly viscous liquids (see Sec. II.A.3). One may be inclined to attribute the quantitative discrepancy between hydrodynamic theory and experiment to the neglect of higher-order correction terms, or even a complete failure of the asymptotic expansions that are formally valid only in the limit $Ca \ll 1$ [24]. Apparent dynamic contact angles obtained from finite-element solutions of the complete set of Navier–Stokes equations and free-surface boundary conditions [227] indeed agree better with Hoffman's universal curve (3) than ϑ_{m0} values calculated from Eq. (26), but above $\theta_D \geq 150°$, predicted angles are still larger than angles measured for $\lambda_\eta \ll 1$. Evidently, purely hydrodynamic theories with ad hoc boundary conditions at a putative wetting fail to accurately describe the θ_D versus Ca curve at capillary numbers larger than $Ca > O(0.1)$.

For small apparent dynamic contact angles on completely wettable surfaces ($\vartheta_w = \theta_0 = 0$), Eq. (26) simplifies to

$$\vartheta_{m0}^3 = 9\, Ca \ln(\epsilon_s^{-1}) \tag{33}$$

This simple form arises because $g(\vartheta, \lambda)$ is closely approximated by $\vartheta^3/9 + \lambda\vartheta^4/8\pi + O(\vartheta^5)$ when $\vartheta \ll \pi/2$ [71]. Equation (33), and of course also Eq. (32a) with $\vartheta_\delta = 0$, are particular versions of the *Hoffman–Voinov–Tanner law* (5) that, in turn, agrees with experiments (see Figs. 4 and 13). For $10^{-6} < \epsilon_s < 10^{-2}$, the constant prefactor $c_T = 9 \ln(\epsilon_s^{-1})$ in Eq. (33) falls in the range $40 < c_T < 130$, which brackets the value $c_T \approx 90$ that provides the best fit through the data in Fig. 4. For $\lambda_\eta \ll 1$, the approximation $g(\vartheta, \lambda) \approx \vartheta^3/9$ remains valid up to $\vartheta \le 135°$ and accords with Voinov's [67] original result (23a) that does not account for viscous effects in the receding airflow. Voinov's [67] and Cox's [24] analyses rationalize why the Hoffman-Voinov-Tanner law (5) competently describes the macroscopic wetting behavior of completely wetting liquid/solid systems up to much higher values of Ca (see Fig. 4) than theories based on the lubrication-flow approximation might suggest [6,14,66].

Tanner [66] derived the relationship $\vartheta_D \sim Ca^{1/3}$ directly from a lubrication-flow approximation in which the ordinary differential equation

$$\frac{d^3h}{dx^3} = 3\, Ca \frac{1}{h^2} \tag{34}$$

describes the profile of a thin liquid front advancing over a dry surface. Tanner avoided the immediate proximity of the wetting line by truncating the analysis where solutions to (34) exhibit a minimum and calculated an apparent contact angle ϑ_β from the slope at the inflection point where the angle of inclination of the advancing meniscus is largest (see Fig. 16b). Solutions of (34) with small curvature at large distances $h(x)$ from the solid assume the limiting form $h(x) \to -x[Ca \ln(x/x_\delta)]^{1/3}$ where x_δ is a cut-off distance [6]. This form suggests that the interface assumes the shape of a wedge of nearly constant slope between a sharply curved meniscus close to the solid and a quasistatic macroscopic meniscus [6], in agreement with Eqs. (24) and (25) as well as detailed measurements of meniscus profiles of spreading drops [73,132]. Tanner [66], and others after him [187,245,246], deduced also that droplets of completely wetting liquids spread according to

$$R \sim t^{1/10} \tag{35}$$

if small enough to retain the shape of a spherical cap away from the immediate proximity of the solid. Like Eq. (33), the simple power law (35) agrees well with experimental data (see Sec. II.D.1).

For liquid/solid systems with partial wetting (i.e., $\vartheta_w = \theta_0 > 0$), Voinov's result (31) and also Cox's leading-order result (26) have the same mathematical structure as the universal form (6b) first proposed by Hoffman [29], and hence, they adequately describe experimental data for selected liquid/solid systems that are not susceptible to peculiar low-speed behavior (see Fig. 6). Numerical solutions of the complete Navier–Stokes equations with the ad hoc boundary conditions $\vartheta_w = \theta_0 \neq 0°$ and $\epsilon_s \ll 1$ also predict that apparent dynamic contact angles collapse onto a single curve when a uniform shift factor $g(\theta_0, \lambda_\eta)/\ln(\epsilon_s^{-1})$ is added to Ca [97,227]. For $\lambda_\eta \ll 1$ and $\vartheta_w \leq \vartheta_{m0} < 135°$, Eq. (26) assumes the limiting form

$$\vartheta_{m0}^3 - \vartheta_w^3 = 9\, Ca \ln(\epsilon_s^{-1}) \tag{36}$$

which is the same as Eq. (32a). Both Eqs. (32a) and (36) provide a theoretical basis for the modified Hoffman–Voinov–Tanner law (7). Results of the form (36) emerge also from leading-order asymptotic solutions of lubrication-flow approximations for thin, two-dimensional drops spreading under the action of capillarity [244,245] and numerical integration of third-order film profile equations like (34) [173].

Even though simple hydrodynamic theories are able to capture the dynamic behavior of $\theta_D = f(Ca, \theta_0)$ for selected liquid/solid systems, the ad hoc assumptions $\vartheta_w = \theta_0$ and $\epsilon_s = \text{const.}$ exclude the influence of specific material effects a priori and fail to account for the systematic deviations and scatter characteristic of θ_D data on partially wettable surfaces. In particular, the theories discussed so far presuppose that the meniscus advances in a steady motion over a smooth, homogeneous surface and, thus, cannot describe peculiar low-speed behavior such as temporary contact line pinning or stick-slip resulting from it. Nevertheless, if the consequences of low-speed anomalies can be absorbed in a pseudostatic angle θ_0^* that possibly may be velocity-dependent itself (see Sec. II.A.4), hydrodynamic theories with $\vartheta_w = \theta_0^*(U)$ account for a significant portion of the dynamic variations of the apparent contact angle at higher speeds. Progress toward modeling low-speed wetting on inhomogeneous, partially wettable surfaces is summarized in Sec. III.A.7.

de Gennes [94] considered the limiting regime of small, yet finite static contact angles on smooth surfaces (i.e., $0 < \theta_E \ll \pi/2$) and exceedingly small capillary numbers (i.e., $Ca \leq O(\theta_E^3)$]. In this regime, the lubrication approximation applies and, away from the immediate vicinity of the wetting line, the deviations in meniscus shape from a straight wedge remain logarithmically small [6,247]. de Gennes [94] assumed that the meniscus is a straight wedge, and its slope $\vartheta_a > \theta_E$ is dictated by a balance between the viscous dissipation Φ_v at distances larger than L_δ from the solid [see Eq. (20b)] and the entropy $F_Y U$ generated by the wetting force $F_Y = \gamma_{12}$

$(\cos \theta_E - \cos \vartheta_a)$. Hydrodynamic theories that assume $\vartheta_w = \theta_0$, in contrast, permit viscous bending in the "wedge region." As a consequence, de Gennes' [94] result

$$\vartheta_a(\vartheta_a^2 - \theta_E^2) = 6\, Ca \ln(\epsilon_\delta^{-1}) \tag{37}$$

is similar but not identical to the *generalized Hoffman–Voinov–Tanner laws* (32a) and (36). In particular, Eq. (37) does not have the mathematical structure $g(\vartheta_a) - g(\theta_E) = f(Ca)$ that Hoffman [29] first postulated and other hydrodynamic theories corroborate. In addition, the coefficient 6 $\ln(\epsilon_\delta^{-1})$ in (37) is different from 9 $\ln(\epsilon_\delta^{-1})$ in (32a) and (36). When $(\vartheta_a - \theta_E)/\theta_E \ll 1$, nonetheless, Eq. (37) closely approximates the low-speed limit (12) and, like the generalized Hoffman-Voinov-Tanner laws (32a) and (36), predicts a linear increase of the unbalanced Young's force $F_Y = \gamma_{12}$ (cos θ_E − cos ϑ_a) with displacement speed U. For complete wetting (i.e., $\theta_E = 0$), Eq. (37) reduces to the familiar form $\vartheta_a \sim Ca^{1/3}$ of the Hoffman-Voinov-Tanner law, even though de Gennes' [94] derivation does not apply rigorously when $S > 0$ (see Sec. III.B.1).

5. Local Wetting Dynamics: Success or Failure of Ad Hoc Assumptions?

Encouraged by apparent agreement between experimental data and hydrodynamic models with the ad hoc boundary conditions $\vartheta_w = \theta_0$ and $\epsilon_\delta = \text{const.} \ll 1$, several authors [23,24,164,174] argued that the contact angle on a submicroscopic scale retains its static value θ_0, even under dynamic conditions. Furthermore, many investigators inferred operative slip lengths L_s from comparisons between theoretical predictions of ϑ_m and experimental θ_D data and relied on the values obtained to speculate about specific wetting mechanisms at work [23,24,31,34,71,97]. Inferred slip lengths typically fall in the range 1 nm $< L_s <$ 0.1 μm. They are shorter than the nearly macroscopic length scales that have been suggested for apparent slip due to roughness [18,95] or submicroscopic air entrainment [56], but longer than length scales of a few Å that are often postulated for slip by molecular mechanisms on smooth surfaces [6,164, 174,175,191]. Hocking and Rivers [23] emphasized that the first-order correction terms $Q_i = O(1)$ and $Q_0 = O(1)$ in Eq. (30) are not substantially smaller than $\ln(\epsilon_s^{-1}) = O(10)$ and must be included when effective slip lengths are deduced from comparisons between ϑ_m and θ_D. Hocking and Rivers [23] calculated $Q_i(\vartheta)$ and $Q_0(\vartheta)$ for the particular problem of spreading drops and found good agreement with their own data for molten glass on platinum when selecting extrapolation lengths in the range 1 nm $< L_\beta <$ 0.1 μm. Finite-element solutions for capillary displacement [227], which implicitly account for higher-order corrections, suggest $L_\beta \approx$ 10 nm for the best fit with Hoffman's universal curve (6). This value is indeed smaller than the slip length $L_s \approx 0.1$ μm for which Cox's [24] leading-order result (26) fits experimental data best (see Fig. 17). Here, L is taken to be 1 mm, which is representative of most experimental configurations. Lowndes [174] assumed a priori that $L_\beta = 1$ nm and systematically overestimated apparent dynamic contact angles in capillary displacement.

The assertion that the actual contact angle at the solid surface does not vary with displacement speed lacks unequivocal scientific verification. The crux is that most experiments probe the meniscus profile at distances of several tens of microns or even further from the solid where numerous models for the cutoff region yield the same results at leading order, including refined models that seek to describe the local wetting dynamics. Therefore, agreement between theoretical predictions and experimental observations, such as apparent contact angles, rates of spontaneous spreading, or forces needed to advance a liquid front over a solid, does not validate the assumptions made for the details within the cutoff region and cannot be invoked to identify the mechanisms by which a liquid advances over a solid [3,25,45,238]. For instance, various functional forms for velocity-dependent values of ϑ_w, ϵ_s, and q_i could be substituted into Eqs. (29–31) to construct the same dynamic behavior of $\vartheta_m = f(Ca, \theta_0)$ [25,45]. Figure 18 illustrates that slip on molecular length scales (e.g., $\epsilon_s \leq 10^{-6}$) and small static angles yield comparable apparent angles as slip on nearly macroscopic length scales (e.g., $\epsilon_s \geq 10^{-2}$) and local angles much larger than 90°. On the other hand, the steep slope of the curves in Fig. 18 toward $\vartheta_w \to 0°$ indicates that the assumption $\vartheta_w = \theta_0$ is not all that critical to the predicted values of ϑ_m when $Ca > g(\vartheta_w)/\ln(\epsilon_s^{-1})$ [177]. Thus, even if macroscopic observations were able to provide insights into the local wetting dynamics, they would not constitute a sensitive test at high displacement speeds.

Further doubts about the validity of the ad hoc assumption $\vartheta_w = \theta_0$ arise from the exceedingly small cutoff lengths that the assumption suggests. At distances from the solid smaller than $L_\delta < 0.1$ μm, van der Waals forces or other long-range interactions influence and possibly even dominate the interface profile [6,191]. Thus, capillary hydrodynamics alone cannot describe the interfacial profile in the cutoff region. For $L_\delta < 30$ nm, short-range interaction forces come into play and, even under static conditions, the local angle of inclination of the interface may deviate from θ_0 measured on larger scales [6,19,21,74,214,248,249]. At molecular length scales, the concepts of interfaces and wetting lines becomes merely mathematical substitutes for highly dynamic molecular events, yet those are often the scales of the inner region where mathematical analyses insist that the boundary condition $\vartheta_w = \theta_0$ be satisfied.

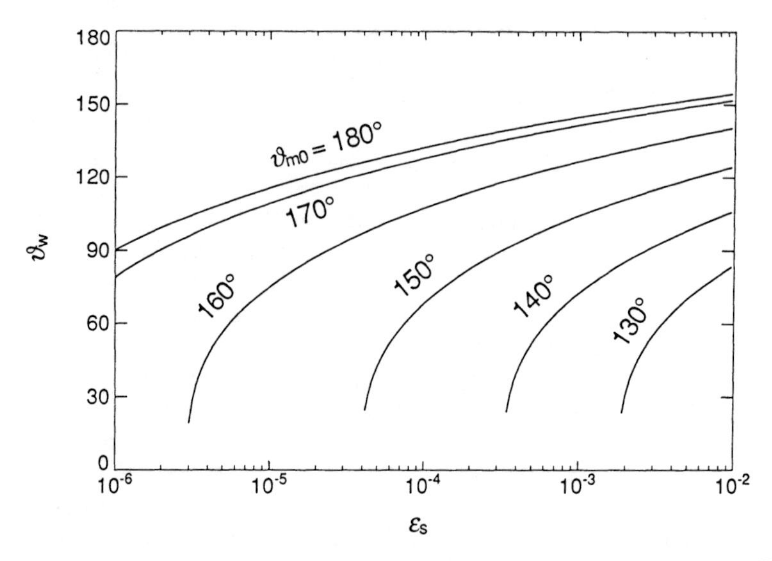

FIG. 18 Contours of constant apparent dynamic contact angle ϑ_{m0} [Eq. (26)] as a function of slip length ϵ_s and contact angle ϑ_w imposed at the solid surface ($Ca = 0.2$, $\lambda_\eta = 10^{-3}$).

To validate any model for the local wetting dynamics, measurements are needed at the local length scales that the model tries to describe. Except for systems that form precursor films at very low rates of spontaneous spreading (see Sec. II.D.2 and III.B.1), such measurements have not been realized. Even at distances of tens of microns from the solid, accurate measurements of the meniscus profile are rare [45,56,73,132]. The profiles measured by Dussan V. et al. [45] confirm that the meniscus can bend substantially at length scales accessible with optical microscopes, as hydrodynamic theories predict. The profiles agree with Eq. (24) that, in turn, is consistent with the ad hoc assumptions $\vartheta_w = \theta_0$ and a fixed value of $\epsilon_s \ll 1$. Hence, the limited experimental evidence available to date does not refute the appropriateness of the ad hoc assumptions. On the contrary, for liquid/gas/solid systems that follow the universal behavior (2) or (6), calculations based on the assumptions $\vartheta_w = \theta_0$ and $\epsilon_s = \text{const.} \ll 1$ must correctly predict the dynamic behavior of the intermediate angle ϑ_R identified by Dussan V. and colleagues [25,45,238] [see Eq. (22)]. Otherwise, calculated predictions of the macroscopically observable contact angle ϑ_m would not agree with experimental observations, unless $\epsilon_s \ll 1$ was not satisfied. The ad hoc assumptions $\vartheta_w =$

θ_0 and ϵ_s = const. provide therefore a useful working hypothesis for calculating macroscopic flow fields and free surface shapes. In particular, the hypothesis can be applied to assess the influence of macroscopic flow fields on apparent dynamic wetting behavior (see Sec. III.A.6).

The success of strictly hydrodynamic theories in predicting ϑ_R for selected liquid/solid systems does not contradict the notion that ϑ_R is a material function, as demanded by Dussan V. and co-workers [25,45,238]. It suggests, however, that liquid viscosity η_1, surface tension γ_{12}, and static contact angle θ_0 are predominant material parameters. It thereby furnishes additional, compelling reasons to speculate that for many liquid/solid systems, especially those with complete static wetting, the dynamic increase of ϑ_R—and hence also θ_D—is dominated by visco-capillary mechanisms away but not too far from the wetting line, rather than specific material effects right at the three-phase juncture.

Refined theories for the cutoff region furnish additional, albeit indirect, support for the notion that the mechanisms which control the local displacement process are quite distinct from those that set the apparent dynamic contact angle—at least at low and moderate displacement speeds. Molecular dynamics simulations of one Lennard-Jones fluid displacing another corroborate that the local contact angle remains close to its static value [207]. Augmented continuum theories that account for long-range interactions also indicate that within the cutoff region, the interfacial shape is not altered much by the macroscopic flow [6,7,14,191]. de Gennes et al. [191], for instance, included van der Waals forces in a lubrication flow analysis of a plate being dragged at a small angle $\alpha = \theta_E \ll \pi/2$ into a shallow pool of liquid. Even though van der Waals forces dominate the calculated meniscus profile in the cutoff region, which is quite large for $\theta_E \ll \pi/2$ (see Sec. III.A.2), capillary pressure and Poiseuille friction control the dynamic variations of the apparent contact angle in an intermediate region further away from the solid where the slope of the interface varies logarithmically slowly as a function of distance from the wetting line [191]—as it does in purely hydrodynamic models [see Eqs. (24) and (25)]. Right at the wetting line, of course, the augmented continuum theory of de Gennes et al. [191] breaks down and cannot elucidate the physics of the local contact angle.

Experimental observations at macroscopic length scales provide further, circumstantial evidence that an intermediate region dominates the dynamic wetting behavior visible at macroscopic length scales. Material effects are often less important than might be expected (see Sec. II.A.3, II.A.4, and II.D.1), consistent with hydrodynamic theories which predict that the dynamic variations of apparent contact angles are insensitive to the wetting mechanisms in the cutoff region. On the other hand, the local meniscus deformation escapes detection through low-powered microscopes (see Sec. II.A.1 and II.A.3), in accord with the theoretical result that measurable bending occurs at length scales significantly smaller than the macroscopic meniscus. Contrary to claims made by some [51,96], quasistatic meniscus shapes observed at macroscopic length scales do not prove that dynamic variations in the contact angle arise from mechanisms other than ordinary visco-capillary effects. Hydrodynamic theories [24,244] confirm that macroscopic menisci retain a nearly static shape at capillary numbers as large as $Ca \leq 0.1$, as seen in experiments [29,47], yet suggest that purely viscous meniscus bending can induce measurable variations in θ_D at capillary numbers as low as $Ca \gtrsim 10^{-4}$.

Calculated meniscus profiles whose slope varies logarithmically slowly as the macroscopic meniscus is approached are consistent with the experimental observation that apparent dynamic contact angles are insensitive to the technique by which the angles are measured (see Sec. II.A.3) and thereby lend additional credibility to hydrodynamic theory. To illustrate this point, Fig. 19 compares computed predictions of several apparent contact angles for capillary displacement between parallel plates [227]. The angle $\vartheta_{D,M}$ that is calculated from the apex height of the meniscus is within 5° from the angle $\vartheta_{D,\text{poly}}$ that is inferred when a fourth-

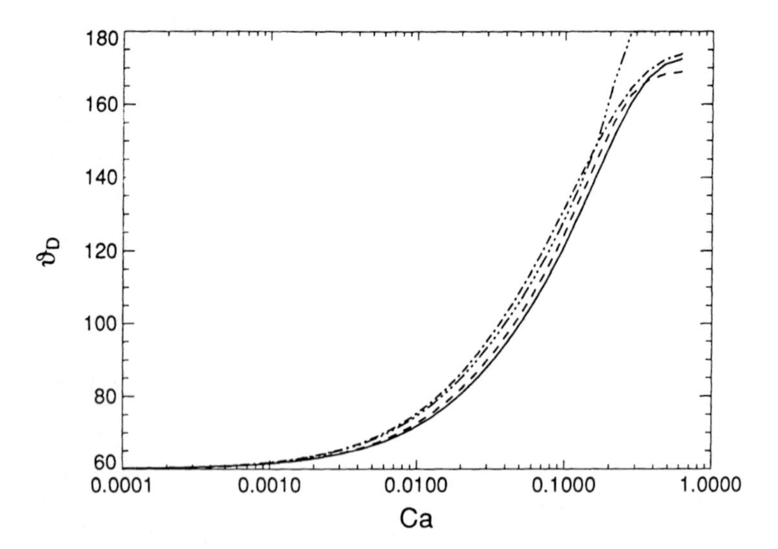

FIG. 19 Finite-element predictions of apparent dynamic contact angles for capillary displacement between parallel plates [227] (—— $\vartheta_{D,T}$, ------ $\vartheta_{D,\text{poly}}$, —·— $\vartheta_{D,M}$, —···— $\vartheta_{D,\Delta P}$; $Re = 0$, $Bo = 0$, $\vartheta_w = 60°$, $\epsilon_\beta = 10^{-6}$, $\lambda_\eta = 0$).

order polynomial is fit to the meniscus profile at distances greater than 20 μm from the solid and extrapolated to the solid surface. The angle of inclination $\vartheta_{D,T}$ of the computed interface at 5 μm from the solid surface is still at most 10° smaller than $\vartheta_{D,M}$ over the entire range of *Ca*. Even the angle $\vartheta_{D,\Delta P}$ that is deduced from the total pressure drop according to West's formula agrees with the other angles quite well up to $Ca \leq O(0.1)$. The calculations make plain that agreement between optical and force measurements [62,72] does not prove that measured θ_D values reflect the contact angle right at the solid surface, as was noted some time ago [238]. In the computed solutions [227], $\vartheta_{D,\Delta P}$ is close to the other apparent angles even though the surface tension force acting on the solid does not change direction because the excess viscous shear stress close to the wetting line becomes a significant portion of the total drag on the solid and nearly makes up for the difference $\gamma_{12}(\cos\vartheta_w - \cos\vartheta_D)$ between the "true" and apparent surface tension force.

6. Effect of Macroscopic Flow

The preeminent influence of the visco-capillary stress balance in the intermediate region rationalizes why apparent dynamic contact angles are insensitive to the flow configuration in which they are measured (see Sec. II.A.3). In particular, the length scale of the intermediate region is so small that inertia and gravity have a negligible effect, as Huh and Scriven [2] first suspected. In hydrodynamic theories for $Ca \ll 1$ and $\epsilon_\delta \ll 1$, the outer flow influences the apparent angle ϑ_m only through higher-order correction terms, that is, $q_0/f(\vartheta_m, \lambda_\eta)$ in Eq. (29) or $Q_0(\vartheta_m)$ in Eq. (30). Voinov [67,68,176] calculated Q_0 to be zero for a flat wall, 1 for a spherical drop, 1.5 for a narrow slot between two parallel plates, and 1.83 for a capillary tube. In many flows of practical interest, higher-order correction terms are cumbersome if not impossible to calculate with methods of matched asymptotic expansion, especially when inertia is important in the outer flow. Numerical analysis of the full Navier–Stokes equations, on the other hand, implicitly accounts for higher-order corrections. Figure 20 summarizes finite-element predictions of apparent dynamic contact angles in a plunge tank [227], calculated under the ad hoc assumptions $\vartheta_w = \theta_0$ and $\epsilon_s = \text{const.} \ll 1$. The values of the property parameter $N_P \equiv (\gamma_{12}^3\rho_1/(\eta_1^4 g))^{1/3}$ were chosen so that inertia and gravity are important at macroscopic length scales, yet the computed curves for ϑ_D versus *Ca* in a plunge tank are insensitive to N_P. The curves are indeed close to the universal curve (6) first identified by Hoffman [29] for capillary displacement and agree therefore with selected data from plunging-tape experiments (see Figs. 5 and 6). de Gennes et al. [6] also found that gravity has no effect on calculated apparent dynamic contact angles, even when capillary statics control the meniscus shape away from the wetting line.

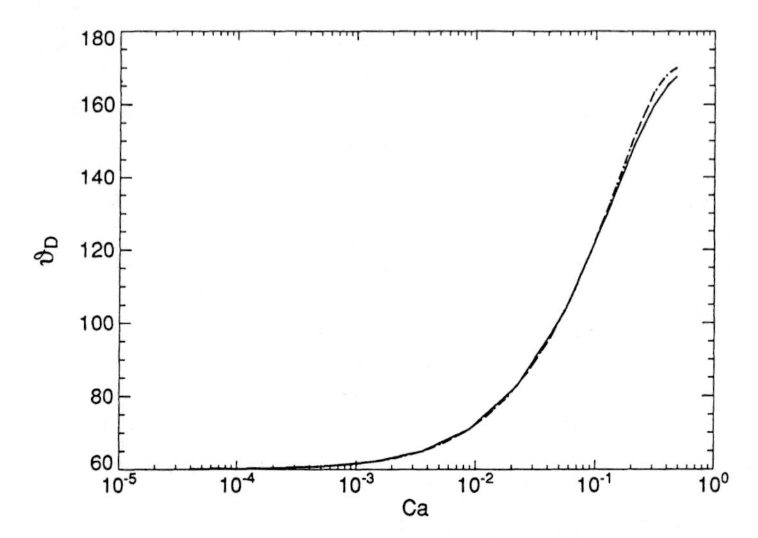

FIG. 20 Finite-element predictions of apparent dynamic contact angles in plunge tanks [227] (—— $N_P \equiv (\gamma_{12}^3\rho_1/\eta_1^4 g)^{1/3} = 57.41$; ------ $N_P = 2.655$; — — — $N_P = 0.124$; for $\rho_1 = 1000$ kg/m^3, $\gamma_{12} = 0.065$ N/m, and $g = 9.81$ m/s, the values of N_P chosen correspond to viscosities of 0.02, 0.2, and 2 Pa·s, respectively; $\vartheta_w = 60°$, $\epsilon_\beta = 10^{-6}$, $\lambda_\eta = 0$, and $W/\sqrt{\gamma_{12}/\rho_1 g} = 8.7$, where W is the width between the plunging tape and confining vertical pool wall).

Hydrodynamic theories of dynamic wetting explain furthermore why the θ_D versus Ca behavior observed for spontaneous spreading is the same as that found for forced wetting (see Sec. II.D.1). The effects of unsteadiness are negligible when the shape of the interface changes on a characteristic time scale T larger than [24]

$$T \geq O\left(\frac{L}{U}\right) \tag{38}$$

For spreading drops, the inequality (38) should be satisfied soon after the drop is placed on the solid [71]. The wetting hydrodynamics in the intermediate region are then the same for spontaneous spreading and forced wetting. Experimental evidence comes from the validity of the Hoffman–Voinov–Tanner law (5) for both modes of dynamic wetting (see Figs. 4 and 13).

The three-region structure predicted by purely hydrodynamic theories—and also refined theories that are augmented with microstructural forces—furnishes formal justification for a simple approach to modeling flows with dynamic wetting lines which assumes that the meniscus retains a static shape over its entire extent and prescribes a velocity-dependent contact angle from an empirical correlation [51,134,250,251]. For capillary rise and capillary imbibition, for instance, the approach has led to much improved agreement with experiments, as compared to the classical Washburn–Rideal–Lucas model that relies on a static angle [134,251]. However, the approach is not genuinely predictive and provides no fundamental insights into the physics of dynamic wetting. It merely validates an empirical θ_D versus U correlation from one set of experiments for different liquid/solid systems or alternative flow configurations.

Even though hydrodynamic theories insinuate that the geometric design of the flow device is unimportant, the term $Ca \ln(\epsilon_\delta^{-1})$ in Eqs. (26), (29), (30), and (31) allows for a weak influence of the size of the device. It is tempting [24,238] to relate the size effect that has been documented in experiments [25,27,30,37,70,71,126] (see Sec. II.A.3 and II.D.1) to changes in the ratio ϵ_δ^{-1} between a characteristic dimension L of the macroscopic flow and the size L_δ of the cutoff region. Figure 21 shows that Cox's result (26) captures the trend of experimental data measured by

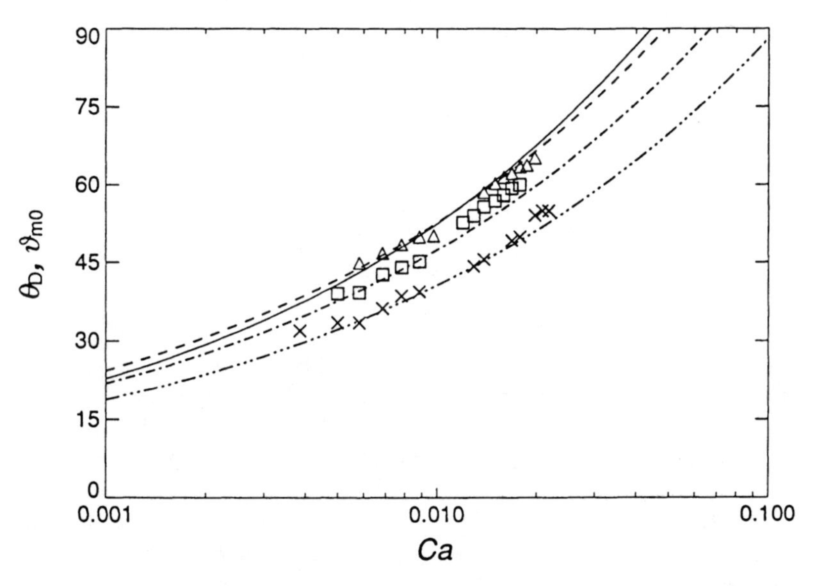

FIG. 21 Size effect for capillary displacement between glass plates: comparison between experimental data by Ngan and Dussan [37] (△ gap = 1.2 mm, □ gap = 0.7 mm, × gap = 0.1 mm), and Cox's [24] leading-order result (26) for ϑ_w = 0° and λ_η = 0 [------ $\epsilon_s = 2\cdot10^{-4}$, —·— $\epsilon_s = 2\cdot10^{-3}$, —···— $\epsilon_s = 2\cdot10^{-2}$; —— Hoffman's universal correlation, Eq. (3)].

Ngan and Dussan V. [37] at three different spacings between two glass plates. However, in conjunction with the assumptions that the slip length L_s remains unaffected by the macroscopic flow and that the local contact angle ϑ_w retains its static value θ_0, the term $Ca \ln(\epsilon_s^{-1})$ cannot account for the entire effect of L on $\theta_{D,M}$ that the data in Fig. 21 imply [37]. Finite-element computations with systematic mesh refinement close to the wetting line [227] confirm that ϑ_D is a weak, logarithmic function of ϵ_s when ϑ_w is set to θ_0. Ngan and Dussan V. [25] relied on the intermediate angle ϑ_R [see Eq. (22)] rather than ad hoc assumptions at the wetting line and also failed to fully predict the measured size effect. Ngan and Dussan V. [25] derived a perturbation solution for $\epsilon_s \ll 1$ and $Ca \ln(\epsilon_s^{-1}) \to 0$ as $Ca \to 0$ and thought that the inadequacies of the two-region expansion for these restrictive limits might explain why the analysis fell short. Finite-element solutions of the complete Navier–Stokes system [227], however, make plain that—as long as ϑ_R is taken to be a material function—hydrodynamic theory alone cannot explain the size effect reported by Ngan and Dussan V. [25,37]. Bach and Hassager [189], on the other hand, argued that their Lagrangian finite-element calculations replicate the rise in $\theta_{D,M}$ from 54 to 63° that Ngan and Dussan V. [37] observed upon increasing the spacing from 0.1 to 0.7 mm at one particular displacement speed ($Ca = 0.02$). However, Bach and Hassager [189] did not calculate $\vartheta_{D,M}$ at the third gap (1.2 mm), nor did they explore the sensitivity of their results to the element size at the wetting line. In their numerical algorithm, the element size set the effective slip length. For the rather coarse meshes used, Bach and Hassager specified $\vartheta_w = 35°$ in order to fit the two data points, even though the silicone oils used by Ngan and Dussan V. [25,37] presumably wetted the glass slides.

Part of the size effect reported by Ngan and Dussan V. [25,37] could possibly be related to the formation of a thin, invisible film ahead of the apparent wetting line [8,252]. Starov et al. [252] predicted that the apparent dynamic contact angle of a meniscus advancing in a prewet capillary depends on both the diameter and microstructural forces. The latter were lumped into a disjoining pressure term [253], and the thickness of the preexisting film was assumed to be given by its equilibrium with a static meniscus in a capillary tube [253,254]. In an alternative explanation, Legait and Sourieau [30] attributed the impact of the capillary radius on $\theta_{D,\Delta P}$ for liquid/liquid displacement to chemical surface heterogeneities.

Apart from a weak size effect, purely hydrodynamic theories that insist on the ad hoc assumptions $\vartheta_w = \theta_0$ and $\epsilon_\delta = \text{const.} \ll 1$ fail to predict any major influence of the global flow field, especially when the theories assume that the liquid displaces a gas of negligible viscosity (i.e., $\lambda_\eta \equiv \eta_2/\eta_1 \ll 1$). Finite-element simulations of single-phase liquid flow in slide coating [227], for instance, yield calculated angles ϑ_D that are close to Hoffman's [29] universal curve $\theta_D = f(Ca, \theta_0)$. The predicted angles are slightly below the data of Muës et al. [56] for water/glycerine/surfactant mixtures on a PET film that had a subbing layer of dried gelatine, but well above the data of Chen [57] for water/glycerine on a prewet metal surface. Finite-element computations [227] furthermore predict that ϑ_D is insensitive to the small pressure difference, $p^A - p^B \ll p^A$, that is typically applied to stabilize the coating bead (see Sec. II.A.4). This result concurs with the few data points that Muës et al. [56] reported at three levels of $p^A - p^B$ ($Ca = 0.16$), but contradicts Chen's finding that above $Ca > O(0.1)$, the apparent contact angle θ_D diminishes as the pressure difference $p^A - p^B$ increases.

Clearly, there is a need to identify the conditions under which apparent dynamic contact angles can be altered by seemingly minor changes in flow parameters other than displacement speed and liquid properties, or by switching from one flow configuration to another—as industrial coating experience suggests [75,107]. In addition, there is a need to identify the mechanisms that prevent the ad hoc assumptions $\vartheta_w = \theta_0$ and $\epsilon_\delta =$ const. $\ll 1$, which for $Ca \leq O(0.1)$ work rather well, from capturing the macroscopic consequences of the unresolved local dynamics toward higher capillary numbers. One such mechanism may be the entrainment and subsequent collapse of an invisibly thin film of air, which would drastically alter the local wetting dynamics from those active during complete low-speed displacement [8,10,17,117]. In particular, invisible air entrainment might give rise to apparent slip over distances that are considerably longer [56] than the near-molecular cutoff lengths commonly thought to enable successful low-speed displacement. Preliminary analyses [8,117] indicate that the onset of the entrainment/collapse regime is sensitive to fluid/solid interaction forces, as well as the radius over which the outer flow forces the macroscopic meniscus to bend upstream of the wetting line (see Sec. III.B.2). The influence of the macroscopic flow at high displacements speeds may, however, also be related to the impact of the ejected airflow at intermediate distances from the apparent wetting line (see Sec. III.A.9).

7. Influence of Specific Material Effects, Surface Roughness, and Heterogeneity

Liquid/solid systems that deviate from the universal behavior $\theta_D = f(Ca, \theta_0)$ at low and moderate capillary numbers provide a reminder that specific material effects can also have a major impact on macroscopic features of displacement flows, especially for systems with partial wetting. Because predicted values of ϑ_D depend only weakly on the cutoff length ϵ_δ, one is tempted to attribute deviations from the universal curve to dynamic variations of the local contact angle [24,67,97]. Dynamic wetting models based on the theory of molecular rate processes appear to be particularly well suited to describe changes in local contact angle at low speeds [255] (see also Sec. III.C.1). In one of the earliest theories of dynamic wetting, Hansen and Miotto [256] proposed that molecular reorientation in the three-phase region may give rise to changes in contact angle at very low displacement speeds. The hypothesis of Hansen and Miotto was later adopted [22,38] to explain intermediate plateaus of slightly elevated yet constant dynamic contact angles at low speeds (see Sec. II.A.4), but the validity of the hypothesis remains unresolved. Systematic deviations from the universal wetting behavior may stem from not only dynamic variations of $\vartheta_w = f(U)$, but also specific fluid/solid interactions. Thompson and Robbins [207,257], for example, showed by means of molecular dynamics simulations that the impact of specific material interactions could be lumped into a constant analogous to the term $q_i/f(\vartheta_w, \lambda_\eta)$ in (29) (see Sec. III.C.2). On surfaces with roughness, chemical heterogeneities, or nonuniform surface charge distribution, most of the deviations from the universal curve are probably related to an unsteady "stick-slip" motion by which the interface advances, especially when θ_0 is well above 0° and $Ca \ll 1$ (see Sec. II.A.4).

Jansons [95] pointed out that, at leading order in Ca, the motion of the macroscopic liquid front over a heterogeneous surface can be described without resolving the complicated, unsteady motion close to the solid—or the local wetting dynamics in the immediate vicinity of the contact line. When a liquid advances slowly over a solid with one-dimensional periodic ridges that occupy only a small fraction of the total surface area, the intermittent "jumps" of the contact line occur on time scales much shorter than the time the macroscopic interface takes to advance the same distance [95]. When $Ca \ll (\cos \theta_E - \cos \theta_A]/\ln[L_D/L_\delta)$ (where L_D is the wavelength of the roughness on the solid surface), the unsteady motion is confined to scales smaller than L_D/Ca and provides in effect a cutoff for the force singularity that arises in conventional steady-state analyses. The singularity persists of course on a local scale, but is isolated from the macroscopic motion by the meniscus jumps. Janson's analysis [95] predicts that, if temporary contact line pinning is an important mechanism, the apparent "slip length" can be much larger than the roughness dimensions, but will diminish with increasing displacement speed U. This prediction agrees with experiments in which stick-slip is typically most noticeable at low speeds, but more difficult to detect at higher speeds (see Sec. II.A.4). For large-area fractions of surface defects or at higher capillary numbers, the disparity between the time scales of the individual jumps and the averaged macroscopic flow disappears, and the effective cutoff length becomes comparable to the extrapolation length on rough surfaces [195]. Steady-state models for liquids advancing over rough surfaces suggest that the extrapolation length ought to be comparable to the depth of the surface irregularities if shallow and to their spacing if deep [18].

Attempts to predict how additional viscous dissipation due to surface heterogeneity influences apparent dynamic contact angles have been restricted to solid surfaces with idealized patterns of heterogeneity, namely, a single isolated chemical defect [258] or one-dimensional periodic variations in surface wettability parallel to the wetting line [247]. Analyses of both cases balance local variations in spreading coefficient with restoring forces from surface tension and viscous dissipation due to contact line motion [247,258]. The analyses assume that the static contact angles are small, that is, $\theta_E(x, y) \ll \pi/2$; the heterogeneity is a small perturbation from a uniform spreading coefficient S_0, that is, $| S(x, y) - S_0 | \ll | S_0 |$, where $S(x, y)$ is the local spreading coefficient $\gamma_{S2}(x, y) - \gamma_{S1}(x, y) - \gamma_{12}$; the cutoff length is much smaller than the heterogeneity (i.e., $L_\delta \ll L_D$); and the variations in dynamic contact angle are small (i.e., $\vartheta_a - \theta_E \ll \theta_E$). The calculations furthermore postulate that the local dynamic contact angle ϑ_a at which the interface glides over the solid depends on the instantaneous speed in the same manner as on a smooth surface [247], that is,

$$\cos \theta_E - \cos \vartheta_a \approx \tfrac{1}{2}(\vartheta_a^2 - \theta_E^2) \sim \frac{Ca}{\theta_E} \tag{39}$$

For $\vartheta_a - \theta_E \ll \theta_E$, Eq. (39) is, in essence, the same as (37). When a fixed force $F = \gamma_{12} \cos \vartheta_D$ drives a liquid front over a heterogeneous surface, the average velocity $\overline{U}$ increases according to [247,258]

$$\frac{F - F_c}{\gamma_{12}} = \cos \vartheta_D - \cos \vartheta_A \sim \left(\frac{\eta \overline{U}}{\gamma_{12}}\right)^2 \tag{40}$$

where $F_c = \gamma_{12} \cos \vartheta_A$ is the critical force needed to exceed the advancing contact angle ϑ_A. The key mechanism in the constant-force regime is the slowing down of the contact line when it advances over poorly wetting regions. On the other hand, when the liquid is forced to advance at a fixed speed U and there is strong pinning [6,259], the dynamics are dominated by the short periods during which the contact line jumps. In this regime, the average force $\overline{F}$ increases as [247,258]

$$\frac{\overline{F} - F_U}{\gamma_{12}} \sim Ca^{2/3} \tag{41}$$

where F_U is the threshold force at which the wetting line unpins from the heterogeneity. The distinction between a fixed force and fixed velocity regime is probably an artifact of the particular defects chosen and may be indistinguishable on surfaces with random heterogeneity [247,258]. Equations (40) and (41) both apply only for small displacement forces. For higher forces or weak pinning, the low-speed F versus U relationship approaches the ideal surface behavior (39) with average wetting properties.

In accord with empirical correlations of the form (11) that capture the overall trend of low-speed data, the force vs. speed relationships (40) and (41) for dynamic wetting on heterogeneous surfaces predict a nonlinear increase of the unbalanced wetting force $\gamma_{12}(\cos \vartheta_A - \cos \vartheta_D)$ with displacement speed. The exponents $a = 2$ and $a = \frac{2}{3}$, however, are well above those inferred from experiments at low Ca, which typically fall in the range $0.35 < a < 0.55$ [32,61]. Zhou and Sheng [97] advanced an alternative model for the impact of low-speed stick-slip behavior on dynamic contact angles that agrees much better with experiments

$$\frac{\Delta F}{\gamma_{12}} = \cos \theta_0 - \cos \vartheta_w(U) = B\ Ca^{0.5} \tag{42}$$

Equation (42) derives from an analysis of the excess viscous dissipation $\Delta F \cdot U$ due to capillary waves that are induced by the jerky motions of the advancing meniscus. Earlier, Rillaerts and Joos [87] invoked a capillary-wave-dissipation mechanism in a semiheuristic argument and also arrived at the simple result $(\cos \theta_0 - \cos \theta_D) \sim \sqrt{Ca}$. Incidentally, the exponent $a = 0.5$ arises in another model for dynamic wetting which assumes that a mono-molecular layer of the advancing liquid is adsorbed on the solid ahead of the advancing wetting line, behaves as a perfect gaseous film, and provides the friction that dissipates the unbalanced Young's force [96].

On rough and porous surfaces, dynamic wetting may proceed by alternative mechanisms that have not been analyzed in detail. At low displacement speeds, the surface imperfections form a network of channels into which liquid is drawn from the macroscopic front by capillary imbibition, even for systems with $\theta_0 \neq 0$ [11,137,161,187]. The macroscopic front then slides over a prewetted, smoothed surface in a manner akin to spreading over a precursor film [9] (see Sec. III.B.1). At higher displacement speeds, the macroscopic front may advance without any local contact line motion, for the liquid may lay down in a rolling motion across successive crests, trapping the displaced fluid in the hollows [18]. The subsequent wetting of individual hollows, promoted by the collapse of the gas pockets due to the Kelvin effect and intermolecular forces, could occur after the apparent wetting line has passed [17]. Some of the experimental findings of Yasui and Tanaka [114], and even those of Miyamoto [118], might be a consequence of such a mode of dynamic wetting (see Sec. II.B). Entrapment and the subsequent collapse of small air pockets may make "air entrainment" tolerable on substrates with textured surfaces under conditions under which it is catastrophic on smooth surfaces and would explain why maximum speeds of successful coating on paper can exceed those on impervious substrates [17].

8. Effect of Preexisting Liquid Film on Solid Surface

When a uniform film of liquid at least 0.1 μm thick covers the solid surface ahead of an advancing liquid front (as sketched in Fig. 22), conventional continuum theory can accurately describe the local wetting dynamics without taking recourse to ad hoc boundary conditions [66,173,176,246, 252,260–265]: The no-slip hypothesis can be applied over the entire length of the solid because the liquid film provides a physical cutoff for the logarithmic force singularity that plagues hydrodynamic models of wetting on dry surfaces; the profile of the advancing meniscus joins smoothly into the preexisting film, which in effect sets the local contact angle to $\vartheta_w \approx 0°$. The film could have been left behind by a receding meniscus [8,260,266], but is often assumed to be of an arbitrary thickness H_∞ that results from an unspecified coating process. Films much thinner than 1 μm may also form when the solid is in contact with a large body of a wetting liquid that is bounded by a curved meniscus [252], a precursor spreads ahead of a macroscopic front (see Sec. III.B.1), or liquid adsorbs from a vapor phase [7,96]. The behavior of thin films is affected by their

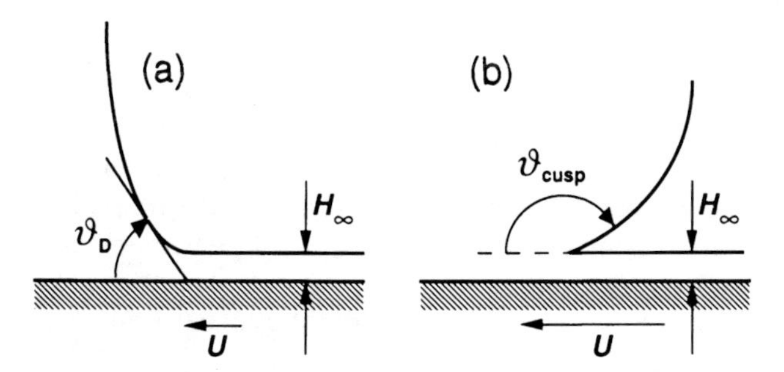

FIG. 22 Schematic of dynamic wetting over a preexisting liquid film: (a) continuously curved meniscus at low and moderate speeds and (b) apparent cusp at high speeds.

proximity to the solid, and an analysis requires accounting for intermolecular forces [8,176,252] (see also Sec. III.B.1 and III.B.2).

When fluid/solid interactions are ignored, lubrication-flow theory yields the profile equation

$$\frac{d^3h}{dx^3} = 3\,Ca\,\frac{h - h_\infty}{h^3} \tag{43}$$

for a liquid front advancing in a steady motion over a thick bulk liquid film [260,267]. Here, $h_\infty = H_\infty/L$ is the dimensionless thickness of the preexisting film. Ahead of the apparent wetting line, solutions to (43) exhibit a characteristic set of rapidly decaying standing waves [260,261]. Away from the film, the solutions approach a static shape that can be extrapolated to the solid to compute an apparent dynamic contact angle (see Fig. 22a). Variations in this angle arise exclusively from hydrodynamic effects, yet follow the same $\vartheta_D \sim Ca^{1/3}$ behavior [66,176,246,261] as the Hoffman–Voinov–Tanner law (33) suggests for dynamic wetting on dry substrates with $S > 0$. For a prewet substrate ($S = 0$) and $\vartheta_D \ll \pi/2$, the power law $\vartheta_D \sim Ca^{1/3}$ can be derived directly by balancing the capillary driving force for wetting, $F_\gamma = \gamma_{12}(1 - \cos\vartheta_D) \approx \gamma_{12}\vartheta_D{}^2/2$, with the viscous force F_v in an advancing liquid wedge [see Eq. (20a)] [6,144]. There is no consensus among different investigators as to the appropriate value of the proportionality constant in $\vartheta_D \sim Ca^{1/3}$. Some authors even lost track of the thickness H_∞ of the preexisting film [246,261] that is essential because it provides the cutoff length L_δ [6,66,173,176]. The calculated dependence of ϑ_D on H_∞ rationalizes why spontaneous spreading on prewetted surfaces proceeds faster than on dry surfaces, why forced wetting over preexisting film yields lower apparent dynamic contact angles than over "dry" surfaces [53,57] and why, for a given film thickness H_∞, apparent dynamic contact angles depend on the characteristics size L of the macroscopic flow geometry [28].

At higher displacement speeds at which $\vartheta_D > 90°$, the meniscus connecting the advancing liquid front with the preexisting film becomes increasingly sharply curved. Above a critical capillary number, the meniscus appears to exhibit a sharp corner that some investigators have referred to as a *cusp* [115,268]. Palmquist and Kistler [269] replicated the transition from a rounded meniscus to an apparent cusp with finite-element calculations. They insisted that, on a local scale that was resolved with extensive mesh refinement, the meniscus remains smoothly curved instead of forming a corner singularity, as was first suggested by Richardson [270]. The results indicate that what appears on a visible scale to be a cusplike corner is merely a region of extremely high curvature, and that the thresholdlike onset of apparent cusp formation arises from an exponential increase of the localized curvature with rising Ca. Critical capillary numbers inferred from the calculations are close to those measured by Joseph et al. [268] for several Newtonian liquids ($Ca \approx 2.5$). Past the onset of apparent cusp formation, the angle ϑ_{cusp} measured through the liquid between the advancing meniscus and the extrapolated film surface (see Fig. 22b) increases in a manner analogous to an apparent contact angle for dynamic wetting on a dry surface. This is illustrated in Fig. 23 for three film thicknesses that, as anticipated by low-speed lubrication-flow analyses [6,66,176], lead to significant differences in apparent cusp angle. To date, no systematic measurements of cusp angles at high capillary numbers have been published. The results in Fig. 23 provide additional if only indirect evidence that dramatic changes in an apparent contact angle can arise from hydrodynamic effects alone (see Sec. III.A.4 and III.A.5).

At sufficiently high Ca, air begins to be entrained in visible amounts at apparent cusps. Capillary numbers at the onset of air entrainment [115] ($Ca_{AE} \leq 10$) are consistently higher than those at the onset of apparent cusp formation [268] ($Ca_{cusp} \leq 8$) and are well above those at the onset of visible air entrainment on dry surfaces ($Ca_{AE} \leq 1.2$; see Fig. 10). The finite-element theory by Palmquist and Kistler [269] predicts an asymptotic approach of the cusp angle toward 180° (see Fig. 23), rather than a critical capillary number at which $\vartheta_{cusp} = 180°$, most likely because the theory neglects airflow effects (see Sec. III.A.9 and III.A.10). The limit of a genuine cusp where two interfaces meet at a common tangent, which

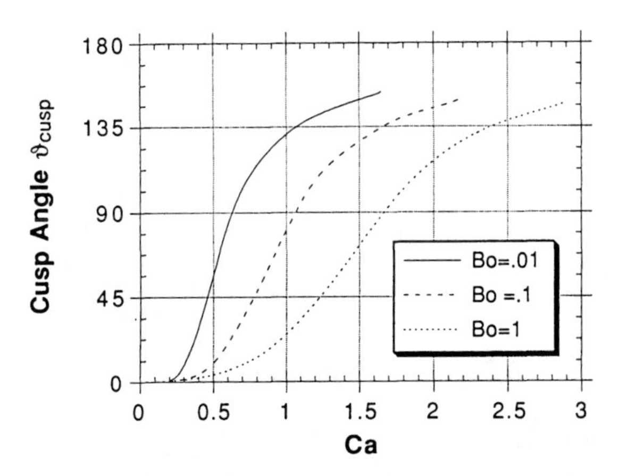

FIG. 23 Formation of cusped interface by liquid film plunging into a pool of the same liquid: apparent cusp angle (defined in Fig. 22) as a function of film thickness (which is measured here by the Bond number $Bo \equiv \rho_1 g H_\infty^2/\gamma_{12}$; $Re = 0$) [269].

in single-phase hydrodynamic analyses is associated with an unphysical singularity [268,270], is probably never reached because the gas phase in the narrow wedge between the converging interfaces builds sufficient forces to separate the interfaces, causing entrainment of the displaced phase [269,271].

9. Effect of Displaced Airflow

Hydrodynamic two-phase flow theories suggest that viscous effects in the phase to be displaced can have a significant impact on apparent dynamic contact angles [24,76]. Figure 24 shows predictions from Eq. (26). The ejected flow of a receding a gas or vapor ($\lambda_\eta \equiv \eta_2/\eta_1 < 0.01$) becomes important when confined to a wedge between the solid and advancing meniscus, that is, $\vartheta_{m0} > 120°$ or $Ca > 0.1$. Low-viscosity liquids (i.e., $\lambda_\eta > 10^{-5}$) are very susceptible to the separating pressures building in the receding gas. The effect becomes self-reinforcing and causes ϑ_{m0} to rapidly head toward 180° with small increases in Ca. High-viscosity liquids, in contrast, are nearly immune to the lubricating action of the ejected gas and ϑ_{m0} approaches 180° slowly, as is predicted when viscous effects in

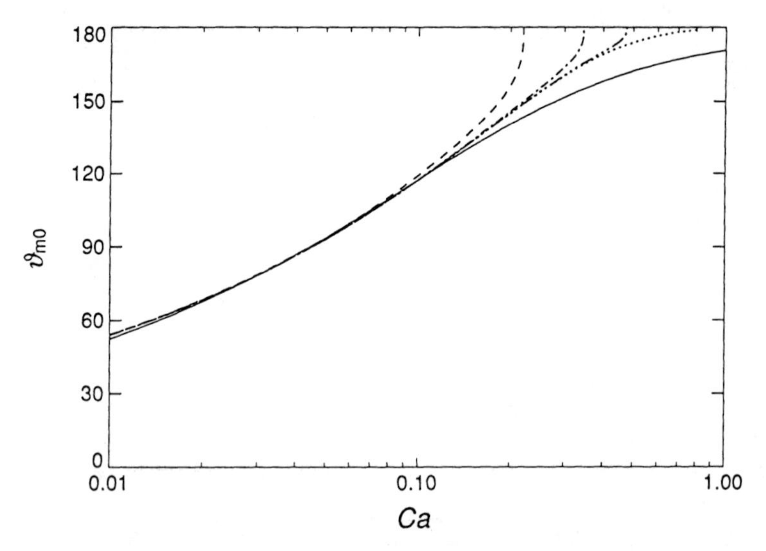

FIG. 24 Effect of the viscosity ratio $\lambda_\eta \equiv \eta_2/\eta_1$ on apparent dynamic contact angles as calculated from Cox's [24] leading-order result (26) for $\vartheta_w = 0$ and $\epsilon_s = 10^{-4}$: — — — $\lambda_\eta = 10^{-2}$, —·— $\lambda_\eta = 10^{-3}$, —···— $\lambda_\eta = 10^{-4}$, ------ $\lambda_\eta = 0$; —— Hoffman's universal correlation (3) for comparison. For air with $\eta_2 = 1.83^{-5}$ Pa·s, the values of $\lambda_\eta > 0$ plotted cover liquid viscosities in the range 1.83 mPa·s $\leq \eta_1 \leq$ 183 mPa·s.

the displaced phase are ignored altogether. Cox [24] made the unnecessary assumption that the slip length ϵ_s is the same for both fluids. Close to the wetting line, the displaced gas phase may be better modeled as Knudson flow, which would implicitly account for slippage. Extensions of Voinov's [67] approximate analysis of the interfacial shape in the intermediate region yield the same predictions for the influence of the second phase [76] as Cox's [24] formal three-region expansion procedure.

Figure 25 shows that experimental data for completely wetting liquid/solid systems corroborate the distinct dynamic wetting behavior of low- and high-viscosity liquids. The quantitative agreement between the low-viscosity data and Cox's leading-order result is probably coincidental, since Eq. (26) with $\lambda_\eta = 0$ deviates systematically from the universal curve (3) above $\theta_D \geq 135°$ (see Fig. 24). Nevertheless, the rapid rise of θ_D toward 180° is characteristic of most data for high-speed wetting with

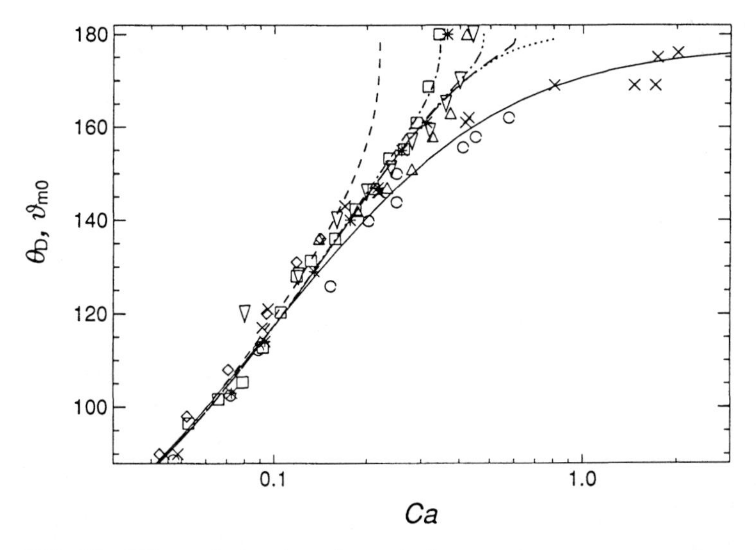

FIG. 25 Effect of the viscosity ratio $\lambda_\eta \equiv \eta_2/\eta_1$ on apparent dynamic contact angles: comparison between theory [Eq. (26) with $\vartheta_w = 0$, $\epsilon_s = 10^{-4}$; — — $\lambda_\eta = 10^{-2}$, —·— $\lambda_\eta = 10^{-3}$, —···— $\lambda_\eta = 10^{-4}$, — — $\lambda_\eta = 10^{-5}$, ------ $\lambda_\eta = 0$] and experimental data for capillary displacement with high-viscosity silicone fluids (× 98.8 Pa·s [29], ○ 5.72 Pa·s [31]), PET tapes plunging into a low-viscosity silicone oil (□ 5 mPa·s [59]), and syringe-needle extrusion coating with low-viscosity fluids (▽ 25.6 mPa·s oleic acid, * 4.7 mPa·s isobutyl alcohol, ◇ 2 mPa·s and △ 20 mPa·s silicone fluids [59]). The solid line —— is Hoffman's universal correlation (3) for comparison.

low-viscosity liquids (1 mPa·s < η_l < 100 mPa·s) [16,56,59]. Experimental data for high-viscosity silicone oils [29,34], also included in Fig. 25, exemplify the slow approach of θ_D toward 180° that is typical of published data for high-viscosity liquids that can reach high capillary numbers ($Ca > 1$) at very low displacement speeds ($U \leq 0.01$ m/s) [29,34,42,43]. The asymptotic approach of θ_D toward 180° is still fast enough to justify the ad hoc boundary condition $\vartheta_w = 180°$ in numerical simulations of polymer processing operations [224,225,242], which typically proceed at $Ca \gg 1$.

The results in Fig. 25 provide compelling evidence that viscous effects in the displaced gas phase can cause substantial deviations from Hoffman's universal curve $\theta_D = f(Ca, \theta_0)$. The importance of the receding air phase has been pointed out before, especially in the context of entrainment and collapse of thin films below the onset of catastrophic air entrainment [8,13,17,117,239]. However, the influence of the second phase was not recognized by most experimentalists working with liquids of low or moderate viscosity [16,47,49–51]. Airflow effects are a primary reason for the failure of empirical correlations to properly draw together data in the range of $120° < \theta_D < 180°$. Published correlations typically ignore λ_η altogether, or have the wrong mathematical form for the dependence of θ_D on λ_η (see Sec. II.A.5). Empirical correlations also overlook the possibility of a size effect. Hydrodynamic theories affirm that when airflow effects accelerate the approach of ϑ_D toward 180°, the size effect could be particularly consequential because small shifts in the steep ϑ_D versus Ca curve would drastically alter the apparent contact angle. The compounded influence of airflow and size effects could be especially important in industrial coating processes that appear to be quite sensitive to small changes in operating conditions near the onset of massive air entrainment [57,75].

Once the ejected airflow is confined to a narrow wedge (i.e., $150° < \theta_D < 180°$), additional mechanisms that Cox's two-phase creeping flow theory does not account for might further alter the θ_D versus Ca curve. Escape of air into rough or porous substrates, for instance, could weaken the impact of the receding phase [17]. The growth of an extended, wedge-shaped air tongue (Fig. 26) might also be delayed by adverse pressure gradients near the wetting line. Such gradients could be generated by inertia or other macroscopic flow effects. Much of the influence of the global flow field in industrial coating processes [75] may indeed stem from changes in growth and stability of the air tongue that tends to form near $\theta_D \rightarrow 180°$. Electrostatic forces and possibly other long-range interactions may also help "squeeze out" the air to be displaced and thereby assist dynamic wetting. Electrostatic forces reach, in fact, sufficiently far into

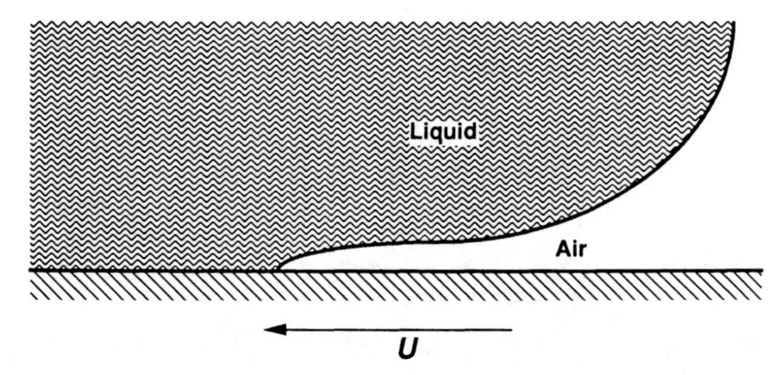

FIG. 26 Schematic of an air tongue penetrating into an advancing front of a low-viscosity liquid near the onset of catastrophic air entrainment.

the intermediate and even outer region to significantly alter apparent contact angles at low speeds at which airflow effects are less important [272].

If specific mechanisms of the sort just listed assist dynamic wetting and permit capillary numbers larger than $Ca \geq O(1)$ without bringing θ_D close to 180°, the macroscopic dynamics of the advancing liquid front become increasingly disjointed from the local wetting physics [78,240], for upstream influence is transmitted primarily by the action of capillarity [3,185]. At high capillary numbers, apparent dynamic contact angles become dominated by the global flow away from the wetting line. In the limit of $Ca \gg 1$, finite-element simulations of liquid flow in curtain coating predict that ϑ_D not only increases with η_1 and U as might be expected, but also diminishes with increasing flow rate or curtain height [78,240]. Predicted flow conditions at which ϑ_D reaches 180° agree qualitatively with those measured at the onset of catastrophic air entrainment [78,118] in high-speed coating in which the wetting line is drawn downstream of the impinging curtain, but fail to capture a second regime of air entrainment that upsets successful coating at lower speeds or high flow rates at which the impinging curtain forms a heel [78,107].

10. Onset of Visible Air Entrainment

Hydrodynamic two-phase flow theories predict that for $\lambda_\eta > 0$, the apparent dynamic contact angle reaches 180° at a finite threshold speed $U_{180°}$ [24,76]. If the limit $\vartheta_D \rightarrow 180°$ is admitted as an ad hoc criterion for the onset of catastrophic air entrainment, purely hydrodynamic theories predict a maximum speed of successful dynamic wetting. For $\lambda_\eta \ll 1$, which is satisfied for all liquids when the displaced phase is air, Cox's [24] leading order result can be approximated by

$$Ca_{180°} = \frac{\frac{1}{6}\pi \ln\left(\frac{4}{3\pi\lambda_\eta}\right) - g(\vartheta_w) + O(\lambda_\eta)}{\ln(\epsilon_s^{-1})} + O(\ln \epsilon_s^{-1})^{-2} \tag{44}$$

where $Ca_{180°} \equiv \eta_1 U_{180°}/\gamma_{12}$. Figure 27 shows that the simple result (44) agrees surprisingly well with experimental data for the onset of visible air entrainment in plunge tanks. The calculated values of $Ca_{180°}$ for $10^{-6} < \epsilon_s < 10^{-3}$ fall in the range $0.2 < Ca_{AE} < 1.0$ covered by the data points. Moreover, at a fixed value of ϵ_s, Eq. (44) replicates the experimental finding that the critical capillary number drops as the viscosity of the advancing liquid is reduced. In conjunction with the ad hoc assumption $\vartheta_w = \theta_0$, Eq. (44) corroborates furthermore the observation that for $\lambda_\eta < 10^{-2}$ and $\theta_0 < 90°$, Ca_{AE} is insensitive to the wettability of the solid surface [43,47,49–51,64,104,113]. Eq. (44) thereby rationalizes why em-

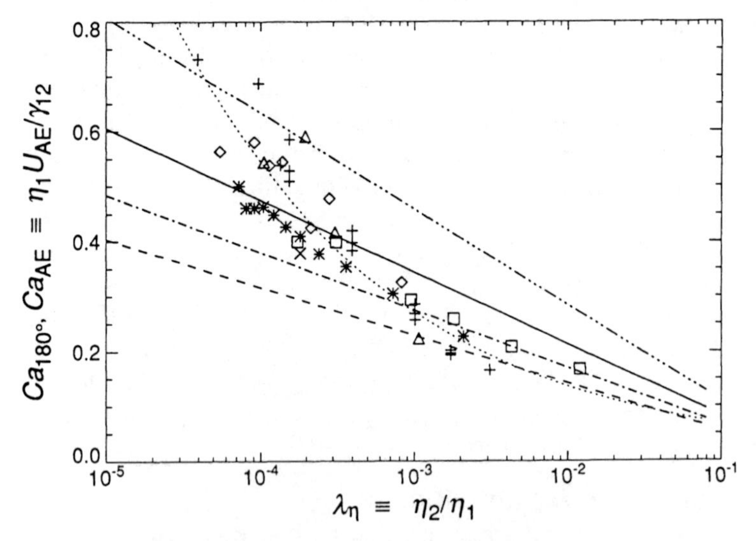

FIG. 27 Comparison between the critical capillary number at which $\vartheta_{m0} = 180°$ as calculated from Eq. (44) [24] (—···— $\epsilon_s = 10^{-3}$, —— $\epsilon_s = 10^{-4}$, —·— $\epsilon_s = 10^{-5}$, — — $\epsilon_s = 10^{-6}$; $\vartheta_w = 0°$) and the critical capillary number Ca_{AE} at the onset of visible air entrainment in plunge-tank experiments. The data are the same as in Fig. 10; the symbols are defined in the legend of that figure. The dotted line ······ is a plot of the empirical correlation (17).

pirical correlations for Ca_{AE} can adequately represent measured data without accounting for θ_0 (see Sec. II.B).

Cox's [24] result (44) provides a theoretical foundation for the claim made in Sec. II.B that the onset of visible air entrainment is dominated by not only capillarity and viscous stresses in the advancing liquid, but also viscous effects in the receding air. As for apparent dynamic contact angles, however, agreement between experiment and theory does not validate the assumptions made in regard to the local wetting dynamics. In fact, Cox's [24] formal expansion procedure breaks down close to 180° (refer to condition 8.11b in Cox's paper). Moreover, the connection between the limit $\vartheta_{m0} \rightarrow 180°$ and the onset of visible air entrainment remains tentative (see Sec. II.B). Further reservations arise because values of $Ca_{180°}$ calculated for $\vartheta_w = \theta_0$ and ϵ_s = const. are less sensitive to λ_η than experimental data for Ca_{AE}. The quantitative discrepancy between theory and experiment is probably related to the failure of Eq. (26) to accurately predict θ_D versus Ca above 135° for $\lambda_\eta \rightarrow 0$ (see Fig. 24).

In spite of serious shortcomings, purely hydrodynamic theories are more successful in predicting the critical conditions at the onset of visible air entrainment than any other attempt published to date. The apparent success of hydrodynamic two-phase flow theory suggests that air entrainment may arise not only from a failure of the liquid to displace sufficient air from the solid at submicroscopic length scales, as has been postulated by some [8,17,56,117], but also from air engulfment induced at larger length scales. Experiments confirm that dynamic wetting can be unsteady below the critical speed for visible entrainment [42,50,54,103] and that catastrophic coating failure may stem from an alternating cycle between successful wetting and entrapment of air bubbles [102,103]. However, neither wavering wetting lines and air entrapment nor serrated wetting lines that sometimes precede or accompany the onset of massive air entrainment have been predicted from first principles. A rigorous analysis of the stability of the air tongue as it penetrates into the advancing liquid front remains one of the most prominent unresolved issues in wetting hydrodynamics. Such an analysis would help establish to what extent dynamic wetting failure is controlled by two-phase flow mechanics at intermediate length scales that Cox's theory accounts for; macroscopic hydrodynamic effects such as the impact of the freely falling sheet in curtain coating [46,78,107]; nonhydrodynamic mechanisms like those listed in Sec. III.A.9 above; and local displacement mechanisms of the sort addressed in Sec. III.B.2 below.

Irregular dynamic wetting lines are much better understood in the low-speed regime. The most prominent example is the well-known fingering instability in Hele–Shaw cells [273–276]. Nearly periodic fingering instabilities have also been observed when a liquid tongue is driven over a previously dry solid by gravity [169,277] or centrifugal forces [278,279], shear stresses induced by a jet of air [278], surface tension gradients [150,280,281], or a positive spreading coefficient [66,150]. Even though reminiscent of Saffman–Taylor instability, periodic wetting line instabilities arise from a different mechanism, for the pressure gradients in the film behind the front are insignificant. The instabilities have been attributed to the breakup of the bulge of liquid that forms at the leading edge [234,264,281]. The mechanism is akin to the Rayleigh instability of a liquid column breaking into drops, but a complete analysis of the nonlinear development is still outstanding [234,282]. An alternative explanation [150] attributes the instability to an increased wetting force that pulls on protrusions of the liquid front. Although some studies suggest that the onset of fingering is insensitive to the dynamics near the wetting line [169,277], fingering can be suppressed by a preexisting film of liquid on the solid or a precursor spreading ahead of the macroscopic front [279]. Once the fingers have reached a finite amplitude, their fate depends strongly on the interactions between the liquid and solid at the wetting line [277]. For liquid/solid systems with complete wetting or low static contact angles, the fingers evolve into a series of nearly periodic triangles with sharply curved leading tips [169,277]. The forms are reminiscent of the sawtooth-shaped wetting lines associated with the visible onset of air entrainment. For higher static contact angles, on the other hand, the advancing liquid may fail to cover the entire solid surface. The fingers become parallel-sided rivulets emerging from a front that remains stationary [169,277]. Sufficiently far downstream, triangular tongues might also evolve into rivulets [234].

11. Liquid/Liquid Displacement

If the second phase is an immiscible liquid, viscous stresses in the receding liquid have a significant effect on the apparent dynamic contact angle even when ϑ_m is much less than 180° [24]. Figure 28 shows that for $\vartheta_w = \theta_0$ and ϵ_s = const., values of ϑ_{m0} calculated from Eq. (26) rapidly rise toward 180° at capillary numbers well below those at which liquids successfully displace air. Experimental observations of liquid/liquid displacement corroborate the strong impact of $\lambda_\eta \equiv \eta_2/\eta_1$ that Eq. (26) predicts [31,33,34,71,97]. Slip lengths inferred from comparisons between theory and experiment are comparable to those for liquid/air displacement (10 nm $< L_s <$ 10 μm) [71,97]. The quantitative agreement between ϑ_{m0} and θ_D, however, is not as good as it is for liquid/air displacement. Measured contact angles sometimes increase abruptly at very low displacement

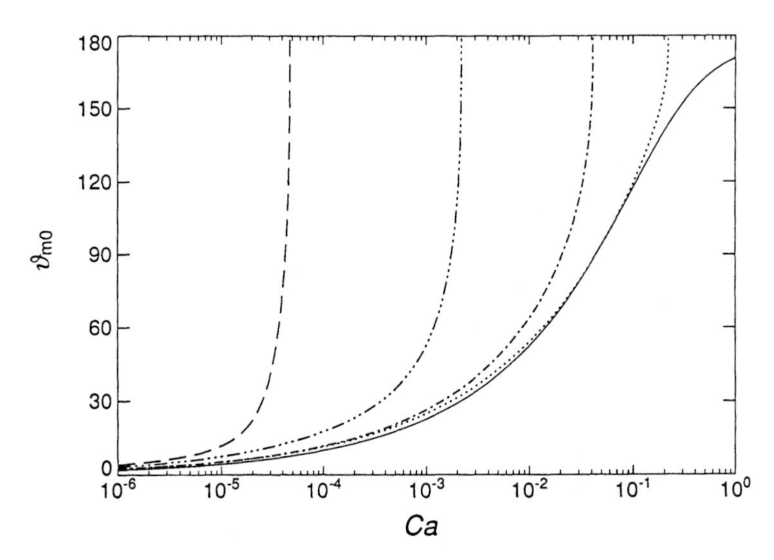

FIG. 28 Effect of the viscosity of displaced fluid on apparent dynamic contact angle as predicted by Cox's [24] leading-order result (26) for $\epsilon_s = 10^{-4}$ and $\vartheta_w = 0°$: ------ $\lambda_\eta = 10^{-2}$, —·— $\lambda_\eta = 1$, —···— $\lambda_\eta = 10^{+2}$, — — $\lambda_\eta = 10^{+4}$; —— is a plot of the empirical correlation (3) for $\lambda_\eta = 0$.

speeds ($Ca < 10^{-5}$) [33], exhibit significant scatter below $Ca \leq O(0.01)$ [34], and often rise more rapidly toward higher *Ca* [31,33,34,71,97,133] than theories suggest that ascribe the dynamic increase in ϑ_m exclusively to viscous bending of the interface [24,97]. Sizable deviations of experimental data from Eq. (26) arise even when the viscosity ratio is as low as it is in liquid/air displacement, that is, $\lambda_\eta < 10^{-2}$. Nonetheless, measured values of the critical capillary number at the onset of visible entrainment of the receding liquid follow the trend of rapidly diminishing $Ca_{180°}$ with increasing $\lambda_\eta \equiv \eta_2/\eta_1$ that Cox's [24] result (26) predicts, as is shown in Fig. 29. Evidently, both θ_D and $Ca_{180°}$ in liquid/liquid displacement are influenced by the same visco-capillary flow effects that presumably dominate liquid/air displacement. The quantitative discrepancies and scatter in Fig. 29 reveal, however, that other nonhydrodynamic mechanisms must also affect the displacement of one liquid by another from a solid surface.

Zhou and Sheng [97] argued that the accelerated increase of θ_D with *Ca* must arise from the velocity dependence of ϑ_w. They found much better agreement between Eq. (26) and experimental data upon introduc-

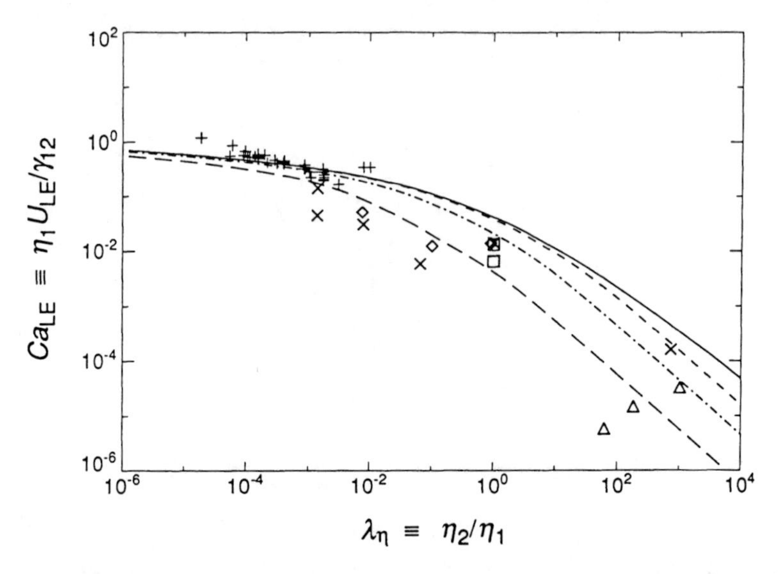

FIG. 29 Critical capillary number at the onset of visible entrainment of a liquid by another liquid: comparison between Cox's [24] leading-order result (26) for $\vartheta_{m0} = 180°$ and $\epsilon_s = 10^{-4}$ (—— $\vartheta_w = 0°$, ------ $\vartheta_w = 45°$, —·— $\vartheta_w = 90°$, — — $\vartheta_w = 135°$) and the experimental data first shown in Fig. 11 (the symbols are the same as in Fig. 11).

ing a heuristic extension of the capillary-wave-dissipation model (42) [97], namely,

$$\frac{\Delta F}{\gamma_{12}} = \cos\theta_0 - \cos\vartheta_w(U) = B\,Ca^a \tag{45}$$

Equation (45) is, in essence, the same as the empirical correlation (11). Zhou and Sheng [97] deduced exponents in the range $0.33 \leq a \leq 0.5$ that agree quite well with $a = 0.4 \pm 0.05$ measured directly by Stokes et al. [32]. Mumley et al. [133] also identified an excess force ΔF^* in liquid/liquid capillary rise that increases roughly as $\Delta F^* \sim Ca^{0.5}$. Fermigier and Jenffer [34], in contrast, were unable to determine the power a in (11) because of scatter in their data. Plots of $|\cos\theta_D - \cos\vartheta_{m0}|$ versus Ca, with ϑ_{m0} calculated from (26), exhibited even more scatter and thus made direct validation of (45) impossible. Nonetheless, stick-slip resulting from temporary contact line pinning is a plausible explanation for the rapid rise of θ_D under dynamic conditions [33,34], since contact-angle hysteresis is typically large in liquid/liquid systems [34]. Other causes for accelerated variations in contact angle, however, cannot be ruled out. In sharp contrast to gases—which are very mobile, have a mean free path quite large compared to the size of the cutoff region, and thus may have little impact on the local wetting physics—liquids previously in contact with the solid have a density comparable to that of the advancing liquid, may be strongly bonded to the solid surface, and thus can offer considerable resistance to displacement. Therefore, the local dynamics of a liquid pushing away another liquid might be quite different from those of a liquid replacing a gas [98] and could lead to significant excess dissipation in the three-phase juncture that would alter the contact angle close to the solid. In particular, adsorption–desorption mechanisms of the sort first proposed by Blake and Haynes [88] and Cherry and Holmes [89] (see Sec. III.C.1) could affect $\vartheta_w(U)$ more strongly in liquid/liquid displacement than liquid/gas displacement and account for some of the discrepancies between Cox's hydrodynamic theory with the ad hoc assumption $\vartheta_w = \theta_0$ and experimental data [33]. Additional discrepancies may arise from the term q_i in Eq. (29) that can account for specific liquid/solid interactions [33,34,207], but is dropped in the leading-order result (26). Further discrepancies might also be related to the presence of thin films of one of the liquid phases [8,33,34,133].

B. Wetting Hydrodynamics of Thin Films

1. Precursor Films of Advancing Liquid

Liquid/solid systems that form precursor films (see Sec. II.D.2) offer an ideal opportunity to solve for the macroscopic wetting dynamics without taking recourse to ad hoc boundary conditions at a putative wetting line. The local displacement process takes place at the leading edge of the precursor, whereas the macroscopic front merely acts as a reservoir for the protruding film and slides over a "prewet" surface. With the exception of rather thick secondary films that may arise from surface tension gradients, precursor films are usually thin enough ($H < 1$ μm) to be strongly influenced by microstructural forces, such as dispersion or electrostatic forces [283]. The controlling influence of such forces on the local wetting dynamics had been recognized for quite some time [2,90,167,175,206, 246,284], but only in the last decade have fluid/solid interactions been incorporated into comprehensive theories that describe low-speed spreading with precursor films and resolve essential aspects of the local wetting mechanisms. Several reviews have appeared [6,7,9,14,285], and this section highlights only a few key results.

Most published analyses of dynamic wetting via precursor films focus on simple, nonvolatile liquids spreading on structureless surfaces. The analyses furthermore assume that long-range fluid/solid interactions furnish the sole driving force for the precursor film, and that the film remains sufficiently thick so that the molecular structure of the interfaces is unimportant and continuum theory remains applicable ($H > 30$ Å). In this case, the net aggregate of all fluid/solid and fluid/fluid interactions as a function of film thickness H can lumped into the *disjoining pressure* $\Pi(H)$ that was first introduced by Derjaguin [286,287]. In lubrication-flow analyses of thin films, $\Pi(H)$ gives rise to an additional pressure gradient that competes with capillary pressure and viscous resistance to flow [6,7, 14,288]

$$\frac{d}{dx}\left[\frac{d^2h}{dx^2} + \frac{H_0\Pi(H)}{\gamma_{12}}\right] = 3\,Ca\,\frac{1}{h^2} \tag{46}$$

Equation (46) is an augmented version of (34) for a thin liquid tongue with small slope in a steady-state advance (H_0 is a reference film thickness). The most commonly used form of $\Pi(H)$ is that for simple, nonretarded van der Waals forces of the sort induced by dipolar interactions in nonionic systems [283], that is,

$$\Pi_{\mathrm{vdW}}(H) = \frac{A_{S12}}{6\pi H^3} \tag{47}$$

Here, A_{S12} denotes the negative of the effective Hamaker constant for the interaction between a solid and gas through a thin liquid film. For completely wetting systems, A_{S12} is positive and yields a repulsive, or "disjoining," force between the solid and gas. Many real solid/liquid/gas systems, especially ionic systems susceptible to electrical double-layer effects, require more complicated disjoining pressure functions [8,289]. For very thin films, $\Pi(H)$ needs to be modified further to account for structural contributions from short-range forces [21,290].

For the macroscopic wetting dynamics of nonvolatile liquid/solid systems with $S > 0$, results from precursor-film theories agree with experimental observations of a nearly universal spreading behavior. At leading order, the calculated apparent contact angle ϑ_a varies according to [6,14,288]

$$\vartheta_a^3 = 9\,Ca\,\ln\left(\frac{X}{X_0}\right) \tag{48}$$

This result emerges from a three-region expansion of the meniscus shape that includes the nearly flat precursor film, the macroscopic drop, and a third region in between in which the meniscus is sharply curved where it joins the precursor but forms a nearly flat, advancing wedge at larger

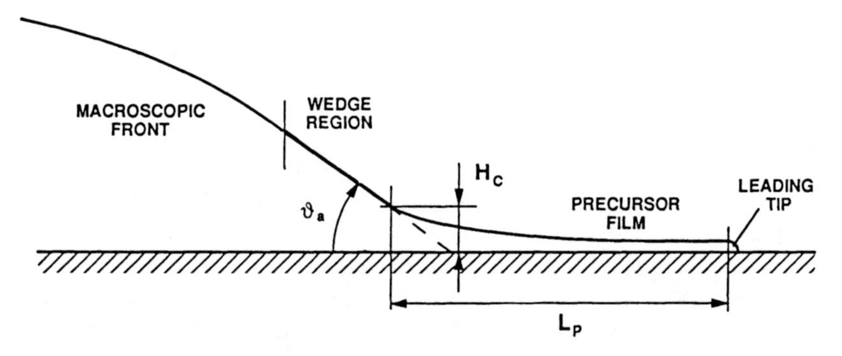

FIG. 30 Schematic of a precursor film of nonvolatile liquid spreading ahead of a macroscopic front (not to scale).

distances from the solid (see Fig. 30). The distance X is measured from the crossover between the wedge region and precursor to the location at which the wedge region needs to be matched to the quasistatic, curved meniscus that caps the spreading droplet. Thus, X is related to the macroscopic size R of the droplet. On the other hand, $X_0 \sim L_a/\vartheta_a^2$ is a short-distance cutoff provided by the precursor. The characteristic length $L_a \equiv \sqrt{A_{S12}/6\pi\gamma_{12}}$ is typically of molecular dimensions.

Equation (48) is, of course, another version of the Hoffman–Voinov–Tanner law (33) that concurs with the power laws $R \sim t^{1/10}$ and $\theta_D \sim t^{-1/3}$ for the macroscopic spreading dynamics of small droplets that are immune to gravity [6,14,66]. A $R \sim t^{1/10}$ behavior results also from numerical integration of drop-profile evolution equations that account for intermolecular forces and describe both the macroscopic drop and the precursor film [7]. For drops large enough to be dominated by gravity, Eq. (48) is consistent with the power law $R \sim t^{1/8}$ [146] (see Sec. II.D.1). Hydrodynamic theory suggests that medium-sized drops whose shape is influenced but not dominated by gravity might not follow a simple power law because the evolving drop shapes are not self-similar [146]. Experimental data nonetheless obey the "pseudoscaling law" $R \sim t^n$ with $n \approx 0.127$, which is very close to $\frac{1}{8}$ [145]. Via the quantity $X \sim R$, Eq. (48) predicts that a smaller droplet will expand slightly faster than a larger drop at the same apparent dynamic contact angle, as some experiments suggest [3,70,126]. In addition, because of the logarithmic term and other higher-order correction terms not included in Eq. (48), the exponent in the power law $\vartheta_a^m \sim Ca$ may deviate slightly from $m = 3$ [6]. Such a deviation from the Hoffman-Voinov-Tanner law is borne out by some θ_D

data for spreading droplets (see Fig. 13). Via the Hamaker constant A_{S12} contained in X_0, Eq. (48) also admits that apparent dynamic contact angles may weakly depend on specific intermolecular forces.

In addition to replicating essential aspects of the macroscopic spreading behavior as it is observed in experiments, augmented lubrication-flow theories furnish valuable insights into the submicroscopic wetting dynamics. In particular, such theories provide a physical explanation for the marked difference between apparent contact angles and much smaller angles of interface inclination close to the solid. The theories furthermore elucidate how microstructural forces alleviate the unbounded force singularity that arises in conventional hydrodynamic theory. Microstructural forces yield a cutoff ($X_0 \sim L_a/\vartheta_a^2$) that dominates over the cutoff that would arise if slippage were important (i.e., L_β/ϑ_a), provided that the extrapolation length L_β is of molecular dimensions [6]. Most important, precursor-film theories explain why the macroscopic wetting dynamics are rather insensitive to the submicroscopic displacement mechanisms. The free energy contribution $S \cdot U$ that arises from the spreading coefficient $S > 0$ is, at leading order, dissipated within the precursor film [6,288]. Consequently, only the second term $\gamma_{12}(1 - \cos \vartheta_a) \approx \gamma_{12}\vartheta_a^2/2$ in the total driving force for wetting, $F = S + \gamma_{12}(1 - \cos \vartheta_a)$, contributes to the meniscus deformation visible on a macroscopic scale. Therefore, the macroscopic wetting dynamics for $S > 0$ are the same as on a preexisting liquid film, that is, $S = 0$, even when $S \gg \gamma_{12}\vartheta_a^2/2$ [6]. For small droplets, bulk spreading is mostly driven by capillary pressure in the region where the interface bends sharply to merge the precursor film with the advancing wedge. In this region, the influence of long-range intermolecular forces decays rapidly as the film thickness grows [6,14,175,288]. In the precursor film, on the other hand, intermolecular forces dominate over capillary pressure and control the thickness and rate of expansion of the film [6,14,288]. The distinct regions of influence of fluid/solid interactions and capillary pressure rationalize the seemingly paradoxical observation that, for nonvolatile liquid/solid systems with $S > 0$, the dynamic variations of ϑ_a are not much affected by changes in surface texture [34,131] or chemistry [42,131] (see also Figs. 4, 5, and 13).

The results from precursor-film theories elucidate furthermore why, for systems with complete wetting, strictly hydrodynamic theories that do not incorporate the proper physics close to the solid can adequately describe the wetting dynamics on larger length scales. Away from the immediate vicinity of the precursor film, augmented lubrication-flow analyses predict the same asymptotic structure of the meniscus shape as purely hydrodynamic theories that take recourse to ad hoc boundary conditions, that is, ϵ_δ = const. and $\vartheta_w = 0$ (see Sec. III.A.4), or modified theories that assume a preexisting, thick liquid film (see Sec. III.A.8). Equation (48), like Eqs. (24) and (25), suggests that the slope of the meniscus varies slowly as a function of the distance X from the crossover to the precursor film. Evidently, the wedge region in precursor-film theories plays a pivotal role that parallels the importance of the intermediate region in strictly hydrodynamic theories. For liquid/solid systems that form precursor films, laser-light interference microscopy verifies that there is a wedge-shaped region where the slope is nearly constant and agrees with Eq. (48) [73,132]. The detailed measurements of meniscus profiles explain further why experimental θ_D data are rather insensitive to the particular definition of an apparent contact angle chosen and also the distance from the solid that conventional optical techniques can resolve (see Ref. [73]; refer also to Sec. III.A.5).

Despite the apparent success of precursor-film theories in describing essential aspects of dynamic wetting, such theories are of limited practical relevance because precursor films are not ubiquitous, even when $S > 0$. Nonvolatile liquids cannot form precursor films unless the rate of film expansion surpasses the speed of the macroscopic front. For forced wetting at a constant rate, theory predicts that the length L_P of the precursor is controlled by uniform "drift" at the displacement speed U and shrinks according to $L_P = L_a/Ca\,\sqrt{S/\gamma_{12}}$ as U increases [6,14,288,291]. The dependence of L_P on U has been corroborated by experiments for the early stage of spontaneous spreading [147]. Sizable precursors form only at very low displacement speeds or small apparent contact angles, even when $S/\gamma_{12} \geq O(1)$ [9]. In the absence of an extended precursor film, the local displacement process is no longer disjoint from the apparent wetting process of the macroscopic front. Therefore, unless $\vartheta_a \ll \pi/2$, resolving the local wetting dynamics for nonvolatile systems with $S > 0$ requires matching conventional continuum theory directly with a refined theory for the displacement process at near-molecular length scales, as is necessary for systems with $S < 0$ and, say, $\theta_E > 10°$ [191].

Even when a precursor extends beyond the macroscopic front, lubrication-flow theory augmented with a simple disjoining pressure term often does not accurately describe the shape of the film. The simple profile $H \sim 1/(X - X_0)$ that theory predicts [6,14] sometimes adequately describes a limited region of the film not too far from the macroscopic front [147,152,158] where the profile is dominated by "drift" at the velocity imposed by the spreading front [291]. Measurements also corroborate the prediction [6,14,288] that precursor films ought to be longer and thinner for large S and extend more rapidly for small ϑ_a [152]. However, not all experiments confirm the simple result $H_c \approx L_a/\vartheta_a$ [6,14] for the thickness at the transition from the wedge to the precursor [147] (see Fig. 30).

Moreover, ahead of the macroscopic front or in later stages of spreading, precursor films often spread faster than predicted by the augmented continuum theory [152], and measured thickness profiles deviate from calculated ones [147,152,158]. In this regime, the shape of the precursor is dominated by diffusive expansion of the film itself [291] and continuum theory may no longer be applicable even when augmented with microstructural forces.

Augmented lubrication-flow theory fails altogether to describe how the leading edge of a precursor film advances over a solid surface and thus cannot resolve the physics of how the first few molecules propagate. de Gennes and Joanny [6,14] restricted their analysis to "dry spreading" of completely nonvolatile liquids and assumed that the liquid advances exclusively by viscous flow induced by colloidal forces. de Gennes and Joanny contended that the viscous stress singularity that persists at the leading edge of the precursor is of no consequence because the local meniscus profile is dominated by capillarity and disjoining pressure. The calculated profile, however, intersects the solid at a 90° angle as it does in conventional lubrication-flow analyses when the unbounded viscous force singularity is not removed (see Sec. III.A.1). In augmented precursor-film analyses, the 90° angle is an artifact of the singular form of the disjoining pressure function (47) that does not account for short-range forces [21]. The 90° angle violates the small-slope approximation that is essential in lubrication-flow analysis. It also disagrees with experimental measurements which indicate that precursor films decay to molecular thicknesses at the leading edge [147,152,154] and sometimes exhibit layered molecular structures [135,155,159]. The layering is probably controlled by short-range intermolecular forces and dynamic molecular events and cannot be captured by augmented continuum theories.

The wetting dynamics in films of molecular thickness have been described by surface-diffusion models [7,12,135,291], Eyring-type models for molecular rate processes [158], and statistical-mechanics models relying on Langevin theory [292], as well as Monte-Carlo simulations [293,294]. Surface diffusion had been proposed a long time ago as the primary wetting mechanism for nonvolatile liquids [11], and diffusion models for the local wetting dynamics can be found in the early literature [88,89,206]. In an attempt to replicate data by Bascom et al. [11] for squalane spreading on metal, Teletzke et al. [7] blended a surface diffusion model at the leading edge of the wetted area with an augmented lubrication-flow theory behind. The calculations suggest that the rate of advance of the primary film by surface diffusion limits the rate of macroscopic spreading. The experimental data [11], nevertheless, seem to be consistent with the universal $R \sim t^{1/8}$ power law for large drops [167].

The meniscus profiles computed by Teletzke et al. [7] exhibit small foot structures at the transition from the macroscopic front to the primary layer. Beaglehole [147] measured similar features for silicone oils, but they appear to be smaller than those predicted by Teletzke et al. [7].

For volatile liquids, transport through the vapor phase provides an alternative mechanism for primary wetting, as was first suggested a long time ago by Hardy [15]. When the liquid wets the solid and the displaced phase is loaded with vapor of the advancing liquid, the primary film may be rather thick, depending on external control parameters such as relative humidity, temperature of the solid surface, or vertical position in the direction of gravity [6,253,289]. In this case, the spreading coefficient S "locks" automatically to zero, for the macroscopic front advances over a surface of identical chemical composition as it does when the solid is prewet with a thick liquid film (see Sec. III.A.8). When the displaced phase is an inert gas, the thickness of the primary film may be limited by the additional resistance due to diffusion through the noncondensing component. Wayner [143] concluded—from an order-of-magnitude estimate of the relative importance of fluid flow and change-of-phase heat and mass transfer in a thin spreading film—that evaporation from the thicker part of the film and condensation on the solid can be an essential mechanism of dynamic wetting, even for liquids of moderate volatility. The reason is that the area available for evaporation and condensation is much greater than that for surface diffusion [143]. Although the impact of evaporation on the meniscus profile and state of stress in the contact line region is fairly well understood for static configurations in near equilibrium (see, for instance, [295] and the extensive literature citations therein), comprehensive theories for dynamic wetting via volatile components are only now being developed [143,215]. Wayner and Schonberg [215] proposed a simple spreading model for a thin liquid wedge that is driven by a long-range intermolecular force field and advances by viscous flow, as well as evaporation, followed by multilayer adsorption. The model suggests that the dimensionless ratio $N_V \equiv \rho_i D \eta_1 X / \pi A_{S12} \rho_1$ measures the relative importance of vapor diffusion to viscous flow in the thin film along the solid (here, ρ_i is the density of the diffusing material in the vapor, D the diffusion coefficient, and X is the distance along the advancing film). The ratio N_V can be used to quantify the modifiers "volatile" and "nonvolatile" [215].

Teletzke et al. [7] analyzed the "moist spreading" of water on glass. They assumed that a *primary film* of water is adsorbed on the solid surface and employed a composite disjoining pressure function for water on glass [296]. Computed predictions [7] replicate observations by Marmur and Lelah [160] that the rate of spreading of a water drop on a glass slide depends on the size of the slide. The computations show that this curious effect arises from the rapid expansion of a *secondary film* that reaches the edge of the slide and transmits upstream influence back to the macroscopic drop [7]. The macroscopic spreading behavior depends only weakly on the thickness of the preadsorbed film, independently of whether there is an obstacle [7] or not [6,7].

Evaporation promotes temperature and, for mixtures or systems with impurities, concentration gradients. Both lead to surface tension gradients that, if directed toward the wetting line, may give rise to significant driving forces for rather thick secondary precursor films [7,11,297]. Such films, in turn, may significantly alter the apparent dynamic wetting behavior on a macroscopic scale.

2. Entrained Films of Receding Fluid

The hypothesis of entrainment and collapse of a thin air film [8,17,72, 100,116] at high speeds of forced wetting suggests another regime of dynamic wetting that can be analyzed without taking recourse to ad hoc boundary conditions at a putative wetting line (see Fig. 31). In a hypothetical flow state of steady, uniform film entrainment, the lubricating action of the film provides a physical mechanism for apparent slippage of the liquid over the solid; the interfacial profile bends to assume an angle of inclination close to $\vartheta_w \approx 180°$, where it approaches the solid and becomes strongly affected by microstructural forces.

Theoretical support for the hypothesis that small amounts of the receding phase may be left behind at speeds well below visible entrainment

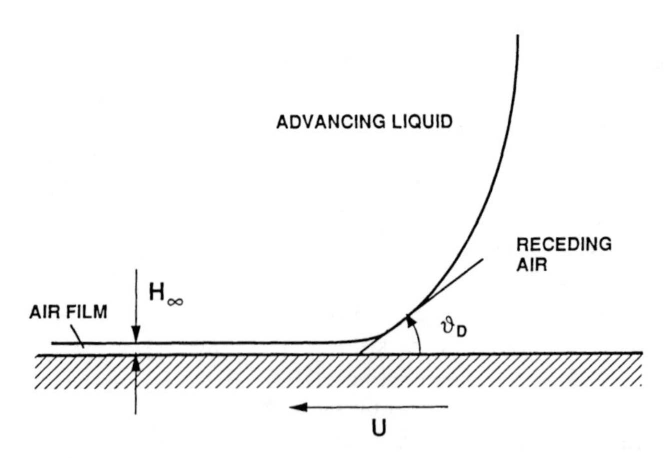

FIG. 31 Schematic of a thin, entrained film of the displaced phase (the film is shown to be uniform as in Teletzke's [8,13] analysis, but is unstable and breaks up into patches and bubbles in reality).

comes from extensions [8,124] of Bretherton's [260] analysis of how an inviscid bubble displaces a liquid in a capillary. Bretherton's original analysis ignored fluid/solid interaction forces and suggested that a liquid film is always left behind. Derjaguin and Levi [100] were the first to consider the effect of intermolecular forces on an entrained layer of air. Prutow and Ostrach [124] accounted for long-range fluid/solid interactions, as well as viscous effects in both phases, and predicted that a nonwetting receding phase can be entrained only above a critical speed. Teletzke et al. [8,13], apparently unaware of Prutow and Ostrach's work, modeled low-speed two-phase flow in a capillary by means of the augmented film profile equation

$$\frac{d}{dx}\left[\frac{h_{xx}}{(1 + h_x^2)^{3/2}} + \frac{H_0\Pi(H)}{\gamma_{12}}\right] = 3\,Ca\,\frac{h - h_\infty}{h^3}\,F_2(h; \lambda_\eta, h_\infty) \tag{49}$$

where F_2 is a function of the film profile $h(x)$, the final thickness of the entrained film h_∞, and the viscosity ratio of the two fluids, $\lambda_\eta \equiv \eta_2/\eta_1$. For $1/\lambda_\eta = \eta_1/\eta_2 \to 0$, the function reduces to $F_2 = 1$, which is the same as in Eq. (43). For thin gas films, the microstructural forces add up to a negative *conjoining pressure* [8,13,17] with $d\Pi(H)/dH > 0$, which tends to collapse the films. Steady-state solutions to Eq. (49) with the boundary condition $dh/dx = 0$ as $x \to -\infty$ confirm that a thin gas film can be left behind by the advancing liquid front above a critical speed U_F. However, such solutions do not prove that a second flow state with complete displacement cannot coexist at $U > U_F$. In augmented film entrainment theories, incipient entrainment is triggered when viscous stresses in the receding phase overwhelm conjoining fluid/solid interactions. Predicted critical capillary numbers $Ca_{2,F} \equiv \eta_2 U_F/\gamma_{12}$ formed with the viscosity η_2 of the receding phase depend strongly on the interaction parameters that enter the conjoining pressure $\Pi(H)$ [8,124]; they are also influenced by the radius of the capillary and, to a lesser extent, the viscosity ratio λ_η.

According to the hypothesis advanced by Scriven and co-workers [8,10,13,17,117], successful forced wetting at high coating speeds is not limited so much by incomplete expulsion of air from the three-phase juncture as it is by insufficient breakdown, trapping, and dissolution of the residual air film that commences to be entrained at speeds well below those at which catastrophic coating failure becomes visible. Steady-state analyses [8,13,124] predict that h_∞ is submicroscopic and influenced by microstructural forces at speeds slightly above U_F, but rapidly approaches the power law $h_\infty \sim Ca^{2/3}$ that is well-established for macroscopic films at higher speeds [260,266]. Entrained gas films are unstable [10,17, 117,298]. The breakup of thin, incipient films into patches and bubbles is driven primarily by conjoining microstructural forces [10,17,117,124]. As the films become thicker at higher speeds, the high shear in the air film [10,117,298] and viscosity discontinuity across the liquid/gas interface [298] drive purely hydrodynamic instabilities, some of the Kelvin–Helmholz type. The threshold of catastrophic entrainment of visible amounts of air has been attributed to the slow breakdown of thick air films and the long dissolution times of large bubbles that result from thick films [10,117].

The hypothesis of entrainment and subsequent collapse of invisibly thin air layers suggests a script for complete, microhydrodynamic theories of how a liquid advances over a previously dry solid at high speeds. Such theories hold promise of delivering insights into essential aspects of dynamic wetting, especially the impact of the size and other characteristics of the global flow. The influence of the capillary radius on $Ca_{2,F}$ in simple lubrication-flow analysis [8,13] indicates that the radius over which the macroscopic flow forces the meniscus to bend upstream of the apparent wetting line affects incipient air entrainment [77]. In bead-coating devices, this radius can be altered by changes in subambient pressure $p^A - p^B$ (see Sec. II.A.4). One may speculate that pressure gradients downstream of the apparent contact line or other hydrodynamic forces could also affect the thickness and fate of the entrained film and thereby assist dynamic wetting [77]. In addition, theories that model the entrainment and collapse of thin air films account for specific liquid/solid interactions, including electrostatic forces, and could be modified to incorporate surface roughness and porosity.

Unfortunately, entrained-air-film theories have not been advanced to the point where direct comparisons with experimental data can be made. Preliminary finite-element studies of two-phase flow in slide coating [239] yield flow fields that resemble flow-visualization photos at macroscopic length scales [55]. However, quantitative predictions of apparent dynamic contact angles or velocities at the onset of visible air entrainment have not been published to date. Teletzke's [8,13] results for capillary displacement with $L = O(1 \text{ mm})$ suggest that the capillary number at which the apparent contact angle reaches 180° can at most be $Ca_{2,180°} \equiv \eta_2 U_{180°}/\gamma_{12} \leq O(10^{-5})$ because, above this value, intermolecular forces become negligible. Experimental data for high-viscosity liquids in capillaries [29,34] ($\eta_1 \geq 100$ Pa·s) fall in the plausible range $10^{-8} < Ca_{2,180°} < 10^{-5}$, but do not unequivocally validate that thin air films are entrained at lower speeds. Plunge-tank data for low and moderate viscosities (1 mPa·s $\leq \eta_1 \leq$ 1 Pa·s), on the other hand, cover the range $2 \cdot 10^{-5} \leq Ca_{2,AE} = \eta_2 U_{AE}/\gamma_{12} \leq 2 \cdot 10^{-3}$ (see Fig. 10b). These values seem to corroborate the notion that an invisibly thin air film must be entrained at speeds well below catastrophic coating failure [8,17]. Circumstantial experimental evidence in support of this notion has also been assembled [10,56,118,119] (see Sec. II.B and II.C).

Nonetheless, unresolved inconsistencies remain between the entrainment/collapse theory in its preliminary form and experimental observations. The wide spread of $Ca_{2,180°}$ calculated from experimental data for high-viscosity liquids ($10^{-8} < Ca_{2,180°} < 10^{-5}$) is incongruent with the claim that Ca_2 based on η_2 rather than the usual capillary number Ca based on η_1 ought to be the preeminent parameter [8]. In fact, for highly viscous liquids displacing air, the liquid viscosity appears to be most important: The values of Ca at which the apparent contact angle comes close to 180° fall in the narrow range $3 \leq Ca_{180°} \leq 10$ (see Fig. 3). Experimental data for low-viscosity liquids also seem to be inconsistent with Teletzke's [8] theory. Measured speeds U_{AE} at the onset of visible air entrainment can exceed predicted speeds U_F at incipient entrainment of a thin film by several orders of magnitude. The theory predicts that apparent contact angles rapidly approach 180° as soon as $U > U_F$ and thereby suggests that apparent contact angles would have to stay near 180° over wide ranges in Ca. Experimental data, in contrast, reveal a rapid drop in θ_D below 180° for low-viscosity liquids when $U < U_{AE}$ (see Fig. 25). Therefore, entrainment and subsequent collapse of a residual air film would have to be confined to a narrow range ($U_F < U < U_{AE}$) just below the onset of visible entrainment. Below that range, the incipient air film presumably breaks down right where it is being formed, but the relevant mechanisms are then difficult to distinguish from "conventional" wetting mechanisms at "true" wetting lines. In addition, entrapment of air in the microstructure of rough or porous surfaces [18], or spasmodic jumps of the advancing meniscus induced by a cyclical collapse of the receding air pocket at larger length scales [111], might also yield submicroscopic amounts of entrained air, as have been detected in experiments [10,118]. An adsorbed layer of the receding phase may stay bonded to the solid surface, even at very low speeds and on smooth surfaces [16,17,196]. Evidently, more refined models and more systematic experiments are needed to establish the parameter ranges over which the theory of entrainment and collapse of submicroscopically thin air films models physical reality.

C. Molecular Dynamics of Fluid/Fluid Displacement

1. Stress-Modified Molecular Rate Processes

To elucidate the mechanisms by which the very leading edge of a liquid front traverses a smooth solid surface under conditions of complete displacement, theories are needed that describe dynamic molecular events within a three-phase zone that may undergo intense fluctuations about some mean state [16]. One such theory adapts Eyring's theory of stress-modified activated-rate processes to the vicinity of the moving wetting line [16,69,88,89] (for an extensive review, the reader is referred to Chap. 5 by Blake in this volume). The theory supposes that the out-of-balance interfacial tension $\gamma_{12}(\cos\theta_0 - \cos\vartheta)$ provides a shearing force that biases the energy barriers to molecule displacements in the direction of the wetting line advance. When retarding interactions between fluid molecules and the solid surface, as well as viscous interactions among fluid molecules themselves, are accounted for, the main result is [16,69]

$$\sinh\left[\frac{\gamma_{12}}{2nkT}(\cos\theta_0 - \cos\vartheta)\right] = \frac{\gamma_{12}\nu}{2K_s^0\lambda h}\, Ca \tag{50}$$

Equation (50) relates the dynamic contact angle ϑ for liquid/gas displacement not only to the static angle θ_0, the displacement speed U, the liquid viscosity η_1, and the interfacial tension γ_{12}, but also to a quasiequilibrium rate constant for the surface wetting process K_s^0, the length of each molecular displacement λ, the number of adsorption sites per unit area n, and the volume of each molecular unit ν (h, k, and T have their usual meaning).

Thanks to the adjustability provided by the molecular parameters $K_s^0\lambda h/\nu$ and nkT, Eq. (50) fits θ_D data for a wide variety of liquid/solid systems in numerous flow configurations [16,36,42,44,69,88]. The parameters $K_s^0\lambda h/\nu$ and nkT that yield the best fit assume reasonable values that agree with simple estimates [16,69] and thereby lend further credibility to the molecular-kinetic theory. As for any theory of dynamic wetting, however, agreement with experimental data obtained from macroscopic measurements is no proof that the theory accurately describes the submicroscopic wetting physics. In addition, because of the rather strong sensitivity of ϑ to $K_s^0\lambda h/\nu$ and nkT, Eq. (50) has limited predictive capabilities.

The wetting physics that simple molecular-kinetic models suggest are fundamentally different from the local wetting dynamics that conventional hydrodynamic models assume a priori and that continuum theories augmented with microstructural forces describe in more detail. The treatment of dynamic wetting as an activated-rate process ascribes the variations in dynamic contact angle ϑ to a material-specific adsorption–desorption process causing friction at molecular distances from the solid surface, and hence, it asserts that $\vartheta = f(U)$ is a material function that needs to be measured for each liquid/solid system. In addition, the derivation of Eq. (50) implies that the angle ϑ represents the slope of the interface very close to the solid, yet calculated values of ϑ are usually compared with apparent dynamic contact angles θ_D that have been measured at macroscopic length scales [16,36,42,44,69,88]. Such comparisons suggest that the angles ϑ and θ_D are the same [16,69] and make no allowance for viscous bending of the interface near the solid. Hydrodynamic theories of dynamic wetting, in contrast, suggest that the dynamic variations in apparent contact angle arise primarily from visco-capillary bending of the meniscus away from the innermost region where material-specific microstructural forces dominate; that meniscus bending gives rise to a universal $\theta_D = f(Ca, \theta_0)$ curve; and that only systematic deviations from this curve arise from material-specific mechanisms or global flow effects. Wetting models based on the theory of stress-modified molecular rate processes cannot disprove the assumptions made in hydrodynamic models because the simple molecular models lump statistical fluctuations into quasiequilibrium quantities and employ a macroscopic concept themselves, namely, the unbalanced wetting force $\gamma_{12}(\cos\theta_0 - \cos\vartheta)$. Experimental observations available to date cannot settle the conflict between the molecular-kinetic and hydrodynamic point of views either. However, evidence for meniscus bending as far as several tens of microns from the solid is undisputed for systems that form precursor films (see Sec. II.D.2 and III.B.1) and is mounting for other regimes of dynamic wetting (see Sec. II.A.3 and III.A.5). Molecular-kinetic theories also deny the possibility of a size effect. Observations of such an effect lend further support to the notion that the local contact angle must be different from the angle measured with conventional optical techniques [25,37]. Toward high displacement speeds, finally, theories based on the concept of activated-rate processes anticipate that there is a maximum speed of wetting at which ϑ reaches 180°, but fail to describe the systematic difference between high- and low-viscosity liquids that is observed in experiments (see Sec. II.A.3 and III.A.9). Equation (50) predicts a rapid approach toward 180° as is observed for low-viscosity liquids, but does not capture the slow asymptotic approach for high-viscosity liquids. Attempts to augment the theory in order to remedy the discrepancy near 180° invoke a "tank-tread" motion due to attractive intermolecular forces [69] or "additional viscous effects" in the liquid [16]. Mathematical models of both mechanisms are designed to retard the increase of ϑ near 180°. In hydrodynamic theories [24,76], in contrast, the asymptotic approach toward 180° is the standard result for high-viscosity liquids, and the accelerated rise in ϑ_m for low-viscosity liquids stems from viscous effects in the displaced air phase (see Sec.II.A.9).

Based on the apparent success of hydrodynamic theories—which in conjunction with the ad hoc assumptions $\vartheta_w = \theta_0$ and $\epsilon_s =$ const. predict universal θ_D versus Ca behavior—one may be tempted to assign molecular-kinetic models the role of describing systematic deviations from the universal behavior that might arise from changes in the local contact angle ϑ_w due to excess dissipation at molecular length scales. When a liquid displaces a gas, deviations from the universal curve are particularly prominent at low speeds (see Sec. II.A.4). For $\gamma_{12}(\cos\theta_0 - \cos\vartheta)/2nkT \ll 1$, Eq. (50) specializes to the linear form $(\cos\theta_0 - \cos\vartheta) \sim Ca$. For systems with partial wetting, this form is the same as the low Ca limit (12) of the modified Hoffman–Voinov–Tanner law (32a). It might be useful to describe a rapid rise in θ_D at extremely low speeds, but like (12) overestimates the rate of increase of $(\cos\theta_0 - \cos\theta_D)$ with Ca in comparison to most low-speed data available (see Sec. II.A.5). At higher speeds, Eq. (50) reduces to the logarithmic form

$$(\cos\theta_0 - \cos\vartheta) \sim \log(Ca) \tag{51}$$

which appears to be better suited to describe anomalous low-speed data [16,44,255]. Nevertheless, evidence that peculiar low-speed behavior arises from adsorption–desorption mechanisms rather than stick-slip related to contact-angle hysteresis remains inconclusive. Blake's [16] suggestion that stick-slip may be related to molecular events also remains in doubt.

For liquids that completely wet the solid, the low-speed limit of Eq. (50) is $\vartheta \sim Ca^{0.5}$, which predicts a more rapid rise of ϑ with Ca than the Hoffman-Voinov-Tanner law $\vartheta \sim Ca^{1/3}$ [see Eq. (33)]. Except perhaps at extremely small Ca, the power law $\vartheta \sim Ca^{0.5}$ agrees poorly with experimental data for completely wetting liquid/solid systems (see Figs. 4 and 13).

2. Molecular Dynamics Simulations

An alternative approach to resolving the physics in the immediate vicinity of a moving apparent contact line is to explicitly compute the positions and velocities of individual molecular particles by integrating Newton's equation of motion for each particle. Molecular dynamics simulations of dynamic wetting have relied on simple Lennard-Jones-type interactions to model immiscible liquid/liquid displacement in either Poiseuille [196,299] or Couette flow [207,257]. The simulations have furthermore been confined to very small systems containing at most a few thousand particles, which limits the characteristic dimensions to 10–15 molecular spacings and leads to unrealistically high deformation rates.

In spite of serious physical and computational limitations, molecular dynamics simulations provide some interesting insights into the local displacement process that, by and large, corroborate the ad hoc assumptions of local slip and a fixed static contact angle that often enter conventional hydrodynamic theories. In particular, the simulations suggest that the Navier–Stokes equations for incompressible viscous flow remain valid outside a three-phase region only a few molecular diameters in size. Inside that region, velocity gradients vary on molecular length scales, continuum hydrodynamics break down, and molecular fluctuations yield slippage of both liquids over the solid [196,207,299]. Despite the breakdown of continuum theory and limited statistical resolution, averaged flow fields that result from the molecular computations [196] are consistent with the multivalued flow fields that are expected from kinematic arguments [121] and have been observed in experiments [122]. Simulated flow fields are also consistent with numerical solutions of the Navier-Stokes system that presuppose an exponentially decaying slip velocity boundary condition [207]. However, molecular events provide a cutoff in the molecular-dynamics simulations and, hence, averaged flow fields do not suffer from the unphysical singularity that persists when conventional continuum theory is modified with an ad hoc slip boundary condition at molecular length scales (see Sec. III.A.2).

Dynamic variations in contact angle inferred from molecular dynamics simulations are also consistent with the assumptions and predictions of conventional hydrodynamic theories. The simulations suggest that, within the uncertainty that arises from statistical fluctuations ($\pm 5°$), the contact angle at a molecular level retains its static value [207]. However, in simulations published to date, the interactions of both fluids with the solid were chosen to be identical, and so the angle was restricted to $\vartheta_w = 90°$. Predicted values of an apparent dynamic contact angle ϑ_D at a constant distance of 6.38 molecular diameters from the solid agree with Cox's [219] result (29) for $\vartheta_w = 90°$, $\lambda_\eta = 1$, and $\ln(\epsilon_s^{-1}) + Q = \text{const.}$, where $Q \equiv q_i/f(\vartheta_w, \lambda_\eta) - q_0/f(\vartheta_m, \lambda_\eta)$. The term Q is important because of the large dimensionless cutoff length $\epsilon_s \approx O(0.1)$ in molecular-dynamics simulations of small systems. The simulations indicate that changes in the interaction potentials between the fluids and solid can significantly alter the apparent contact angle ϑ_D even though the local angle retains its static value [257]. Above $Ca \geq O(0.1)$, molecular-dynamics simulations of very small liquid/liquid systems with $\lambda_\eta = 1$ predict that a film of the receding liquid is left behind on the solid [196,207].

IV. SUMMARY

This chapter gives an overview of what is known about the hydrodynamics of wetting. Many unresolved issues remain, related mostly to the submicroscopic mechanisms by which a liquid displaces another fluid from a solid surface. Nonetheless, considerable progress has been made in the past 15 years, and several unifying concepts emerge that may not be readily apparent upon casual reading of selected papers in the vast and sometimes controversial literature on dynamic wetting.

Most significant progress has been made in synthesizing a unified view of the key mechanisms that affect the macroscopic wetting dynamics, especially for nonvolatile liquids that completely wet the solid surface when allowed to reach static equilibrium. For such systems, apparent dynamic contact angles are insensitive to specific liquid/solid interactions. Measured angles depend primarily on capillary number, $Ca \equiv \eta_1 U/\gamma_{12}$, and gather close to a universal curve $\theta_D = f_{\text{Hoff}}(Ca)$. Up to $Ca \leq 0.1$, the curve is independent of the flow configuration in which θ_D is measured and is well approximated by the much celebrated $\theta_D \sim Ca^{1/3}$ law. For systems with partial equilibrium wetting, on the other hand, the choice of materials can strongly affect measured apparent dynamic contact angles. Nevertheless, for selected liquid/solid combinations—especially nonpolar, nonvolatile liquids advancing over smooth and clean surfaces—the static contact angle θ_0 successfully accounts for much of the influence of surface wettability under dynamic conditions, and the generalized curve $\theta_D = f_{\text{Hoff}}[Ca + f_{\text{Hoff}}^{-1}(\theta_0)]$ correlates experimental data quite well. The mathematical structure of the generalized correlation indicates that surface wettability is important at low *Ca*, but its influence weakens at higher displacement speeds.

Even though material-specific mechanisms can give rise to significant deviations from the universal wetting behavior for either completely wetting or partially wetting systems, the predominant role of *Ca* in correlating θ_D for many liquid/solid systems suggests that the angle variations detectable at visible length scales arise primarily from hydrodynamic mechanisms. Purely hydrodynamic models indeed predict the universal curve—even though conventional continuum theory cannot resolve the physics of how a liquid displaces another fluid from a solid. Most models employ slip or a related ad hoc boundary condition to truncate the unphysical force singularity that arises in conventional continuum theory. Other models ignore the immediate vicinity of the putative wetting line altogether. Many hydrodynamic models furthermore insist that the contact angle close to the solid retain its static value θ_0 and thus attribute variations in apparent contact angle exclusively to viscous bending of the interface close to the solid. The bending arises at length scales much smaller than the macroscopic flow, yet larger than the submicroscopic cutoff region where the unresolved wetting physics dominate. Up to $Ca \leq 0.2$, hydrodynamic models replicate the functional form of $\theta_D = f_{\text{Hoff}}(Ca)$. In particular, the models predict the low-speed behavior $\theta_D \sim$

$Ca^{1/3}$ for complete static wetting and corroborate the generalized universal curve $\theta_D = f_{\text{Hoff}}[Ca + f_{\text{Hoff}}^{-1}(\theta_0)]$ for partial wetting. Above $Ca \geq 0.2$, results from hydrodynamic models with ad hoc boundary conditions begin to deviate from the universal curve, reflecting a more rapid rise of θ_D near 180° than is observed in experiments.

The apparent success of hydrodynamic models with ad hoc boundary conditions intimates that a complete knowledge of the local displacement process is not required to predict macroscopic wetting dynamics. Refined models that incorporate long-range fluid/solid interactions to describe the meniscus profile close to the solid confirm that the free-surface profile outside the cutoff region is independent of the local wetting mechanism chosen, and that apparent contact angles vary as a universal function of *Ca*. Purely hydrodynamic models which assume that the solid is prewet by a thin liquid film yield analogous results.

Because predicted macroscopic wetting dynamics are insensitive to the boundary conditions imposed at the solid, agreement between theory and experiment on macroscopic scales does not validate the assumptions made for the local wetting dynamics. Unequivocal verification of *any* theory of dynamic wetting, including refined theories that attempt to describe the local displacement process, requires experimental measurements on the length scales that the theory tries to resolve. To date, such verification is available only for the special case of precursor films that expand ahead of the macroscopic front when a completely wetting liquid spreads slowly. Detailed measurements of the interfacial profile between the precursor film and macroscopic meniscus confirm that apparent dynamic contact angles detected by the naked eye or measured through low-power microscopes can deviate substantially from the angle of inclination of the interface at submicroscopic distances from the solid. Direct comparisons between measured and calculated profiles verify also that viscocapillarity dominates the contact angle at which the quasistatic meniscus appears to intersect the solid, even when microstructural forces control the thickness and extent of the precursor film by which primary wetting occurs. Both theory and experiment show furthermore that the angle of inclination varies slowly upstream from where the precursor film joins the macroscopic meniscus and thereby rationalize why different techniques for measuring θ_D yield similar values. Even for the much studied case of slow spreading on top of precursor films, however, the mechanisms by which the first few molecular layers advance over the solid are not fully resolved. Surface diffusion is a plausible mechanism, but transport of volatile components through the vapor phase is probably more important than is commonly appreciated.

At larger apparent contact angles that arise either from higher displacement speeds or incomplete wetting, precursor films cannot form and the reasons for the success of purely hydrodynamic theories with ad hoc boundary conditions need to be clarified further. Molecular-dynamics simulations corroborate that slip provides a physical cutoff mechanism, at least at molecular length scales, and that the local angle of interface inclination remains close to its static value. Simulations published to date, however, employ highly idealized interaction potentials and are confined to unrealistically small systems. Augmented hydrodynamic continuum theories further support the notion that microstructural forces control the interfacial profile close to the solid as they do in the static case, yet the dynamic variations of the apparent contact angle arise from viscous bending at larger distances from the solid where the influence of long-range fluid/solid interactions is negligible. These same theories also suggest that microstructural forces may dominate over slip in providing a cutoff for the unbounded viscous stress singularity. However, the predictions from augmented continuum theories and molecular dynamics simulations—and also the ad hoc assumptions that enter conventional hydrodynamic models—are squarely at odds with other models of dynamic wetting that attribute the variations of the dynamic contact angle to excess dissipation due to molecular events right at the three-phase juncture and that fit experimental θ_D data for numerous liquid/systems quite well. Local measurements of the interface profile substantiate that the meniscus bends in the direction opposite of the wetting line motion as hydrodynamic theories predict, but data are available only at distances larger than 20 μm from the solid and are restricted to $\theta_0 \approx 0°$ and $Ca \leq 0.03$. To fully elucidate the physics of the local displacement process and verify refined wetting theories, measurements are needed at length scales of a few μm or even less—a formidable challenge with no straightforward solution in sight.

Additional experiments and refined theories are also needed to identify and predict specific material effects that cause systematic deviations from the universal curve $\theta_D = f(Ca, \theta_0)$. For completely wetting systems, topography and chemical composition of the solid surface appear to be of minor consequence. Exceptions may arise at low displacement speeds at which capillary imbibition on rough surfaces may enhance wetting rates and specific liquid/solid interactions may also come into play. For partially wetting systems, on the other hand, the observed dynamic wetting behavior is susceptible to intricate details of surface texture and chemistry. Deviations from the universal curve $\theta_D = f_{\text{Hoff}}[Ca + f_{\text{Hoff}}^{-1}(\theta_0)]$ can be dramatic and seem to persist over wide ranges of Ca. Nonetheless, the deviations can sometimes be adsorbed into a pseudostatic contact angle, $\theta_0^* > \theta_0$, which indicates that the "anomalous" behavior of a partially wetting liquid/solid system may stem from peculiar low-speed phenomena. Stick-slip arising from surface roughness and heterogeneity is a particularly plausible explanation, for individual data points often exhibit scatter, but other material-specific mechanisms may also produce excess friction in the immediate vicinity of the three-phase juncture and cause θ_D to rise at lower speeds than the universal correlation suggests. Simple models for distinct low-speed regimes have been advanced, but agree only qualitatively with experimental data. The subsequent rise of θ_D at higher speeds is presumably governed by the same hydrodynamic mechanisms that dominate dynamic wetting for $\theta_0 = 0$. This needs to be verified, however, with careful low-speed experiments involving state-of-the-art analysis of the solid surface and quantitative measurement of specific liquid/solid interactions. Careful experiments and accompanying theories are also needed to quantify the consequences of surfactants and contaminants, volatile components, non-Newtonian rheology, soluble priming layers, and last, but not least, electrostatic surface charges that all might yield measurable deviations from the universal correlation for either partial or complete static wetting.

Systematic deviations from the universal θ_D versus Ca curve can also be induced by the global flow field, of which the wetting process is part, especially at capillary numbers $Ca \geq O(0.1)$ above which the macroscopic meniscus may deviate substantially from its static shape. Changes in global hydrodynamics can be used to great advantage to assist dynamic wetting in industrial coating practice—as, for instance, in curtain coating—but a sizable effect of the macroscopic flow on θ_D has not been documented in the scientific literature for well-characterized liquid/solid systems. Neither have hydrodynamic models been advanced that predict the effect, except in the limiting regime of $Ca \to \infty$ where the apparent dynamic contact angle is dominated entirely by the hydrodynamics away from the wetting line. Refined models for the local wetting dynamics at high speeds of forced displacement, which suppose that an invisibly thin film of air is entrained and successfully collapses below the onset of catastrophic air entrainment, suggest that the curvature of the macroscopic meniscus upstream of the wetting line, as well as pressure gradients imposed by the flow field downstream, might have a significant effect. Such models, however, need to be incorporated into global flow analyses and the results verified with experiments. Circumstantial evidence for a high-speed entrainment/collapse regime comes from microscopic bubbles detected in coated layers and laser-Doppler velocimetry measurements downstream of the apparent wetting line.

At capillary numbers less than $O(0.1)$, the influence of the macroscopic flow enters primarily through the size of the flow geometry. The flow configuration in which the apparent dynamic contact angle is measured is unimportant. In particular, spontaneous spreading and forced wetting experiments yield comparable θ_D versus *Ca* curves. Hydrodynamic forces acting on macroscopic length scales, such as gravity or inertia, are inconsequential—consistent with theories that attribute dynamic contact-angle variations to a visco-capillary stress balance at small, yet finite distances from the solid surface. Such theories suggest that a size effect arises from the ratio of the characteristic length of the macroscopic flow to the cutoff length, within which local wetting mechanisms dominate. Predicted size effects, however, agree only qualitatively with experimental data for silicone oils advancing between parallel glass plates, the only system for which a size effect has been documented systematically. More experiments with other liquid/solid combinations and different flow geometries are needed to establish whether a size effect is ubiquitous and, if so, to verify that changes in the ratio between the macroscopic length and cutoff length account for the effect.

Further deviations from the universal θ_D versus *Ca* curve arise from viscous stresses—and possibly other hydrodynamic forces—in the receding phase. The viscosity ratio $\lambda_\eta \equiv \eta_2/\eta_1$ is a key parameter in dynamic wetting, second in importance only to *Ca* and, at low speeds, θ_0. If the receding phase is a gas, its influence is measurable only when the gas flow is confined to a wedge between the advancing meniscus and solid surface, that is, above $\theta_D \geq 120°$. For low-viscosity liquids, lubricating pressures building in the displaced gas cause θ_D to rapidly head toward 180°. For liquids with viscosities above 100 Pa · s, on the other hand, the displacement flow on length scales accessible by conventional optical means is nearly immune to hydrodynamic separating forces in the displaced gas, and capillary numbers well above $Ca \geq O(1)$ can be reached before θ_D comes close to 180°. Hydrodynamic theories that account for viscous flow in the receding phase corroborate the distinct high-speed wetting behavior of low- and high-viscosity liquids, even though such theories assume complete displacement and rely on ad hoc boundary conditions at the wetting line. In refined theories that describe the entrainment and collapse of a thin air layer, viscous dissipation in the second phase is paramount and competes with long-range fluid/solid interaction forces in controlling the local wetting dynamics.

The influence and technological control of local airflow effects ought to become a major focus of future experimental and theoretical studies of dynamic wetting—for it is key to successful high-speed coating. In particular, the flow in the receding phase dictates the maximum speed beyond which successful displacement fails in a catastrophic manner visible by the unaided eye. Empirical correlations for the threshold of visible coating failure indicate that the critical capillary number diminishes with increasing viscosity ratio $\lambda_\eta \equiv \eta_2/\eta_1$. When the limit $\theta_D \rightarrow 180°$ is admitted as an ad hoc criterion for the onset of visible entrainment, strictly hydrodynamic two-phase-flow theories corroborate the dependence of Ca_{AE} on $\lambda_\eta \equiv \eta_2/\eta_1$ and thereby suggest that the macroscopic failure of dynamic wetting is dominated by hydrodynamic separating forces that build in the receding gas away from the immediate proximity of the wetting line. In contrast, theories that postulate the entrainment and collapse of an invisibly thin air film at subcritical speeds contend that the catastrophic limit of visible entrainment arises from the failure of microstructural forces to exclude sufficient air and from a strong sensitivity of the air film stability to film thickness. Neither theory predicts the sawtooth-shaped wetting line that sometimes accompanies air entrainment, nor the tenuous unsteady motion that the wetting line may undergo before the onset of visible entrainment. Analyzing these phenomena and related modes of coating failure remains a formidable challenge in wetting hydrodynamics. Complete theories would have to consider both spatial and temporal modes of two-phase flow instability. In addition, theories for air entrainment would have to account for not only the evidently dominant role of capillary pressure and viscous stresses in both phases, but also the potential influence of inertia and other global flow effects, as well as specific long-range fluid/solid interactions, surface roughness, porosity, electrostatic charges, and possibly other characteristics of the solid.

SYMBOLS

A	Hamaker constant (J)
Bo	Bond number ($Bo \equiv \rho_1 g L^2/\gamma_{12}$)
Ca	Capillary number ($Ca \equiv \eta_1 U/\gamma_{12}$)
Ca_{AE}	Critical capillary number at onset of air entrainment ($Ca_{AE} \equiv \eta_1 U_{AE}/\gamma_{12}$)
h	Dimensionless film thickness ($h \equiv H/H_0$)
H	Film thickness (m)
H_0	Reference film thickness (m)
L	Characteristic, macroscopic length scale (m)
L_a	Molecular length scale (m)
L_s	Slip length (m)
L_β	Extrapolation length in Navier's slip boundary condition ($L_\beta \equiv \beta\eta_1$) (m)
L_δ	Cutoff length for unbounded force singularity at apparent dynamic wetting line (m)
L_γ	Capillary length ($L_\gamma \equiv \sqrt{\gamma_{12}/\rho_1 g}$) (m)

R	Radius of spreading droplet (m)
Re	Reynolds number ($Re \equiv \rho_1 UL/\eta_1$)
S	Spreading coefficient ($S \equiv \gamma_{S2} - \gamma_{S1} - \gamma_{12}$) (N/m)
U	Displacement speed of apparent wetting line (m/s)
U_{AE}	Critical displacement speed at onset of air entrainment (m/s)
We	Weber number ($\rho_1 U^2 L/\gamma_{12}$)
β	Slip coefficient (m^2s/kg)
ϵ_i	Dimensionless intermediate length scale
ϵ_β	Dimensionless extrapolation length ($\epsilon_\beta \equiv L_\beta/L = \beta\eta_1/L$)
ϵ_δ	Dimensionless cutoff length ($\epsilon_\delta \equiv L_\delta/L$)
ϵ_s	Dimensionless slip length ($\epsilon_s \equiv L_s/L$)
γ_{12}	Interfacial tension between advancing and receding fluid (N/m)
η_1	Viscosity of advancing liquid (Pa · s)
η_2	Viscosity of receding fluid (Pa · s)
λ_η	Viscosity ratio ($\lambda_\eta \equiv \eta_2/\eta_1$)
λ_ρ	Density ratio ($\lambda_\rho \equiv \rho_2/\rho_1$)
ρ_1	Density of advancing liquid (kg/m^3)
ρ_2	Density of receding fluid (kg/m^3)
Π	Disjoining pressure (N/m^2)
θ	Contact angle measured in experiment
θ_D	Apparent dynamic contact angle
θ_E	Equilibrium contact angle
θ_0	Apparent static contact angle
ϑ	Contact angle in mathematical model
ϑ_a	Macroscopic angle of inclination of liquid wedge advancing over solid
ϑ_D	Calculated apparent dynamic contact angle
ϑ_m	Apparent dynamic contact angle calculated from asymptotic expansion method
ϑ_{m0}	Apparent dynamic contact angle calculated from leading-order solution
ϑ_w	Local contact angle imposed as mathematical boundary condition

REFERENCES

1. G. E. P. Elliot, and A. C. Riddiford, in *Recent Progress in Surface Science*, Vol. 2, (J. F. Danielli, K. G. A. Pankhurst, and A. C. Riddiford, eds.), Academic Press, New York, 1964, pp. 111–128.
2. C. Huh, and L. E. Scriven, *J. Coll. Inter. Sci. 35:*85 (1971).
3. E. B. Dussan V., *Ann. Rev. Fluid Mech. 11:*371 (1979).
4. S. H. Davis, *J. Appl. Mech.* (*Trans.* ASME) *50:*977 (1983).

5. A. Marmur, *Adv. Coll. Inter. Sci. 19:*75 (1983).
6. P. G. de Gennes, *Rev. Mod. Phys. 57:*827 (1985).
7. G. F. Teletzke, H. T. Davis, and L. E. Scriven, *Chem. Eng. Comm. 55:* 41 (1987).
8. G. F. Teletzke, H. T. Davis, and L. E. Scriven, *Rev. Phys. Appl. 23:*989 (1988).
9. A. M. Cazabat, *Contemp. Phys. 28:*347 (1987).
10. K. Miyamoto, *Ind. Coating Res. 1:*71 (1991).
11. W. D. Bascom, R. L. Cottington, and C. R. Singleterry, in *Contact Angle, Wettability and Adhesion*, Advances in Chemistry Series, Vol. 43 (F. M. Fowkes, ed.), ACS, Washington, DC, 1964, pp. 355–379.
12. A. M. Cazabat, N. Fraysse, F. Heslot, and P. Carles, *J. Phys. Chem. 94:* 7581 (1990).
13. G. F. Teletzke, Ph.D. thesis, Univ. Minn., Minneapolis, 1983.
14. J. F. Joanny, *J. Theor. Appl. Mech. 23 (Numéro Special):*249 (1986).
15. W. P. Hardy, *Philos. Mag. 38:*49 (1919).
16. T. D. Blake, AIChE Spring National Meeting, New Orleans, LA, 1988 (see also chap. 5 of this volume).
17. L. E. Scriven, AIChE Spring National Meeting, Orlando, FL, 1982.
18. L. M. Hocking, *J. Fluid Mech. 76:*801 (1976).
19. R. E. Benner, H. T. Davis, and L. E. Scriven, in *Structure of the Interface Region*, Faraday Symposium No. 16, Royal Soc., London, 1982, pp. 169–190.
20. J. S. Rowlinson, and B. Widom, *Molecular Theory of Capillarity*, Calarendon Press, Oxford, 1982.
21. F. Brochard-Wyart, J. M. di Meglio, D. Quéré, and P. G. de Gennes, *Langmuir 7:*335 (1991).
22. R. J. Hansen, and T. Y. Toong, *J. Coll. Inter. Sci. 37:*196 (1971).
23. L. M. Hocking, and A. D. Rivers, *J. Fluid Mech. 121:*425 (1982).
24. R. G. Cox, *J. Fluid Mech. 168:*169 (1986).
25. C. G. Ngan, and E. B. Dussan V., *J. Fluid Mech. 209:*191 (1989).
26. W. Rose, and R. W. Heins, *J. Coll. Sci. 17:*39 (1962).
27. R. J. Hansen, and T. Y. Toong, *J. Coll. Inter. Sci. 36:*410 (1971).
28. B. V. Zheleznyi, *Dokl. Acad. Nauk SSSR 207:*647 (1972).
29. R. L. Hoffman, *J. Coll. Inter. Sci. 50:*228 (1975).
30. B. Legait, and P. Sourieau, *J. Coll. Inter. Sci. 107:*14 (1985).
31. M. Fermigier, and P. Jenffer, *Ann. Physique 13:*37 (1988).
32. J. P. Stokes, M. J. Higgins, A. P. Kushnick, S. Bhattacharya, and M. O. Robbins, *Phys. Rev. Lett. 65:*1885 (1990).
33. A. Calvo, I. Paterson, R. Chertcoff, M. Rosen, and J. P. Hulin, *J. Coll. Inter. Sci. 141:*384 (1991).
34. M. Fermigier, and P. Jenffer, *J. Coll. Inter. Sci. 146:*226 (1991).
35. T. A. Coney, and W. J. Masica, NASA Rept. TN D-5115, 1969.
36. E. V. Gribanova, and L. I. Molchanova, *Coll. J. USSR* (English trans.) *40:*22 (1978).

37. C. G. Ngan, and E. B. Dussan V., *J. Fluid Mech. 118:*27 (1982).
38. G. E. P. Elliot, and A. C. Riddiford, *J. Coll. Inter. Sci. 23:*389 (1967).
39. R. E. Johnson, R. H. Dettre, and D. A. Brandreth, *J. Coll. Inter. Sci. 62:* 205 (1977).
40. E. Zimmermann, and J. Siekmann, *Appl. Microgravity Tech. II 2:*96 (1989).
41. J. Siekmann, and E. Zimmermann, *Z. Angew. Math. Mech. 70:*T351 (1990).
42. G. Ström, M. Fredriksson, P. Stenius, and P. Radoev, *J. Coll. Inter. Sci. 134:*107 (1990).
43. G. Inverarity, *Br. Polym. J. 1:*245 (1969).
44. A. M. Schwartz, and S. B. Tejada, *J. Coll. Inter. Sci. 38:*359 (1972).
45. E. B. Dussan V., E. Ramé, and S. Garoff, *J. Fluid Mech. 230:*97 (1991).
46. R. T. Perry, Ph.D. thesis, Univ. Minn., Minneapolis, 1967.
47. R. Burley, and B. S. Kennedy, *Br. Polym. J. 8:*140 (1976).
48. T. D. Blake, and K. J. Ruschak, *Nature 282:*489 (1979).
49. E. B. Gutoff, and C. E. Kendrick, *AIChE J. 28:*459 (1982).
50. R. Burley, and R. P. S. Jolly, *Chem. Eng. Sci. 39:*1357 (1984).
51. M. Bracke, F. de Voeght, and P. Joos, *Prog. Coll. Polym. Sci. 79:*142 (1989).
52. R. Ablett, *Philos. Mag. 46:*244 (1923).
53. M. N. Esmail, and M. T. Ghannam, *Canad. J. Chem. Eng. 68:*197 (1990).
54. M. T. Ghannam, and M. N. Esmail, *AIChE J. 36:*1283 (1990).
55. P. M. Schweizer, *J. Fluid Mech. 193:*285 (1988).
56. W. Mues, J. Hens, and L. Boiy, *AIChE J. 35:*1521 (1989).
57. K. S. A. Chen, Ph.D. thesis, Univ. Minn., Minneapolis, 1992.
58. G. R. Zvan, L. J. Douglas, and S. F. Kistler, AIChE Spring National Meeting, New Orleans, LA, 1992.
59. L. J. Douglas, S. F. Kistler, and N. K. Nelson, IS&T 44th Annual Conference, St. Paul, MN, 1991.
60. E. Bayramli, T. G. M. van de Ven, and S. G. Mason, *Coll. Surf. 3:*131 (1981).
61. J. E. Seebergh, and J. C. Berg, *Chem. Eng. Sci. 47:*4455 (1992).
62. J. E. Seebergh, and J. C. Berg, *Chem. Eng. Sci. 47:*4468 (1992).
63. G. D. Towell, and L. B. Rothfeld, *AIChE J. 12:*972 (1966).
64. R. Burley, and B. S. Kennedy, *Chem. Eng. Sci. 31:*901 (1976).
65. T.-S. Jiang, S.-G. Oh, and J. C. Slattery, *J. Coll. Inter. Sci. 69:*74 (1979).
66. L. H. Tanner, *J. Phys. D: Appl. Phys. 12:*1473 (1979).
67. O. V. Voinov, *Fluid. Dyn. 11:*714 (1976).
68. O. V. Voinov, *Sov. Phys. Dokl.* (English trans.) *23:*891 (1978).
69. R. L. Hoffman, *J. Coll. Inter. Sci. 94:*470 (1983).
70. J.-D. Chen, *J. Coll. Inter. Sci. 122:*60 (1988).
71. R. T. Foister, *J. Coll. Inter. Sci. 136:*266 (1990).
72. T. D. Blake, Ph.D. thesis, Univ. Bristol, 1968.
73. J.-D. Chen, and N. Wada, *J. Coll. Inter. Sci. 148:*207 (1992).
74. L. R. White, *J. Chem. Soc. Faraday 73:*390 (1977).
75. K. J. Ruschak, *Ann. Rev. Fluid Mech. 17:*65 (1985).

76. L. Grader, *Coll. Polym. Sci. 264:*719 (1986).
77. L. E. Scriven, AIChE Spring National Meeting, Orlando, FL, 1990.
78. S. F. Kistler, Ph.D. thesis, Univ. Minn., Minneapolis, 1984.
79. R. E. Johnson, and R. H. Dettre, in *Surface and Colloid Science*, Vol. 2 (E. Matijevic, ed.), Wiley-Interscience, New York, 1969, pp. 85–153.
80. G. D. Yarnold, *Proc. Phys. Soc. Lond. 50:*540 (1938).
81. J. B. Cain, D. W. Francis, R. D. Venter, and A. W. Neumann, *J. Coll. Inter. Sci. 94:*123 (1983).
82. W. H. Seward, M.S. thesis, Univ. Minn., Minneapolis, 1967.
83. J. F. Oliver, and S. G. Mason, *Coll. Surf. 1:*79 (1980).
84. M. A. Cohen Stuart, and A. M. Cazabat, *Prog. Coll. Polym. Sci. 74:*64 (1987).
85. J. G. E. M. Fraaije, M. Cazabat, X. Hua, and A. M. Cazabat, *Coll. Surf. 41:*77 (1989).
86. H. M. Princen, A. M. Cazabat, M. A. Cohen Stuart, F. Heslot, and S. Nicolet, *J. Coll. Inter. Sci. 126:*84 (1988).
87. E. Rillaerts, and P. Joos, *Chem. Eng. Sci. 35:*883 (1980).
88. T. D. Blake, and J. M. Haynes, *J. Coll. Inter. Sci. 30:*421 (1969).
89. B. W. Cherry, and C. M. Holmes, *J. Coll. Inter. Sci. 29:*174 (1969).
90. E. Ruckenstein, *Chem. Eng. Sci. 24:*1223 (1969).
91. H. van Oene, Y. F. Chang, and S. Newman, *J. Adhesion 1:*54 (1969).
92. W. W. Y. Lau, and C. M. Burns, *J. Coll. Inter. Sci. 45:*295 (1973).
93. V. A. Ogarev, T. N. Timonina, V. V. Arslanov, and A. A. Trapeznikov, *J. Adhesion 6:*337 (1974).
94. P. G. de Gennes, *Coll. Polym. Sci. 264:*463 (1986).
95. K. M. Jansons, *J. Fluid Mech. 167:*393 (1986).
96. K. Ishimi, H. Hikita, and M. N. Esmail, *AIChE J. 32:*486 (1986).
97. M.-Y. Zhou, and P. Sheng, *Phys. Rev. Lett. 64:*882 (1990).
98. G. Ström, M. Fredriksson, and P. Stenius, *J. Coll. Inter. Sci. 134:*117 (1990).
99. H. L. Goldsmith, and S. G. Mason, *J. Coll. Sci. 18:*237 (1963).
100. B. V. Derjaguin, and S. M. Levi, *Film Coating Theory*, Focal Press, London and New York, 1964.
101. W. L. Wilkinson, *Chem. Eng. Sci. 30:*1227 (1975).
102. P. J. Veverka, and C. K. Aidun, *TAPPI J. 74:*203 (Sept. 1991).
103. A. O'Connel, Ph.D. thesis, Heriot-Watt Univ., Edinburgh, 1989.
104. K. Ishizaki, N. Chino, and S. Fuchigami, AIChE Spring National Meeting, Orlando, FL, 1982.
105. D. R. Brown, *J. Fluid Mech. 10:*297 (1961).
106. J. F. Greiller, U.S. Patent 3,632,374 to Eastman Kodak Company, 1972.
107. T. D. Blake, A. Clarke, and K. J. Ruschak, *J. Fluid Mech.* (1993) (submitted for publication).
108. E. B. Gutoff, and C. E. Kendrick, *AIChE J. 33:*141 (1987).
109. S. M. Levi, *Zh. Nauchn. Prikl. Fotogr. Kenematogr. 11:*401 (1966).
110. S. M. Levi, and V. I. Akulov, *Zh. Nauchn. Prikl. Fotogr. Kenematogr. 9:*124 (1964).

111. J. N. Tilton, Ph.D. thesis, Univ. Houston, 1985.
112. D. J. Coyle, C. W. Macosko, and L. E. Scriven, *AIChE J. 36:*161 (1990).
113. R. A. Buonopane, E. B. Gutoff, and M. M. T. Rimore, *AIChE J. 32:*682 (1986).
114. Y. Yasui, and T. Tanaka, AIChE Spring National Meeting, Orlando, FL, 1990.
115. B. Bolton, and S. Middleman, *Chem. Eng. Sci. 35:*597 (1980).
116. J. Bataille, *C. R. Acad. Sci. Paris Ser. A 262:*843 (1966).
117. K. Miyamoto, and L. E. Scriven, AIChE Annual Meeting, Los Angeles, CA, 1982.
118. K. Miyamoto, AIChE Spring National Meeting, New Orleans, LA, 1986.
119. K. Miyamoto, U.S. Patent 4,842,900 to Fuji Photo Film Co., Ltd., 1989.
120. A. M. Schwartz, C. A. Rader, and E. Huey, in *Contact Angle, Wettability, and Adhesion*, Advances in Chemistry Series, Vol. 43 (F. M. Fowkes, ed.), ACS, Washington DC, 1964, pp. 250–267.
121. E. B. Dussan V., and S. H. Davis, *J. Fluid Mech. 65:*71 (1974).
122. E. B. Dussan V., *AIChE J. 23:*131 (1977).
123. B. C. Sakiadis, *AIChE J. 7:*26 (1961).
124. R. J. Prutow, and S. Ostrach, AFOSR Scientif. Rept. 70-2882TR, 1971.
125. J. Hyypia, *Anal. Chem. 20:*1039 (1948).
126. H. Schonhorn, H. L. Frisch, and T. K. Kwei, *J. Appl. Phys. 37:*4967 (1966).
127. W. Radigan, H. Ghirardella, H. L. Frisch, H. Schonhorn, and T. K. Kwei, *J. Coll. Inter. Sci. 49:*241 (1974).
128. A. R. Karnik, *J. Photograph. Sci. 25:*197 (1977).
129. G. C. Sawicki, in *Wetting, Spreading, and Adhesion* (J. F. Padday, ed.), Academic Press, London, 1978, pp. 361–375.
130. M. D. Lelah, and A. Marmur, *J. Coll. Inter. Sci. 82:*518 (1981).
131. D. Ausserré, A. M. Picard, and L. Léger, *Phys. Rev. Lett. 57:*2671 (1986).
132. J.-D. Chen, and N. Wada, *Phys. Rev. Lett. 62:*3050 (1989).
133. T. E. Mumley, C. J. Radke, and M. C. Williams, *J. Coll. Inter. Sci. 109:* 398 (1986).
134. P. Joos, P. van Remoortere, and M. Bracke, *J. Coll. Inter. Sci. 136:*189 (1990).
135. A. M. Cazabat, N. Fraysse, and F. Heslot, *Coll. Surf. 52:*1 (1991).
136. C. Allain, D. Ausserré, and F. Rondelez, *J. Coll. Inter. Sci. 107:*5 (1985).
137. A. M. Cazabat, and M. A. Cohen Stuart, *J. Phys. Chem. 90:*5845 (1986).
138. D. Bangham, and Z. Saweris, *Trans. Faraday Soc. 34:*554 (1938).
139. D. Pesach, and A. Marmur, *Langmuir 3:*519 (1987).
140. P. Carles, and A. M. Cazabat, *Coll. Surf. 41:*97 (1989).
141. P. Carles, and A. M. Cazabat, *Prog. Coll. Polym. Sci. 82:*76 (1990).
142. R. L. Cottington, C. M. Murphy, and C. R. Singleterry, in *Contact Angles, Wettability and Adhesion*, Advances in Chemistry Series, Vol. 43 (F. M. Fowkes, ed.), Washington, DC, 1964, pp. 341–354.
143. P. C. Wayner, *Coll. Surf. 52:*71 (1991).
144. P. Levinson, A. M. Cazabat, M. A. Cohen Stuart, F. Heslot, and S. Nicolet, *Rev. Phys. Appl. 23:*1009 (1988).

145. C. Redon, F. Brochard-Wyart, H. Hervet, and F. Rondelez, *J. Coll. Inter. Sci. 149:*580 (1992).
146. F. Brochard-Wyart, H. Hervet, C. Redon, and F. Rondelez, *J. Coll. Inter. Sci. 142:*518 (1991).
147. D. Beaglehole, *J. Phys. Chem. 93:*893 (1989).
148. P. Silberzan, and L. Léger, *Macromolecules 25:*1267 (1992).
149. W. B. Hardy, and I. Doubleday, *Proc. Roy. Soc. Ser. A. 100:*550 (1922).
150. R. Williams, *Nature 266:*153 (1977).
151. F. Meyer, and G. J. Loyen, *Acta Electronica 18:*33 (1975).
152. L. Léger, M. Erman, A. M. Guinet-Picard, D. Ausserré, and C. Strazielle, *Phys. Rev. Lett. 60:*2390 (1988).
153. L. Léger, M. Erman, A. M. Guinet-Picart, D. Ausserre, C. Strazielle, J. J. Benattar, F. Rieutord, J. Daillant, and L. Bosio, *Rev. Phys. Appl. 23:* 1047 (1988).
154. F. Heslot, A. M. Cazabat, and P. Levinson, *Phys. Rev. Lett. 62:*1286 (1989).
155. F. Heslot, N. Fraysse, and A. M. Cazabat, *Nature 338:*640 (1989).
156. F. Heslot, A. M. Cazabat, and N. Fraysse, *J. Phys.: Condens. Matter 1:* 5793 (1989).
157. H. Ghirardella, W. Radigan, and H. L. Frisch, *J. Coll. Inter. Sci. 51:*522 (1975).
158. J. Daillant, J. J. Benattar, and L. Léger, *Phys. Rev. A 41:*1963 (1990).
159. F. Heslot, A. M. Cazabat, P. Levinson, and N. Fraysse, *Phys. Rev. Lett. 65:*599 (1990).
160. A. Marmur, and M. D. Lelah, *J. Coll. Inter. Sci. 78:*262 (1980).
161. J. F. Oliver, and S. G. Mason, *J. Coll. Inter. Sci. 60:*480 (1977).
162. H. K. Moffat, *J. Fluid Mech. 18:*1 (1964).
163. S. Bhattacharji, and P. Savic, in *Proceedings of Heat Transfer Fluid Mechanics Institute* (A. F. Charwat, ed.), Stanford Univ. Press, CA, 1965, pp. 248–262.
164. C. Huh, and S. G. Mason, *J. Fluid Mech. 81:*401 (1977).
165. S. H. Smith, *Z. Angew. Math. Phys. 20:*556 (1969).
166. C. Nakaya, *J. Phys. Soc. Japan 37:*539 (1974).
167. J. Lopez, C. A. Miller, and E. Ruckenstein, *J. Coll. Inter. Sci. 56:*460 (1976).
168. J. Buckmaster, *J. Fluid Mech. 81:*735 (1977).
169. H. E. Huppert, *Nature 300:*427 (1982).
170. H. E. Huppert, J. B. Shepherd, H. Sigurdsson, and R. S. J. Sparks, *J. Volcan. Geotherm. Res. 14:*199 (1982).
171. L. M. Hocking, *Quar. J. Mech. Appl. Math 36:*55 (1983).
172. R. Goodwin, and G. M. Homsy, *Phys. Fluids A 3:*515 (1991).
173. E. O. Tuck, and L. W. Schwartz, *SIAM Rev. 32:*453 (1990).
174. J. Lowndes, *J. Fluid Mech. 101:*631 (1980).
175. P. Neogi, and C. A. Miller, *J. Coll. Inter. Sci. 86:*525 (1982).
176. O. V. Voinov, *Zh. Prikl. Mekh. i. Tekh. Fiz. 18:*92 (1976).

177. W. Boender, A. K. Chesters, and A. J. J. van der Zanden, *Int. J. Multiphase Flow 17:*661 (1991).
178. C. L. M. H. Navier, *Mém. Acad. R. Sci. Inst. Fr. 6:*389 (1827).
179. L. M. Hocking, *J. Fluid Mech. 79:*209 (1977).
180. W. J. Silliman, Ph.D. thesis, Univ. Minn., Minneapolis, 1979.
181. H. P. Greenspan, *J. Fluid Mech 84:*125 (1978).
182. P. J. Haley, and M. J. Miksis, *J. Fluid Mech. 223:*57 (1991).
183. P. J. Haley, and M. J. Miksis, *Phys. Fluids A 3:*487 (1991).
184. N. van Quy, *Int. J. Eng. Sci. 9:*101 (1971).
185. E. B. Dussan V., *J. Fluid Mech. 77:*665 (1976).
186. P. A. Durbin, *J. Fluid Mech. 197:*157 (1988).
187. P. Neogi, and C. A. Miller, *J. Coll. Inter. Sci. 92:*338 (1983).
188. V. V. Kalinin, and V. M. Starov, *Colloid J. USSR* (English trans.) *51:*744 (1989).
189. P. Bach, and O. Hassager, *J. Fluid Mech. 152:*173 (1985).
190. C. Baiocchi, and V. V. Pukhnachev, *J. Appl. Mech. Techn. Phys. USSR 2:*185 (1990).
191. P. G. de Gennes, X. Hua, and P. Levinson, *J. Fluid Mech. 212:*55 (1990).
192. D. Kröner, *Z. Angew. Math. Mech. 67:*T303 (1987).
193. S. Goldstein, *Modern Developments in Fluid Dynamics*, Calarendon Press, Oxford, 1938; Dover Reprint, Dover, New York, 1965.
194. S. Richardson, *J. Fluid Mech. 59:*707 (1973).
195. K. M. Jansons, *Phys. Fluids 31:*15 (1988).
196. J. Koplik, J. R. Banavar, and J. F. Willemsen, *Phys. Fluids A 1:*781 (1989).
197. U. Heinbuch, and J. Fischer, *Phys. Rev. A. 40:*1144 (1989).
198. P. A. Thompson, and M. O. Robbins, *Phys. Rev. A. 41:*6830 (1990).
199. T. D. Blake, *Coll. Surf. 47:*135 (1990).
200. J. Galt, and B. Maxwell, *Mod. Plas. 42:*115 (1964).
201. D. S. Kalika, and M. M. Denn, *J. Rheol. 31:*815 (1987).
202. W. R. Schowalter, *J. Non-Newtonian Fluid Mech. 29:*25 (1988).
203. C. J. S. Petrie, and M. M. Denn, *AIChE J. 22:*209 (1976).
204. A. V. Ramamurthy, *J. Rheol. 30:*337 (1986).
205. G. D. Yarnold, and B. Mason, *Proc. Phys. Soc. Lond. B 62:*121 (1949).
206. E. Ruckenstein, and C. S. Dunn, *J. Coll. Inter. Sci. 59:*135 (1977).
207. P. A. Thompson, and M. O. Robbins, *Phys. Rev. Lett. 63:*766 (1989).
208. P. G. de Gennes, *C. R. Acad. Sci. Paris Ser. B. 288:*219 (1979).
209. F. Brochard, and P. G. de Gennes, *J. Physique Lett. 45:*L597 (1984).
210. R. H. Dettre, and R. E. Johnson, *J. Adhesion 2:*61 (1970).
211. R. Bruinsma, *Macromolecules 23:*276 (1990).
212. F. W. Bruns, *J. Coll. Inter. Sci. 74:*341 (1980).
213. D. H. Michael, *Mathematika 5:*82 (1958).
214. P. C. Wayner, *J. Coll. Inter. Sci. 88:*294 (1982).
215. P. C. Wayner, and J. Schonberg, *J. Coll. Inter. Sci. 152:*507 (1992).
216. M. E. R. Shanahan, *J. Phys. D: Appl. Phys. 21:*981 (1988).

217. S. Rosenblat, and S. H. Davis, in *Frontiers in Fluid Mechanics* (S. H. Davis, and J. L. Lumley, eds.), Springer-Verlag, Berlin and New York, 1985, pp. 171–183.
218. P. R. Schunk, and L. E. Scriven, *J. Rheol. 34:*1085 (1990).
219. R. G. Cox, *J. Fluid Mech. 168:*195 (1986).
220. P. R. Schunk, Ph.D. thesis, Univ. Minn., Minneapolis, 1989.
221. D. J. Benney, and W. J. Timson, *Stud. Appl. Math. 63:*93 (1980).
222. L. M. Pismen, and A. Nir, *Phys. Fluids 25:*3 (1982).
223. V. V. Pukhnachev, and V. A. Solonnikov, *PMM USSR* (English trans.) *46:*771 (1983).
224. D. J. Coyle, J. W. Blake, and C. W. Macosko, *AIChE J. 33:*1168 (1987).
225. H. Mavridis, A. N. Hrymak, and J. Vlachopoulos, *AIChE J. 34:*403 (1988).
226. J. N. Tilton, *Chem. Eng. Sci. 43:*1371 (1988).
227. S. F. Kistler, and G. R. Zvan, IS&T 44th Annual Conference, St. Paul, MN, 1991.
228. H. P. Greenspan, and B. M. McCay, *Stud. Appl. Math. 64:*95 (1981).
229. S. H. Davis, *J. Fluid Mech. 98:*225 (1980).
230. E. B. Dussan V., and R. T.-P. Chow, *J. Fluid Mech. 137:*1 (1983).
231. G. W. Young, and S. H. Davis, *J. Fluid Mech. 174:*327 (1987).
232. L. M. Hocking, *J. Fluid Mech. 179:*253 (1987).
233. L. M. Hocking, *J. Fluid Mech. 179:*267 (1987).
234. L. M. Hocking, *J. Fluid Mech. 211:*373 (1990).
235. P. Ehrhard, and S. H. Davis, *J. Fluid Mech. 229:*365 (1991).
236. K. N. Christodoulou, and L. E. Scriven, *J. Fluid Mech. 208:*321 (1989).
237. M. Viriyayuthakorn, and R. V. Deboo, *SPE ANTEC Tech. Papers 41:*178 (1983).
238. F. Y. Kafka, and E. B. Dussan V., *J. Fluid Mech. 95:*539 (1979).
239. K. N. Christodoulou, and L. E. Scriven, AIChE Annual Meeting, San Francisco, CA, 1984.
240. S. F. Kistler, and L. E. Scriven, *Phys. Fluids A* (1993) (to be submitted).
241. H. Mavridis, A. N. Hrymak, and J. Vlachopoulos, *Polym. Eng. Sci. 26:* 449 (1986).
242. R. A. Behrens, M. J. Crochet, C. D. Denson, and A. B. Metzner, *AIChE J. 33:*1178 (1987).
243. L. M. Hocking, *Quar. J. Mech. Appl. Math. 34:*37 (1981).
244. L. M. Hocking, in *Proceedings of 2nd International Colloquium on Drops and Bubbles* JPL Publ. 82-7, NASA-JPL, Pasadena, CA, 1982, pp. 315–321.
245. A. A. Lacey, *Stud. Appl. Math. 67:*217 (1982).
246. V. M. Starov, *Coll. J. USSR* (English trans.) *45:*1009 (1983).
247. J. F. Joanny, and M. O. Robbins, *J. Chem. Phys. 92:*3206 (1990).
248. K. K. Mohanty, H. T. Davis, and L. E. Scriven, in *Surface Phenomena in Enhanced Oil Recovery* (D. O. Shah, ed.), Plenum, New York, 1981, pp. 595–609.

249. G. J. Merchant, and J. B. Keller, *Phys. Fluids A 4:*477 (1992).
250. T. D. Blake, D. H. Everett, and J. M. Haynes, in *Wetting*, S.C.I. Monograph No. 25, Society of Chemical Industry, London, and Gordon & Breach, New York, 1967, pp. 164–173.
251. P. van Remoortere, and P. Joos, *J. Coll. Inter. Sci. 141:*348 (1991).
252. V. M. Starov, N. V. Churaev, and A. G. Khvorostyanov, *Coll. J. USSR* (English trans.) *39:*176 (1977).
253. B. V. Derjaguin, *Acta Physicochim. USSR* (English trans.) *12:*181 (1940).
254. B. V. Derjaguin, and N. Churaev, *Coll. J. USSR* (English trans.) *38:*402 (1976).
255. P. G. Petrov, and J. G. Petrov, *Coll. Surf. 61:*227 (1991).
256. R. S. Hansen, and M. Miotto, *J. Am. Chem. Soc. 79:*1765 (1957).
257. M. O. Robbins, and P. A. Thompson, *M.R.S. Symp. Proc. 177:*411 (1989).
258. E. Raphaël, and P. G. de Gennes, *J. Chem. Phys. 90:*7577 (1989).
259. J.-F. Joanny, and D. Andelman, *J. Coll. Inter. Sci. 119:*451 (1987).
260. F. P. Bretherton, *J. Fluid Mech. 10:*166 (1961).
261. G. Friz, *Z. Angew. Phys. 19:*374 (1965).
262. V. Ludviksson, and E. N. Lightfoot, *AIChE J. 14:*674 (1968).
263. T. E. Mumley, C. J. Radke, and M. C. Williams, *J. Coll. Inter. Sci. 109:*413 (1986).
264. S. M. Troian, E. Herbholzheimer, S. A. Safran, and J. F. Joanny, *Europhys. Lett. 10:*25 (1989).
265. J. A. Moriarty, L. W. Schwartz, and E. O. Tuck, *Phys. Fluids A 3:*733 (1991).
266. C.-W. Park, and G. M. Homsy, *J. Fluid Mech. 139:*291 (1984).
267. L. Landau, and B. Levich, *Acta Phyicochim. USSR* (English trans.) *17:*42 (1942).
268. D. D. Joseph, J. Nelson, M. Renardy, and Y. Renardy, *J. Fluid Mech. 223:*383 (1991).
269. K. E. Palmquist, and S. F. Kistler, AIChE Spring National Meeting, Orlando, FL, 1990.
270. S. Richardson, *J. Fluid Mech. 33:*475 (1968).
271. O. Hassager, *Nature 279:*402 (1979).
272. J. Q. Feng, and L. E. Scriven, AIChE Spring National Meeting, New Orleans, LA, 1992.
273. P. G. Saffman, and G. I. Taylor, *Proc. R. Soc. Lond. A 245:*312 (1958).
274. S. D. R. Wilson, *J. Coll. Inter. Sci. 51:*532 (1975).
275. K. M. Jansons, *J. Fluid Mech. 163:*59 (1986).
276. J. F. Joanny, *Physica Scripta T29:*270 (1989).
277. N. Silvi, and E. B. Dussan V., *Phys. Fluids 28:*5 (1985).
278. L. H. Tanner, *La Recherche 17:*182 (1986).
279. F. Melo, J. F. Joanny, and S. Fauve, *Phys. Rev. Lett. 63:*1958 (1989).
280. V. Ludviksson, and E. N. Lightfoot, *AIChE J. 17:*1166 (1971).
281. A. M. Cazabat, F. Heslot, S. M. Troian, and P. Carles, *Nature 346:*824 (1990).

282. L. W. Schwartz, *Phys. Fluids A 1:*443 (1989).
283. J. N. Israelachvili, *Intermolecular and Surface Forces*, 2nd ed., Academic Press, London and San Diego, 1992.
284. C. A. Miller, and E. Ruckenstein, *J. Coll. Inter. Sci. 48:*368 (1974).
285. B. D. Summ, V. S. Yushchenko, and E. D. Shchukin, *Coll. Surf. 27:*43 (1987).
286. B. V. Derjaguin, and M. M. Kussakov, *Acta Physicochim. USSR* (English trans.) *10:*25 (1939).
287. B. V. Derjaguin, *Colloid J. USSR* (English trans.) *17:*191 (1955).
288. H. Hervet, and P. G. de Gennes, *C. R. Acad. Sci. Paris 299 II:*499 (1984).
289. I. E. Dzyaloshinskii, E. M. Lifshitz, and L. P. Pitaevskii, *Adv. Phys. 10:* 165 (1961).
290. G. F. Teletzke, L. E. Scriven, and H. T. Davis, *J. Coll. Inter. Sci. 87:*550 (1982).
291. J. F. Joanny, and P. G. de Gennes, *J. Phys.* (*Paris*) *47:*121 (1986).
292. D. B. Abraham, P. Collet, J. de Coninck, and F. Dunlop, *Phys. Rev. Lett. 65:*195 (1990).
293. D. B. Abraham, J. Heiniö, and K. Kaski, *J. Phys. A: Math. Gen. 24:*L309 (1991).
294. J. Cook, and D. E. Wolf, *J. Phys. A: Math. Gen. 24:*L351 (1991).
295. M. Sujanani, and P. C. Wayner, *J. Coll. Inter. Sci. 143:*472 (1991).
296. R. M. Pashley, *J. Coll. Inter. Sci. 78:*246 (1980).
297. P. Neogi, *J. Coll. Inter. Sci. 105:*94 (1985).
298. A. M. Lezzi, and A. Prosperetti, *J. Fluid Mech. 226:*319 (1991).
299. J. Koplik, J. R. Banavar, and J. F. Willemsen, *Phys. Rev. Lett. 60:*1282 (1988).
300. E. Ramé, Ph.D. thesis, Univ. Pennsylvania, Philadelphia, 1988.

7

Role of Solid/Solid Wetting in Catalysis

EDMOND I. KO Department of Chemical Engineering, Carnegie Mellon University, Pittsburgh, Pennsylvania

I. INTRODUCTION

Wetting is a prevalent issue in heterogeneous catalysis since a large number of catalysts involve the contact between two solid phases. Often the ease with which one solid wets another under specific pretreatment and reaction conditions dictates the structure and, in turn, the catalytic behavior of the system. A common example is the change in the dispersion of supported metal catalysts through sintering and redispersion. In this case, wetting behavior affects the number of surface atoms available for catalysis. Another example is the wettability of one oxide onto another, as the chemical reactivity is often different between a dispersed surface phase and supported bulk phase.

Despite its importance, the role of solid/solid wetting in catalysis has received relatively little attention in the literature. One reason, perhaps, is that many people still associate wetting with liquid/solid interfaces only. In fact, one may even question whether the term wetting in its strictest sense can be used to describe solids that have low atomic mobility at ambient temperature. The purpose of this article is to show that the basic concepts of liquid/solid wetting are indeed useful for a qualitative understanding of solid catalysts under typical thermal treatments. We start by presenting a general formalism for the thermodynamics of wetting and showing how it can be used to explain relevant experimental observations. We will also review the techniques that have been most widely used to study wetting in catalysis.

Our survey of experimental results will focus on systems consisting of metal on oxide (supported metal) and oxide on oxide (supported oxide), since many commercial catalysts belong to one of these two types. We will briefly discuss the wetting of oxide on metal within the context of strong metal-support interactions (SMSI). Readers who are interested in the deposition of metal on metal in relation to bimetallic catalysts should consult the work of Dodson [1].

II. THERMODYNAMICS OF WETTING

The treatment in this section is a synopsis of the detailed derivations that can be found elsewhere [2,3]. Our intent is to present the general relationships that are useful in guiding a qualitative understanding of the wetting of a substrate by a film of varying thicknesses. Several issues particularly relevant to supported catalysts will then be highlighted.

A. Formalism

1. Spreading Condition

When a thick film is transferred from a large reservoir to a uniform substrate, the specific free energy of formation is given by

$$\sigma_{\infty} = \sigma_{mg} + \sigma_{ms} - \sigma_{gs} \tag{1}$$

where σ_{ij} denotes the interfacial free energy between phases i and j, and subscripts m, g, and s denote film, gas, and substrate, respectively.

The interfacial free energy between film and substrate is given by

$$\sigma_{ms} = \sigma_m + \sigma_s - U_{ms} \tag{2}$$

where U_{ms} incorporates two contributions: the interaction energy per unit area of film/substrate interface that decreases the interfacial free energy, and the strain energy per unit area that increases it. Substitution of Eq. (2) into Eq. (1) gives

$$\sigma_\infty = \sigma_{mg} + \sigma_m + \sigma_s - \sigma_{gs} - U_{ms} \tag{3}$$

Since $\sigma_{mg} \approx \sigma_m$ and $\sigma_{gs} \approx \sigma_s$, Eq. (3) can be approximated as

$$\sigma_\infty = 2\,\sigma_{mg} - U_{ms} \tag{4}$$

At the wetting condition, $\sigma_\infty < 0$ and the film spreads over the substrate because this causes a decrease of the free energy of the system. This condition can be restated as $U_{ms} > 2\,\sigma_{mg}$ with Eq. (4). When the reverse is true, that is, $\sigma_\infty > 0$ or $U_{ms} < 2\,\sigma_{mg}$, the film does not wet the substrate and instead coalesces into an island. The island forms an angle θ with the substrate described by Young's equation

$$\sigma_{mg} \cos\theta = \sigma_{gs} - \sigma_{ms} \quad \text{or} \quad \sigma_m \cos\theta = \sigma_s - \sigma_{ms} \tag{5}$$

Let us briefly examine the physical insight provided by Eq. (4). Wetting represents a tradeoff between the surface free energy of the film and the interaction energy at the interface. In the case of an oxide-supported metal, wetting is not expected to occur due to the large surface free energy of metal and weak metal-oxide interactions. For the same oxide-metal combination, however, it will be easier for the oxide to wet the metal substrate because of its generally lower surface free energy. Finally, one oxide could wet another oxide even with a moderate interfacial interaction due to the generally lower surface free energies of oxides.

2. Effect of Film Thickness

The discussion above is valid only for a film that is thick enough to be considered a bulk phase. When the range of the interaction forces between one atom at the surface of the film and the substrate is comparable to the thickness of the film, the specific free energy of formation of the film becomes a function of thickness. Using a Lennard-Jones potential and assuming pairwise additivity, Ruckenstein [3] showed that

$$\sigma(h) = \sigma_\infty + \frac{\alpha}{h^2} - \frac{\beta}{h^6} \tag{6}$$

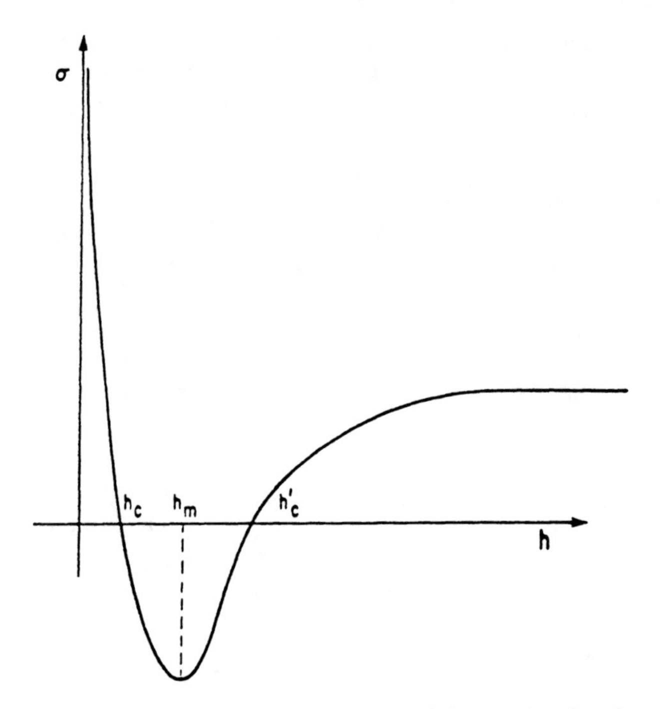

FIG. 1 Plot of free energy vs. film thickness showing the case in which a thick film does not wet the substrate but a very thin film does. (Reproduced from Ref. 3 with permission.)

where h is the thickness of the film and α and β depend on Lennard-Jones parameters and atomic densities.

The derivation of this equation is based on a continuum approach and disregards the strain energy between film and substrate. Nonetheless, it shows that the variation of σ with h has a minimum when $h = h_m$ (see Fig. 1). It is thus possible to have $\sigma_\infty > 0$ but $\sigma < 0$. Physically, this corresponds to the interesting situation in which a thick film does not wet the substrate, but a thin film of thickness h_m does. The thickness h_m is of the order of the distance between two atoms, or a monolayer.

3. Catalyst Loading

We have seen so far that wetting depends on the surface free energy of the film, interaction between film and substrate, and film thickness. Experimental data for the first two parameters are scarce, even though qualitative trends can often be established between different systems. On the other hand, the loading of a supported catalyst, defined usually in terms

TABLE 1 Summary of Qualitative Guidelines

σ_{mg}	U_{ms}	σ_{∞}	σ	Remark
Large	small	large and positive	positive for all h	no wetting
Small	small	small and positive	positive for large h, could be negative for small h	a film of less than monolayer coverage wets, larger than monolayer does not
Small	large	large and negative	negative for all h	wetting always occurs

of weight percent (wt.%), can be converted readily into film thickness. One common approach is to calculate the loading necessary to achieve what is known as a theoretical monolayer coverage, which corresponds to the limit of perfect wetting. By assuming an atom density of 10^{15} atoms/cm^2 (which is equivalent to saying each atom occupies 10 Å^2), we have

$$\text{loading required for monolayer coverage, in wt.\%} = \frac{1.66 \times 10^{-5}\, S_A M_W}{1 + 1.66 \times 10^{-5}\, S_A M_W} \times 100\% \tag{7}$$

where

S_A = surface area of the support in m^2/g
M_W = molecular weight of the supported species

For example, a nickel loading of 22.6 wt.% is necessary to completely wet a SiO_2 support with a surface area of 300 m^2/g. The loading of many supported metal catalysts is below the theoretical monolayer coverage or not much above it. On the other hand, the loading of supported oxides often exceeds the monolayer coverage.

Table 1 summarizes several situations that we are likely to encounter in supported catalysts. We will discuss specific examples in Sec. IV within this framework but, before we do so, there are several important issues that need to be addressed.

B. Other Considerations

1. Kinetic Constraint

Our discussion so far has implicitly assumed that the catalytic system is at equilibrium. Such is not necessarily the case for a solid phase with low

TABLE 2 Bulk Melting Temperatures (T_m) of Some Representative Materials

Metals		Oxides	
Material	T_m(K)	Material	T_m(K)
Ag	1235	V_2O_5	963
Cu	1356	Nb_2O_5	1758
Ni	1726	SiO_2	1996
Fe	1808	TiO_2	2113
Pt	2045	Al_2O_3	2345
Ir	2683	ZrO_2	2988
Mo	2890	MgO	3125

Data for metals are taken from Ref. 2 and those for oxides from Ref. 5.

mobility of atoms. For example, it is possible to have a sintering process proceeding at such a slow rate that the thermodynamically favored configuration is never achieved within the experimental time scale.

One good measure of atomic mobility is the Tammann temperature, which is equal to a fraction of the melting temperature of the bulk solid in degrees Kelvin. The mobility of surface atoms, expressed in terms of their surface diffusion, becomes significantly larger above this temperature. Ruckenstein [4] suggested that the Tammann temperature is associated with a two-dimensional melting of the surface of the solid and discussed ways by which it can be calculated. In general, the Tammann temperature can be estimated as half the bulk melting temperature. Table 2 lists the bulk melting temperatures of some commonly used metals and oxides in catalysis, arranged in the order of increasing melting temperature. It is of interest to note that many catalyst pretreatments are done in the temperature range of 600–800K and sintering studies in one of 800–1300K. Under these conditions, solid atoms are of sufficient mobility for wetting (or dewetting) to occur.

2. Effects of Pretreatment and Impurities

The preparation of a supported catalyst usually involves the introduction of a precursor and the subsequent activation in a particular gaseous environment. Sometimes the support is also heated in air or vacuum prior to preparation. For oxide supports, the removal of surface hydroxyl groups can lead to two important consequences. First, dehydroxylation can significantly change the surface free energy of the substrate. Second, if OH groups react with the precursor, then their absence will weaken the interfacial interaction [the term U_{ms} in Eq. (4)].

The choice of a precursor can also affect wetting in two ways. In cases where the interaction between precursor and support involves chemical bonding, it is important to pick a precursor with the right constituent group. At the same time, it is undesirable to have a precursor that will leave behind impurities. The presence of an impurity in the film, at the film/substrate or the film/gas interface, can change the wetting angle [6]. Experimental data supporting this view include the addition of titanium to liquid nickel drops on α-Al_2O_3 [7] and the effect of sulfur, oxygen, nitrogen, and carbon on the interfacial energy between liquid iron and Al_2O_3 [8].

Finally, the type of gas used in the pretreatment or activation step needs to be considered. The choice of a gas can affect wetting by (1) reacting with and thus removing an impurity, (2) reacting with the supported species, and (3) adsorbing on the film and/or substrate. The last two possibilities will change the surface free energies.

3. Effect of Curvature

Another assumption in our derivation in Sec. II.A is that the substrate is flat. This does not pose a severe limitation because, out of convenience, many direct experimental observations on the wetting behavior of supported catalysts have been made with low-surface-area, flat substrates. However, many commercial catalysts consist of metals or oxides supported on high-surface-area, microporous substrates that have curved surfaces. Surface curvature affects surface free energy and, in particular, the growth of particles through particle migration and interparticle transport. These issues were discussed by Wynblatt and Ahn [9].

III. EXPERIMENTAL TECHNIQUES

Experimental measurements of surface free energy in solids involve both direct and indirect techniques [10,11]. Although some data are available for metals [12] and oxides [13], information remains scarce for a quantitative prediction of wetting behavior. This deficiency is particularly true for systems containing adsorbed species and impurities. Thus, experimental studies of wetting in catalysis have been phenomenological in nature. In this section, we discuss the techniques most suitable for this purpose. Our focus will be on how these techniques can determine whether a supported species wets a substrate and what they can tell us about the structure of the supported phase. Detailed descriptions of the principles and other applications of these techniques can be found elsewhere [14,15].

A. X-Ray Line Broadening

The relationship between the size of small crystallites and the breadth of the diffraction lines they give rise to was recognized by Scherrer in 1918. In its simplest form, x-ray line broadening can give the mean dimension L of the crystallites by

$$L = \frac{0.9\ \lambda}{\beta \cos \theta} \tag{8}$$

where

λ = wavelength
β = diffraction line breadth (calculated from the experimental line breadth and instrumental broadening)
θ = Bragg angle

The technique is applicable to crystallites in the size range of 5–50 nm. In addition to obtaining the mean crystallite size, a detailed analysis of the x-ray line profile can give the distribution of crystallite sizes [16,17].

The major limitation of x-ray line broadening is that it will not detect the really small crystallites. With a shorter wavelength (e.g., using Mo radiation instead of Cu) and careful experimentation, it is possible to extend the lower limit of detection to 2–3 nm, which is still too large if a film wets the substrate perfectly. The technique is also unable to detect amorphous particles.

Despite these limitations, x-ray line broadening has been widely used because of its simplicity. The technique is particularly useful in identifying the onset of crystallization of either metal or oxide that usually accompanies "dewetting." Most experiments done to date employed ex situ heat treatments, but in situ studies are possible in a diffractometer equipped with a heating stage. Recently, Liu et al. [18] described a method by which they calculated the wt.% of crystalline Mo_2O_3 in a series of $Mo_2O_3/\gamma\text{-}Al_2O_3$ samples. The addition of a reference allows a quantitative analysis of diffraction peak intensities.

B. Transmission Electron Microscopy

Transmission electron microscopy (TEM) is probably the most common technique used in studying the growth of metal particles on flat substrates. The main objective is to prepare a thin film that is a model of the supported catalyst so that a direct observation with the microscope can be made. There are several ways to prepare such a sample. Wynblatt et al. [19,20] prepared a film by first depositing alumina (~10 nm thick) onto a silica microscope slide, followed by deposition of a thin film of platinum (~1 nm thick). Both films were deposited by radio-frequency (rf) sputtering in argon from high-purity targets and their thicknesses measured by a vibrating crystal thickness monitor. Ruckenstein and Malhotra [21] prepared a 30-nm alumina film by anodizing a high-purity aluminum foil and stripping away the unoxidized aluminum. The alumina film was then put on a gold microscope grid, dried in air, and used as a substrate for the deposition of Pd by vacuum evaporation [22]. Wang and Schmidt [23] prepared silica substrates by depositing in vacuum a 20-nm film of Si on NaCl, floating Si flakes off in water, catching a flake on a gold grid, and oxidizing to 800°C in air for several days. Chojnacki and Schmidt [24] later described a simpler procedure in which silicon was deposited on a gold microscope grid precovered with a thin Formvar film. A subsequent high-temperature oxidation step burned off the Formvar and converted silicon to silica.

In general, the thickness of an oxide film is kept below 100 nm for ease of imaging. Once a stable oxide is formed, it can be used as the substrate for another oxide. For example, both Wynblatt et al. [19,20] and Bukhardt and Schmidt [25] have prepared silica-supported alumina for the subsequent deposition of a metal film. Lately, the same preparation technique has been applied to the study of supported oxides themselves. Hayden and Dumesic [26] prepared a nonporous amorphous alumina film by anodization and then deposited onto it molybdenum oxide by vacuum evaporation. Weissman et al. [27] followed the procedure of Wang and Schmidt [23] to prepare an amorphous silica film and used it as a substrate for rf-sputter-deposited niobia.

Thin films prepared by any of the above methods can be directly imaged in a TEM to give, for example, the distribution of particle sizes as a function of heat treatment. With enhanced capabilities such as high-resolution imaging, microdiffraction, and microanalysis, much can be learned about the wetting behavior of metal or oxide on oxide, as we shall see in the next section. However, we need to be aware of the limitations of these experiments.

The first limitation is that many heat treatments are done outside the microscope in a conventional furnace. Even though the sample is usually stable after the heat treatment and not expected to change during the transfer, it is desirable to have a heating stage in the microscope, or better still, an environmental cell. Derouane et al. [28] have used a controlled atmosphere electron microscope to observe directly the wetting and spreading of copper particles on magnesium oxide.

Another limitation is that a flat substrate may not be a perfect model of a microporous, high-surface-area support. Other than the effect of curvature as mentioned in Sec. II.B.3, different preparation methods may also affect wetting behavior [27]. One way to overcome this difficulty is to examine in a single study similarly prepared low- and high-surface-area materials. Unfortunately, such an approach is not always possible due to experimental complexities. For example, image contrast is usually more of a problem in examining high-surface-area samples. However, researchers have successfully studied on high-surface-area supports the sintering of supported nickel [29] and platinum [30] catalysts, as well as the dispersion of vanadia on silica [31] and alumina [32].

Finally, we need to realize that thin films only provide a two-dimensional view of the supported particles. Careful experimentation (e.g., tilting the sample) may yield information on the thickness, but it is not always possible to do so [33]. Burkhardt and Schmidt [25] showed that when Si or Al was oxidized on a gold grid, portions of the resulting SiO_2 or Al_2O_3 extended from the grid. Metal particles subsequently deposited onto these edges can then be imaged from the side. Recently, Datye et al. described an alternate approach in which nonporous spherical particles are used as model supports that permit direct edge-on viewing. This approach has been applied to systems including Rh on SiO_2 [34], Ru on SiO_2 and MgO [35], and Rh on TiO_2 [36]. Datta et al. [37] have also used nonporous silica spheres as model supports for MoO_3. In the same study, these authors examined the spreading of MoO_3 on a planar SiO_2 substrate, as well as with in situ electron microscopy.

C. Laser Raman Spectroscopy

When one oxide is deposited onto another oxide substrate, the interfacial interaction can be sufficiently strong to stabilize a surface structure which is different from that of the bulk. Since a unique vibrational spectrum can usually be related to each structure, laser Raman spectroscopy (LRS) is a powerful technique to characterize supported oxides. With the appropriate reference compounds, the technique can provide direct information on the structure of each molecular state, in addition to differentiating among several molecular states. Unlike XRD, LRS detects both amorphous and crystalline phases. In situ experiments (see [38] for a typical experimental setup) are also possible for studying the effects of temperature and water vapor on the transformations of various molecular states.

The work of Chan et al. [38] illustrates some of these capabilities of LRS. In the alumina-supported tungsten oxide systems, these authors found that below monolayer coverage, the tungsten oxide phase is present as a highly dispersed and amorphous surface complex. This complex forms a close-packed monolayer as the surface area of the alumina decreases at high calcination temperatures. At still higher temperatures a $Al_2(WO_4)_3$ phase is formed from the reaction between the two oxides. Chan et al. concluded that the surface density of tungsten oxide is the controlling parameter for these observations.

The successful applications of LRS in identifying different supported oxide phases and following their transformations have now been demonstrated for numerous systems. A representative list includes vanadia/titania [39], nickel oxide/γ-alumina [40], chromia/alumina [41,42], molybdenum oxide/γ-alumina [43], and titania/silica [44,45]. In the last case, Reichman and Bell used LRS to characterize the state of titanium throughout the impregnation and calcination steps. Their results shed light on the chemistry by which small particles of titania are formed.

We mention in passing that extended x-ray absorption fine structure (EXAFS) is a complementary technique to provide structural information on an atomic scale. Bond-length data have been obtained for supported oxides such as V_2O_5/TiO_2 [46] and Nb_2O_5/SiO_2 [47].

D. X-Ray Photoelectron Spectroscopy and Auger Electron Spectroscopy

Both x-ray photoelectron spectroscopy (XPS) and Auger electron spectroscopy (AES) are surface-sensitive techniques. Since the sampling depth is in general a few atomic layers thick, a particle with a dimension larger than the sampling depth will give a smaller signal than that given by the same amount of material spread out as a two-dimensional phase. It is thus possible to determine whether a film wets a substrate or, in a similar vein, to differentiate between different growth modes. Figure 2 shows the variation of Auger signals with time as a metal film is deposited onto a substrate [48]. If there is two-dimensional layer-by-layer growth (known as Frank-van der Merwe growth), then the intensity vs. time plot consists of several linear segments [panel (a)]. Each break between two segments corresponds to the completion of a layer. If growth proceeds by the formation of the first layer followed by the growth of three-dimensional particles (known as Stranski-Krastanov growth), then the intensity vs. time plot is linear up to the point when the first layer is complete. In reality, the breaks are not always as sharp as those indicated in Fig. 2, but the qualitative features remain the same.

As mentioned in Sec. II.A.3, the loading of supported catalysts is usually below or at monolayer coverage. The interest thus lies on determining whether a supported species wets the substrate as a two-dimensional phase or exists as a three-dimensional particle. In the catalytic literature, the parameter most often used in characterizing these situations is dis-

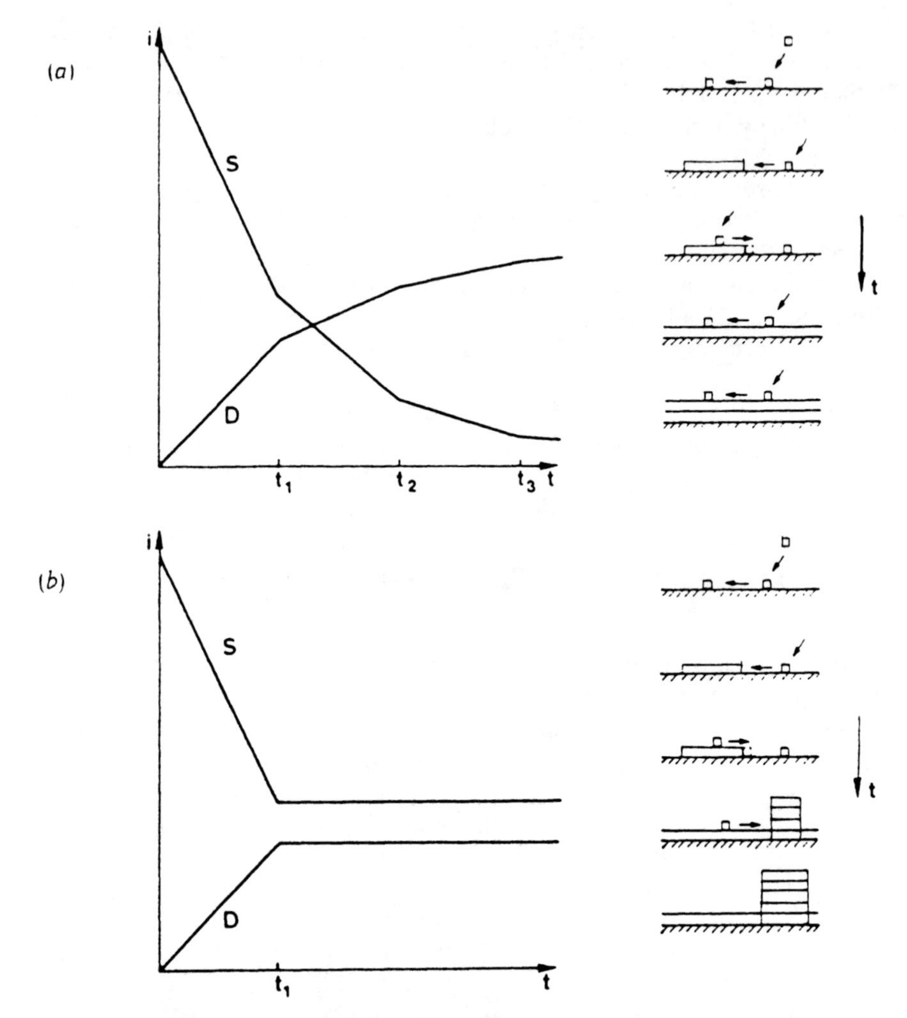

FIG. 2 Schematic plots of the Auger signals for the substrate S and the deposited metal D vs. deposition time for the two ultrathin film growth modes that include monolayer formation. (a) Layer-by-layer growth: the nth layer is complete at t_n. (b) Monolayer followed by three-dimensional clusters or crystallites. Right: schematic presentations of the growth mechanisms. (Reproduced from Ref. 48 with permission.)

persion, defined as the ratio of the number of surface atoms to the number of total atoms. A phase that wets can be thought of as highly dispersed, and a phase that does not, as poorly dispersed.

Defosse [49] presented a detailed discussion on the use of XPS in determining dispersion, although the quantitative correlation between XPS intensity ratio and dispersion is complex [49,50]. The simple idea that a high dispersion gives a high ratio is still qualitatively correct. This point is well illustrated by Houalla et al. [51,52] in their study of silica-supported nickel catalysts. They found that the Ni/Si intensity ratio increased linearly with Ni wt.% for samples prepared by ion exchange, but was rather insensitive to loading for samples prepared by impregnation. The higher dispersion of ion-exchanged samples was confirmed by TEM and reduction studies. As discussed by Xie and Tang [53], XPS is equally applicable for the determination of dispersion in supported oxides. They cited an example in which MoO_3 supported as a monolayer on TiO_2 gives a much stronger XPS signal than a poorly dispersed MoO_3.

Besides identifying an element and measuring its intensity, XPS can provide information on the oxidation state. A particular important application of this capability is to study the reducibility of supported oxides, as has been shown for WO_3/Al_2O_3 [54,55]. The interfacial interaction between two oxides that affects the wetting behavior often results in different reduction characteristics between the dispersed and bulk phase.

E. Chemical Characterization

The ultimate interest in studying wetting in catalysts is to relate chemical reactivity to dispersion. For supported metals, the issue of structure sensitivity is usually studied via a variation of particle size, or dispersion [56]. For supported oxides, the strong interfacial interaction often leads to a well-dispersed phase over the substrate. The fundamental issue then becomes whether the well-dispersed phase behaves chemically different from the corresponding bulk crystallite. To address these issues, chemical characterization becomes a necessity to supplement the physical techniques described above. By chemical characterization, we mean the study of adsorption and reaction of chemical species, an area that is too vast to receive a comprehensive review here. Our intent is merely to cite a few examples to highlight some of the interesting features.

The use of chemisorption in measuring the dispersion of supported metals has been well established [14,15]. The idea is to pick an adsorbate that adsorbs strongly on the metal but weakly on the support. Dispersion can then be calculated from the amount adsorbed with the appropriate assumptions on particle geometry and adsorption stoichiometry, etc. For supported metals, very good agreement has been found for dispersion measured by H_2 chemisorption and TEM.

The situation is less clearcut for supported oxides since it is more difficult to find a selective adsorbate, even though there are criteria for choosing one [57]. Rethwisch and Dumesic [58] used NO chemisorption to determine the dispersion of iron oxide supported on SiO_2, Al_2O_3, TiO_2, MgO, and ZnO. With the exception of ZnO, the supports adsorbed small amounts of NO that could be subtracted from the total uptakes of the supported samples. Since the goal is achieving selective adsorption, an adsorbate that adsorbs on the support but not on the supported oxide will also work. Niwa et al. [59] demonstrated this approach with the adsorption of benzoate species on V_2O_5 supported on alumina and titania. At 523K the benzoate species adsorbed on the two supports but not on V_2O_5 at all, which allowed the exposed surface area of the support to be calculated.

Another adsorbate commonly used to characterize supported oxides is pyridine for probing surface acidity. Dumesic and co-workers [60,61] have used pyridine adsorption in conjunction with other techniques to characterize the acidic properties of binary oxides. For some oxides, such as vanadia and molybdena, more acid sites were found when they are supported on silica. These authors further proposed models to explain the generation of acidity when one oxide is dispersed onto another. At present, it is still difficult to relate acidic properties directly to dispersion. However, it is significant to note that for many systems, the surface acidity increases linearly with the oxide loading and levels off at the point when monolayer coverage is reached [53].

As noted earlier, a two-dimensional oxide phase interacting with a substrate may reduce differently from the corresponding bulk phase. Temperature-programmed reduction (TPR) can thus be used to separate in favorable circumstances a highly dispersed phase from a poorly dispersed one. For a coverage of one monolayer or less, WO_3 supported on Al_2O_3 [54,55] and Nb_2O_5 supported on SiO_2 [62] were found to be more difficult to reduce than bulk WO_3 and Nb_2O_5, respectively, whereas V_2O_5 supported on TiO_2 reduces more readily than unsupported V_2O_5 [39]. When the supported oxide exceeds monolayer coverage, crystallites formed on top of the monolayer may, as in the case of WO_3/Al_2O_3, or may not, as in the cases of Nb_2O_5/SiO_2 and V_2O_5/TiO_2, reduce like the bulk phase. The important point is that once the reduction behavior of the various supported species has been characterized, TPR can provide a fingerprint of their distribution.

We close this section with two examples on how a dispersed, two-dimensional oxide phase alters catalytic chemistry. Wachs et al. [39,63]

showed that supporting vanadia on titania has two favorable effects on the oxidation of o-xylene to phthalic anhydride: (1) A surface vanadia species is more active and selective than crystalline vanadia, and (2) surface vanadia covers up the titania support, which leads to complete combustion of the partial oxidation products. Ko and co-workers showed that dispersing either titania [64] or niobia [65] as a monolayer on silica produces an interacting support for Ni catalysts. The interactions as characterized by several chemical probes are not as strong as those exhibited by crystalline titania or niobia supports. These results show a promising avenue of introducing a chemical modifier via the wetting of an oxide substrate.

IV. SURVEY OF LITERATURE DATA

A. Supported Metals

Supported metal catalysts, which are prevalent in many industrial processes, usually consist of metal particles deposited onto an inorganic support. The support serves the role of stabilizing small particles of the metal, so that the amount of metal atoms on the surface, and hence its surface area, can be maximized. However, the metal surface area tends to decrease during use as metal particles grow in size in a process known as sintering. When the loss of surface area becomes severe, it is necessary to restore the performance of the catalyst by a process known as redispersion.

The preparation, sintering, and redispersion of a supported metal catalyst involve heat treatment of the sample under different gaseous environments. The dispersion of the metal particles, defined as the ratio of surface to total metal atoms, is governed by the thermodynamic driving force to minimize free energy that, in turn, is related to the surface and interfacial energies. It is in this context that we will discuss these phenomena in terms of the criteria of wetting and spreading. Furthermore, we will draw on literature results to show how the same framework can be used to understand qualitatively the effect of sintering and redispersion on the microstructure of metal particles, as well as the phenomenon of metal-support interactions.

1. Sintering and Redispersion

A supported metal catalyst is typically prepared by introducing a metal precursor and subsequently reducing it. The loading, or the amount of metal introduced, is usually low such that only a fraction of the support is covered. Thus, if dispersion is the only concern, then it would be desirable to have the metal wetting the support. As shown in Eq. (4), the spreading condition is to have U_{ms} greater than $2\sigma_{mg}$. This is equivalent to saying the ratio $U_{ms}/2\sigma_{mg}$ is a good indicator of the wettability of the metal particles. In other words, the larger this ratio is, the higher the dispersion will be.

Physically, this criterion makes sense as it simply states that a metal particle which interacts strongly with a substrate wets it well. Unfortunately, it is of limited utility since values of U_{ms} are rarely available. Values of σ_{gs} (the interfacial free energy between gas and support), however, are available for many oxides and, on the basis of Eq. (1), it is tempting to use σ_{gs} as a guideline to predict the stability of metal on different supports. In fact, Boudart et al. [66] reported that iron has a higher dispersion on MgO than SiO_2 and Al_2O_3, an observation that is consistent with the higher surface free energy of MgO [2]. The danger of using such a guideline, though, lies in the assumption of a constant U_{ms}. As shown in Eq. (2), σ_{ms} changes with U_{ms} that, as stated earlier, reflects the strength of interaction between the metal and support. In general, the interactions between a metal and an oxide are complex and can be dispersive, polar, and covalent in nature. However, in most cases the interactions are not covalent and hence not very strong. It is for this reason that metal particles grow in size, or sinter, when they are heated at high temperatures.

There are two mechanisms by which metal particles sinter. The first one involves the migration and coalescence of particles, and the second the emission of atoms by small particles and their capture by larger particles. The second mechanism can further be divided into Ostwald ripening and direct ripening [67]. In Ostwald ripening, the small particles lose atoms to a two-dimensional phase dispersed over the support from which the large particles capture atoms. In direct ripening, a small particle releases atoms directly to a neighboring large particle. For brevity, we will refer below to these two possibilities simply as the ripening mechanism.

Both the migration of particles and the ripening mechanism can occur in a given system. In fact, whether one mechanism dominates over the other depends on the interactions involved. Consider, for example, the interactions between the particles and support. If such interactions are strong, then there is a driving force for phase transition, leading to a surface film in coexistence with the particles. The resulting spreading of the particles decreases the likelihood of particle migration and increases that of ripening. Conversely, if such interactions are weak, then the migration of particles is likely to play a major role in sintering.

The effect of chemical atmosphere on the sintering process can now be understood in terms of the pertinent surface and interfacial free energies. Chen and Ruckenstein [22,68] observed that alumina-supported Pd particles sinter differently in hydrogen and oxygen. During heating in oxygen, PdO particles form and wets the substrate. Wetting leads to the spreading and even tearing and fragmentation of the particle with more severe oxygen heat treatments. In this case, there are two contributions to the driving force for wetting: the lower surface free energy of PdO compared to Pd, and the stronger interactions between PdO and alumina compared to between Pd and alumina. This corresponds to a lower σ_m and larger U_{ms} in Eq. (2). The magnitude of U_{ms} can be significant in cases when oxide-oxide interactions result in compound formation, as found for alumina-supported Ni [69] and Fe [70]. The formation of aluminates in these systems leads to enhanced surface spreading phenomena and unusual particle shapes such as torus. Figure 3 illustrates some of these phenomena for alumina-supported iron catalysts [70].

The spreading of a metal particle from oxygen heat treatment and its ensuing breakup into smaller fragments are the physical basis behind the redispersion phenomenon. The small fragments, which are metal oxide, can then be reduced to metal. Thus, most industrial regeneration techniques involve an oxygen-hydrogen cycle. During heating in oxygen, the temperature has to be sufficiently high to ensure complete oxidation. The wetting and spreading of the particles will then proceed according to the principles described above. During heating in hydrogen, the temperature

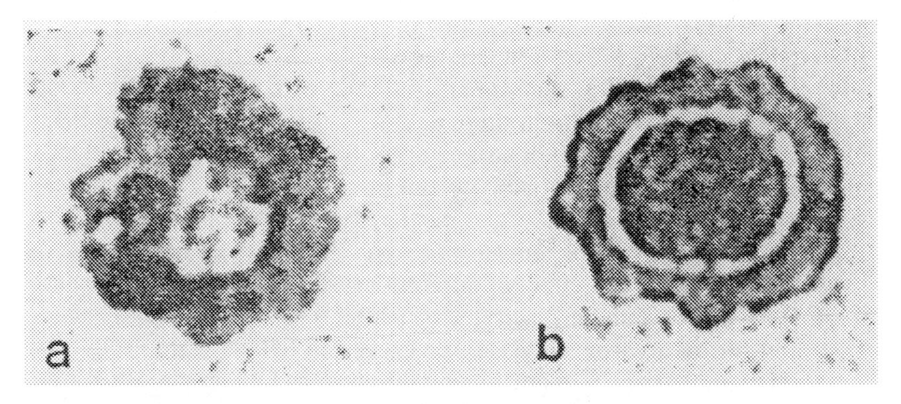

FIG. 3 TEM micrographs showing (a) torus and (b) core-and-ring structures of iron on alumina following heat treatment in hydrogen. (Reproduced from Ref. 70 with permission.)

has to be high enough for reduction, but not so high that sintering can again take place. Finally, redispersion can also occur by the emission of atoms from the particles to a two-dimensional phase on the substrate.

2. Microstructure of Supported Metals

We have thus far discussed the sintering and redispersion of supported metal catalysts in terms of dispersion, which reflects the size of metal particles. However, particle size alone is not always a good predictor of catalytic performance. Certain catalytic reactions are known to depend on the finer details of particle surfaces, such as the types of planes exposed and their relative abundance. In this section we use the term microstructure to refer to these features and, within the context of wetting, discuss how heat treatment in different gaseous environments affects the microstructure of supported metals.

Lee and Schmidt [71] prepared planar SiO_2-supported Rh samples and heated them in oxygen-hydrogen cycles. They observed that all Rh particles spread on the substrate upon oxidation, and the oxide particles split into smaller particles upon heating in H_2 at 300°C for 1 h. Subsequent hydrogen treatment at higher temperatures results in the sintering of the reduced Rh particles. These observations can be readily understood on the basis of our earlier discussion on redispersion and sintering. When kinetic data are obtained for a high-surface-area, similarly prepared Rh/SiO_2 catalyst, Lee and Schmidt noted that the activity of ethane hydrogenolysis significantly decreases (close to three orders of magnitude) as the temperature of hydrogen treatment increases (300–600°C). This decline in activity, expressed as the rate of CH_4 formation, can be reversed by oxidizing the catalyst and rereducing it at 300°C, as shown in Fig. 4. In essence, each oxidation–reduction cycle corresponds to the redispersion of the sample and its subsequent sintering with increasing reduction temperatures. Since the variation of activity is too large to be accounted for by a change in the metal surface area, these authors suggested that high-temperature annealing leads to the preponderance of (111) planes. As supporting evidence, they quoted the study of Goodman [72], who showed that ethane hydrogenolysis proceeds 1000 times faster on Ni (100) than (111) surfaces. The effect of different crystal planes induced by heat treatments on the catalytic activity of Rh/SiO_2 was subsequently demonstrated for the hydrogenolysis of propane and butane [73].

When a comparison is made among different metals on the same SiO_2 support, Gao and Schmidt [74] showed that the variation in activity between low- and high-temperature reduction (following oxidation) correlates with the ability of each metal to retain the high-activity, low-coordination sites. Metals with lower melting points, such as Ni and Pd, anneal

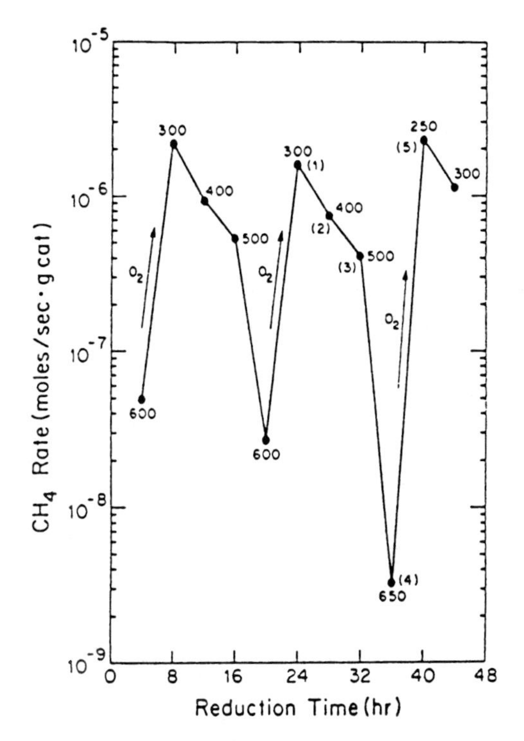

FIG. 4 Effect of heat treatment on methane formation rate from the hydrogenolysis of ethane over Rh/SiO_2. For each oxidation–reduction, the sample was oxidized in flowing air at 500°C for 4 h and then reduced in H_2 for another 4 h at the temperature indicated. Each oxidation-reduction is indicated by an O_2 and arrow next to the line between those points. (Reproduced from Ref. 71 with permission.)

easily at low temperatures and have small variations in activity with heat treatment. In contrast, the more refractory metals, such as Ru, Rh, and Ir, retain their high activities in hydrogenolysis after low-temperature annealing. Pt is unique in its inability to form the oxide and thus shows no morphological or phase changes upon heating in oxygen at 500°C.

The behavior of different metals toward oxidation and annealing hints at the interesting and complex morphology of multimetal catalysts. As an example, Chen et al. [75] showed for SiO_2-supported Pt-Rh that heating in air above 400°C produces a rhodium oxide layer on the surface of the alloy particles and that subsequent low-temperature reduction produces an Rh-enriched surface layer. These morphologies are different from those of supported Pt-Pd alloys [76], for which PdO nucleates at the side of the metal particles instead of on top. Finally, heating the Pt-Rh alloy particles in hydrogen above 600°C leads to a two-phase structure [24].

The support also plays a role in the spearing of metal particles upon oxidation. Burkhardt and Schmidt [25] found that Rh_2O_3 spreads more readily on Al_2O_3 than SiO_2 and ascribed this more extensive spreading to a stronger Rh_2O_3-Al_2O_3 interaction. Even though this is qualitatively consistent with the wetting criterion, it is important to note that both SiO_2 and Al_2O_3 are generally considered to be weakly interacting supports. The next section discusses the behavior of oxide supports that interact more strongly with metals.

3. Strong Metal-Support Interactions

The term strong metal-support interactions (SMSI) was originally coined by Tauster et al. [77] to describe the suppression of the chemisorption ability of group VIII metals supported on TiO_2 after high-temperature reduction. Tauster and Fung [78] later reported the same phenomenon for Ir supported on other oxides such as Ta_2O_5, V_2O_3, and Nb_2O_5. These reports led to a flurry of research activities in subsequent years aimed at understanding the mechanism of SMSI.

The role of wetting in a SMSI system was recognized early by Baker et al. [79,80]. These authors showed that when the Pt/TiO_2 system is heated in hydrogen at 825K or higher, the Pt particles assume a pillbox morphology. At the same time, the TiO_2 support transforms to a lower oxide, Ti_4O_7. Treatment of this sample with oxygen at 875K causes the Pt particles to increase in size and attain a globular morphology. At first glance, these results seem to be inconsistent with the trends reported by Ruckenstein and Chu [81] for Pt/Al_2O_3 under oxidation–reduction treatments. However, as Baker [82] pointed out, the different support behavior can be understood in terms of (1) the lower surface free energy of TiO_2 compared to Al_2O_3 [13] and (2) the high surface free energy of the non-stoichiometric oxide Ti_4O_7. Furthermore, the interaction between Pt and Ti_4O_7 could facilitate the wetting of Pt particles into the observed pillbox morphology. For these reasons, wetting occurs in Pt/Al_2O_3 but not Pt/TiO_2 under an oxidizing environment, whereas the reverse is true under a reducing environment (see Fig. 5).

Stoneham [83] proposed a correlation between SMSI and wetting by making the following empirical observations. Of the oxides studied by Tauster and Fung [78], those that exhibit SMSI behavior all have a refractive index roughly equal to or larger than 5. Since Stoneham found the refractive index to be a good gauge on whether a particular liquid will wet a given oxide substrate, he went on to suggest that SMSI systems are those in which the metal wets the substrate.

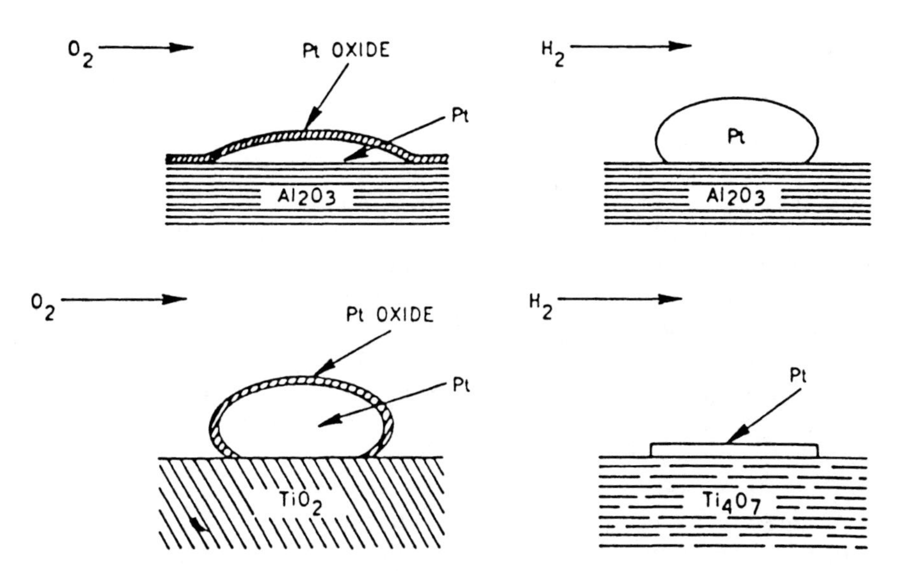

FIG. 5 Schematic representation of Pt catalysts supported on alumina and titania heat-treated in O_2 and H_2 environments. (Reproduced from Ref. 82 with permission.)

Shortly after the work of Baker et al. and Stoneham, it became clear that one important feature of SMSI is that a reduced oxide moiety migrates onto the metal particle in what is commonly referred to as a decoration model [84]. Thus, it is necessary to think in terms of the metal wetting the oxide, as well as the oxide wetting the metal. The two issues are, of course, related through the interfacial free energy between the two species. For the wetting of an oxide over a metal surface, Ruckenstein [85] showed that, similar to the derivation of Eq. (4), the inequality

$$U_{\text{metal-oxide}} > 2\sigma_{\text{oxide-gas}}$$

must be satisfied. In the case of TiO_2, the strong interaction between the reduced oxide and metal would result in a large $U_{\text{metal-oxide}}$. Along this line, it is noteworthy that Tauster and Fung [78] suggested that SMSI behavior is related to the reducibility of the oxide. Apparently, the stronger interaction between the reduced oxide and metal is sufficient to drive the migration process, even though the reduced oxide could have a higher surface free energy than the unreduced one. The spreading of the metal particle itself over the substrate could also be driven by such an interaction, in addition to the lower surface free energy of an oxide-covered surface.

One other key characteristic of the SMSI behavior is its reversibility. In other words, oxidizing a SMSI catalyst and then reducing it at low temperatures can restore its chemisorption capability and catalytic activity. The work of Bond et al. [86] suggested that for Ru/TiO_2 catalysts, this reversibility can also be understood in terms of morphological changes due to wetting. They found that after a Ru/TiO_2 catalyst attains the SMSI state, oxidation leads to the spreading of ruthenium oxide over the support. A subsequent low-temperature reduction produces highly dispersed Ru particles that give higher activities for the hydrogenolysis of n-butane. These authors did not mention any migrating species in their study. In general, the fate of any such species during an oxidation–reduction treatment is unclear. The restoration of chemisorption capability and catalytic activity argues that the metal surface is not covered by the decorating oxide. Thus, one possibility could be the decorating oxide becomes fully oxidized and returns to the substrate. Using high-resolution transmission electron microscopy, Braunschweig et al. [87] provided direct evidence of an amorphous overlayer on Rh particles after a model Rh/TiO_2 catalyst has been reduced at 773K. Oxidation at 573K followed by reduction at 473K leads to the partial removal of the overlayer and a roughening of the Rh metal surfaces. At the same time, the H_2 chemisorption uptake and activity of n-butane hydrogenolysis are almost completely restored.

B. Supported Oxides

Supported oxides, consisting of one oxide deposited onto the surface of another, are an important class of catalytic materials. Since a supported oxide often possesses structural and chemical properties that are different from the unsupported bulk phase, it is of great interest to study the dispersion of the supported phase that is governed by how well it wets the substrate. Other than catalytic applications, the wetting of one oxide over the surface of another is also relevant in any reactions involving a solid/solid interface [88,89]. Our intent in this section is not to provide a comprehensive review of the literature, but to use selected examples to illustrate the most interesting observations and trends. Additional examples can be found in other review articles [53,63,88,90,91].

One system that has received a lot of attention in recent years is V_2O_5/TiO_2 [92], a combination known to be active in the selective oxidation and ammoxidation of aromatics. As shown in Table 3, the morphology of supported V_2O_5 can be broadly classified into two groups: the well-dispersed species at below a monolayer coverage and the crystalline ones above it. Such a transition turns out not to be a rare occurrence. In fact,

TABLE 3 Summary of Results of V_2O_5/TiO_2

Support	Preparation[a]	Morphology	Ref.
Anatase	$VOCl_3$/benzene (5, 450)	monolayer coverage	94
Anatase	V_2O_5/aq. oxalic acid (2, 450)	moderate crystallinity	63
Anatase	$VOCl_3/CCl_4$ (5, 350)	monolayer coverage destabilizes above 450°C	95
Anatase and rutile	NH_4VO_3/aq. oxalic acid (3, 500)	well-dispersed monolayer coverage	96
Anatase	aq. V-oxalate (2, 500), Li_2O promoted	V_6O_{13} crystals anatase transformed to rutile	97
Anatase	aq. NH_4VO_3 (3, 450), $VOCl_3$/benzene (3, 450), $VOCl_2 \cdot 2H_2O$/isopropyl alcohol (3, 450)	surface V-O species	98
Anatase	$VOCl_3$ in CCl_4 (5, 350), vanadyl triisopropoxide in aq. alcohol (5, 350)	amorphous, intrinsic disorder	99
Anatase and rutile	NH_4O_3/aq. oxalic acid, $VOCl_3$ gas	uniform monolayer	100
Anatase	V-oxalate/aq. oxalic acid (4, 450)	small crystals	101
Anatase	NH_4VO_3/aq. oxalic acid (5, 500)	small V_2O_5 crystals, areas of TiO_2 exposed	102
Anatase	NH_4VO_3/aq. oxalic acid (3, 500)	V_2O_5 crystals	96
Anatase	NH_4VO_3 (3, 450), $VOCl_3$/benzene (3, 450), $VOCl_2 \cdot 2H_2O$/isopropyl alcohol (3, 450)	bulk on top of dispersed monolayer	98
Anatase and rutile	NH_4VO_3/aq. oxalic acid, $VOCl_3$ gas	crystals covering a small fraction of stable monolayer	100
Anatase	V-oxalate/aq. oxalic acid (4, 450)	crystalline and amorphous V_2O_5	101
Anatase	V_2O_5 vapor	destabilized surface phase	103

[a] Calcining conditions given as (x,y), where x is in hours and y in °C.

the work of Xie and co-workers [53,93] has shown that a large number of oxides disperse spontaneously onto the surface of oxide supports to form a monolayer. They referred to this process as solid/solid adsorption and explained it in terms of thermodynamic and kinetic driving forces [93]. Even though it is useful to think of supported oxides in terms of such a monolayer threshold, a couple of cautionary remarks are in order. Xie and co-workers inferred a dispersed phase from the lack of x-ray diffraction peaks. Since x-ray diffraction cannot detect particles smaller than 3–5 nm, it is possible that not the entire support surface is covered with a complete overlayer. The limitations and implications of having a perfect monolayer have been noted by Bond et al. [90]. A closer examination of Table 3 also reveals that other factors, such as the nature of the support, the precursor used, and the pretreatment conditions, can affect the final morphology.

In the following sections, we will present literature results that demonstrate directly the spreading process and, by comparing the behavior of differently prepared samples, identify the key parameters in affecting solid/solid wetting.

1. Model Studies

Haber et al. [89] studied the wetting of V_2O_5 on TiO_2 by placing a pellet of V_2O_5 in the center of a larger TiO_2 pellet. After the sample was heated in air at 923K for 48 h or at 823K for 24 days, two regions (one in the center and one near the edge of the TiO_2 pellet) were analyzed by x-ray diffraction, XPS, and ESR spectroscopy. Their results showed that under these conditions, V_2O_5 migrates over the surface of anatase TiO_2 but not rutile. Even on anatase, the spreading is nonuniform as part of the TiO_2 surface remains free of V_2O_5. Apparently, the interaction between the two oxides depends on the polymorphic modification (anatase vs. rutile) and the different crystal planes of anatase. Such is not the case of MoO_3, which is found to wet both anatase and rutile in a similar study of Haber et al. [104].

Although MoO_3 wets TiO_2 readily, it has been found not to wet SiO_2 [91,104]. However, here again, the specific form of MoO_3 may be a factor. In most studies, the impregnated precursor is calcined at or near 500°C, leading to orthorhombic MoO_3 crystallites that once formed, are difficult to redisperse on SiO_2 [105]. Datta et al. [37] confirmed the stability of orthorhombic MoO_3 upon calcination in air at 500°C on both nonporous SiO_2 spheres and planar SiO_2 substrates using TEM and in situ electron microscopy. Furthermore, these authors showed that hexagonal MoO_3 crystallites, formed by calcining the precursor at 300°C, spread readily onto these SiO_2 substrates. The independence of this wetting behavior on the atmosphere suggests strong chemical interactions between hexagonal MoO_3 and SiO_2.

We have now seen two examples in which even for the same oxide/oxide combination, the structure of one of the oxides, be it the surface or supporting phase, significantly affects the wetting behavior. It is thus not surprising that solid/solid wetting, in general, is system-specific. Murrell et al. [106] studied the spreading of WO_3 on Al_2O_3 and SiO_2 model thin films. Using a mask, they vapor-deposited WO_3 onto the various substrates and used a scanning Auger microprobe to quantify the distribution of WO_3 concentration after the samples had been heated in air at 500°C. Their results showed that WO_3 spreads onto both θ- and γ-Al_2O_3 and the rate appears to be the same on the two Al_2O_3 surfaces. On the other hand, no WO_3 signal was detected on SiO_2 after the heat treatment, suggesting that either WO_3 had diffused into SiO_2 or SiO_2 had diffused onto WO_3.

2. Effects of Preparation Methods

The samples used in the model studies all have very low surface area. Many oxide supports, however, are of high surface area and supported oxides are usually introduced via the impregnation of a solution containing the appropriate precursor. Furthermore, many studies on oxide/oxide wetting employ physical mixtures that are mechanically put together initially [53,91,93]. It is thus of interest to see if and how the preparation method affects the dispersion of the supported oxide.

For the V_2O_5/TiO_2 system, Haber [88] showed that when a mechanical mixture of V_2O_5 and anatase TiO_2 is heated in a reactor, the activity for the oxidation of o-xylene increases with time on stream. On the other hand, the activity remains low and invariant with time when a mixture of V_2O_5 and rutile TiO_2 is heated under the same conditions. These results, shown in Fig. 6, are totally consistent with the spreading of V_2O_5 over anatase but not rutile [89]. Leyrer et al. [91] also reported the formation of a surface vanadia species when a physical mixture of V_2O_5/TiO_2 is heated at 770K for 48 h in a flow of moist oxygen. The TiO_2 used is Degussa P25 that is composed of 80% anatase and 20% rutile. When the same mixture is heated at 870K, the surface vanadia species forms in dry as well as moist conditions and TiO_2 transforms completely into rutile.

Table 3 shows several examples in which impregnation techniques lead to well-dispersed V_2O_5 species on TiO_2. Recently, Bond and Flamery [107] showed that V_2O_5/TiO_2 catalysts can be prepared either by wet impregnation with NH_4VO_3-oxalic acid solution or by drafting with either $VOCl_3$ or $VO(O^iBu)_3$. Both methods can give a single monolayer of V_2O_5, even though they differ in the transition from the dried precursor to the final calcined state.

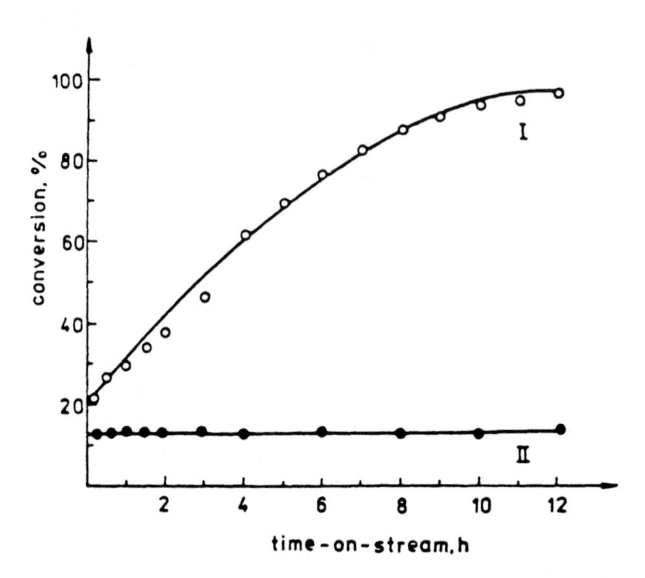

FIG. 6 Time-on-stream activity data of O-xylene oxidation for V_2O_5-anatase (curve I) and V_2O_5-rutile (curve II) mechanical mixtures. (Reproduced from Ref. 88 with permission.)

Alumina-supported molybdena is another system that has received a lot of attention. Xie et al. [93] showed that when a physical mixture of MoO_3 and γ-Al_2O_3 is heated at 450°C for 24 h, the x-ray diffraction pattern of MoO_3 disappears up to a loading of 0.22 g of MoO_3 per g of γ-Al_2O_3. This observation is interpreted as the spontaneous spreading of MoO_3 to form a monolayer until the threshold coverage is reached. Using Raman spectroscopy [91], ion scattering spectroscopy [108], and EXAFS analysis [109], Leyrer and co-workers also showed that MoO_3 spreads on the surface of γ-Al_2O_3 when a physical mixture of the two components is heated. The spreading process itself occurs independent of the water content in the gas atmosphere, but water is necessary for the formation of a surface polymolybdate. These results were confirmed in another study in which Raman microscopy was used to probe special specimens consisting of an alumina wafer in contact with a molybdenum trioxide wafer [110]. The wetting behavior of MoO_3/Al_2O_3 samples was also related to their activity in thiophene hydrodesulfurization [111], demonstrating again the important relationship between structure and reactivity.

Stampfl et al. [105] used Raman spectroscopy and Fourier transform infrared spectroscopy to study the interactions of MoO_3 with TiO_2, SnO_2,

γ-Al_2O_3, MgO, and SiO_2 and reported the thermally induced spreading of the bulk MoO_3 over the first four oxides. By comparing their results with literature data, these authors suggested that the nature of the final calcined sample could be independent of the method used to load the support. In fact, the spreading of MoO_3 to form a monolayer on γ-Al_2O_3 has been found for samples prepared by the incipient wetness impregnation of the support with an ammonium heptamolybdate solution [112,113].

The spreading of WO_3 on γ-Al_2O_3 also seems to be independent of the preparation method. Murrell et al. [106] prepared two ball-milled mixtures of WO_3 and γ-Al_2O_3 containing 10 and 25% WO_3. After calcination at 700°C, bulk WO_3 x-ray diffraction peaks disappeared for the 10% sample and reduced significantly for the 25% sample. This observation is similar to that of Leyrer et al. [91] and consistent with the results of the thin-film samples. Furthermore, the 10% sample showed an acid cracking activity that was identical to a sample prepared from the impregnation of ammonium meta-tungstate.

At this point, the readers may be tempted to conclude that the preparation method does not affect the spreading and hence the dispersion of supported oxides. However, we must caution that this statement seems to be true only for systems with strong oxide-oxide interactions. If a support interacts weakly with the supported oxide, then a preparation method that would maximize such an interaction is more likely to favor the wetting of the supported phase. As an example, it is well known that oxides such as MoO_3, WO_3, and V_2O_5 do not wet SiO_2 [91,105]. When a physical mixture of MoO_3 and SiO_2 is heated, the crystalline MoO_3 x-ray diffraction peaks remain. However, Cheng and Schrader [114] reported the evidence of a surface MoO_3 phase on SiO_2 for a sample prepared by the impregnation of an ammonium heptamolybdate solution. The preferential wetting of the hexagonal to the orthorhombic form of MoO_3 on SiO_2 has already been discussed [37]. Another example is the Nb_2O_5/SiO_2 system. When Nb_2O_5 is introduced by impregnating SiO_2 with a hexane solution of niobium ethoxide, no x-ray diffraction peaks of Nb_2O_5 are observed after the sample has been calcined at 600°C [65]. However, when Nb_2O_5 is sputter-deposited onto SiO_2 thin films, crystalline Nb_2O_5 is detected after a similar heat treatment [27]. Apparently, the wet preparation gives a stronger Nb_2O_5-SiO_2 interaction through the reaction between the precursor and surface hydroxyl group, resulting in a better wetting of Nb_2O_5.

3. Physical Basis of Interfacial Interactions

We have thus far referred to the strength of oxide-oxide interactions vaguely as "strong" or "weak" as values for the interaction energy, a major driving force for wetting, are rarely available. Thus, the description of solid/solid wetting is by necessity phenomenological in nature and the microscopic mechanism remains unknown. Still, it is instructive to consider the physical basis of some potential interfacial interactions to arrive at a qualitative understanding of the experimental observations.

Compound formation represents a strong interaction between two oxides. Leyrer et al. [91] noted that MoO_3 reacts with Al_2O_3 to form $Al_2(MoO_4)_3$ and with TiO_2 to form $Ti(MoO_4)_2$. Since MoO_3 also wets Al_2O_3 and TiO_2 readily, these authors speculated that the interaction energies in these pairs contain "chemical" contributions from the compounds. Even though wetting occurs at temperatures lower than those for solid-state reactions to take place, the existence of a stable compound could be viewed as an affinity of the two oxides to interact strongly and consequently to favor wetting. The difficulty of MoO_3 in wetting SiO_2 can then be understood in terms of the lack of a chemical contribution to the interaction energy of these two oxides.

Another factor important in interfacial interactions is surface hydroxyl groups (OH). In many preparation methods, the impregnation of a support with a solution containing the precursor involves the interaction of the precursor with surface hydroxyl groups, a process often referred to as "grafting." Since grafting leads to actual chemical bonds, in the form of cation-oxygen-cation linkages, the interactions are sufficiently strong to induce wetting. Bond and co-workers [90,107] have shown that reactions between surface hydroxyl groups and compounds such as $VOCl_3$ and $VO(OR)_3$, where R is iso-C_4H_9, for example, are effective in forming V_2O_5 monolayers on TiO_2 through formation of V-O-Ti bonds. Kijenski et al. [115] also used the reactions of vanadyl triisobutoxide with surface hydroxyl groups to prepare monolayers of V_2O_5 on Al_2O_3, SiO_2, MgO, and TiO_2 supports. In fact, a stoichiometric reaction between metal alkoxides and surface hydroxyl groups has been quantified for certain systems [116,117]. As noted earlier, Ko et al. [65] suggested that a similar reaction between niobium ethoxide and surface hydroxyl groups is what causes a monolayer of Nb_2O_5 to wet SiO_2. In support of this claim is the observation that Nb_2O_5 monolayer is less stable on a SiO_2 thin-film support that has a lower surface hydroxyl concentration [27]. Apparently, in this system Nb-O-Si linkages provide the driving force for wetting to occur, and their relative propensity accounts for the different behaviors of samples prepared from wet impregnation vs. dry deposition.

V. SUMMARY

Many catalysts consist of one solid phase supported on another. The particle size of the supported phase, or its dispersion, is often related to how well it wets the substrate. Over the years, many experimental techniques have been developed to characterize the dispersion of a supported catalyst and relate the dispersion to chemical properties. At present, we are still not in a position to accurately predict the dispersion of a particular system under certain conditions. However, what we have attempted to show in this article is that simple arguments based on the thermodynamics of wetting can explain many experimental observations, as well as establish qualitative trends.

The two key parameters in solid/solid wetting are the surface free energy of the supported phase and the interaction energy between the two solids. Metals, which have high surface free energies and interact weakly with oxides, do not wet oxide supports well and sinter upon heating. In contrast, oxides wet oxides well and this is the basis of using oxidation to redisperse supported metal catalysts. The same principles also apply to the migration of an oxide species over metal particles in systems that exhibit the SMSI behavior.

For supported oxides, the spreading of the supported phase to form a monolayer over the substrate seems to be the rule rather than the exception. In fact, the final state is independent of how the two solids are brought into contact for systems with strong oxide-oxide interactions. For systems with weaker interactions, however, factors such as the type of precursor used and the nature and pretreatment of the support become important as it is necessary to optimize a set of conditions to favor wetting. For certain systems, the specific structure of either the supporting or supported oxide could also be important. Any comparisons among literature results should be made with these factors in mind.

Obviously, more accurate data on surface free energies and interaction energies are needed before a quantitative description of wetting in catalysis becomes possible. Short of attaining this goal, it would still be desirable to identify the physical basis of interfacial interactions in as many systems as possible and perform studies in which one or more of the pertinent parameters are systematically varied. We hope that this article will generate sufficient interests toward building a larger database in this important but often overlooked area of catalysis.

ACKNOWLEDGMENT

We wish to thank Professor Eli Ruckenstein for making us aware of the importance of solid/solid wetting in catalysis. Many ideas in this article (especially those in Sec. II) come from his extensive work in the area.

REFERENCES

1. B. W. Dodson, *Surf. Sci. 184:*1 (1987).
2. E. Ruckenstein, in *Metal-Support Interactions in Catalysis, Sintering, and Redispersion* (S. A. Stevenson, J. A. Dumesic, R. T. K. Baker, and E. Ruckenstein, eds.), Van Nostrand Reinhold, New York, 1987, Chap. 13.
3. E. Ruckenstein, *J. Cryst. Growth 47:*666 (1979).
4. E. Ruckenstein, *Mater. Sci. Res. 16:*199 (1984).
5. R. C. Weast, and M. J. Astle, eds., *CRC Handbook of Chemistry and Physics*, 60th ed., CRC Press, Boca Raton, FL, 1979.
6. E. Ruckenstein, and B. Pulvermacher, *J. Catal. 29:*224 (1973).
7. H. Sutton, and E. Feingold, in *Materials Science Research*, Vol. 3, (W. W. Kreiger, and H. Palmour III, eds.), Plenum, New York, 1966.
8. F. A. Halden, and W. D. Kingery, *J. Phys. Chem. 59:*557 (1955).
9. P. Wynblatt, and T. M. Ahn, in *Sintering and Catalysis* (G. C. Kuczynski, ed.), Plenum, New York, 1975, p. 83.
10. J. M. Blakely, in *Introduction to the Properties of Crystal Surfaces*, Pergamon Press, New York, 1973, Chap. 3.
11. A. W. Adamson, in *Physical Chemistry of Surfaces*, 3rd ed., Wiley, New York, 1976, Chap. V.
12. J. R. Anderson, in *Structure of Metallic Catalysts*, Academic Press, New York, 1975, Appendix 1.
13. S. H. Overbury, P. A. Bertrand, and G. A. Somorjai, *Chem. Rev. 75:*547 (1975).
14. R. B. Anderson, and P. T. Dawson, eds., in *Experimental Methods in Catalytic Research, vols. II and III*, Academic Press, New York, 1976.
15. F. Delannay, ed., in *Characterization of Heterogeneous Catalysts*, Marcel Dekker, New York, 1984.
16. B. E. Warren, and B. L. Averbach, *J. Appl. Phys. 21:*595 (1950).
17. S. R. Sashital, J. B. Cohen, R. L. Burwell, and J. B. Butt, *J. Catal. 50:* 479 (1977).
18. Y. J. Liu, Y. C. Xie, J. Ming, J. Liu, and Y. Q. Tang, *Chinese J. Catal. 3:*262 (1982).
19. P. Wynblatt, and N. A. Gjostein, *Scripta Met. 9:*969 (1973).
20. P. Wynblatt, *Acta Metallurgical 24:*1175 (1976).
21. E. Ruckenstein, and M. L. Malhotra, *J. Catal. 41:*303 (1976).
22. J. J. Chen, and E. Ruckenstein, *J. Catal. 69:*254 (1981).
23. T. Wang, and L. D. Schmidt, *J. Catal. 70:*187 (1981); *71:*411 (1981); *78:* 306 (1982).
24. T. P. Chojnacki, and L. D. Schmidt, *J. Catal. 115:*473 (1989).
25. J. Burkhardt, and L. D. Schmidt, *J. Catal. 116:*240 (1989).
26. T. F. Hayden, and J. A. Dumesic, *J. Catal. 103:*366 (1987).
27. J. G. Weissman, E. I. Ko, and P. Wynblatt, *J. Catal. 108:*383 (1987).
28. E. G. Derouane, J. J. Chludzinski, and R. T. K. Baker, *J. Catal. 85:*187 (1984).
29. D. G. Mustard, and C. H. Bartholomew, *J. Catal. 67:*186 (1981).

30. D. White, T. Baird, J. R. Fryer, L. A. Freeman, D. J. Smith, and M. Day, *J. Catal. 81:*119 (1983).
31. A Baiker, P. Dollenmeier, M. Glinski, A. Reller, and V. K. Sharma, *J. Catal. 111:*273 (1988).
32. G. Bergeret, P. Gallezot, K. V. R. Chary, B. Rama Bao, and V. S. Subrahmanyam, *Appl. Catal. 40:*191 (1988).
33. J. V. Sanders, in *Catalysis Science and Technology*, Vol. 7 (J. R. Anderson, and M. Boudart, eds.), Springer-Verlag, New York, 1985, Chap. 2.
34. S. Chakraborti, A. K. Datye, and N. L. Long, *J. Catal. 108:*444 (1987).
35. A. K. Datye, A. D. Logan, and N. L. Long, *J. Catal. 109:*76 (1988).
36. E. J. Braunschweig, A. D. Login, A. K. Datye, and D. J. Smith, *J. Catal. 118:*227 (1989).
37. A. Datta, J. R. Regalbuto, and C. W. Allen, *Ultramicroscopy 29:*233 (1989).
38. S. S. Chan, I. E. Wachs, L. L. Murrell, and N. C. Dispenziere, Jr., *J. Catal. 92:*1 (1985).
39. I. E. Wachs, R. Y. Saleh, S. S. Chan, and C. C. Cherish, *Appl. Catal. 15:* 339 (1985).
40. S. S. Chan, and I. E. Wachs, *J. Catal. 103:*224 (1987).
41. I. E. Wachs, F. D. Hardcastle, and S. S. Chan, *Spetctroscopy 1:*30 (1986).
42. M. I. Zaki, N. E. Fouad, J. Leyer, and H. Knözinger, *Appl. Catal. 21:*359 (1986).
43. J. Leyrer, M. I. Zaki, and H. Knözinger, *J. Phys. Chem. 90:*4775 (1986).
44. M. G. Reichmann, and A. T. Bell, *Langmuir 3:*111 (1987); *3:*563 (1987).
45. M. G. Reichman, and A. T. Bell, *Appl. Catal. 32:*315 (1987).
46. R. Kozlowski, R. F. Pettifer, and J. M. Thomas, *J. Phys. Chem. 87:*5176 (1983).
47. K. Asakura, and Y. Iwasawa, *Chem. Lett:*633 (1988).
48. G. E. Rhead, *Contemp. Phys. 24(6):*535 (1983).
49. C. Defosse, in *Characterization of Heterogeneous Catalysts* (F. Delanny, ed.), Marcel Dekker, New York, 1984, Chap. 6.
50. S. C. Fung, *J. Catal. 58:*454 (1979).
51. M. Houalla, F. Delannay, I. Matsuura, and B. Delmon, *J. Chem. Soc., Faraday Trans. I 76:*2128 (1980).
52. M. Houalla, and B. Delmon, *Surf. Inter. Anal. 3:*103 (1981).
53. Y. C. Xie, and Y. Q. Tang, *Adv. Catal. 37:*1 (1990).
54. L. Salvati, J. M. Makovsky, J. M. Stencel, F. R. Brown, and D. M. Hercules, *J. Phys. Chem. 85:*3700 (1981).
55. I. E. Wachs, C. C. Chersich, and J. H. Hardenbergh, *Appl. Catal. 13:*335 (1985).
56. M. Boudart, and G. Djéga-Mariadassou, in *Kinetics of Heterogeneous Catalytic Reactions*, Princeton Univ. Press, NJ, 1984, Chap. 5.
57. H. Knözinger, in *Advances in Catalysis*, Vol. 25 (D. D. Eley, H. Pines, and P. B. Weisz, eds.), Academic Press, New York, 1976, p. 184.
58. D. G. Rethwisch, and J. A. Dumesic, *J. Phys. Chem. 90:*1863 (1986).
59. M. Niwa, S. Inagaki, and Y. Murakami, *J. Phys. Chem. 89:*3869 (1985).

60. G. Connell, and J. A. Dumesic, *J. Catal. 101:*103 (1986); *102:*216 (1986); *105:*285 (1987).
61. T. Kataoka, and J. A. Dumesic, *J. Catal. 112:*66 (1988).
62. P. A. Burke, J. G. Weissman, E. I. Ko, and P. Wynblatt, in *Catalysis 1987* (J. W. Ward, ed.), Elsevier, New York, 1988, p. 457.
63. I. E. Wachs, R. Y. Saleh, S. S. Chan, and C. Chersich, *Chemtech* :756 December 1985.
64. E. I. Ko, and F. H. Rogan, *Chem. Eng. Comm. 55:*139 (1987).
65. E. I. Ko, R. Bafrali, N. T. Nuhfur, and N. J. Wagner, *J. Catal. 95:*260 (1985).
66. M. Boudart, A. Delbouille, J. A. Dumesic, S. Khammanouma, and H. Topsoe, *J. Catal. 37:*486 (1975).
67. E. Ruckenstein, in *Sintering and Heterogeneous Catalysis* (G. C. Kuczynski, A. E. Miller, and G. A. Sargent, eds.), Plenum, New York, 1984, p. 199.
68. J. J. Chen, and E. Ruckenstein, *J. Phys. Chem. 85:*1606 (1981).
69. E. Ruckenstein, and P. S. Lee, *Surf. Sci. 52:*298 (1975).
70. I. Shushuma, and E. Ruckenstein, *J. Catal. 94:*239 (1985).
71. C. Lee, and L. D. Schmidt, *J. Catal. 101:*123 (1986).
72. D. W. Goodman, *Surf. Sci. 123:*L679 (1982).
73. S. Gao, and L. D. Schmidt, *J. Catal. 111:*210 (1988).
74. S. Gao, and L. D. Schmidt, *J. Catal. 115:*356 (1989).
75. M. Chen, T. Wang, and L. D. Schmidt, *J. Catal. 60:*356 (1979).
76. M. Chen, and L. D. Schmidt, *J. Catal. 56:*198 (1979).
77. S. J. Tauster, S. C. Fung, and R. L. Garten, *J. Am. Chem. Soc. 100:*170 (1978).
78. S. J. Tauster, and S. C. Fung, *J. Catal. 55:*29 (1978).
79. R. T. K. Baker, E. B. Prestridge, and R. L. Garten, *J. Catal. 56:*390 (1979).
80. R. T. K. Baker, E. B. Prestridge, and R. L. Garten, *J. Catal. 59:*293 (1979).
81. E. Ruckenstein, and Y. F. Chu, *J. Catal. 59:*109 (1979).
82. R. T. K. Baker, *J. Catal. 63:*523 (1980).
83. A. M. Stoneham, *Appl. Surf. Sci. 14:*249 (1982).
84. A. J. Simoens, R. T. K. Baker, D. J. Dwyer, C. R. F. Lund, and R. J. Madon, *J. Catal. 86:*359 (1984).
85. E. Ruckenstein, in *Strong Metal-Support Interactions* (R. T. K. Baker, S. J. Tauster, and J. A. Dumesic, eds.), ACS Symp. Series 298, Washington, DC, 1986, p. 152.
86. G. C. Bond, and X. Yide, *J. Chem. Soc., Faraday Trans. I 80:*3103 (1984).
87. E. J. Braunschweig, A. D. Logan, A. K. Datye, and D. J. Smith, *J. Catal. 118:*227 (1989).
88. J. Haber, *Pure Appl. Chem. 56(12):*1663 (1984).
89. J. Haber, T. Machej, and T. Czeppe, *Surf. Sci. 151:*301 (1985).
90. G. C. Bond, S. Flamerz, and R. Shukri, *Faraday Discuss. Chem. Soc. 87:* 65 (1989).
91. J. Leyrer, R. Margraf, E. Taglauer, and H. Knözinger, *Surf. Sci. 201:*603 (1988).

92. G. C. Bond, *Appl. Catal. 71*:1 (1991).
93. Y. Xie, L. Gui, Y. Liu, Y. Zhang, B. Zhao, N. Yang, Q. Guo, L. Duan, H. Huang, X. Cai, and Y. Tang, in *Adsorption and Catalysis on Oxide Surfaces* (M. Che and G. C. Bond, eds.), Elsevier, Amsterdam, 1985, p. 139.
94. G. C. Bond, and K. Bruckman, *Faraday Discuss. Chem. Soc. 72*:235 (1981).
95. J. Haber, A. Kozlowska, and R. Kozlowski, *J. Catal. 102*:52 (1986).
96. M. Inomata, K. Mori, A. Miyamoto, T. Ui, and Y. Murakami, *J. Phys. Chem. 87*:754 (1983).
97. R. Hasse, U. Illger, J. Scheve, and I. W. Schulz, *Reac. Kinet. Catal. Lett. 28*:395 (1985).
98. G. Busca, G. Centi, L. Marchetti, and F. Trifiro, *Langmuir 2*:568 (1986).
99. R. Kozlowski, P. F. Pettifer, and S. M. Thomas, *J. Phys. Chem. 87*:5176 (1983).
100. G. C. Bond, J. P. Zurita, and S. Flamerz, *Appl. Catal. 27*:353 (1986).
101. Z. C. Kang, and Q. X. Bao, *Appl. Catal. 26*:251 (1986).
102. M. del Arco, M. Holgado, C. Martin, and V. Rives, *J. Catal. 99*:19 (1986).
103. A. Vejux, and P. Courtine, *J. Solid State Chem. 63*:179 (1986).
104. J. Haber, T. Machej, and R. Grabowski, *Solid State Ionics 32/33*:887 (1989).
105. S. R. Stampfl, Y. Chen, J. A. Dumesic, C. Niu, and C. G. Hill, Jr., *J. Catal. 105*:445 (1987).
106. L. L. Murrell, and N. C. Dispenziere, Jr., paper 273 presented at 64th Colloid and Surface Science Symposium, Lehigh Univ., Bethlehem, PA 1990.
107. G. C. Bond, and S. Flamerz, *Appl. Catal. 46*:89, (1989).
108. R. Margraf, J. Leyrer, H. Knözinger, and E. Taglauer, *Surf. Sci. 189/190*: 842 (1987).
109. G. Kisfaludi, J. Leyrer, H. Knözinger, and R. Prins, *J. Catal. 130*:192 (1991).
110. J. Leyrer, D. Mey, and H. Knözinger, *J. Catal. 124*:349 (1990).
111. T. I. Koranyi, Z. Paal, J. Leyrer, and H. Knözinger, *Appl. Catal. 64*:L5 (1990).
112. J. M. Stencel, L. E. Makovsky, T. A. Surpus, J. de Vries, R. Thomas, and J. A. Moulijn, *J. Catal. 90*:314 (1984).
113. S. Kasztelan, J. Grimblot, and J. P. Bonnelle, *J. Phys. Chem. 91*:1503 (1987).
114. C. P. Cheng, and G. L. Schrader, *J. Catal. 60*:276 (1979).
115. J. Kijenski, A. Baiker, M. Glinski, P. Dollenmeier, and A. Wokaum, *J. Catal. 101*:1 (1986).
116. R. Hombeck, J. Kijenski, and S. Malinowski, in *Preparation of Catalysts, Vol. II* (B. Delmon, P. Grange, P. Jacobs, and G. Poncelet, eds.), Elsevier, Amsterdam, 1979, p. 595.
117. M. Glinski, and J. Kijenski, in *Preparation of Catalysts*, Vol. III (P. Grange, and P. A. Jacobs, eds.), Elsevier, Amsterdam, 1983, p. 553.

8

High-Temperature Wetting Behavior of Inorganic Liquids

BRIAN J. J. ZELINSKI, JOHN PAUL CRONIN,* MATTHEW DENESUK, and DONALD R. UHLMANN Department of Materials Science and Engineering, University of Arizona, Tucson, Arizona

I. INTRODUCTION

In all natural and technological systems that consist of mixtures of solid and liquid phases, the interplay between interphase boundary energies

* *Current affiliation:* Advanced Technology Center, Donnelly Corporation, Tucson, Arizona.

affects the extent to which the solid and liquid phases come into contact. It is these surface energy interactions that produce the macroscopically observable wetting behavior. Enhanced wetting promotes such everyday occurrences as the heterogeneous nucleation of rain drops on airborne dust particles or the use of detergents to improve the cleansing action of water in washing machines, whereas wetting prevention is the objective of every good car wax. Because the wetting behavior of liquids subtly or overtly influences both everyday phenomena and technologically sophisticated manufacturing processes, it has been extensively studied. Many excellent reviews on this topic exist, a few of which are included in the reference list [1–4].

By comparison with organic liquids and aqueous solutions, the wetting behavior of high-temperature oxide liquids remains largely unexplored, despite the importance of such phenomena in many manufacturing processes. As examples, the formation of strong glass-to-metal seals requires good spreading of the molten glass and the development of adhesion upon cooling; during the firing of electronic packages, component densification is often controlled by the presence of a wetting liquid phase; the development of interlayer adhesive strength in multilayer structures relies on the spreading characteristics of the liquid bonding agent; the poor wetting of molten glass by molten metals, such as tin, is exploited in the float glass process to manufacture large panes of architectural glass; the wetting of refractories by molten glass promotes undesirable corrosion of the walls of glass melters; and the manufacturing of ceramic-ceramic and metal-ceramic composites requires careful control of wetting and bonding relationships between the matrix and fibers to ensure optimization of mechanical properties.

This chapter focuses on the wetting behavior of inorganic liquids on solid surfaces. Whereas factors such as surface roughness, composite interfaces (interfaces of nonuniform surface energy), and precursor films can impact wetting behavior [1,3], few studies have been conducted to determine their effect on the wetting of high-temperature liquids. The discussion begins with a brief theoretical description of wetting phenomena, in which the importance of interfacial energies and interfacial reactions will be reviewed. It is pointed out that small changes in the interfacial energies can result in large variations in wetting phenomena. Studies from the literature on the wetting behavior of inorganic liquids on metal and ceramic substrates are then reviewed and discussed. Almost all wetting phenomena have a transient phase in that some time is required before steady-state contact angles are achieved. Modeling of the kinetics of spreading will be presented. Finally, the suitability of Zisman's critical surface tension concept to high-energy surfaces, such as ceramics and metals, at elevated temperatures will be discussed.

II. THEORY

A. Wetting Phenomena

Several different parameters are used to characterize the wetting of a liquid on a solid. The most commonly used parameter is the contact angle θ. This is the angle, through the liquid, between the substrate/liquid interface and the tangent to the droplet liquid/vapor interface at the solid/liquid/vapor junction (see Fig. 1). By balancing the horizontal force components at the three-phase line, Young [5] developed the following expression, which relates the contact angle θ to the interfacial energies:

$$\cos\theta = \frac{\gamma_{SV} - \gamma_{SL}}{\gamma_{LV}} \tag{1}$$

where γ_{SV}, γ_{LV}, and γ_{SL} are the equilibrium solid/vapor, liquid/vapor, and solid/liquid interfacial energies, respectively. As seen in Eq. (1), the contact angles formed by a given liquid on different solids are determined by $\gamma_{SV} - \gamma_{SL}$, a factor called the driving force for wetting.

The contact angle is a directly observable parameter describing the compatibility of a solid and a liquid in equilibrium with a vapor. A liquid is said to wet a solid if $\theta < 90°$; otherwise, it is nonwetting. If $\theta = 0°$, the liquid completely spreads over the surface of the solid. Although the contact angle may be measured with a number of different geometries, the method of choice for investigation of high-temperature systems is the sessile drop method [e.g., 7,8], where a small liquid droplet is brought into contact with a flat, solid surface and the contact angle is directly measured.

The degree of wetting is controlled by the relative values of the three interfacial energies. These interfacial energies are associated with unsatisfied bonding states at the surfaces of the liquid and solid and at the solid/liquid interface. In order to explore regions of wetting behavior, consider the case where two phases, 1 and 2, are placed in contact with each other. γ_1 is the interfacial energy of the phase 1/vacuum interface, γ_2 is the interfacial energy of the phase 2/vacuum interface, $\gamma_1 > \gamma_2$, and γ_{12} is the interfacial energy of the phase 1/phase 2 interface. If the excess bond energy of the surface atoms of phase 2 satisfies very little of the unsatisfied bonding requirements of surface atoms in the adjoining phase 1, then γ_{12} is greater than both γ_1 and γ_2, the surface of neither phase has lost much

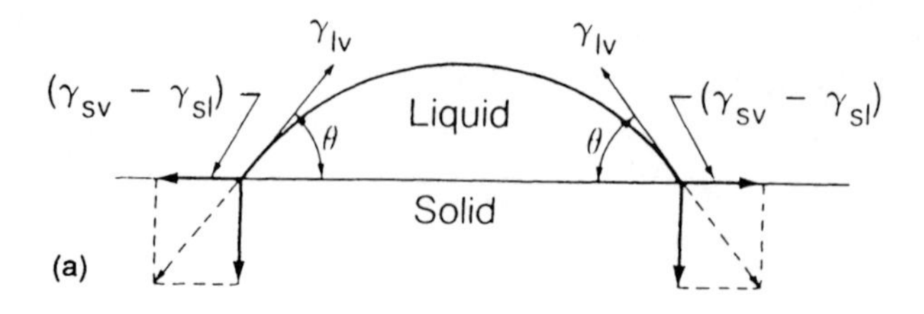

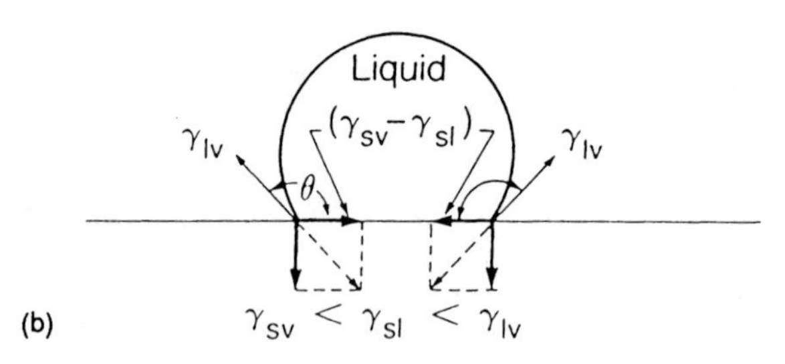

FIG. 1 Equilibrium forces at the solid/liquid/vapor junction (three-phase line) for an (a) acute contact angle and (b) obtuse contact angle. (From Ref. 5, by permission.)

of its individual identity, and a true interface actually has not formed [9]. In this case, the contact angle approaches 180° as γ_{12} approaches the value $\gamma_1 + \gamma_2$.

In contrast, in the absence of a chemical reaction, a true interface is considered to form when the presence of the surface atoms in the adjoining phases leads to a significant reduction in the number and/or strength of the unsatisfied bonds of the atoms on both surfaces. This occurs when chemical bonds develop across the phase 1/phase 2 interface. Under these conditions, γ_{12} will lie between γ_1 and γ_2. In the limit, all the surface bonding requirements of phase 2 may be met by the adjoining phase 1. In this case, phase 2 brings no broken or unsatisfied bonds to the interface and the magnitude of the total unsatisfied bond energy of phase 1 is reduced by an amount equal to γ_2. Based on this simplified argument, the value of γ_{12} cannot be less than the limiting value of $\gamma_1 - \gamma_2$.

Returning to solid/liquid/vapor interfaces, when $\gamma_{SV} < \gamma_{SL} < \gamma_{LV}$, as is the case for a metallic liquid droplet on a ceramic substrate, the droplet contact angle will be obtuse. When $\gamma_{LV} < \gamma_{SL} < \gamma_{SV}$, such as for an oxide liquid droplet on a metallic substrate, the contact angle will be acute. As the driving force for wetting approaches the value of γ_{LV} (i.e., $\gamma_{SL} \rightarrow \gamma_{SV} - \gamma_{LV}$), the contact angle approaches zero. Hence, in the absence of chemical reactions at the interface, no spreading or unrestrained extension of the liquid surface occurs. Dispersion interactions, however, can serve to reduce γ_{12} below $\gamma_1 - \gamma_2$. The magnitude of typical dispersion contributions to interfacial energies is ~25 mN/m [10]. Consequently, spreading may be expected to occur in systems where full or nearly full compensation of the unsatisfied bonding states at the liquid/vapor interface occurs upon formation of the solid/liquid interface.

According to Pask [11], if the solid phase actively participates in a reaction at the interface, the specific Gibbs energy of the reaction at the droplet periphery per unit area and unit time dG_R can contribute to the driving force for wetting. In this case, the driving force for wetting becomes

$$\gamma_{SL} - (\gamma_{SL} + dG_r) \quad (2)$$

As a result, the contact angle decreases, and if the driving force exceeds γ_{LV}, spreading will occur.

In some cases, it is useful to view a solid/liquid interface in terms of its energy relative to a hypothetical state where the solid and liquid are separated over a long distance with an intervening vapor phase. The energy difference between these two states can be seen as a measure of the compatibility of the solid and liquid and is often termed the thermodynamic work of adhesion W_a (first introduced by Dupré [12])

$$W_a = \gamma_{SV} + \gamma_{LV} - \gamma_{SL} \quad (3)$$

γ_{SV} is often decomposed into a solid/vacuum component γ_{SO} and a quantity termed the spreading pressure π_e, which represents the reduction in surface energy of the bare solid surface on exposure to the vapor ($\gamma_{SV} = \gamma_{SO} - \pi_e$).

Wettability is also discussed with reference to the spreading coefficient S, typically defined as

$$S \equiv \gamma_{SO} - \gamma_{SL} - \gamma_{LV} \quad (4)$$

Provided that π_e is negligible, the condition $S \geq 0$ ensures a zero contact angle and indicates strong compatibility between the liquid and solid (complete wetting); the condition $S < 0$ indicates incomplete wetting and the formation of a finite contact angle of the liquid on the solid. The latter case is separated further into two more cases: partial wetting, where $-\gamma_{LV} < S < 0$ and the contact angle is between 0 and 90°, indicates a weaker compatibility between the liquid and solid, and nonwetting, where $-\gamma_{LV} > S$ and the contact angle is between 90 and 180°, indicates a relative incompatibility of the solid and liquid. If π_e is significant, it may be added to S to maintain the validity of the above classifications.

B. Interfacial Energy Considerations

The energies of interfaces, particularly interfaces in contact with the vapor phase, are highly sensitive to environmental factors. This sensitivity can have great impact on the wetting behavior. Figure 2 illustrates the relationship between the contact angle and the driving force for wetting for a liquid with $\gamma_{LV} = 300$ mN/m. This value for the liquid/vapor interfacial energy is typical for molten oxide glasses. As seen in the figure, the contact angle is very sensitive to small variations in the driving force, particularly for small contact angles. A change in the driving force of 30 mN/m reduces the contact angle from 25–0°. Since a typical value γ_{SV} for metals or oxides is in the range of 1000–2000 mN/m, variations in γ_{SV} (or

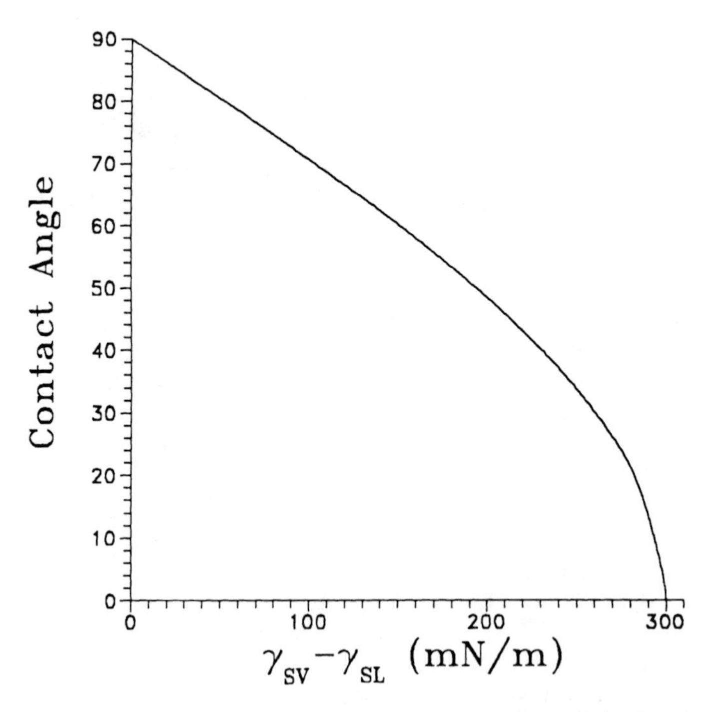

FIG. 2 Relationship between the contact angle and driving force for wetting, $\gamma_{SV} - \gamma_{SV}$, with $\gamma_{LV} = 300$ mN/m.

γ_{SL}) of only a few percent can drastically change wetting interactions. Similar changes in wetting are brought about by ~10% changes in γ_{LV} if one assumes a constant driving force.

Changes in surface energies of this magnitude can be induced by modest variations in the temperature, atmosphere, or chemistry of a wetting system. For example, although the surface tensions of most liquids decrease with increasing temperature, γ_{LV} for silicate liquids can exhibit a wide range of behavior. As seen in Fig. 3, from the work of Cronin et al. [13]

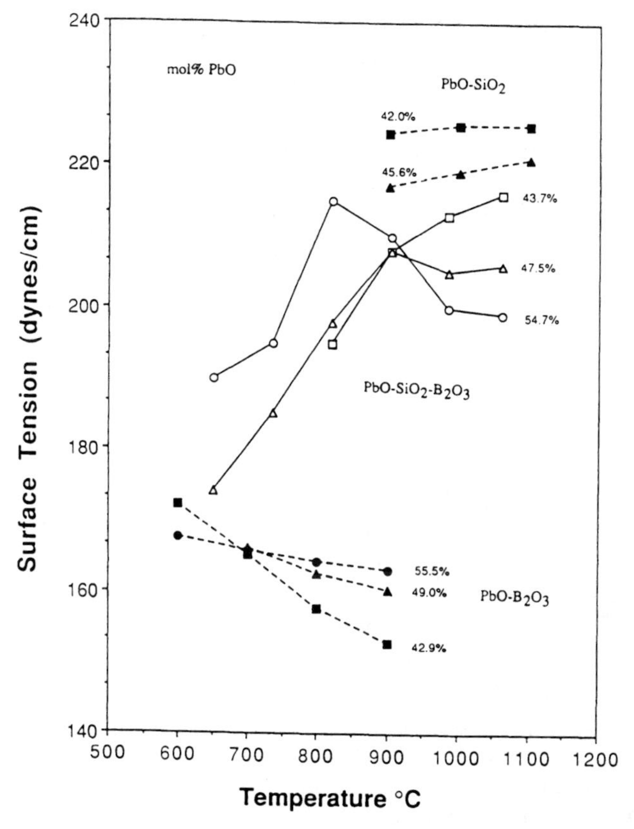

FIG. 3 Temperature dependence of the surface tension for molten $PbO-SiO_2$ liquids, $PbO-B_2O_3$ liquids, and $PbO-B_2O_3-SiO_2$ liquids [13]. The B_2O_3/SiO_2 molar ratio is 0.34 in ternary liquids. Percentages indicate mol. % of PbO in all liquids. (Binary data from Ref. 14, by permission.)

in the temperature range of 600–900°C, $PbO\text{-}B_2O_3$ liquids demonstrate the behavior common to most liquids. The surface tension decreases with increasing temperature and is well described by an equation of the form

$$\gamma_{LV} = \gamma_0 + AT \tag{5}$$

where γ_0 is a material constant and the temperature coefficient of surface tension A is negative. This is consistent with the requirement that the surface tension vanish at the critical point.

In contrast, $PbO\text{-}SiO_2$ liquids exhibit positive temperature coefficients of surface tension in the 700–1100°C range. Over a limited temperature range, $PbO\text{-}B_2O_3\text{-}SiO_2$ liquids exhibit a strongly positive temperature dependence; and one of the liquids exhibits a pronounced maximum in the γ_{LV} vs. T relationship. Positive temperature coefficients can be exhibited by liquids when increasing temperatures reduce the concentration of species that preferentially segregate at the surface to reduce the surface tension. Alternatively, increasing temperatures may destabilize structural states on the surface of the liquid that minimize the surface tension [13].*

The adsorption of gases on the surfaces of solids and liquids can also lead to a large reduction in their surface energies. Figure 4 shows the variation in γ_{SV} of silver as a function of oxygen partial pressure in the gas phase. Here the solid/vapor interfacial energy is reduced via the adsorption of a monolayer of oxygen on the surface of silver. Parikh [16] has shown that the adsorption of water vapor can drastically reduce γ_{LV} of molten $Na_2O\text{-}CaO\text{-}SiO_2$ glass. In this liquid, the adsorption of a monolayer of OH^- ions lowers γ_{LV} by approximately 50 mN/m. Similar effects may be observed in other systems via the absorption of gaseous species such as CO, CO_2, etc.; and atmospheric effects would be anticipated for solid and liquid oxide surfaces if species in the phases are easily reduced or oxidized. Changes in the oxidation state of the metal species can produce chemical and/or structural changes in the liquid or solid surfaces and cause a modification of the surface energy.

Minor changes in chemistry can cause large changes in γ_{LV} as well. The effects of various oxides, added in small quantities, on the surface tension of a $Na_2O\text{-}CaO\text{-}SiO_2$ liquid are shown in Fig. 5, after the work of Badger et al. [17]. The surface tension of this liquid has been increased by as much as 15 mN/m and decreased by 70 mN/m in this way.

Taken together, these examples indicate that great care must be exerted during the measurement and interpretation of wetting behavior. If pos-

* It is expected that at sufficiently high temperatures, all these liquids would show a negative temperature coefficient of surface tension.

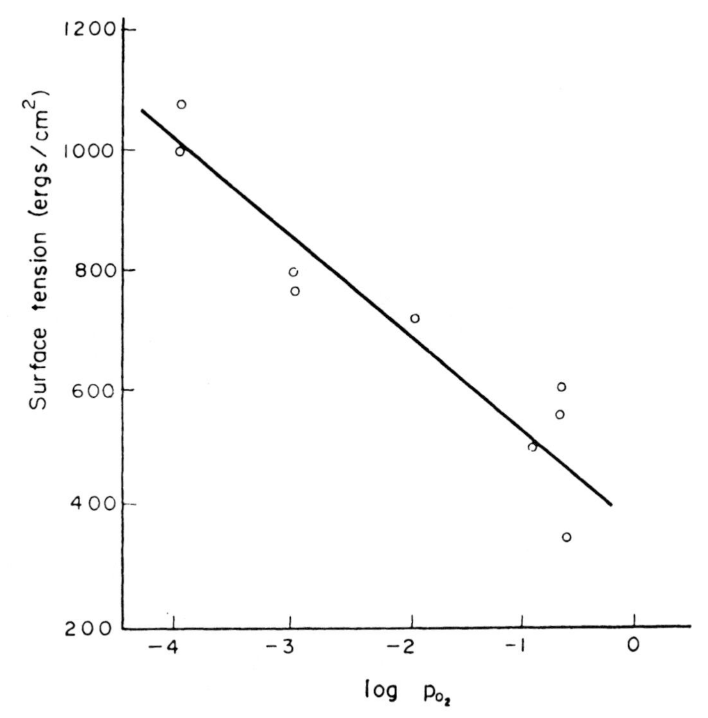

FIG. 4 Variation of γ_{SV} with oxygen partial pressure. (From Ref. 15, by permission, after F. H. Buttner, R. R. Funk, and H. Udin, *J. Phys. Chem. 56*:657 (1952)).

sible, values of interfacial energies should be measured via independent techniques under experimental conditions (e.g., atmosphere, temperature) identical to those employed in the wetting study. Also, if substrate dissolution, or the condensation of volatile liquid species on the solid surface, is occurring, the impact of these chemical changes on the appropriate interfacial energies must be assessed.

C. Adhesion

In many applications, the degree of wetting is secondary to the primary goal of achieving good adherence between the substrate and solidified liquid (glass). According to Hoge et al. [18], maximum adherence of a glass to metal occurs when the regions of both the glass and metal at the interface are saturated with the oxide of the lowest-valence cation of the

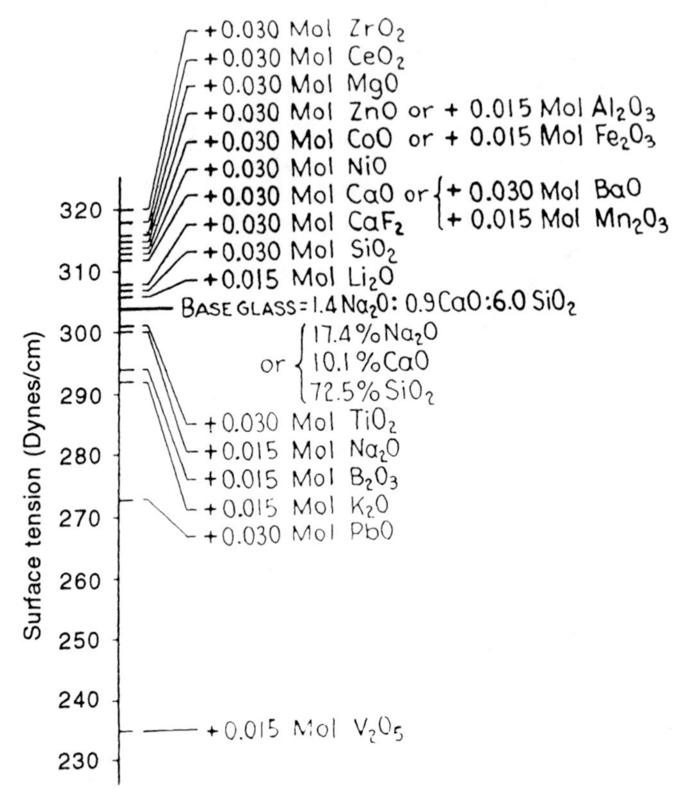

FIG. 5 Impact of oxide additions on γ_{LV} of a base glass having the composition 17 mol. % Na_2O-11% CaO-72% SiO_2. For the base liquid, γ_{LV} = 304 mN/m. (From Ref. 17, by permission.)

substrate metal. This equilibrium condition leads to a balance of the attractive forces exerted by the glass and metal on oxygen and metal ions in the interfacial region.

As discussed by Pask [19], chemical equilibrium implies the existence of an intermediate oxide layer at the interface, whose metal atoms are part of the metal structure and oxygen ions are part of the glass structure. This oxide layer may be of molecular dimensions and therefore difficult to detect. Saturation with respect to the liquid is required because, in its absence, the oxide layer may be dissolved by the liquid, resulting in the loss of adhesion.

As discussed earlier, the development of good chemical bonding at the solid/liquid interface occurs when a large fraction of the available bonding states at the liquid/vapor interface satisfy unsaturated bonds of the solid/vapor interface. In the limit of maximum compensation, in the absence of reaction, the contact angle decreases to zero. This means that small contact angles correspond to good bonding across the interface and should lead to good adhesion. Pask [19] notes that the adherence of glass to metals is good when the contact angle is less than ~30° under good vacuum conditions. This corresponds to a reduction of the surface energy of the metal by the glass by an amount equivalent to ~85% or more of the surface energy of the molten glass.

If one or both of the phases is not saturated with the intermediate oxide material, then poor adhesion results. However, under appropriate conditions, chemical reactions can occur, such as the reduction of species in the liquid, to produce this intermediate oxide. If the rate of reaction is faster than the diffusion of oxygen into the substrate or metal cation into the liquid, then local saturation at the interface will occur, resulting in good adhesion (and wetting). However, if chemical diffusion is fast, then saturation with the substrate/metal oxide may not occur until the entire droplet or substrate becomes saturated.

Also, physical factors may enhance adhesion. As examples, these factors include the development of surface roughness, where the improved adhesion results from an increase in the contact area, and the formation of interlocking dendrites, or other crystal shapes, of an intermediate compound at the interface.

III. REVIEW OF WETTING BEHAVIOR

A. Wetting on Metal Substrates

1. Nonreactive Systems

(a) Platinum. Early results obtained for the wetting of $Na_2O{\cdot}SiO_2$ on Pt provide a good example of both the influence of adsorbed gases on wetting behavior and the care with which the planning and execution of wetting studies need to be conducted. The work of Ellefson and Taylor [8,20] on the wetting characteristics of a molten $Na_2O{\cdot}SiO_2$ liquid on Pt showed that the equilibrium contact angle at 1000°C is dependent on atmosphere. A marked decrease in contact angle from 65–0° was observed when the atmosphere was changed from vacuum to oxygen.

In a study by Fulrath et al. [21], the pressure dependence of the wetting behavior of $Na_2O{\cdot}2SiO_2$ liquid on Pt exhibited an unusual pressure dependence. At 10^{-3} Pa, the contact angle was 15° from 900–1300°C. On increasing the pressure from 10^{-3}–10^{-1} Pa, the contact angle rose to 55°. Decreasing the pressure again resulted in the angle returning to 15°. At

TABLE 1 Effects of Atmosphere on Contact Angle of $Na_2O \cdot 2SiO_2$ on Platinum

Gas	1 Pa	10 Pa	100 Pa
O_2	15°	13°	10°
N_2	22°	21°	32°
CO_2	20°	29°	—
Air	17°[a,b]	—	37°[a], 22°[b]
He	—	—	24°[b]
H_2O	58°	63°	—
CO	40°	55°	66°
H_2	56°	51°	49°

Contact angle in vacuum (10^{-3} Pa) = 22° ± 3°.
[a] Absence of zirconium.
[b] Presence of zirconium.
Source: Ref. 24.

a pressure of 9×10^4 Pa, this same group had earlier reported a contact angle of 0° [22].

Volpe et al. [23] later clarified these results in an investigation of the effects of atmosphere on the wetting behavior of $Na_2O \cdot 2SiO_2$ on Pt and Au. By using a modified experimental apparatus, contact angles of 22° ± 3° were observed for molten $Na_2O \cdot 2SiO_2$ in vacuum on Pt at 1000°C. In the presence of water vapor, CO and H_2, at pressures of 1–100 Pa, the contact angle increased to values between 56 and 66°. This increase was attributed to a reduction in γ_{SV} via the adsorption of the gases on the Pt surface. When O_2 was introduced into the system, it caused a rapid decrease in the contact angle as it displaced the adsorbed gas on the Pt surface. As the O_2 pressure changed from 1–10^5 Pa, the contact angle changed from 15–6°. Apparently, early measurements of the contact angle as a function of pressure were plagued by the adventitious adsorption of water vapor or other gases that reduced γ_{SV}. Adventitious adsorption of O_2, which would have caused a decrease in the contact angle, did not occur because the test apparatus was heated by W and Mo elements, which naturally getter O_2.

The effects of atmosphere on the contact angle of liquid $Na_2O \cdot 2SiO_2$ on Pt are shown in Table 1. Note that the contact angle decreases slightly with increasing O_2 pressure. As noted by Volpe et al., this indicates that O_2 has a larger effect on γ_{SL} than γ_{SV},* since the expected reduction in

* As seen in Fig. 4, Udin [24] showed that oxygen lowered the surface energy of silver. Similar behavior would be expected for Pt since its affinity for oxygen is higher than Ag [21].

γ_{SV} due to O_2 adsorption would tend to increase the contact angle. These authors show that significant quantities of O_2 are absorbed by Pt at 1000°C and suggest that diffusion of this absorbed O_2 through the metal to the liquid/metal interface leads to the required modification of γ_{SL}.

An investigation of the contaminating effects of carbon was conducted by Holmquist and Pask [25], who explored the wetting behavior of sodium borate liquids on Pt. These researchers showed that carbon contamination on Pt increases the observed contact angles by reducing γ_{SV}. The contact angle of anhydrous B_2O_3 on Pt at 900°C in carbon-contaminated vacuum was 68°. As O_2 was admitted by leaking ambient air into the system, the contact angle rapidly decreased to 6°. This reduction was attributed to the removal of carbon via oxidation at the Pt surface and its replacement by adsorbed O_2. Identical results were obtained for 0.94 B_2O_3-0.06 Na_2O and 0.69 B_2O_3-0.31 Na_2O liquids. According to Holmquist and Pask, the lack of variation between liquids indicates that the Na_2O content of the glass does not significantly affect γ_{LV} and γ_{SL}. However, since γ_{LV} is known to more than double as the liquid composition is varied between pure B_2O_3 and the 69% B_2O_3 liquid [26], it is possible that volatilization losses of Na_2O during vacuum annealing drastically reduced the Na_2O content of the liquids. No composition analysis was performed on the sessile drops after wetting, so this explanation cannot be confirmed. It is unlikely that the majority of Na_2O is lost from the droplet, indicating that other explanations for these results may need to be explored.

Using Eq. (1) and assuming that the difference between the contact angle for B_2O_3 on Pt and carbon-contaminated Pt was due to effects on γ_{SV} and not γ_{LV}, Holmquist and Pask calculated the change in γ_{SV} to be 50 mN/m (taking γ_{LV} for B_2O_3 liquid at 900°C to be 80 mN/m). Due to the extremely low solubility of carbon in both the liquid and solid, it is likely that carbon contamination does not alter the value of γ_{SL}. Adherence of the droplet on carbon-contaminated samples was poor, whereas droplets with contact angles of 6° exhibited good adhesion.

Similar effects on Pt were noted by Nagesh et al. [27], who investigated the wetting behavior of a Pb borosilicate liquid (70 wt. % PbO-10% SiO_2-20% B_2O_3) on Pt in air, vacuum, and He. In air, this liquid had a contact angle of 2–3° at 700°C and exhibited excellent adhesion. In carbon-free vacuum, the contact angle was 58°, and in carbon-contaminated vacuum, it was 73°. In the latter cases, adhesion was poor. Nagesh et al. attributed the large contact angles observed in the carbon-contaminated furnace to a reduction in γ_{SV} via adsorption of carbon. Based on the observed contact angle of 73°, the value of γ_{SL} = 1905 mN/m for wetting in carbon-free vacuum and γ_{LV} = 180 mN/m, the reduction in γ_{SV} caused by carbon absorption was calculated to be 43 mN/m. This compares well with the value of 50 mN/m obtained by Holmquist and Pask [25] for carbon contamination in the B_2O_3-Pt system.

The results of Cronin et al. [28], discussed in detail in the next section, for the Pb borosilicate-Au system suggest, however, an alternate explanation. In this study, changes in the wetting behavior of a Pb borosilicate liquid on Au at 574°C, induced by changing the atmosphere from air to forming gas (N_2 + 3% H_2), were shown to be consistent with an increase in γ_{LV} caused by the presence of Pb at the liquid/vapor interface. In the reducing atmosphere, the liquid droplet darkened and developed a metallic sheen, suggesting the presence of metallic Pb at the liquid surface. Using the value of Nagesh et al. of the driving force for wetting of Au in air, and assuming that their driving force value for wetting in a vacuum also applies for wetting in forming gas, the results of Cronin et al. indicate an increase in γ_{LV} from 180 mN/m in air to 312 mN/m in forming gas.

Nagesh et al. had reported that their Pb borosilicate droplets became dark and shiny when melted in the graphite furnace, indicating the presence of metallic Pb in the liquid. If the value of 312 mN/m is used for γ_{LV} and it is assumed that carbon contamination does *not* alter either γ_{SV} or γ_{SL} from their values of 2000 and 1905 mN/m (obtained in the carbon-free vacuum), then the resulting contact angle would be 72°, almost identical to the measured value of 73°. Consequently, the increase in the contact angle of the Pb borosilicate liquid on Pt can be fully accounted for by assuming that the reduced Pb on the surface of the liquid increases γ_{LV}. Clearly, more work needs to be done to clarify the relative contributions of carbon contamination and changes in γ_{LV} to the wetting behavior in this system.

In another study, Holmquist and Pask [25] suggest that when a wetting liquid contains easily reduced cations, reactions can occur with Pt that promote spreading. During sessile-drop experiments conducted in "wet" atmospheric air (water vapor present), sodium borate liquids spread on the Pt substrates and excellent adhesion developed. Spreading and adhesion indicate the occurrence of an interfacial reaction and mutual saturation of the reaction product with the solid and liquid at the interface. The reactions proposed were the reduction of hydroxyls in the liquid structure by Pt and the formation of the metal oxide or hydroxide as

$$x(\mathrm{OH}^-)_{\mathrm{liq}} + \mathrm{Pt} \rightleftarrows \mathrm{PtO}_{x(\mathrm{liq})} + \frac{x}{2}\,\mathrm{H}_{2(\mathrm{liq})} \tag{6}$$

or

$$3(\mathrm{OH}^-)_{\mathrm{liq}} + \mathrm{Pt} \rightleftarrows \mathrm{Pt(OH)}_{2(\mathrm{liq})} + \mathrm{O}^{2-}_{\mathrm{liq}} + \frac{1}{2}\,\mathrm{H}_{2(\mathrm{liq})} \tag{7}$$

Insufficient data were obtained to distinguish between these two reaction schemes.

Since good adhesion develops between the liquid and Pt in water-free air atmospheres, the development of adhesion is not a clear indication of reaction. As noted earlier, Parikh [16] has shown that γ_{LV} of a Na_2O-CaO-SiO_2 liquid can be drastically reduced by the presence of water vapor. Noting that the contact angle obtained by Holmquist and Pask in "dry" air was 6° and assuming that γ_{SL} is not a function of water vapor pressure, it is noted that a reduction in γ_{LV} due to the presence of water vapor of <1 mN/m is required to initiate spreading (assuming γ_{LV} is 80 mN/m in dry air). Consequently, a reduction in γ_{LV} is likely responsible for the observation of spreading in the Holmquist and Pask study. The work of Tummala and Foster [29] on the wetting of borosilicate liquids on 99.5% Al_2O_3 ceramic in wet and dry atmospheres also supports this conclusion.

In a related study, Simhan et al. [30] measured the effects of air, steam, and glycol atmospheres on the wetting behavior of molten E-glass and borosilicate glass on Pt-20% Rh alloy. They attributed modification of the wetting behavior to changes in the negativity of the liquid surface via the introduction or elimination of nonbridging oxygens. The observed changes in the contact angle are, however, consistent with the previously reported variations in γ_{SV} for Pt and γ_{LV} of the glasses as functions of carbon contamination and water vapor pressure.

The results of Copley et al. [31] on the wetting behavior of molten E-glass on Pt group metals in the temperature range 1010–1180°C suggest that the wetting behavior is related to the affinity of the metal substrate for nonbridging oxygens in the melt. In this case, the degree of wetting would be determined by the competition for nonbridging oxygen ions between the substrate metal and the cations in the liquid.

In an extensive study [32], these researchers investigated this effect using contact-angle measurements on Pt of a series of silicate liquids. The liquid compositions maintained a 4:1 mole ratio of SiO_2 to Na_2O and contained 5, 10, or 20% additions of CaO, B_2O_3, TiO_2, ZrO_2, ZnO, Al_2O_3, and PbO_2. The effects of these oxide additions on the contact angles on Pt are shown Fig. 6. Additions of CaO and TiO_2 decreased the contact angle, whereas additions of B_2O_3, ZrO_2, and Al_2O_3 markedly increased θ.

Copley and Rivers attributed these observations to the formation of two types of ionic groups, caused by the breaking of bonds in the silicate liquids. In the broken Si—O— Si bond, the nonbridging oxygen stays with the anionic group, which then has an excess negative charge, whereas the remaining cationic group has excess positive charge.

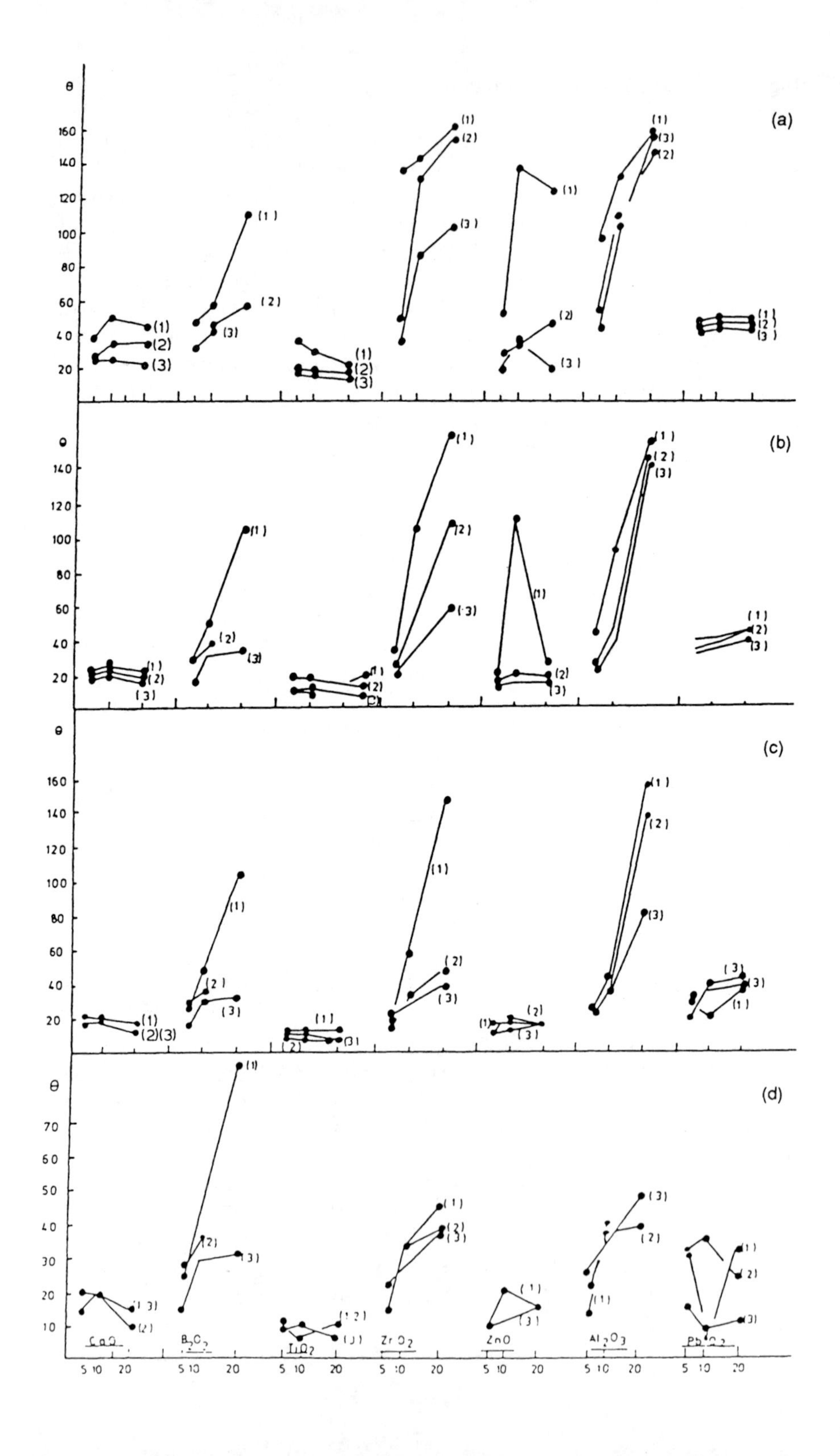
(a)
(b)
(c)
(d)
θ
CaO
B_2O_3
TiO_2
ZrO_2
ZnO
Al_2O_3
PbO_2
5 10 20

When the liquid is brought into contact with the Pt surface, the valence electrons of the Pt atoms seek to form bonding orbitals with the oxygen ions; this demand can best be satisfied by the anionic groups. However, the cations in the liquid also compete for the nonbridging oxygens. The resulting change that occurs in the negativity of the anionic group determines the degree of wetting and the contact angle. For example, when CaO is added to a Na_2O-SiO_2 liquid, the greater ionic field strength of the Ca^{2+}, compared to Na^+, reduces the negativity of the anionic group; hence, the liquid with Ca^{2+} wets the metal less than the glass with Na^+. A reduction in the positivity of the cationic group or an increase in the negativity of the anionic group decreases wetting.

Copley et al. developed equations to describe the degree of wetting in terms of the affinity of the cations within the liquid for oxygen. Figure 7 illustrates the relationship between the measured equilibrium contact angle θ and the affinity for oxygen as derived by these authors. The plot shows that wetting decreases (θ increases) as the demand for oxygen bonding by the cations in the liquid increases. B and Al were exceptions to this behavior.

(b) Gold. In a study of the wetting behavior of Na_2O-SiO_2 liquids on metal surfaces, Ellefson and Taylor measured contact angles of ~55° on Au both in the presence and absence of O_2 [8,20]. Fulrath et al. [21] confirmed this behavior by noting a static equilibrium contact angle of 60° in the same system. This angle was constant with changes in temperature from 900–1040°C and in pressure from 10^{-1}–10^{-3} Pa.

In later work, Volpe et al. [23] measured the contact angles formed by $Na_2O \cdot 2SiO_2$ liquid on Au at 1000°C under various atmospheres at a pressure of 1 Pa. Their results are shown in Table 2. Within experimental error, the contact angles did not change with increasing pressure of the gas atmosphere up to pressures of 100 Pa. The constancy of the contact angle is an indication that Au surfaces are inert and exhibit no tendency to adsorb any of these gases in this pressure range. The data also indicate that the surface tension of $Na_2O \cdot 2SiO_2$ liquid is not appreciably affected by these gases at 1000°C.

←

FIG. 6 Contact angles vs. composition of ternary silicate liquids on Pt. The liquid composition maintained a 1:4 molar ratio of Na_2O to SiO_2. Each column reports the θ values obtained for 5, 10, and 20 mol. % additions of the indicated oxide at temperatures of (curve 1) 1000°C, (curve 2) 1105°C, and (curve 3) 1180°C. Contact angles were measured at (a) 1 min, (b) 10 min, (c) 100 min, and (d) at equilibrium. (From Ref. 32, by permission.)

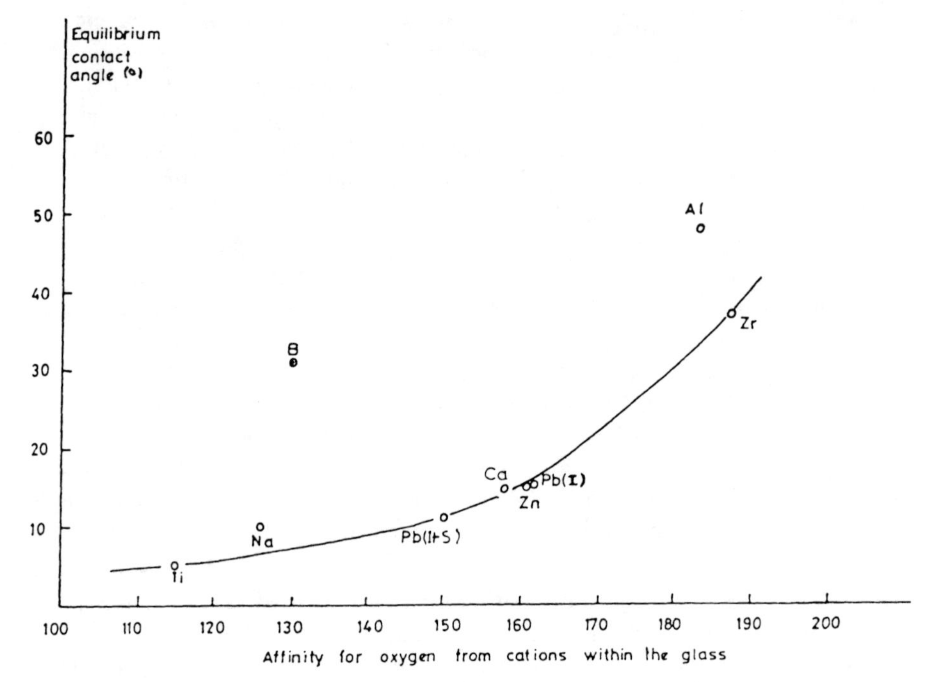

FIG. 7 Empirical relationship between the equilibrium contact angle and the affinity for oxygen from cations with the glass. (From Ref. 32, by permission.)

TABLE 2 Effects of Atmosphere on Contact Angle of $Na_2O{\cdot}2SiO_2$ on Gold

Gas	1 Pa	10 Pa	100 Pa
O_2	65°	65°	64°
CO_2	61°	60°	—
N_2	58°	—	62°
H_2O	66°	65°	65°
He	62°	63°	63°
H_2	64°	63°	66°
Air	62°	63°	62°

Contact angle in vacuum (10^{-3} Pa) = 63 ± 2°.
Source: Ref. 24.

In contrast, Tso and Pask [33] determined that the wetting behavior on Au can be influenced by O_2. They measured a 46° contact angle for a 0.94% B_2O_3-0.06% Na_2O liquid at 900°C on Au in vacuum. The same contact angle was observed in an ambient pressure of 10^5 Pa of N_2. The contact angle in O_2 was also 46° at low pressures. However, at ambient pressures (>2 Pa of O_2), the contact angle was 6°. Since it is known that the chemisorption of oxygen onto the surface of Au reduces γ_{SV} from 1370 mN/m to 1210 mN/m [24], this was taken as an indication that γ_{SL} decreased by 184 mN/m to a value of 1130 mN/m in the O_2 atmosphere.

At high oxygen pressures, when the contact angle was small, the liquid drop migrated until it reached the edge of the Au substrate, leaving behind a very thin film of liquid that adhered strongly to the Au. From this the authors concluded that both the film and Au substrate were saturated with gold oxide. With prolonged heating, the gold oxide diffused into the glass, resulting in loss of saturation of the interface and loss of adhesion.

Tso and Pask also explored the effects of carbon contamination on the wetting behavior of the 0.94% B_2O_3-0.06% Na_2O liquid on Au at 900°C. The Au substrate was found to be inert with respect to carbonaceous contaminants. Samples of Au vigorously pretreated to prevent carbon contamination and samples purposely treated with carbon gave the same contact angle of 46° in vacuum (10^{-3} Pa) at 900°C.

Based on their measurements, Tso and Pask reported values of the driving force for wetting to be 56 mN/m in vacuum and 80 mN/m in oxygen. These values indicate that mutual compensation of unsatisfied surface bonds occurs to a larger extent between Au saturated with gold oxide and the liquid than between metallic Au and the liquid. In oxygen, the driving force has reached the maximum allowable value (in the absence of dispersion energy contributions) of 80 mN/m, equal to γ_{LV}.

It is likely that the results of Volpe et al. [23] concerning the impact of O_2 pressure on the wetting of Au differ from those of Tso and Pask because of variations in experimental conditions. The measurements of Volpe et al. were made in a W-wound furnace whose elements would naturally getter oxygen; hence, this work was possibly conducted at O_2 pressures much less than the recorded values.

Nagesh et al. [27] explored the wetting behavior of a 70 wt. % PbO-10% SiO_2-20% B_2O_3 borosilicate liquid on Au in air, vacuum, and He. At 700°C the liquid formed a contact angle of 18° in air and 7° in vacuum. Although adherence was poor in vacuum, it was good in air, indicating the presence of an oxide layer at the solid/liquid interface. The driving forces for wetting in air and vacuum were calculated to be 171 and 179 mN/m. These results differ considerably from those obtained for Na_2O-B_2O_3 and $Na_2O{\cdot}2SiO_2$ liquids. Here the driving force term for wetting in

TABLE 3 Effect of Glass Composition on Contact Angle with Silver at 900°C

	30.8% soda		33.6% soda		36.9% soda	
Atmosphere	Test value	Avg.	Test value	Avg.	Test value	Avg.
Helium	71	71	66, 74	70	70, 70, 72, 70, 70	70
Hydrogen	73	73	76, 77	76	76, 72, 70, 70	72
Air	0	0	0	0	0	0

Source: Ref. 22.

vacuum is unusually large, larger than the corresponding value obtained in air. Note that both the air and vacuum values are approaching the value of γ_{LV} = 180 mN/m.

In a related study, Cronin et al. [28] demonstrated that the wetting behavior of Pb borosilicate liquids can be sensitive to variations in the atmosphere. The contact angle for a 73.5 wt. % PbO-19% SiO_2-7.5% B_2O_3 liquid on Au in air a 574°C was 17°. When the atmosphere was changed to forming gas, while the sample was at temperature, the three-phase line retracted to an equilibrium position, yielding a new contact angle of 55°. Changing the atmosphere back to air caused a decrease in the contact angle to the original value of 17°. As noted in an earlier section, the retraction of the three-phase line in the forming gas atmosphere is attributed to an increase in γ_{LV} due to the formation of metallic Pb on the surface of the liquid. Assuming that γ_{SV} = 1210 mN/m [24] and γ_{SL} = 1039 mN/m [27] for the Pb borosilicate-Au system in air, and that γ_{SV} = 1370 mN/m and γ_{SL} = 1192 mN/m for this system in forming gas,* γ_{LV} is calculated to increase from 180 mN/m in air to 312 mN/m in forming gas. The surface tension of molten Pb at 574°C is 426 mN/m [34]. Since metallic Pb raises the surface tension of the liquid, it would be expected to migrate away from the liquid/vapor interface. However, the solubility of reduced Pb in the liquid is expected to be quite low. In light of these complications, the value of 312 mN/m for γ_{LV} in a reducing atmosphere compares favorably with the value of γ_{LV} for molten Pb.

(c) Silver. Zackay et al. [22] measured the contact angles observed in He, H_2, and air atmospheres for Na_2O-SiO_2 liquids containing 30.8 wt. %, 33.6%, and 36.9% Na_2O on Ag. Their results are shown in Table 3.

* γ_{SV} = 1370 mN/m was determined by Udin for Au in He atmosphere. γ_{SL} = 1039 mN/m was determined by Nagesh et al. in vacuum.

TABLE 4 Effect of Metal Composition and Atmosphere on Contact Angle with Sodium Silicate Glass at 900°C

Atmosphere	Copper	Nickel	Palladium
Helium	60	55	55
Hydrogen	60	60	40
Air	0	0	25
Oxygen	0	0	20

Source: Ref. 22.

In all cases, spreading is observed for the liquids in air, indicating the presence of Ag_2O at the interface. The lack of a compositional dependence on the contact angles in He and H_2 indicates that neither γ_{SL}, nor γ_{LV} vary much with Na_2O content in the liquid over this composition range. The lack of compositional variability in γ_{LV} is supported by the surface tension measurements of Shartsis and Spinner [35], which indicate that γ_{LV} varies by only 1–2% in this range. In this same study, no dependence was found between the contact angle and grain size of the metal. Single-crystal and polycrystalline specimens of Ag (as well as Au) gave similar contact angles.

(d) Other Metals. Contact angles formed by $Na_2O \cdot 2SiO_2$ liquid, at 900°C, for Cu, Ni, and Pd are shown in Table 4. Because the data in both Tables 3 and 4 were measured using environments that may have contained adventitious gases, caution must be exercised in the use of these data. As noted by Zackay et al. [22], the reduction in the contact angle on Pd when the atmosphere is changed from He to H_2 is particularly interesting in light of the strong affinity of Pd for H_2 and its tendency to form hydrides.

2. Reactive Systems

Considerable experimental evidence indicates that enhanced wetting and spreading are observed when reactions occur at the solid/liquid interface [36]. Because of the larger range of observable contact angles, most data in support of the role of interfacial reactions on wetting behavior have been obtained on liquid metal/oxide substrate systems. Despite the preponderance of data that demonstrate this phenomenon, studies designed to identify the major factors underlying the contribution of interfacial reactions to wetting behavior in oxide liquid/metal substrate systems are scarce.

Pask [11] notes that the Gibbs energy change of the reaction only contributes to the driving force for wetting when the substrate is an active participant in the reaction. To participate actively, the composition of the substrate must change, or it must react with liquid constituents to form a new phase. Figure 8, for example, shows the phase diagram for the Cu-Ag system, with four sessile drop configurations. For this system, Pask proposed that when droplets of composition C or D are placed on pure Cu at 900°C, Ag will diffuse into the substrate and the Gibbs energy of this dissolution reaction will contribute to the driving force for wetting. However, when droplet D is placed on an Ag-saturated Cu substrate of composition B, the Gibbs energy of the dissolution of $Cu(Ag)_{ss}$ into the liquid does not contribute to the driving force for wetting because the substrate plays a passive role in the reaction. Because liquid C and solid B are in equilibrium at 900°C, no reaction occurs when they are placed in contact, and a large contact angle should be observed.

Sharps et al. [36] obtained experimental verification of some of these scenarios using sessile drop experiments in the Cu-Ag system. In this study, a saturated Cu-Ag liquid was observed to spread on pure Cu and pure Ag substrates, but exhibited acute contact angles on saturated $Cu(Ag)_{ss}$ and $Ag(Cu)_{ss}$ substrates. Results representing the case of a reactive liquid on a passive substrate were, however, inconclusive.

Aksay et al. [9] discuss both the thermodynamic and kinetic factors that influence the potential contribution of the interfacial reaction to the driving force. When the substrate is an active participant, dG_R can facilitate spreading only if the flow rate of the spreading drop is faster than the rate at which the reaction products can diffuse through the solid or along the substrate surface. Under these conditions, continuous extension of the three-phase line (spreading) leads to reductions in Gibbs energy in the vicinity of the droplet periphery as the liquid encounters fresh, unreacted substrate. If highly mobile reaction products diffuse into the substrate *ahead* of the droplet, they will stall the interfacial reaction, the dG_R term will no longer contribute to the driving force for wetting, and the contact angle will increase to a new value consistent with the altered interfacial chemistries.

When the substrate is a passive participant the composition of the liquid, not the solid, changes during the reaction. In this case, minor fluctuations in the location of the droplet periphery do not cause significant reductions in Gibbs energy because the substrate regions behind and ahead of the droplet periphery have the same composition. Consequently, the Gibbs energy associated with the ongoing substrate dissolution process does not contribute to the driving force for wetting. As suggested by Aksay et al. [9], however, during the very early stages of wetting, the interfacial reaction may contribute to spreading if *unreacted* liquid can be cycled into the region of the three-phase line as by convection in the

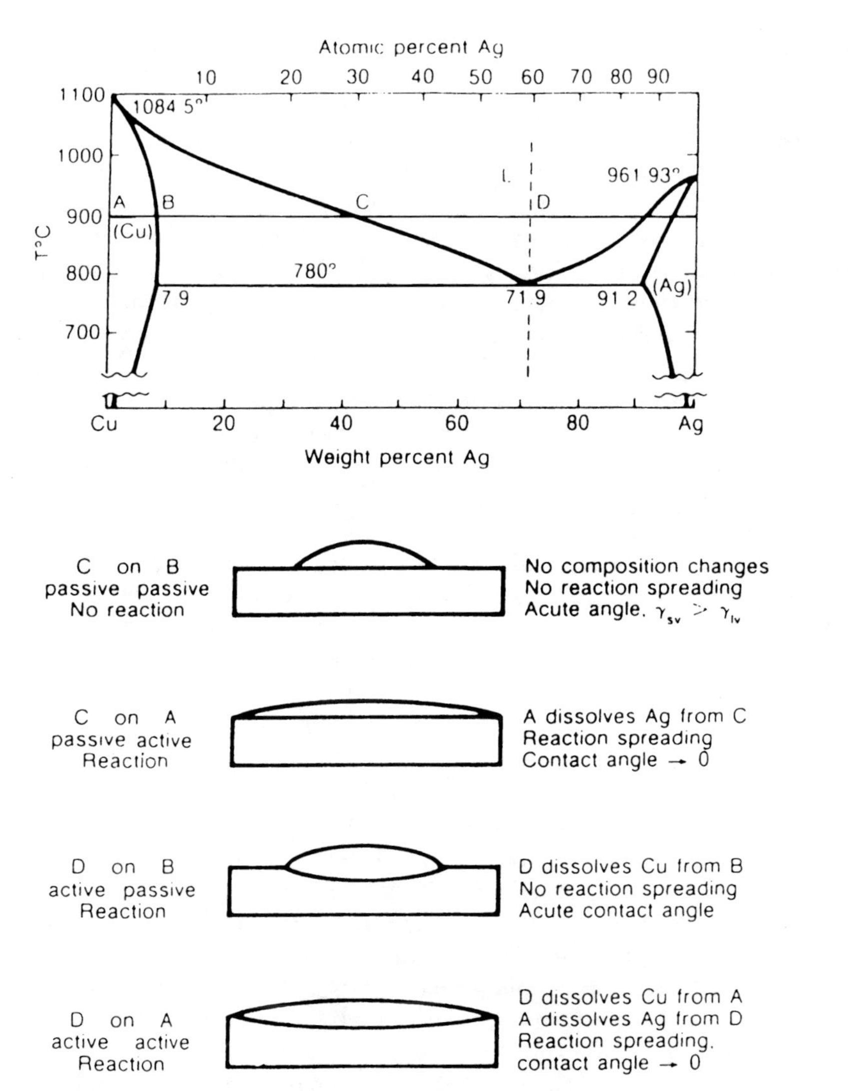

FIG. 8 Stable-phase equilibrium diagram for Cu and Ag with sessile drop examples. In all cases, $\gamma_{SV} > \gamma_{LV}$. See text. (From Ref. 11, by permission.)

liquid. In this case, a short-lived reduction in the contact angle may be observed until diffusion gradients of the reaction products become well established in the liquid phase at the three-phase line.

As noted by Aksay et al., two components contribute to dG_R. Consider a near-surface region of the substrate that has just been exposed to the reactive liquid. Here a localized reduction in the Gibbs energy dG_{chem} occurs due to the specific reaction occurring in the near-surface region. Also present is a contribution due to the modification of the strength or number of unsatisfied bonding states at the surface of the substrate phase dG_γ. With time, the reaction interface moves into the substrate phase, away from the solid/liquid interface, and the dG_{chem} contribution decreases to zero. As a result, dG_R steadily increases and finally stabilizes at dG_γ. Note that this behavior describes the time dependence of dG_R at a constant location in the solid/liquid interface. At the droplet periphery, the liquid continues to encounter fresh substrate; dG_R still includes both contributions to the driving force.

Little insight has been provided by past wetting studies into the relative magnitudes of the dG_{chem} and dG_γ terms. In the case of solution formation at the interface, dG_{chem} could be quite small, compared to its magnitude when the chemical reaction at the interface involves compound formation. For example, dG_{chem} would range between zero and several hundred mN/m for the formation of an ideal solution in an interfacial region that is several nm thick. The magnitude of dG_{chem} for intermediate compound formation may be larger by as much as a factor of 10 [8]. Regardless, dG_γ alone may be sufficient to induce complete spreading.

(a) Iron. The system that provides the greatest insight into the potential impact of substrate/liquid reactions on wetting is the $Fe\text{-}Na_2O\cdot 2SiO_2$ system. This system has been studied extensively by Pask and his associates; their work is reviewed below. These studies also reveal the susceptibility of wetting behavior to atmosphere and substrate effects, including the partial pressures of O_2, $P(O_2)$, and Na vapor, P(Na), and the initial activity of FeO, a(FeO), in the Fe substrate.

In their early work, Cline et al. [37] demonstrated that molten $Na_2O\cdot 2SiO_2$ wets the surface of FeO better than that of metallic Fe. On samples of Armco iron that had been preoxidized to form a 5-μm thick layer of FeO, $Na_2O\cdot 2SiO_2$ initially formed a contact angle of 24° at 900°C in vacuum. As the glass dissolved the oxide and contacted clean metal, the angle rose to ~55°. The molten glass also exhibited contact angles of 55° on unoxidized Armco iron at temperatures between 900 and 1000°C in vacuum. The Armco iron contained small precipitates of FeO; hence, the activity of FeO at the solid/liquid and solid/vapor interfaces was unity in all samples. Also, although not reported, the vacuum atmosphere likely contained a $P(O_2)$ that was slightly larger than that required to reduce FeO (based on data provided in later work by Pask). Consequently, the activity of FeO remained unchanged at these interfaces during the measurements.

It was also observed that the glass developed a greenish-blue color, and a loss in total sample weight occurred, indicating that a reaction had taken place between Fe and the liquid. In a related study, Hagan and Ravitz [38] investigated reactions occurring in couples of metallic Fe and molten $Na_2O{\cdot}2SiO_2$. They determined that under high vacuum, the oxide liquid begins to react with Fe powder, at an appreciable rate, at ~940°C. At 950–1000°C, the Fe metal reduces Na in the glass to produce FeO and Na vapor, which escapes from the liquid and causes the observed weight loss.

The importance of this and other reactions on wetting in this system was further investigated by Tomsia and Pask [39]. Their results show that molten $Na_2O{\cdot}2SiO_2$ exhibits complete wetting (contact angle of 0°) in a graphite furnace and forms small contact angles (around 10°) in an Al_2O_3 furnace on Marz A iron after several hours at 1000°C. The graphite furnace possessed an estimated $PO_2 \approx 10^{-17}$ Pa, below the dissociation pressures of both FeO (1.5×10^{-10} Pa) and Fe_2O_3 (6.4×10^{-10} Pa) at 1000°C. The Al_2O_3 furnace had $PO_2 \approx 10^{-7}$ Pa, which is above the dissociation pressure of both FeO and Fe_2O_3. Details of the observed time dependencies of the contact angle and Na loss at 1000°C are shown in Fig. 9. At all times, the contact angle is lower and the weight loss larger for samples in the graphite furnace compared to those in the Al_2O_3 furnace. Most if not all of the Na in the glass is lost as the wetting behavior evolves.

In a repeat of the earlier work by Cline et al. [37], the contact angle of the molten glass on Armco iron at 1000°C after 1 h in the graphite furnace was 53°; in the Al_2O_3 furnace it was 60°. A Marz A iron sample stored for 8 mon at room temperature in a vacuum desiccator with $P(O_2)$ ~1 Pa also exhibited a contact angle of ~55° in both furnaces. When this same sample was annealed for 18 h at 1100°C in the graphite furnace prior to deposition of the molten glass, the original wetting behavior of Marz A iron (contact angles of 0° in the graphite furnace and small angles in the Al_2O_3 furnace) was restored. As pointed out by Cline et al., these results clearly indicate that variations in the quantity of dissolved O_2 (or activity of FeO) in the Fe substrate can drastically modify wetting behavior.

Tomsia and Pask showed that complete wetting of the Fe substrate is observed when thermodynamic conditions permit the substrate to participate in a chemical reaction with the $Na_2O{\cdot}2SiO_2$ liquid. As previously

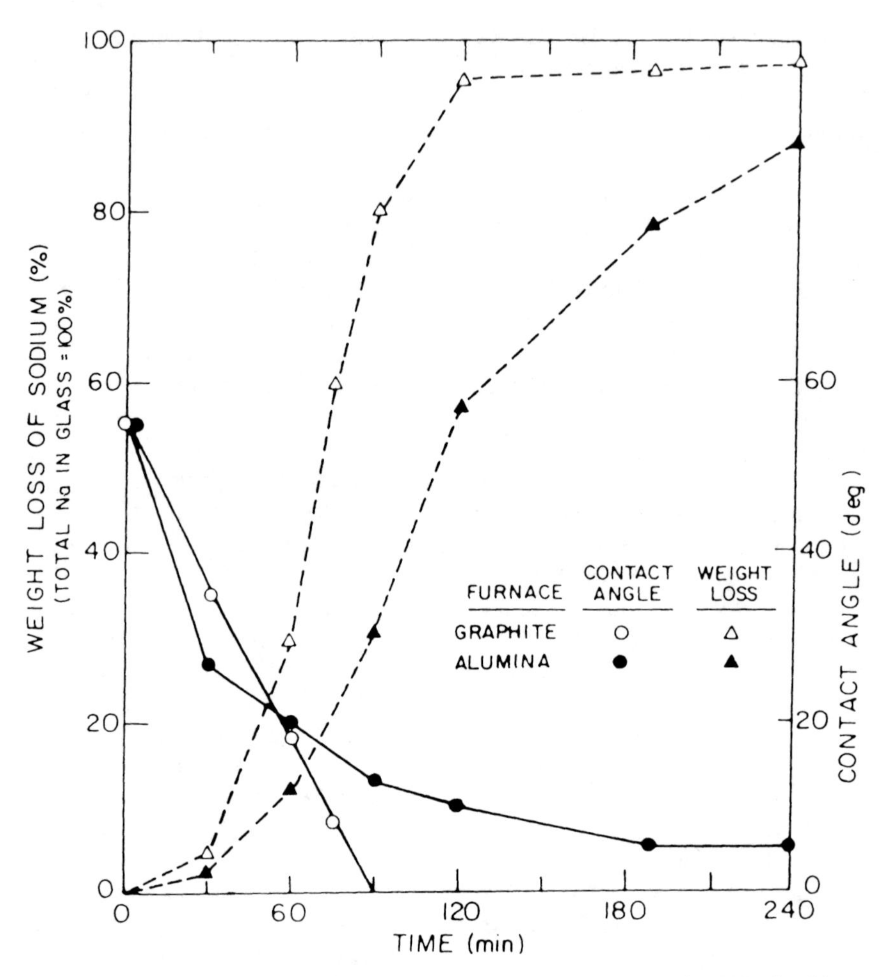

FIG. 9 Contact angles and weight loss of total Na from $Na_2O{\cdot}2SiO_2$ liquid on Marz A iron vs. time from 10–250 min at 1000°C and 2.6×10^{-4} Pa. (From Ref. 39, by permission.)

discussed, the resulting negative Gibbs energy of the reaction would contribute to the driving force for wetting and promote spreading. Tomsia and Pask propose that the substrate is involved in two relevant reactions: (1) oxidation of Fe by reduction of Na in the molten glass and (2) oxidation of ferrous ions in the liquid by reduction of Na in the liquid.

(1) The oxidation of Fe is a two-step process that includes the reduction of Na in the glass to form FeO at the solid/liquid interface and

subsequent mixing of this FeO with the $Na_2O \cdot SiO_2$ liquid to form a Na_2O-FeO-$SiO)_2$ liquid. Tomsia and Pask assert that since the Gibbs energy change in the mixing step is negative, the controlling reaction is the oxidation of Fe to form interfacial FeO

$$x\text{Fe}_s + \text{Na}_2\text{O}_{\text{liq}} \rightleftarrows x\text{FeO}_{\text{int}} + 2x\text{Na}_g \uparrow \tag{8}$$

with the accompanying ΔG

$$\Delta G = \Delta G^\circ + RT \ln \frac{a(\text{FeO})_{\text{int}}^x \, P(\text{Na})^{2x}}{a(\text{Na}_2\text{O})_{\text{liq}}^x} \tag{9}$$

ΔG° at 1000°C is positive, so this reaction can occur only when the equilibrium constant in the above equation is sufficiently smaller than unity. This requires $a(\text{FeO})_{\text{int}}$ and $P(\text{Na})$ to be small and $a(\text{Na}_2\text{O})_{\text{liq}}$ to be large. Under these conditions, oxygen from the FeO formed at the interface may diffuse into the substrate and change its composition. Thus, according to Tomsia and Pask, the substrate plays an active role in the reaction; the resulting Gibbs energy change will contribute to the driving force for wetting. If $a(\text{FeO})_{\text{subs}}$ is unity, then the Fe substrate is saturated with FeO and can no longer change composition upon reaction. Further generation of FeO merely causes dissolution of the substrate.

(2) After the dissolution of FeO in the glass, the ferrous ions can be oxidized as

$$2\text{FeO}_{\text{liq}} + \text{Na}_2\text{O}_{\text{liq}} \leftrightarrows \text{Fe}_2\text{O}_{3(\text{liq})} + 2\text{Na}_g \uparrow \tag{10}$$

with a corresponding ΔG

$$\Delta G = \Delta G^\circ + RT \ln \frac{a(\text{Fe}_2\text{O}_3)_{\text{liq}} \, P(\text{Na})^2}{a(\text{FeO})_{\text{liq}}^2 \, a(\text{Na}_2\text{O})_{\text{liq}}} \tag{11}$$

Again, the standard Gibbs energy change at 1000°C for this reaction is positive, so it will only occur for low $P(\text{Na})$, high $a(\text{Na}_2\text{O})_{\text{liq}}$, and a small ratio of $a(\text{Fe}_2\text{O}_3)_{\text{liq}}/a(\text{FeO})_{\text{liq}}$.

The substrate does not participate in this reaction, so the Gibbs energy change does not contribute to wetting. Once ferric ions are present in the glass, however, they can react with the substrate to produce ferrous ions once again

$$x\text{Fe}_s + x\text{Fe}_2\text{O}_{3(\text{liq})} \rightleftarrows (3x - y)\text{FeO}_{\text{liq}} + y\text{FeO}_{\text{int}} \tag{12}$$

with the accompanying ΔG

$$\Delta G = \Delta G^\circ + RT \ln \frac{a(\text{FeO})_{\text{liq}}^{(3x-y)} \, a(\text{FeO})_{\text{int}}^y}{a(\text{Fe}_2\text{O}_3)_{\text{liq}}^x} \tag{13}$$

The Gibbs energy for this reaction is negative at 1000°C and so $Fe_2O_{3(liq)}$ reacts favorably with the substrate to produce FeO either in the liquid or at the interface. Note that reaction (10) imposes the necessary, but not sufficient, conditions under which the Gibbs energy change of reaction (12) can contribute to the driving force for wetting. An additional condition requires that $a(FeO)_{int} < 1$. Then $y > 0$ in reaction (12), indicating that the substrate incorporates additional oxygen (via solution of FeO) and actively participates in the reaction. When $a(FeO)_{int} = 1$, $y = 0$; substrate dissolution, with no change in composition, occurs upon further reaction. In this case, the Gibbs energy change does not contribute to the driving force for wetting.

Based on these reactions, Tomsia and Pask argue that $Na_2O \cdot 2SiO_2$ liquid forms large contact angles on Armco iron because no Gibbs energy contribution to the driving force for wetting is present. Reaction (8) cannot occur because $a(FeO)_{int}$ is unity due to the presence of FeO precipitates in the substrate. Also, while reactions (10) and (12) occur, causing a gradual replacement of FeO for Na_2O in the liquid, the substrate does not change composition. Consequently, reaction (12) does not promote wetting. Similar behavior is anticipated for other Fe substrates in which a(FeO) in the substrate approaches unity.

Wetting is promoted on the Marz A iron in the Al_2O_3 and graphite furnaces because both reactions (8) and (12) contribute their change in Gibbs energy to the driving force for wetting. [This necessarily assumes that a(FeO) is much less than unity in the Marz A substrate at the beginning of the experiment.] The magnitude of the contribution is principally determined by the P(Na), which was equal in both furnaces, and $a(FeO)_{int}$. Since $P(O_2)$ in the Al_2O_3 furnace is larger than the dissociation pressure of FeO, the activity of O_2 in the substrate, and thus of FeO in the metal and at the interface, is likely larger (but still less than unity) than the oxygen activity of the substrate in the graphite furnace. Hence, the magnitude of the Gibbs energy contribution to the driving force, as determined by Eqs. (9) and (13), is larger for the droplet in the graphite furnace, and complete wetting is achieved in a shorter time.

When 10% FeO or Fe_2O_3 was added to the $Na_2O \cdot 2SiO_2$, the liquid completely spread at a rate that was *faster* than that of the pure liquid, in both the graphite and Al_2O_3 furnaces. Both the direct addition of Fe_2O_3 and formation of Fe_2O_3 via reaction (10) raise $a(Fe_2O_3)$, increase the Gibbs energy change associated with the reduction of Fe_2O_3 at the substrate interface [Eq. (13)], and result in a larger dG_{chem} contribution to the dG_R. Also, the addition of FeO or reduction of Fe_2O_3 additions increase a(FeO) in the substrate at the interface. The presence of O_2 in the

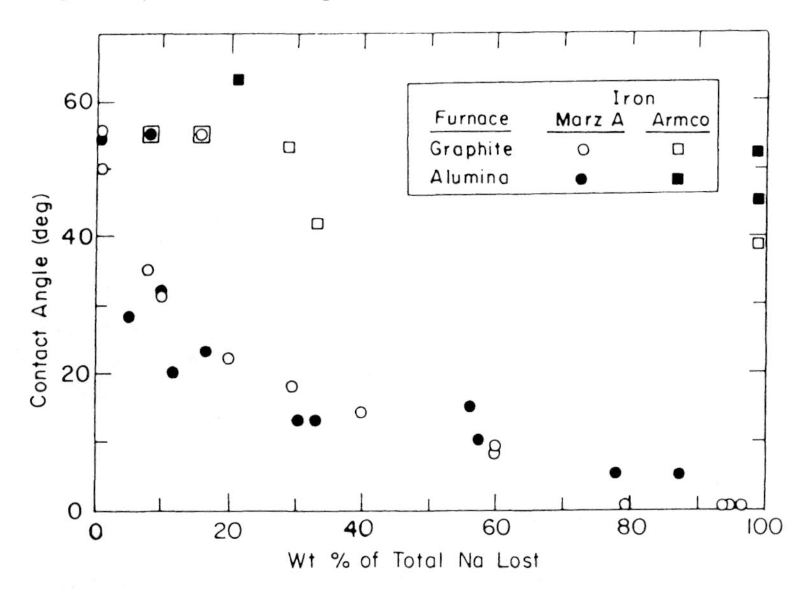

FIG. 10 Contact angles vs. weight loss of total Na for $Na_2O{\cdot}2SiO_2$ liquid on Marz A and Armco irons in the graphite and Al_2O_3 furnaces at 1000°C. (From Ref. 39, by permission.)

substrate under the droplet reduces γ_{SL} but not γ_{SV}, causing an increase in dG_γ.

Insight into the relative contributions of dG_{chem} and dG_γ to dG_R can be obtained by examining the composition dependence of the contact angles for pure $Na_2O{\cdot}2SiO_2$. When the contact-angle data from Fig. 9 are replotted as a function of Na loss (Fig. 10), the points for the graphite and Al_2O_3 furnace fall on a single curve. Tomsia and Pask argue that the change in composition of the glass is the critical factor in determining both the change in driving force for wetting and the magnitude of dG_R contributions to this driving force. However, these results also suggest that the dG_γ component of dG_R may dominate the change in driving force. The magnitude of the dG_{chem} components should be a function of the furnace atmosphere. The presence of a one-to-one correspondence between the contact angles (interfacial energies) and the glass composition, regardless of furnace atmosphere, supports this conjecture.

Despite the development of much understanding in this system, the unusual observations of Hoge et al. [18] remain to be explained. They reported that the addition of 9.1% FeO to $Na_2O{\cdot}2SiO_2$ results in complete spreading on an Fe substrate after 2 h at 1000°C. In contrast, pure $Na_2O \cdot 2SiO_2$ exhibited a 55° contact angle on the same substrate under the same conditions, indicating that $a(FeO)_{sub}$ is unity. Under these conditions, the substrate cannot actively participate in any of the previously discussed reactions, regardless of the FeO content of the liquid. The current model cannot account for the observed spreading in the FeO-containing liquid. Hoge et al. attributed this behavior to variations in $a(Na_2O)_{liq}$ as a function of the O/Si ratio in the liquid. Further work needs to be done to clarify this issue.

In a study of the wetting and adhesion behavior of FeO-containing $Na_2O \cdot SiO_2$ liquids in vacuum at 1000°C, Adams and Pask [40] demonstrated that a glass saturated with FeO when in contact with Fe showed contact angles of ~22° and developed good adherence. The work of Barom and Bhat supports this observation. Barom [41,42] studied the diffusion of Fe in $Na_2O \cdot 2SiO_2$ liquid and showed that no adherence developed at glass/metal interfaces when the concentration of FeO in the liquid was low. In a series of push-off tests, Bhat [43] showed that $Na_2O \cdot SiO_2$ compositions containing low concentrations of FeO exhibited only weak physical adherence to steel, whereas those that were nearly saturated with FeO tended to adhere chemically.

In order to improve adhesion, "adherence oxides" are often added to a glass. Barom et al. [4,45] examined the role of CoO as an adherence oxide addition in the $FeO\text{-}Na_2O \cdot 2SiO_2$ system. Because CoO is easily reduced by Fe to form FeO, a high concentration of FeO is maintained at the interface, and adhesion is greatly improved. The subsequent formation of an $Fe_{(1-x)}Co_x$ alloy consumes the metallic Co, allowing the continuation of the CoO reduction reaction to produce additional FeO.

Precipitation of the $Fe_{(1-x)}Co_x$ alloy in the form of dendrites also provides an electronic transport mechanism for atmospheric oxidation of the metal. The effectiveness of this process depends, however, on the glass composition and degree of preoxidation of the metal. The atmospheric oxidation of the metal produces a rough surface that, in combination with the dendrites, contributes to mechanical adherence and becomes especially important with the loss of phase equilibrium, and thus chemical bonding, at the interface.

Brennan et al. [46] applied these concepts to $Na_2O \cdot 2SiO_2$ containing varying amounts of FeO, CoO, or NiO in contact with Fe, Co, Ni, Ni-Fe, or Ni-Co substrates under several atmospheric conditions. They showed that reactions involving the formation of metallic precipitates (e.g., Co-Fe and Ni-Fe dendrites) or reduction of multivalent cations to a lower valence whose standard free energies are positive (e.g., Fe^{3+} to Fe^{2+}) took place under all experimental conditions. In the absence of reactions involving the adherence oxide, maximum adherence was observed when the activity of the metal oxide, $a(MO)$, at the interface approached unity. In the absence of saturation, all redox reactions contributed to the development of adherence because they introduce substrate cations into the interface, thereby increasing the $a(MO)_{int}$, and lead to the formation of interlocking dendrites at the interface.

(b) Chromium. The results obtained by Tomsia et al. [47] in a study of the wetting and adhesion of $Na_2O{\cdot}2SiO_2$ liquid on Cr are interesting in the differences they offer compared to the wetting behavior of this same liquid on Fe. The time dependencies of the contact angles and associated weight losses for this liquid on Cr substrates at 1000°C in an Al_2O_3 furnace (furnace A − $PO_2 = 1 \times 10^{-5}$ Pa) and graphite furnace (furnace G − $PO_2 = 1 \times 10^{-15}$ Pa) are shown in Fig. 11. In contrast with the results obtained for Fe, better wetting is observed in the Al_2O_3 furnace. Also, because Cr is highly reactive, it can react with either Na_2O or SiO_2, depending on local environments in different regions of the sessile drop.

In all regions of the drop, Cr reduces Na_2O via the following reaction:

$$x\mathrm{Cr}_s + \mathrm{Na_2Si_2O_{5(liq)}} \rightleftarrows \mathrm{Na_{2-2x}CrSi_2O_{5(liq)}} + 2x\mathrm{Na}_g\uparrow \tag{14}$$

The addition of Cr^{2+} ions imparts a blue color to the glass. This reaction consists of two parts: the formation of CrO at the interface and its subsequent diffusion and mixing into the glass. Because Tomsia et al. do not detect adhesion as a result of this reaction, the interface cannot be saturated with CrO. Consequently, the diffusion step must be fast, and reaction (14) is controlled by the reduction process

$$\mathrm{Cr}_s + \mathrm{Na_2O_{liq}} \rightleftarrows \mathrm{CrO_{int}} + 2\mathrm{Na}_g\uparrow \tag{15}$$

where ΔG for the reaction is given by

$$\Delta G = \Delta G^\circ - RT \ln \frac{a(\mathrm{CrO})_{int}\, P(\mathrm{Na})^2}{a(\mathrm{Cr})_s\, a(\mathrm{Na_2O})_{liq}} \tag{16}$$

ΔG° for this reaction is negative at 1000°C and so it is favorable under conditions of unit activities.

In the peripheral regions of the droplet, the Na_g is able to escape because the droplet is thin. This promotes reaction (15), which continues until the CrO content in the liquid is sufficiently large to initiate the following reaction:

$$2\mathrm{CrO_{liq}} + \mathrm{Na_2O_{liq}} \rightleftarrows \mathrm{Cr_2O_{3(liq)}} + 2\mathrm{Na}_g\uparrow \tag{17}$$

Note that low $P(\mathrm{Na})$ also promotes this reaction. The Cr^{3+} ions being

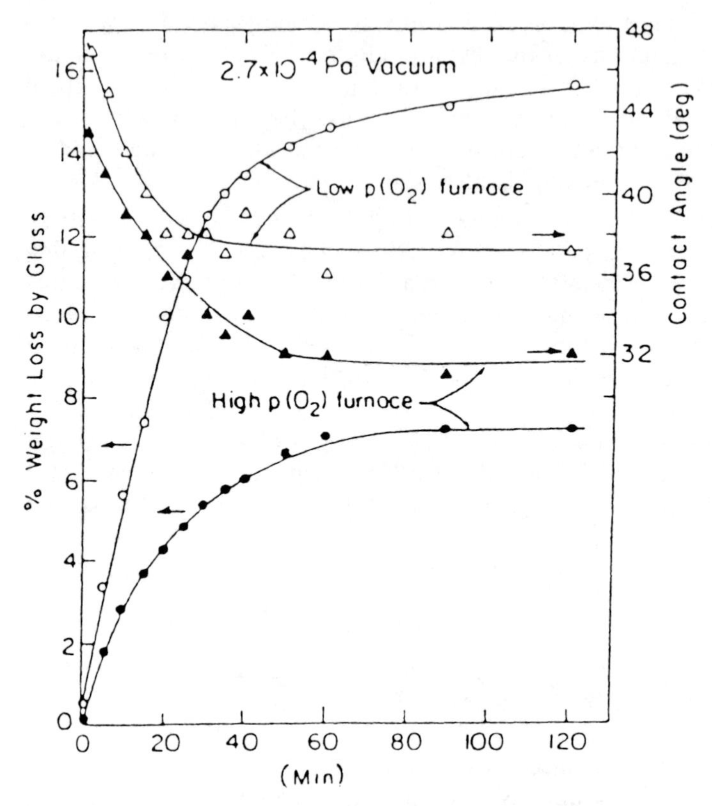

FIG. 11 Contact angles and weight losses of $Na_2O \cdot SiO_2$ liquid on Cr vs. time at 1000°C and 2.7×10^{-4} Pa vacuum in furnaces G and A. (From Ref. 47, by permission.)

formed via this reaction turn the glass green and promote adhesion. Adhesion develops (in the peripheral region) because Cr_2O_3 is much less soluble than CrO and easily saturates both the solid and liquid.

The central portions of the droplet behave differently. Because the droplet is thick in this region, the P(Na) rapidly builds up and stalls reactions (15) and (17). As a result, the Cr begins to reduce SiO_2 and form a CrSi alloy via the reaction

$$(1 + 2x)Cr_s + xSiO_{2(liq)} \rightleftarrows CrSi_{x(s)} + 2xCrO_{liq} \tag{18}$$

Thermodynamic data are not available to calculate the standard Gibbs energy of this reaction $\Delta G°$, but it is considered negative because of the expected negative Gibbs energy of formation for the $CrSi_x$ alloy [47]. The alloy forms metallic-appearing platelets and wirelike crystals at the interface. The surrounding glass is blue, due to the dissolved CrO in the liquid. Also, adhesion is poor under these conditions.

As the CrO content in the drop increases, reaction (18) becomes less favorable. At some concentration of CrO, the appearance of a greenish glass at the interface suggests the formation of Cr^{3+} by the reaction

$$2CrO_{liq} + SiO_{2(liq)} \rightleftarrows Cr_2O_{3(liq)} + SiO_g \uparrow \tag{19}$$

Tomsia et al. [47] calculate ΔG°_{1000} to be 440 kJ/mol under standard conditions of unit activities, but the reaction proceeds because the low ambient pressure in the furnace maintains a low $P(SiO)$. Eventually, Cr_2O_3 precipitates form as nodules on the CrSi phase, and good adhesion is obtained.

As seen in Fig. 11, the droplet in the graphite furnace exhibits more weight loss and attains a steady value for the contact angle at an earlier time than the droplet in the Al_2O_3 furnace. The higher $P(O_2)$ in the Al_2O_3 furnace produces a larger $a(CrO)$ at the interface. This inhibits reaction (15), causing a delay in the formation of and subsequent saturation of the liquid with Cr_2O_3. Good adhesion, caused by saturation of the liquid with Cr_2O_3, was achieved in both furnaces when the contact angle reached steady-state conditions.

(c) Silver. Nagesh et al. [27] explored the wetting and adhesion behavior of a Pb borosilicate liquid of composition 70 wt. % PbO-10 wt. % SiO_2-20 wt. % B_2O_3 on Ag at 700°C. In He the contact angles showed wide variations, in the range of 5–48°, and adhesion was poor. In air the contact angle was nominally zero, spreading occurred, and adhesion was excellent due to the presence of a thin oxide film on the metal surface. In vacuum, contact angles in the range of 2–5° and good adherence were observed due to a redox reaction that the authors postulated to be

$$2Ag_s + PbO_{liq} \rightleftarrows Ag_2O_{int} + Pb_g \uparrow \tag{20}$$

Evidence for this reaction was provided by deposits of Pb metal in the cold parts of the furnace that were directly related to weight losses in the liquid. In terms of the free energy of the system, reaction (20) can be represented by

$$\Delta G = \Delta G^\circ + RT \ln \frac{a(Ag_2O)_{int}\, P(Pb)}{a(PbO)_{liq}} \tag{21}$$

The standard free energy ΔG° for this reaction is positive since the oxidation potential of Ag is lower than that of Pb. Hence, for this reaction to occur, the equilibrium constant in Eq. (21) must be sufficiently smaller than unity to result in a negative ΔG. This is achieved when the $a(Ag_2O)$ is low and Pb formed as vapor is immediately removed from the reaction site. Nagesh et al. showed that reaction (20) was effectively curtailed by changing the atmosphere from vacuum to He at 1 atm. This change in atmosphere resulted in poor adhesion and a high contact angle of 48°. The high ambient pressure raised the value of $P(Pb)$; thus, the equilibrium constant in Eq. (21) was shifted toward a positive ΔG. Prevention of the redox reaction results in no Ag_2O at the interface and, thus, the absence of adherence.

(d) Other Metals. Molten $Na_2O{\cdot}2SiO_2$ on Ta [21] at 10^{-3} Pa and 1000°C exhibited an initial contact angle of 80°, but then a considerable amount of reaction occurred at the interface. Similar reactivity was noted on Ti and Zr substrates. Considerable spreading was noted on Mo and W; the perimeter of the contact area was irregular in shape. This indicates that reactions may be occurring or, as suggested by Fulrath et al. [21], that the retention of surface-adsorbed gases, which alter γ_{SV}, may be sensitive to the crystallographic orientations of the polycrystalline grains of the metal substrate.

For the Ta case, columnar grains of $Na_2Ta_2O_6$ grew from the Ta surface into the liquid droplet. Subsequent contact-angle measurements were not reported due to the occurrence of this reaction. Using thermodynamic analysis, Mitoff [48] proposed three possible reactions to form $Na_2Ta_2O_6$. Each of the reactions involves the reduction and subsequent loss of Na from the glass.

On the basis of the interfacial reactions observed when $Na_2O{\cdot}2SiO_2$ liquids wet a wide variety of metals, Pask and Tomsia [5] proposed the general reaction

$$Me + Na_2Si_2O_{5(liq)} \rightleftarrows MeO_{int} + 2SiO_{2(liq)} + 2Na_g\uparrow \qquad (22)$$

to describe the interaction of the liquid with the metal in an atmosphere of low $P(O_2)$. The general applicability of this equation to a wide range of metals was demonstrated in this reference.

B. Wetting on Ceramic Substrates

1. Oxides

(a) Aluminum Oxide. Most studies of wetting behavior on oxide substrates have involved liquid metals. There are few data in the literature on wetting of such substrates by oxide liquids. Tummala and Foster [29] report that a borosilicate liquid on 99.5% Al_2O_3 at 900°C required approximately 100 min to reach a steady-state value of the contact angle in

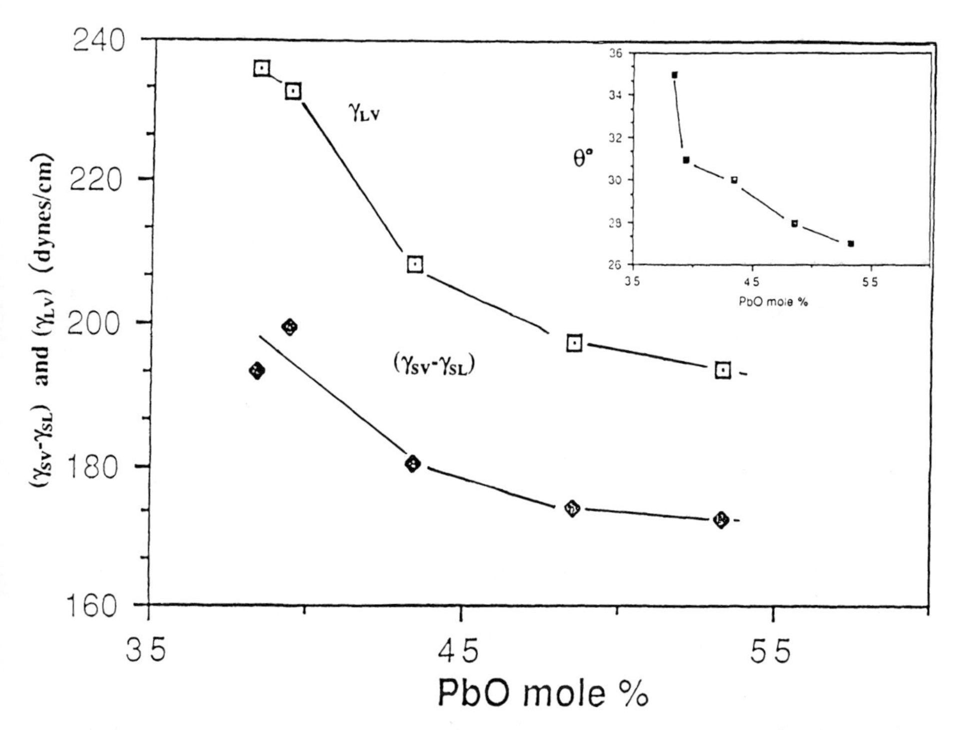

FIG. 12 Surface tension and driving force vs. PbO content for sessile drops of PbO-SiO_2 on Al_2O_3 at 850°C. (γ_{LV} data from Ref. 14, by permission.) Insert shows corresponding contact angle vs. composition relationship. (From Ref. 28, by permission.)

the range of 32–38°. No discussion of the possible impact of reactions between the liquid and substrate was presented.

Cronin et al. [28] investigated the temperature and composition dependence of the wetting behavior of PbO-SiO_2 liquids on Al_2O_3. The insert to Fig. 12 shows the composition dependence of θ at 850°C. The contact angle is seen to decrease as the PbO content of the liquid increases. The composition dependencies of γ_{LV} and the driving force for wetting, $\gamma_{SV} - \gamma_{SL}$, are also shown in Fig. 12. Increases in the PbO content lead to reductions in both the driving force and γ_{LV}. The reductions in driving force must be caused by increases in γ_{SL}. The larger reductions in γ_{LV}, however, dominate the wetting behavior since the contact angle decreases with increasing PbO content.

The temperature dependence of the wetting behavior of a 0.55 PbO-0.45 SiO_2 (molar) liquid on Al_2O_3 is shown in Fig. 13. Here large reduc-

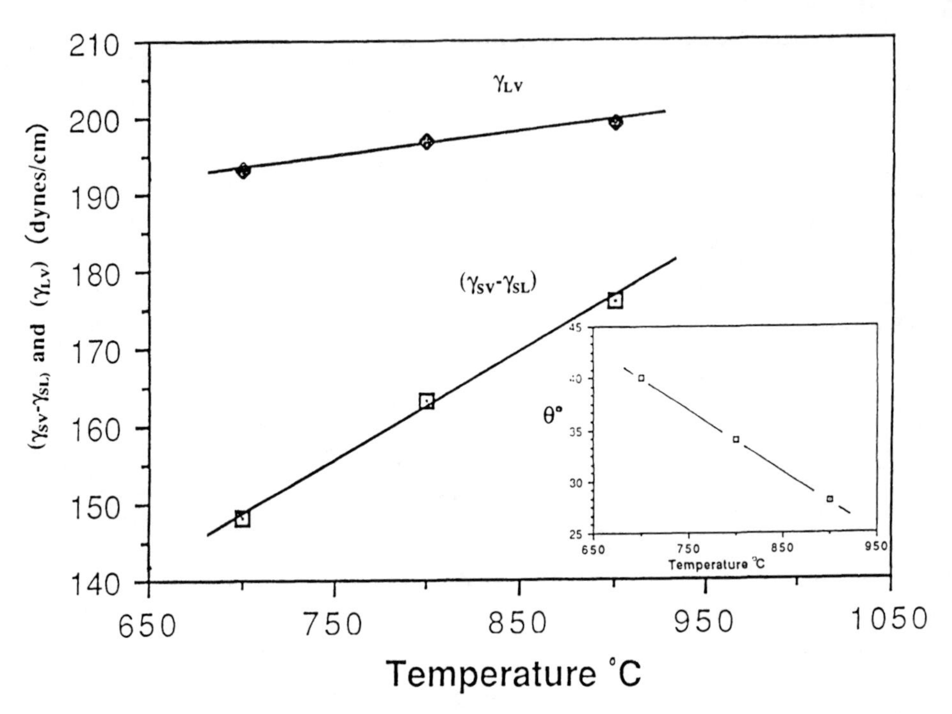

FIG. 13 Surface tension and driving force vs. temperature for sesile drops of composition 0.55 PbO-0.45 SiO_2 on Al_2O_3. (γ_{LV} data from Ref. 14.) Insert shows corresponding contact angle vs. temperature relationship. (From Ref. 28, by permission.)

tions in γ_{SL} overcome the accompanying increases in γ_{LV} and slight reduction in γ_{SV} [34], causing θ to decrease with increasing temperature.

Comparison of the temperature and composition dependencies of the wetting behavior of the PbO-SiO_2 liquids on Al_2O_3 indicates the importance of considering the driving force term when making predictions about wetting behavior. Decreasing values of γ_{LV} in any given system are *not* sufficient to assume that wetting will improve.

Cronin et al. [28] also evaluated the wetting behavior of a ternary Pb-borosilicate liquid on Al_2O_3, as well as AlN. These results are discussed in the aluminum nitride section below.

Comeforo and Hursh [49] investigated the wetting of Al_2O_3-SiO_2 refractories by Na_2O-CaO-SiO_2 liquid. Steady-state contact angles ranging from 0 to >40° were observed. The results are complicated, however, by the possible effects of porosity in and dissolution of the substrate.

(b) Iron Oxide. Cline et al. [37] measured the wetting behavior of $Na_2O{\cdot}2SiO_2$ liquid on a thin film of FeO and on polished Fe_3O_4 at 1000°C in vacuum. On the FeO film, the $Na_2O{\cdot}2SiO_2$ liquid exhibited $\theta = 24°$ at 900°C. The time dependence of the wetting process on Fe_3O_4 is shown in Fig. 14. Within 20 min, the liquid achieves steady-state conditions with a final contact angle of 2°. No reaction is expected to contribute to the driving force for wetting on either FeO or Fe_3O_4. In vacuum, however, the liquid dissolves both oxides. Hence, under steady-state conditions, in each case the droplet contains an unknown quantity of FeO. Despite the uncertainties in experimental conditions, it is interesting to note that the liquid exhibits better wetting behavior on the oxide having the higher valence state. Clearly, many more careful and quantitative studies of the wetting behavior of inorganic liquids on oxide substrates are required in order to develop a suitable understanding of this process.

2. Nonoxides

An important finding from the studies of liquid metals on ceramic substrates was that the surface energies of nonoxide ceramic solids are very sensitive to $P(O_2)$. This was shown by Barsoum et al. [50] in their studies of the wetting of SiC, AlN, and Si_3N_4 by Si. They used a modified version of Eq. (1)

$$\cos\theta = -\frac{RT\,\Gamma\,\ln P(O_2)}{\gamma_{LV}} - \frac{\gamma_{SL}}{\gamma_{LV}} \tag{23}$$

where Γ is the surface excess oxygen adsorbed on the solid surface per unit area and γ_{SV} has been replaced by $-RT\,\Gamma\,\ln P(O_2)$.

Barsoum et al. showed that at 1430°C, the surface energy of SiC decreased by ~85 mN/m for every tenfold increase in O_2 activity. For AlN, the decrease was 115 mN/m and for Si_3N_4, 70 mN/m. Thus, for any wetting experiments on these nonoxide ceramic substrates, a knowledge of $P(O_2)$ and the pressures of other adsorbable gases is an important factor, since these will affect γ_{SV}.

(a) Silicon Nitride. Kossowsky [51] studied the wetting of Si_3N_4 by alkaline oxide-doped $MgO{\cdot}SiO_2$ under nonequilibrium conditions, in an attempt to understand better the sintering of Si_3N_4 with MgO. (To sinter successfully, good wetting of the solid phase by the liquid must occur [52].) Alkaline oxide additions (CaO, Li_2O, Na_2O, and K_2O) were found to improve significantly the wetting of Si_3N_4 by Mg silicate liquids. This is illustrated in Fig. 15, which shows plots of θ vs. temperature for doped $MgO{\cdot}SiO_2$ liquids in N_2 under constant heating rate (10°/min) conditions. In all cases, θ decreases with increasing alkali or alkaline earth oxide in the liquid. For the same addition of alkaline oxide, wetting improved on

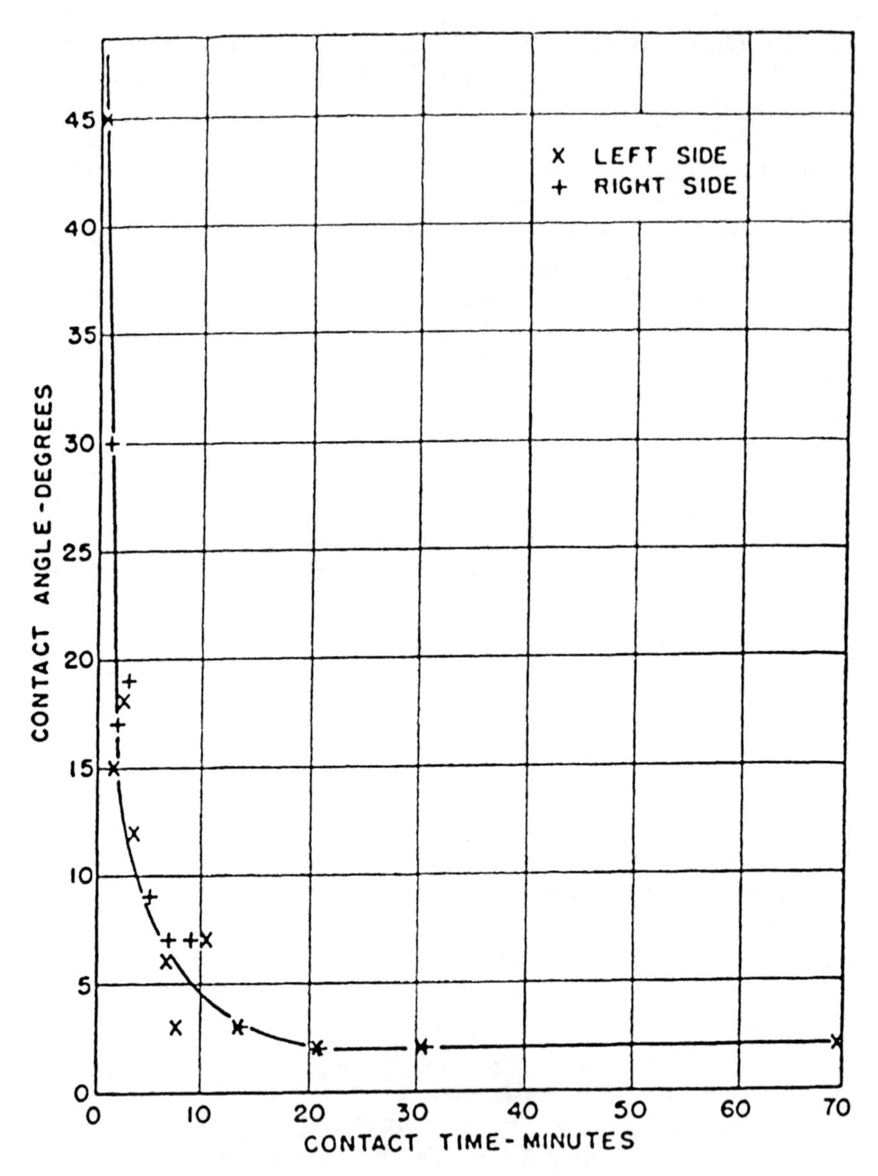

FIG. 14 Kinetics of wetting of $Na_2O \cdot 2SiO_2$ liquid on polished Fe_3O_4 at 1000°C in vacuum. (From Ref. 37, by permission.)

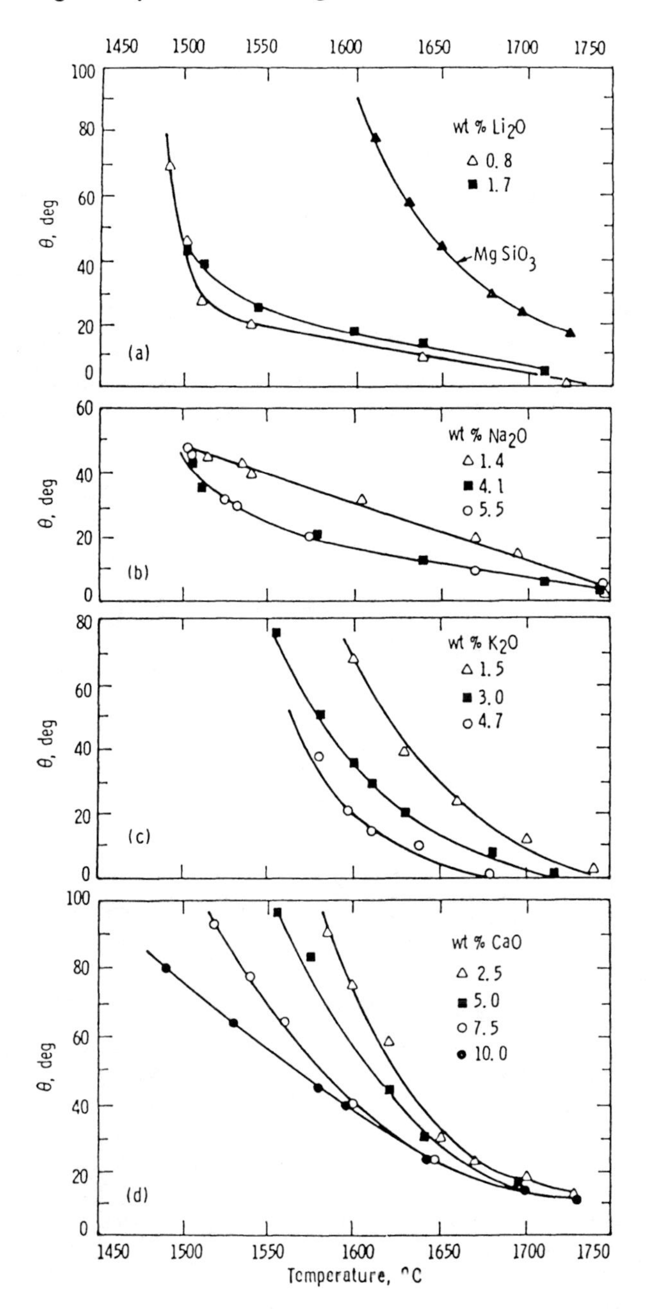

FIG. 15 Wetting of $MgO \cdot SiO_2$ on Si_3N_4 as vs. temperature and oxide additions. Nitrogen atmosphere, heating rate 10°C/min. (From Ref. 51, by permission.)

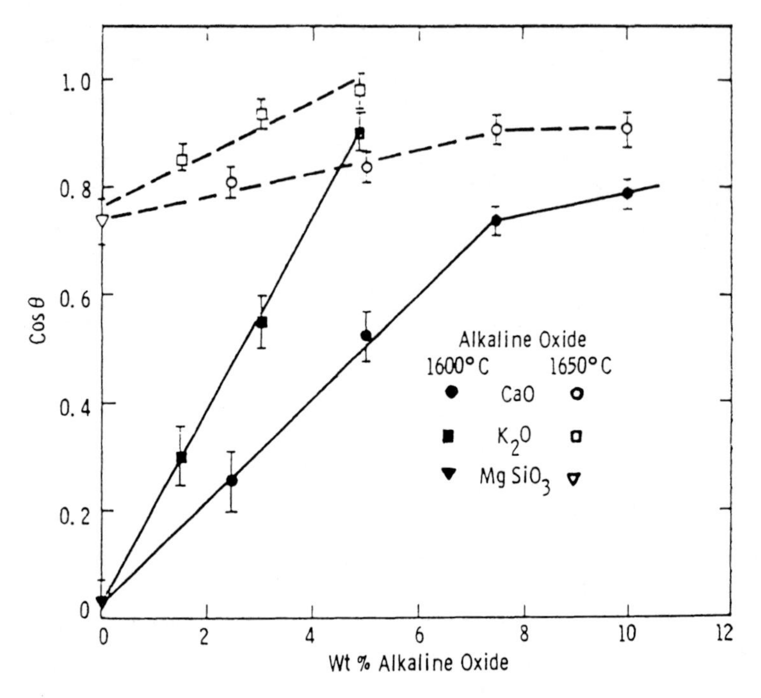

FIG. 16 Cos θ vs. alkaline oxide content of MgO-SiO_2 liquids wetting Si_3O_4 at 1600 and 1650°C. (From Ref. 51, by permission.)

moving from CaO to Li_2O. The authors showed by light and scanning electron microscopy and energy dispersive x-ray analysis that reactions took place at the interface and that the main reaction was the dissolution of Si_3N_4 in the silicate liquid.

Figure 16 shows the cos θ vs. oxide addition relationship for these Mg silicate liquids. The four lines in the figure extrapolate to two common points that agree well with the values of cos θ for pure $MgO \cdot SiO_2$ at the stated temperatures. Thus, from Fig. 16, cos θ can be expressed

$$\cos\theta = A + BX_{\mathrm{alk}} \tag{24}$$

where X_{alk} is the fraction of alkaline addition, A the value of cos θ for pure $MgO \cdot SiO_2$, and B a constant.

The author assumed that additions of alkaline oxides to the liquid should not affect γ_{SV}; Kingery [53] has shown that dilute mixtures of SiO_2 with alkaline oxides do not affect the surface energy of liquid SiO_2. Kossowsky [51] assumed the same behavior for the $MgO \cdot SiO_2$ system with small additions of alkaline oxides. Thus, Eq. (1) can be written

$$\cos \theta = A' - B'\gamma_{SV} \tag{25}$$

Combining Eqs. (23) and (25) yields

$$\gamma_{SL} = \alpha - \beta X_{alk} \tag{26}$$

These results suggest that, for the system Si_3N_4-$MgO \cdot SiO_2$, additions of alkaline oxides to the glass affect mainly the solid/liquid interfacial energy and that this variation controls the wetting behavior.

(b) Silicon Carbide. The wetting of SiC by a series of melts in the family MgO-Li_2O-Al_2O_3-SiO_2 was investigated by Coon et al. [54] over the temperature range 950–1150°C in an air atmosphere. Contact angles in the range of 30–50° were observed. Increasing the MgO/Li_2O ratio or Al_2O_3 content of the liquid led to an increase in the contact angle at a given temperature. At temperatures well above the melting point of the liquids ($T_{liq} > 950$°C) the authors claim that the liquids do not chemically interact with SiC, and that wetting is governed by the balance of interfacial forces alone. Based on Barsoum et al.'s observation of a reduction in γ_{SV} of SiC with oxygen adsorption, it is presumed that much smaller contact angles would be observed for this system if the measurements were conducted in vacuum.

(c) Aluminum Nitride. Cronin et al. [28] compared the wetting behavior of 0.435 PbO-0.565 SiO_2 and 0.435 PbO-0.14 B_2O_3-0.425 SiO_2 liquids on both AlN and Al_2O_3. As seen in Fig. 17, for these liquids on both substrates, θ decreased with increasing temperature. The contact angles of the binary Pb-silicate liquid on AlN were lower than those on Al_2O_3 by ~15°. In contrast, the ternary Pb-borosilicate liquid exhibited much smaller contact angles on Al_2O_3 than AlN. The addition of B_2O_3 to the binary liquid likely causes a significant reduction in γ_{LV} (see Fig. 3), leading to the observed decrease in contact angles on Al_2O_3. Despite the reduction in γ_{LV}, adding B_2O_3 slightly *increased* the contact angles observed on AlN. Cronin et al. suggest that the formation of a BN layer at the interface may account for this behavior (θ = 70° for the ternary liquid on BN at 57°C), but note that the participation of the substrate in a reaction at the interface usually (but not always) leads to enhanced wetting.

Both liquids reacted strongly with the AlN substrate. The reactions produced a gaseous product that formed bubbles in the liquid. Cronin et al. [28] list the thermodynamic reactions that may be occurring at the liquid/AlN interfaces. With the exception of the possible formation of intermediate compounds at the interface, the substrate is not an active

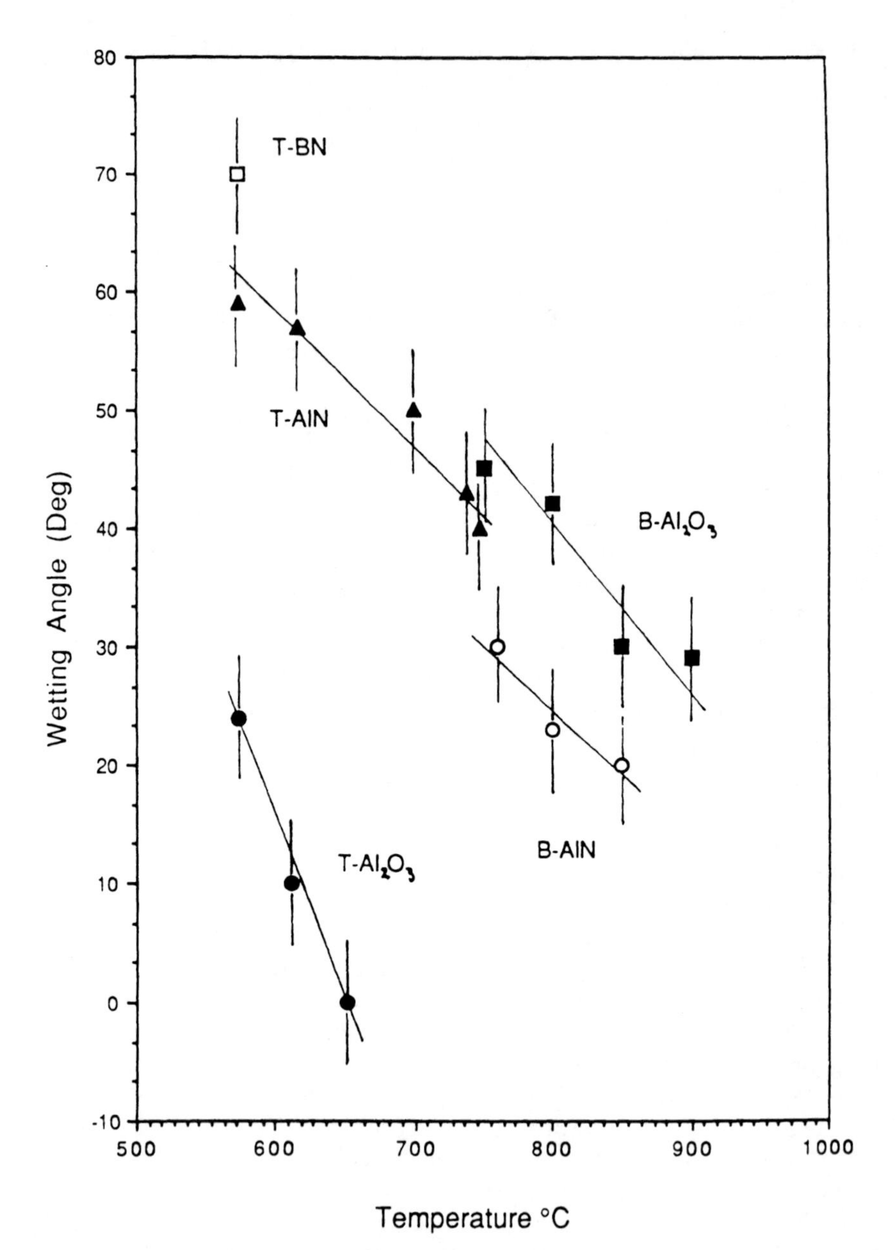

FIG. 17 Contact angle vs. temperature for 0.435 PbO-0.565 SiO_2 (B) and 0.435 PbO-0.14 B_2O_3-0.425 SiO_2 (T) liquids on Al_2O_3 and AlN. Open square is T liquid on BN. (From Ref. 28, by permission.)

participant. As a result, the reactions are not anticipated to impact wetting behavior.

IV. KINETICS OF WETTING AND SPREADING

Time-dependent variations in the contact angle are often observed during high-temperature wetting experiments. Depending on the viscosity of the wetting liquid, the magnitude of the driving force for wetting and contributions due to reactions, θ will change with time to a steady-state value over time scales ranging from seconds to hours. In nonreactive systems, the dynamics of wetting and spreading are generally driven by reductions in interfacial and gravitational energies, with the latter often contributing only negligibly; the spreading is expected to be impeded primarily by viscous dissipation in the liquid. Reactive systems may be described in many cases by adding appropriate temporal and spatial dependencies to the affected material parameters.

Most attention in the literature has been directed to spreading dynamics associated with the small θ regime [1,3,55–58], where the liquid is generally quite thin and nonreactive. A number of models giving some appropriate functional dependencies have been obtained, but always involve the use of at least one arbitrary adjustable parameter. At best, it may be that the adjustable parameter does not vary by too much within a variety of systems [57].

A method for treating general viscous flow processes under the action of surface tension was proposed by Frenkel [59]. He suggested that the energy gained via decreases in surface energy is balanced by the energy lost through viscous deformation of the body. This method was first applied to the analysis of spreading liquids by Strella [60], using a liquid velocity field corresponding to a cylinder undergoing uniaxial compression (which was the same velocity field used by Frenkel in describing the viscous coalescence of two spheres). Denesuk et al. [61] extended this model and applied it to specific systems. Their development is discussed below. Denesuk et al. also applied this model to the case of wetting on porous substrates [62].

Denesuk et al. [61] initially described the energies of spreading in terms of a liquid having the idealized shape of a constant-volume, right circular cylinder expanding on a flat solid. For making quantitative predictions of the dynamic spreading behavior in real systems, the authors proposed the use of a mapping function that correlates model cylinder shapes to more realistic spheric cap shapes with the same interfacial energy equilibria.

Although the description of the interfacial and gravitational driving forces is relatively straightforward, Denesuk et al. note that approxi-

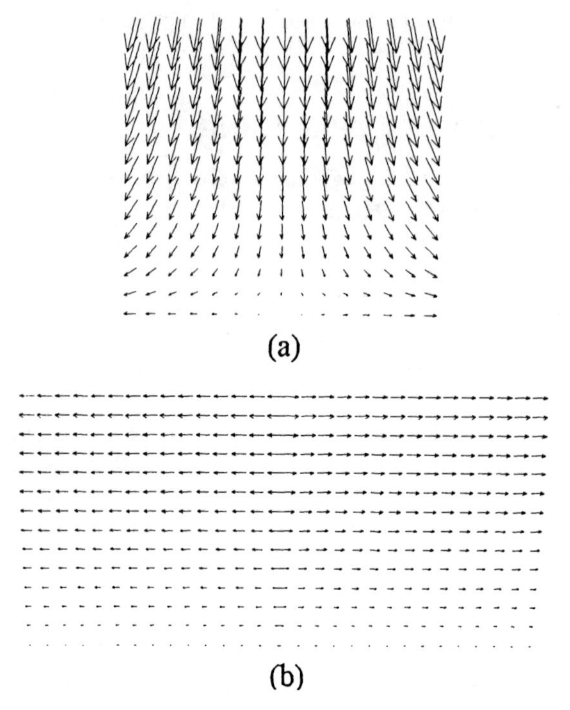

FIG. 18 Velocity fields employed in the spreading models of Ref. 61: (a) bulk-limited flow and (b) thin liquid flow.

mations to the energy lost through viscous dissipation are considerably more uncertain. The dissipation results from not the magnitudes and directions of liquid velocities, but their spatial gradients; appropriate analytic expressions for these gradients are unavailable. Consequently, approximations must be used that maintain appropriate values of the velocity gradients in regions of the volume where the majority of the viscous dissipation occurs. Two different liquid velocity fields were used to describe the energy associated with viscous dissipation in the liquids. The first, which defines the bulk-limited spreading model, corresponds to a cylinder undergoing uniaxial compression (see Fig. 18a); and the second, which defines the thin liquid model, assumes that only radial components in the velocity field are important (see Fig. 18b).

A. Bulk-Limited Spreading

The velocity field employed in this model ignores the expected no-slip (zero tangential velocity) condition at the liquid/solid interface and is expected to apply to spreading situations in which the presence of the solid substrate does not affect the overall liquid velocity field too strongly. The exact, analytical solution to the model is quite unwieldy, but an approximate solution was derived for the time for the droplet to go from a contact angle of θ_0 to a contact angle of θ

$$t(\theta) = \tau g(\theta,\theta_0) \tag{27}$$

where

$$\tau = \frac{9\eta}{\gamma_{LV}}\left(\frac{V_0}{\pi}\right)^{1/3} \tag{28}$$

and

$$g(\theta,\theta_0) = \frac{1}{(1 - \cos\theta)^{1/3}} - \frac{1}{(1 - \cos\theta_0)^{1/3}} \tag{29}$$

and V_0 is the liquid volume. The approximate solution neglects the impact of gravity and assumes that the system is not too close to equilibrium.

The factor $g(\theta,\theta_0)$ typically does not vary much, for example, for θ changing from 90–20°, g changes only from zero to 1.55. The factor τ may thus be seen as a time constant that gives an estimate of the expected time scale of bulk-limited spreading processes; such processes are expected to occur on the scale of a few τ. The quantity $(V_0/\pi)^{1/3}$ may be viewed as a characteristic length associated with the liquid droplet; it is seen that τ is proportional to this characteristic length and the liquid viscosity and is inversely proportional to the liquid surface tension.

Unfortunately, there appear to be few quantitative data in the literature regarding spreading processes that are rate-limited by bulk viscous flow. Most studies have centered on completely wetting systems for which a potential bulk-limited regime was either not of interest, passed by too quickly, or was inseparable from other transient effects. To test the spreading model, Denesuk et al. [61] measured the kinetics of wetting of a 0.43 PbO-0.57 SiO_2 (molar) liquid on Au in air at 710°C. The system was chosen because of the relative inertness and smoothness of the Au, and because of the availability of data on the viscosity and surface energy of PbO-SiO_2 liquids. The data in Fig. 19 show the time dependence of the contact angle in this system. The solid line in Fig. 19 is the behavior predicted by the exact model using appropriate literature values for interfacial energies and liquid viscosity. As seen in the figure, the fit of the model to the experimental data is quite good. Additionally, τ was calculated to be ~200 s, which agrees approximately with the observed time scale of the spreading.

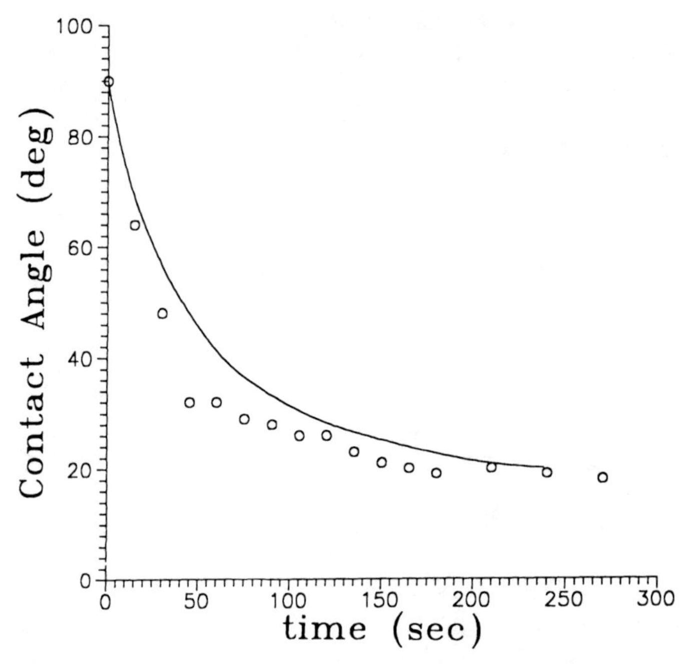

FIG. 19 Spreading data for PbO-SiO_2 liquid on Au at 710°C (t in s). Points are experimental data; solid line is model prediction. (From Ref. 61, by permission.)

Comparison of τ with observed spreading time scales may be used generally as an indication of whether or not a particular spreading process is limited by viscous flow in the bulk of the droplet. For example, consider wetting experiments using $Na_2O \cdot 2SiO_2$ liquid at 1000°C. With $\eta = 1.2 \times 10^2$ Pa/s, $\gamma_{LV} = 300$ mN/m, and $V_0 = 1.5 \times 10^{-8}$ m^3, $\tau = 17$ s. As seen in Fig. 14, significant spreading occurs on the time scale of a new τ for $Na_2O \cdot 2SiO_2$ liquid on Fe_3O_4 (a nonreactive substrate). This indicates that the early stage of the time dependence of the contact angle is likely controlled by viscous flow in the bulk of the droplet. At longer times, the rate of change of the contact angle slows considerably. At such longer times, dissolution of Fe_3O_4 may lead to an increase in liquid viscosity [63] and affect the values of γ_{SL} and γ_{LV}. The implication is that this is a regime in which one or more of these mechanisms is controlling the spreading.

Comparison of τ for $Na_2O \cdot 2SiO_2$ to the kinetic data of Tomsia and Pask [39] for this liquid on Fe at 1000° (shown in Fig. 9) clearly illustrates the dominant role that reactions can play in the kinetics of the spreading process. For these samples, large changes in θ occur on the time scale of hundreds of τ.

These results demonstrate that in high-temperature systems, processes other than viscous flow will often occur that complicate the dynamic spreading behavior. If the impact of the process(es) is confined to the temporal and/or spatial variation of a single material parameter, an appropriate dependency can be added to the parameter. If the process is a chemical reaction (e.g., dissolution of the substrate) that affects a surface energy, the equilibrium configuration will change with time. If the kinetics of the reaction are sufficiently slow, they may even dictate the spreading rate. Dissolution processes may also affect the liquid viscosity, likely in a boundary layer near the substrate.

Spreading models can be used to extract material parameters from kinetic wetting data. For example, with the model of [61], the time to reach a given radius can be expressed

$$t = \eta f(r) \tag{30}$$

Thus, plotting experimentally measured times on the y axis and the corresponding function of the transformed cylinder radii data (obtained, e.g., by applying the inverse of the energy equivalence mapping to the contact-angle data) on the x axis should yield a straight line of slope η. This was done with the experimental data in Fig. 19; the inferred viscosity was found to within a factor of 1.4 of the value expected from the literature (see Fig. 20). This type of analysis may also be used to obtain information on the dynamic behavior of material parameters from dynamic wetting experiments. For example, if the slope in Fig. 20 were found to be increasing, one might propose that the viscosity was increasing in at least some region of the liquid (probably through a dissolution/diffusion mechanism).

B. Thin Liquid Spreading

The second velocity field employed considers only radial velocity components directed in the plane of the solid and incorporates the "no-slip" condition with a parabolic velocity profile away from the solid (see Fig. 18b). This velocity field should be valid if large changes in the radius of the liquid are accompanied by only small changes in the height, that is, if $|dh/dr| \ll 1$. This generally corresponds to thin liquid layers ($|dh/dr| < 1/10$ if $\theta < 26°$).

Again, the exact solution is unwieldy, but an approximate solution is given for the contact radius of the droplet, which is expected to be valid

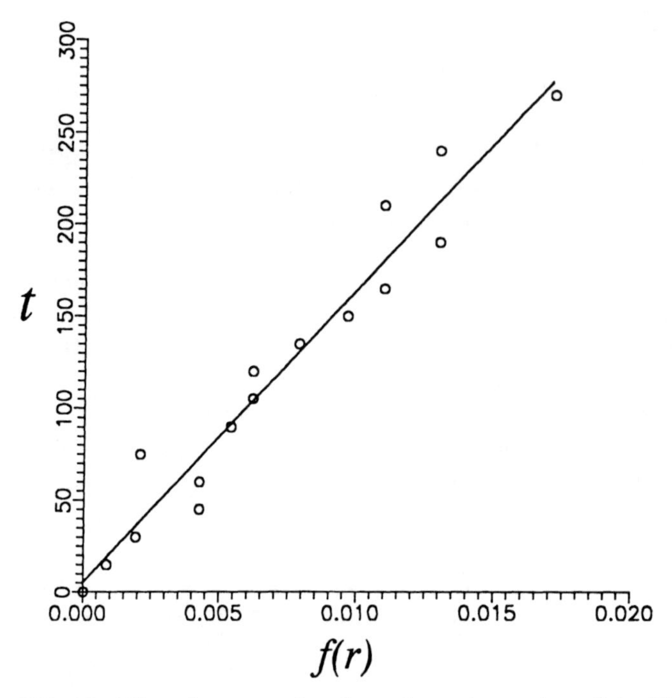

FIG. 20 Viscosity extraction from dynamic wetting of the $PbO\text{-}SiO_2$ liquid on Au at 710°C. Points obtained by calculating $f(r)$ of Ref. 61 using experimental data of Fig. 19. Line is least-squares fit to points. (From Ref. 61, by permission.)

in the small contact-angle limit ($\theta \ll 60°$) for completely wetting systems ($S > 0$), where S is larger than a small fraction of a mN/m

$$r \approx \left(\frac{16S}{3\eta}\right)^{1/8} \left(\frac{8V_0}{\pi}\right)^{7/24} t^{1/8} \tag{31}$$

This model has been found to give reasonable time, volume, viscosity, and surface tension dependencies for a wide class of liquid/solid systems, and to give reasonable and consistent quantitative predictions for the spreading of several polydimethylsiloxane liquids on a modified oxidized Si solid. It is generally difficult to test quantitative agreement since S is normally difficult to determine when $S \geq 0$.

V. CRITICAL SURFACE TENSION OF METAL AND CERAMIC SUBSTRATES

Zisman introduced the concept of the critical surface tension, γ_c, as a relative measure of the wettability of solid substrates. Using wetting data from organic liquid-low surface energy solids, he showed that plots of cos θ vs. γ_{LV} for liquids in a homologous series on a single solid were linear. Zisman defined γ_c as the value of interfacial energy at the intersection of the plotted line, or its extrapolation, with the line defined by cos θ = 1, the condition for spreading [2]. This relationship led to the following expression for the equilibrium contact angle formed by liquid members of the homologous series as a function of γ_{LV}:

$$\cos \theta = 1 + b(\gamma_c - \gamma_{LV}) \tag{32}$$

Even when data from a variety of nonhomologous liquids were used, the points fell close to a straight line or collected in a narrow band around a line, suggesting that γ_c is a characteristic of the solid surface alone. Consequently, γ_c corresponds to the maximum surface tension a liquid can have and still completely wet the solid. As such, it can serve as a useful empirical parameter for the characterization and comparison of the wettabilities of low-energy surfaces (see Fig. 21).

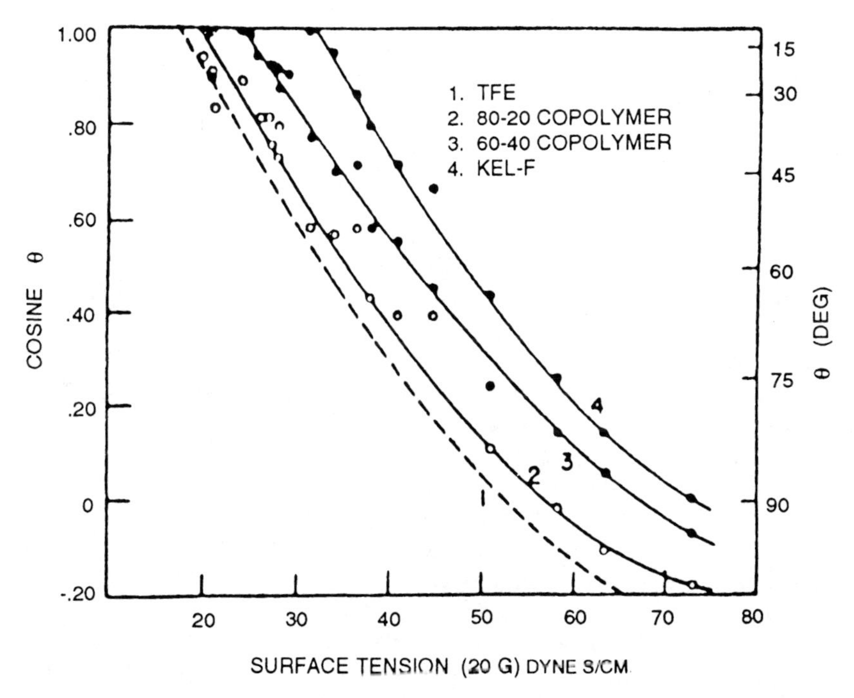

FIG. 21 Wettability of solid copolymers of polytetrafluoroethylene and polychlorotrifluoroethylene. The intersection of the curves with the line cos θ = 1 identifies γ_c for each solid. (From Ref. 64, by permission.)

When using the Zisman construction, caution is advised if approximately linear behavior is not observed, or data are not available over a wide enough range to ensure an accurate extrapolation. In addition, Zisman did not explore the applicability of Eq. (32) for liquids with widely varying bond types (i.e., metallic liquids, inorganic liquids, polymeric liquids). It seems likely that the utility of a given value of γ_c is restricted to a particular class of liquids.

Also, care must be exercised to avoid complications due to surface contamination. For example, Olsen and Osteraas [66] investigated the wetting of alkali borosilicate, lead alkali silicate, fused silica, and soda-lime-silicate glasses by aqueous solutions of K_2CO_3 or $CaCl_2$ at 20°C and by a variety of low-melting metals. The same value of $\gamma_c = 75 \pm 10$ mN/m was extracted from the Zisman plots of both types of liquids. Olsen and Osteraas noted that the similarity of the value of γ_c to γ_{LV} of water suggests that the wetting behavior of the substrates is controlled by an adsorbed water film. The data from the liquid metals indicate that adsorbed water was preexisting on the solid. Adsorbed water layers are expected to be thermally stable, being removed only when the glass is heated to near its softening point or with long exposure to high temperature and reduced pressure [67,68].

Despite these cautions, the utility of the critical surface tension is clear. Having established the values of γ_c and b, we can predict the wetting behavior of a given liquid on the solid from a knowledge of γ_{LV} alone.

In applying the critical surface tension concept to the wetting of high-energy surfaces at room temperature, it was found that molecules of the liquid often adsorb on the solid, creating an effective solid surface with a γ_c that can be appreciably lower than that which would correspond to the pure solid surface [64]. This was offered as an explanation as to why low surface tension liquids do not always completely wet high-energy surfaces, which are expected to have large "intrinsic" γ_c's. Liquids whose surface tension was larger than the corresponding γ_c of their own adsorbed layer were termed "autophobic." Such systems should therefore form finite contact angles. Liquids with surface tension lower than that of their own adsorbed layer should wet completely. In addition, chemical reactions could also occur between the liquid and solid, creating a layer on the solid that has a different γ_c than the unreacted solid.

Eberhart constructed Zisman plots to extract γ_c using literature values for contact angles formed by various alloys on single-crystal Al_2O_3 ($\sim\langle 10\bar{1}2\rangle$ plane) [69] and UO_2 [70]. γ_c was estimated to be 1050 ± 100 mN/m for Al_2O_3 under vacuum, 600 ± 50 mN/m for UO_2 in Ar + O_2, and 1650 ± 100 mN/m for UO_2 in H_2. The data were far from the $\cos\theta = 1$ abscissa, and the extrapolations were extended over a long range, adding appreciable uncertainty to the estimated γ_c's. Additionally, for the Al_2O_3 data, Eberhart made some particularly strong assumptions to justify using the surface tensions of pure metals to correspond to the contact angles of the alloys, and then to extrapolate these surface tensions to the appropriate temperature. Eberhart compared his γ_c for Al_2O_3 with an extrapolation of literature data for polycrystalline Al_2O_3 and found reasonable agreement. Based on the assumptions and extrapolations used, however, the significance of this agreement is questionable.

The Zisman plots for the UO_2 solid gave positive slopes, that is, θ decreases, $\cos\theta$ increases, as γ_{LV} increases. This behavior might be anticipated, since all the contact angles are $>90°$; the liquid/vapor surface area of a spherical cap liquid droplet decreases as the contact angle decreases over the range $180° < \theta < 90°$. (A Zisman plot for systems where all measured contact angles are less than 90° normally has a negative slope; this may be seen as due largely to the fact that the liquid/vapor surface area of the liquid increases as the contact angle decreases.) Experimentally, however, reported slopes have been negative for contact angles both above and below 90° [65,70–73], so the physical significance of Eberhart's positive slope is not clear. Eberhart additionally points out that the reported γ_c in the Ar/O_2 environment is close to estimates of γ_{SV} for UO_2 based on literature data, and that the reported γ_c in the H_2 atmosphere is close to estimates of γ_{SV} for U metal (reduced UO_2) based on literature data.

Rhee has plotted the system temperature vs. $\cos\theta$ and established linear, negative slope relationships for a number of different liquid metals on ceramic substrates [e.g., 72–75]. From the temperature dependence of γ_{LV}, the Rhee T-$\cos\theta$ plots can be translated into Zisman-type γ_{LV} – $\cos\theta$ plots. Caution in interpretation must be exercised, however, since the former type of plot is not expected to be formally equivalent to the latter. Rhee's plots differ from those of Zisman in that the surface tension is varied by changing the temperature rather than the liquid. The relation between the two plots is not clear, and equivalence likely hinges on the effect of temperature on γ_{SV}. Interestingly, if γ_{LV} is linear in temperature, then $(\gamma_{SV} - \gamma_{SL})$ must be roughly quadratic in temperature to give the observed linear plots.

For AlN and TiC substrates [73], Rhee's γ_c was found to depend on the liquid used (Al, Ag, or Cu). Thus for these systems, Rhee's γ_c is a property not of the solid alone, but reflects some specific interaction with the liquid. Although Rhee reports that no severe chemical reactions take place, it should be noted that the wettability can be very sensitive to the nature of the solid/liquid interface. Hence, even reactions of a minor extent can be quite important. Thus, if Rhee's γ_c is expected to be qualitatively similar to Zisman's and reflect a characteristic of the solid, one is led to believe that a minor reaction or an adsorption layer of the liquid on the solid is affecting or controlling the γ_c. For the systems of liquid Al on AlN, TiC, TiB_2, and TiN, it was reported that all the solids were chemically inert with respect to the Al over the temperature range of the experiments, but that some residual TiO_2 in the TiB_2 reacted with the Al, and that the Al did slightly attack the grain boundaries in the TiN. Both these processes would generally be expected to increase γ_c, making the solids more wettable. All of Rhee's plots display the Zisman-type negative slope, even when $\theta > 90°$.

In an extension of Zisman's work, Rhee combined Young's equation with Eq. (32) to establish an approximate relationship between γ_{SV}, γ_c, and b [71]. After making the assumption that the minimum value of γ_{SL} with respect to cos θ is zero, this relationship becomes

$$\gamma_{SV} = \frac{(b\gamma_c + 1)^2}{4b} \tag{33}$$

In addition to the assumption of a minimum γ_{SL} of zero, it is also implicitly assumed that γ_{SV} is constant with respect to variations in cos θ. This latter assumption should be true when cos θ is varied by changing liquids that do not volatilize (and adsorb on the solid), but will generally not be valid when cos θ is varied by changing the temperature (as Rhee does). If the temperature dependence of γ_{SV} is comparably weak, however, this may not be an issue.

Rhee has calculated γ_{SV} from Eq. (33) using literature data for γ_c and b corresponding to systems composed of organic liquids wetting on low-energy (organic) solids [71]. The values obtained were found to be in very good agreement with those given by others. Additionally, literature data on contact angles and liquid surface tensions were used to extract γ_c and b for liquid Ni on MgO, for Cu-Al alloys on UO_2, and for Al-Mg alloys on BeO. For the first case, γ_{LV} was varied by changing the atmosphere, whereas in the latter two cases, the relative composition of the alloy was varied. As discussed earlier, the relationship between such plots is not necessarily clear. In addition, it is argued for the first two cases that $\gamma_{SV} = \gamma_{SO}$ (or $\pi_e = 0$) since γ_{LV} for the two liquids is appreciably larger than γ_{SO} (which, in a rough sense, make adsorption of the vapor of the liquid phase on the solid seem unlikely). In the third case, it is argued that Mg vapor probably adsorbs on the BeO surface since the surface energy of Mg is much less than that of BeO. The resulting computed values for γ_{SO} were found in all three cases to be in good accord with extrapolations of literature data and calculations. Although the agreement is good, it should be noted that a rather liberal use of temperature coefficients is necessary for the comparison.

Additionally, the physical meaning of the assumption that γ_{SL} is zero at its minimum is not clear. In the latter three systems, γ_{SL} at $\theta = 0$ is 10, 333, and 467 mN/m, respectively. Hence, since γ_{SL} is apparently not zero for $\theta > 0$ in these systems, a point where γ_{SL} might be equal to zero, and hence the point of the prediction for γ_{SV}, is not describable in terms of contact angles. For extension into such regimes, one may need to use the spreading parameter S, rather than $\cos\theta$, in the Zisman-type plots. Unfortunately, the measurement of S can be experimentally difficult.

For the special case where the minimum of γ_{SL} is zero and occurs at $\theta = 0$, Eq. (33) reduces to [71]

$$\gamma_{SV} = \gamma_c = \frac{1}{b} \tag{34}$$

and Rhee thus predicts that the critical surface tension should be equal to the solid/vapor surface energy. Rhee [70] found that $1/b$ was approximately equal to γ_c for the wetting of graphite by Al-Mg alloys and of sapphire by Cu-O alloys.

In summary, the limited experimental studies carried out to date suggest that Zisman's concept of the critical surface tension is valid when applied to the wetting behavior of high-energy liquids on high-energy solids. However, to allow for an extrapolation to γ_c, one clearly needs to use liquids with surface tension greater than γ_c. Since the intrinsic γ_c corresponding to pure high-energy surfaces is expected to be quite large, organic liquids (which typically have relatively low γ_{LV}'s) cannot normally be used. An alternative for obtaining an intrinsic γ_c for a high-energy solid using low-energy liquids would involve plotting γ_{LV} vs. the spreading parameter S (which, with certain assumptions, can be estimated from the final thickness of a completely wetting film [1]). Extrapolation to $S = 0$ will give a parameter nominally equivalent to γ_c. However, the applicability of this low-energy liquid value to the interpretation of wetting of high-energy liquids on high-energy solids is uncertain; the estimation of S from the film thickness may be experimentally quite difficult. Such an approach may prove, however, to be a fruitful avenue for future work.

VI. DISCUSSION AND CONCLUSIONS

Wetting and adhesion are complex phenomena that involve often subtle changes in interfacial energies and chemical reactions. These changes are induced by altering the composition of the substrate or wetting liquid (directly or indirectly, as by aging, volatilization, adsorption), or by changing the temperature or atmosphere.

The majority of data and insight into the high-temperature wetting behavior of inorganic liquids has been obtained using $Na_2O \cdot 2SiO_2$ liquid. This liquid is an excellent choice for modeling wetting behavior because its melting point lies in the temperature range used by many ceramic and metal component manufacturing processes; and γ_{LV} is insensitive to changes in atmosphere. Despite the simplicity of this liquid, however, extremely complex and variable behavior has been observed, depending on the choice of substrate and environment. In the Fe-$Na_2O \cdot SiO_2$ system, wetting is controlled by several sequential and/or linked chemical reactions involving the substrate. When Cr is used as a substrate, other additional reactions come into play, with the result that basic behavior, such as the extent of wetting in air vs. vacuum, is drastically modified. Unfortunately, $Na_2O \cdot SiO_2$ is of limited technological importance because it is hygroscopic.

The situation becomes even more complex when the wetting behavior of "practical" liquids is investigated. Consider the surface tension measurements of Cronin et al. [13] for PbO-B_2O_3-SiO_2 liquids.* The investigated compositions exhibit an unusually strong positive temperature dependence (see Fig. 3). This behavior indicates that local changes in the surface chemistry (enrichment in surface tension-reducing elements) and/or structure are occurring. Also, Cronin et al. [28] noted that drastic changes in γ_{LV} of these liquids can be induced upon changing from an oxidizing atmosphere to a reducing one. These changes are occurring at the liquid/vapor interface. To date, no work has been directed toward investigating the corresponding changes that might be occurring at the solid/liquid interface. Since both interfaces are involved in establishing wetting behavior, both must be well characterized. However, a good first step toward advancing the understanding of high-temperature wetting is the investigation of the temperature and atmosphere dependence of γ_{LV} of a wide range of inorganic liquids. Such data would provide an essential building block for the development of the theory of wetting of inorganic liquids.

Despite the limited range of investigated systems, this survey of the literature indicates that the interplay between interfacial energies and reactions leads to three general types of wetting behavior for inorganic liquids on metal and oxide substrates: (1) In the absence of reactions, inorganic liquids tend to exhibit large but acute angles on metal substrates.

* This system is of interest because it is the foundation for many glass-to-metal seals.

In many cases, the metal may be covered by a microscopically thin oxide layer. The dissolution of the oxide layer by the inorganic liquid cleans the metal surface and reveals true oxide liquid/metal behavior. (2) When placed on oxides or thick oxide layers on metal substrates, liquids generally exhibit improved wetting or spreading. In some cases, such as for Cu, Ni, Au, and Ag substrates in air (where an oxide layer is assumed present), complete wetting ($\theta = 0°$) is observed. (3) When the substrate plays an active role in a reaction, improved wetting or spreading is normally observed.

The Fe-$Na_2O{\cdot}2SiO_2$ system provides examples of each of these behaviors. $Na_2O{\cdot}2SiO_2$ liquid forms contact angles of ~55° on O_2-saturated Fe substrates. The initial contact angle observed on unsaturated Fe substrates is also similar to this value. When placed on FeO or Fe_3O_4, contact angles in the range of 2–22° are observed. Finally, under conditions that promote the active participation of the unsaturated Fe substrate in reactions, the contact angle is observed to decrease continuously from 55–0°.

Kingery [76] noted that the observation of better wetting on oxides than metals is contrary to conventional wisdom, which might conclude that the higher values of γ_{SV} for metals should promote wetting. The enhanced wetting on oxides is explained by consideration of $\gamma_{SV} - \gamma_{SL}$, the driving force for wetting. The magnitude of this factor may be viewed as being determined by the fraction of surface bonding states on the oxide liquid that are filled by unsatisfied bonds at the substrate surface. The observation of better wetting on oxide substrates indicates that the degree of mutual bond compensation between the liquid and substrate is larger for inorganic liquid/oxide systems than it is for inorganic liquid/metal systems. Better compensation would be expected when the liquid and solid phase possess the same type of bonding (i.e., ionic, covalent, or metallic).

If we assume that the dispersion energy contribution to γ_{SL} is ~25 mN/m and $\gamma_{LV} = 300$ mN/m, a reduction in the surface energy of the substrate by at least 83% of γ_{LV} must occur to initiate the spreading of inorganic liquids (if we assume no reactions). This result is independent of the actual magnitude of γ_{SV}. It is not surprising, then, that wetting behavior is dominated by the bonding interactions that occur between the solid and liquid phases.

Our understanding of these interactions and how they determine γ_{SL} is in its infancy. What insight has been developed stems from studies of the wetting behavior of a single inorganic liquid on a variety of metal substrates. Studies need to be conducted that focus on the impact of systematic variations in the chemistry and structure of the inorganic liquid on wetting behavior. For example, the results of Copley et al. [31] suggest that interactions between the quantity and electronegativity of nonbridging oxygens in silicate liquids and the metal ions determine the wetting behavior. This concept deserves further attention.

The occurrence of reactions at the interface can impact significantly both the macroscopic wetting behavior and development of adhesion. As the reactions proceed, the contact angles decrease. At sufficiently small contact angles ($\theta < 30°$), adhesion begins to develop. The small contact angles and development of adhesion are both consistent with the development of chemical bonds across the solid/liquid interface. In most cases, this process occurs on a time scale that is much larger than that associated with bulk-viscous flow of the droplet. Consequently, manufacturing processes that focus on the development of adhesion must be designed with this time scale in mind. Alternatively, adhesion-promoting oxides may be added to the liquid to promote spreading and adhesion.

Currently, it is thought that saturation of the interface with the metal oxide must be established and maintained to ensure both good wetting and adhesion. However, the application of this concept is not necessarily straightforward. For example, in the Cr-$Na_2O \cdot SiO_2$ system, under appropriate conditions, Cr reduces species in the liquid to form Cr oxides. After significant dissolution, adhesion has still not been achieved due to a lack of saturation with respect to CrO. Some portion of the CrO, however, is being converted to Cr_2O_3. After a slight increase in CrO content, the conversion reaction causes the concentration of Cr_2O_3 to exceed the solubility limit. The liquid precipitates Cr_2O_3 and good adhesion results. At the current level of understanding of adhesion, it is unlikely that this complex behavior could have been predicted before the fact. This is particularly true for the wetting and adhesion of inorganic liquids on oxide surfaces.

Details of the manner in which chemical reactions at the interface contribute to spreading also remain to be established. Experiments designed to distinguish between the dG_{chem} and dG_γ contributions to dG_R have not been conducted. Experiments of this type will determine if reactions contribute principally by modification of the instantaneous values of the interfacial energies or by local reduction of the Gibbs energy through chemical reaction. In addition, the kinetic conditions at the three-phase line required to promote the dG_R contribution remain to be explored. Research investigating these issues will provide guidelines for selecting liquid/substrate pairs that optimize the dG_R contribution.

As seen in many studies, environmental factors can have a great impact on wetting behavior. As examples, Cline et al. [37] showed that variations in storage conditions and annealing treatments can change the wetting behavior of $Na_2O \cdot 2SiO_2$ liquids on Fe substrates through modification of the activity of FeO in the substrate; similarly, modification of the firing atmosphere promotes or impedes reactions at the interface, which facilitate spreading; because of the high surface energies of metal substrates, gases easily adsorb on their surfaces and reduce γ_{SV}, causing major changes in the wetting behavior (atmospheric contamination of Pt surfaces plagued early measurements of wetting behavior on this substrate). These and other factors must be carefully considered in order to obtain reproducibility of wetting behavior.

The wetting behavior of inorganic liquids on oxide and refractory ceramics remains largely unexplored. Several studies have been conducted that begin to document behaviors, but studies that systematically vary the compositions of both the substrate and liquid remain to be conducted. As noted previously, advances in this area are hampered by the need to measure the compositional dependence of γ_{LV} for a wide range of inorganic liquids. The influence of adsorbed gases on γ_{SV} for oxides remains to be explored as well.

The application of Zisman's concept of the critical surface tension to the wetting behavior of inorganic liquids on oxide and metal substrates could provide valuable insight into this phenomenon. By using liquid metals, γ_c has been determined for a variety of ceramics. The applicability of these values across classes of liquids (oxides, metals, organics) remains, however, to be determined. No investigations have been made on the effect of adsorbed impurities, temperature, or surface roughness on γ_c for inorganic solids. The concept of γ_c has not been applied to metal surfaces; hence, data on the critical surface tensions of metals are completely lacking.

The development of fundamental theory and predictive models of the high-temperature wetting behavior of inorganic liquids is in its infancy, particularly when compared to the level of understanding in the related area of the wetting of organic and polymeric liquids at relatively low temperatures. Also, despite the important role played by high-temperature wetting in both simple and technologically sophisticated manufacturing processes, little progress has been made in this field in recent years. One needs to go back nearly a quarter of a century to find most of the "recent" fundamental studies. Although revealing the complexity of wetting behavior, these studies also proved that systematic and careful measurements can provide significant insight. Clearly, the field would benefit greatly from a resurgence of interest on the part of researchers from a broad spectrum of disciplines including materials science, chemistry, and physics.

ACKNOWLEDGMENTS

The authors wish to express their appreciation to Drs. Kevin Ewsuk, Marc laBranche, Robert Bouchard, Daniel Button, and Norbert Kreidl for stimulating discussions of wetting behavior, and to E.I. du Pont de Nemours & Co. and the U.S. Air Force Office of Scientific Research for their support of research in this area.

REFERENCES

1. P. G. de Gennes, *Rev. Mod. Phys.—Part 1 57:*827 (1985).
2. W. A. Zisman, in *Contact Angle, Wettability and Adhesion* (R. F. Gould, ed.), ACS, Washington, DC, 1964, Chap. 1 and 13.
3. L. Leger, in *Physicochemical Hydrodynamics: Interfacial Phenomena* (M. G. Velarde, ed.), Plenum, New York, 1988, pp. 721–740.
4. S. Wu, *Polymer Interface and Adhesion*, Marcel Dekker, New York, 1982.
5. J. A. Pask, and A. P. Tomsia, *Matr. Sci. Res. 14:*411 (1980).
6. T. Young, "Essay on the Cohesion of Fluids," *Phil. Trans. Roy. Soc. Lond. 95:*65 (1805).
7. N. K. Adam, *Physics and Chemistry of Surfaces*, Oxford Univ. Press, New York, 1941, p. 448.
8. B. S. Ellefson, and N. W. Taylor, *J. Amer. Ceram. Soc. 21:*205 (1938).
9. I. A. Aksay, C. E. Hoge, and J. A. Pask, *J. Phys. Chem. 78:*1178 (1974).
10. J. M. Israelachvili, *Intermolecular and Surface Forces*, Academic Press, San Diego, CA, 1985.
11. J. A. Pask, *Ceram. Bull. 66:*1587 (1987).
12. A. Dupré, *Theorié méchanique de la chaleur*, Gauthier-Villars, Paris, 1869, p. 369.
13. J. P. Cronin, B. J. J. Zelinski, N. J. Kreidl, and D. R. Uhlmann, "Surface Tension of PbO-B_2O_3-SiO_2 Liquids" (to be published).
14. L. Shartsis, S. Spinner, and A. W. Smock, *J. Amer. Ceram. Soc. 31:*23 (1948).
15. Blakely, J. M., *Introduction to the Properties of Crystal Surfaces*, International Series on Materials Science and Technology, Pergamon Press, Oxford, V. 12, 1973, p. 36.
16. N. M. Parikh, *J. Amer. Ceram. Soc. 41:*18 (1958).
17. A. E. Badger, C. W. Parmelee, and A. E. Williams, *J. Amer. Ceram. Soc. 20:*325 (1937).
18. C. E. Hoge, J. J. Brennan, and J. A. Pask, *J. Amer. Ceram. Soc. 56:*51 (1973).
19. J. A. Pask, in *Modern Aspects of the Vitreous State*, Vol. 3 (J. D. Mackenzie, ed.), Butterworths, London, 1964, p. 1.
20. B. S. Ellefson, and N. W. Taylor, *J. Amer. Ceram. Soc. 21:*205 (1938).
21. R. M. Fulrath, S. P. Mitoff, and J. A. Pask, *J. Amer. Ceram. Soc. 40:*269 (1957).

22. V. F. Zackay, D. W. Mitchell, S. P. Mitoff, and J. A. Pask, *J. Amer. Ceram. Soc. 36:*84 (1953).
23. M. L. Volpe, R. M. Fulrath, and J. A. Pask, *J. Amer. Ceram. Soc. 42:*102 (1959).
24. H. Udin, "Metal Interfaces—A Seminar," Amer. Soc. for Metals, Cleveland, OH, 1951, p. 114
25. G. A. Holmquist, and J. A. Pask, *J. Amer. Ceram. Soc. 59:*384 (1976).
26. L. Shartsis, and W. Capps, *J. Amer. Ceram. Soc. 35:*169 (1952).
27. V. K. Nagesh, A. P. Tomsia, and J. A. Pask, *J. Mater. Sci. 18:*2173 (1983).
28. J. P. Cronin, B. J. J. Zelinski, M. Denesuk, N. Kreidl, and D. R. Uhlmann, "Wetting Behavior of PbO-SiO_2 and PbO-B_2O_3-SiO_2 Liquids" (to be published).
29. R. R. Tummala, and B. J. Foster, *J. Mater. Sci. Lett. 10:*905 (1975).
30. R. S. Simhan, L. L. Moore, and P. R. van Gunten, *J. Mater. Sci. 20:*1748 (1985).
31. G. J. Copley, A. D. Rivers, and Smith, *J. Mater. Sci. 10:*1285 (1975).
32. G. J. Copley, and A. D. Rivers, *J. Mater. Sci. 10:*1291 (1975).
33. S. T. Tso, and J. A. Pask, *J. Amer. Ceram. Soc. 62:*543 (1979).
34. A. Bondi, *Chem. Rev. 52:*417 (1952).
35. L. Shartsis, and S. Spinner, *J. Res. NBS 46:*385 (1951).
36. P. R. Sharps, A. P. Tomsia, and J. A. Pask, *Acta Metall. 29:*855 (1981).
37. R. W. Cline, R. M. Fulrath, and J. A. Pask, *J. Amer. Ceram. Soc. 44:*423 (1961).
38. L. G. Hagan, and S. R. Ravitz, *J. Amer. Ceram. Soc. 44:*428 (1961).
39. A. P. Tomsia, and J. A. Pask, *J. Amer. Ceram. Soc. 64:*523 (1981).
40. R. B. Adams, and J. A. Pask, *J. Amer. Ceram. Soc. 44:*430 (1961).
41. M. P. Borom, and J. A. Pask, *Phys. Chem. Glasses 8:*194 (1967).
42. M. P. Borom, *Phys. Chem. Glasses 8:*203 (1967).
43. V. K. Bhat, and C. R. Manning, Jr., *J. Amer. Ceram. Soc. 56:*455 (1973).
44. M. P. Borom, J. Longwell, and J. A. Pask, *J. Amer. Ceram. Soc. 50:*61 (1967).
45. M. P. Borom, and J. A. Pask, *J. Amer. Ceram. Soc. 49:*1 (1966).
46. J. J. Brennan, and J. A. Pask, *J. Amer. Ceram. Soc. 56:*58 (1973).
47. A. P. Tomsia, Z. Feipeng, and J. A. Pask, *J. Amer. Ceram. Soc. 68:*20 (1985).
48. S. P. Mitoff, *J. Amer. Ceram. Soc. 40:*118 (1957).
49. J. E. Comeforo, and R. K. Hursh, *J. Amer. Ceram. Soc. 35*:130 (1952).
50. M. W. Barsoum, and P. D. Ownby, *Mater. Sci. Res. 14:*457 (1980).
51. R. Kossowsky, *J. Mater. Sci. 9:*2025 (1974).
52. T. J. Whalen, and M. Humenik Jun, in *Sintering and Related Phenomena* (G. C. Kuezynski, N. A. Hooton, and G. F. Gibbon, eds.), Gordon and Breach, New York, 1967, p. 715.
53. W. D. Kingery, H. K. Bowen, and D. R. Uhlmann, *Introduction to Ceramics*, Wiley, New York, 1976, p. 183.
54. D. N. Coon, and R. M. Neilson, Jr., *Ceram. Eng. Sci. Proc. 10:*1735 (1989).
55. V. A. Ogarev, T. N. Timonina, V. V. Arslanov, and A. A. Trapeznikov, *J. Adhesion 6:*337 (1974).

56. L. H. Tanner, *J. Phys. D: Appl. Phys. 12:*1473 (1979).
57. F. T. Dodge, *J. Coll. Inter. Sci. 121:*154 (1988).
58. J. Lopez, C. A. Miller, and E. Ruckenstein, *J. Coll. Inter. Sci. 56:*460 (1976).
59. J. Frenkel, *J. Phys. (Moscow) 9:*385 (1945).
60. S. Strella, *J. Appl. Phys. 40:*4242 (1970).
61. M. Denesuk, J. P. Cronin, B. J. J. Zelinski, N. J. Kreidl, and D. R. Uhlmann, "Predictive Modelling of Liquids Spreading on Solid Surfaces" (submitted to *Phys. Chem. Glass*, 1992).
62. M. Denesuk, B. J. J. Zelinski, N. J. Kreidl, and D. R. Uhlmann, "Spreading on Porous Surfaces" (to be sumitted).
63. M. Cukierman, and D. R. Uhlmann, *J. Geophys. Res. 79:*1594 (1974).
64. H. W. Fox, and W. A. Zisman, *J. Coll. Sci. 7:*109 (1952).
65. D. A. Olsen, and A. J. Osteraas, *J. Phys. Chem. 68:*2730 (1964).
66. P. Debye, and L. K. H. van Beek, *J. Chem. Phys. 31:*1595 (1959).
67. T. W. Hickmott, *J. Appl. Phys. 31:*128 (1960).
68. J. G. Eberhart, *J. Phys. Chem. 71:*4125 (1967).
69. J. G. Eberhart, *J. Nucl. Mater. 25:*103 (1968).
70. S. K. Rhee, *J. Amer. Ceram. Soc. 54:*379 (1971).
71. S. K. Rhee, *Mater. Sci. Eng. 11:*311 (1973).
72. S. K. Rhee, *J. Amer. Ceram. Soc. 53:*386 (1970).
73. S. K. Rhee, *J. Amer. Ceram. Soc. 53:*639 (1970).
74. S. K. Rhee, *J. Amer. Ceram. Soc. 54:*332 (1971).
75. S. K. Rhee, *J. Amer. Ceram. Soc. 55:*157 (1972).
76. W. D. Kingery, *Ceram. Bull. 35:*108 (1956).

Index